BIRKHÄUSER

Cornerstones

Series Editors

Emmanuele DiBenedetto

Partial Differential Equations

Second Edition

Birkhäuser
Boston • Basel • Berlin

Emmanuele DiBenedetto
Department of Mathematics
Vanderbilt University
Nashville, TN 37240
USA
em.diben@vanderbilt.edu

ISBN 978-0-8176-4551-9 e-ISBN 978-0-8176-4552-6
DOI 10.1007/978-0-8176-4552-6
Springer New York Dordrecht Heidelberg London

Library of Congress Control Number: 2009938184

Mathematics Subject Classification (2000): 31B05, 31B20, 35A10, 35B45, 35B65, 35D10, 35J05, 35K05, 35L05, 35L60, 35L65, 45A05, 45B05, 45C05, 49J40

Printed on acid-free paper

Birkhäuser Boston is part of Springer Science+Business Media (www.birkhauser.com)

Contents

Preface to the Second Edition

This is a revised and extended version of my 1995 elementary introduction to partial differential equations. The material is essentially the same except for three new chapters. The first (Chapter 8) is about non-linear equations of first order and in particular Hamilton–Jacobi equations. It builds on the continuing idea that PDEs, although a branch of mathematical analysis, are closely related to models of physical phenomena. Such underlying physics in turn provides ideas of solvability. The Hopf variational approach to the Cauchy problem for Hamilton–Jacobi equations is one of the clearest and most incisive examples of such an interplay. The method is a perfect blend of classical mechanics, through the role and properties of the Lagrangian and Hamiltonian, and calculus of variations. A delicate issue is that of identifying "uniqueness classes." An effort has been made to extract the geometrical conditions on the graph of solutions, such as quasi-concavity, for uniqueness to hold.

Chapter 9 is an introduction to weak formulations, Sobolev spaces, and direct variational methods for linear and quasi-linear elliptic equations. While terse, the material on Sobolev spaces is reasonably complete, at least for a PDE user. It includes all the basic embedding theorems, including their proofs, and the theory of traces. Weak formulations of the Dirichlet and Neumann problems build on this material. Related variational and Galerkin methods, as well as eigenvalue problems, are presented within their weak framework. The Neumann problem is not as frequently treated in the literature as the Dirichlet problem; an effort has been made to present the underlying theory as completely as possible. Some attention has been paid to the local behavior of these weak solutions, both for the Dirichlet and Neumann problems. While efficient in terms of existence theory, weak solutions provide limited information on their local behavior. The starting point is a sup bound for the solutions and weak forms of the maximum principle. A further step is their local Hölder continuity.

An introduction to these local methods is in Chapter 10 in the framework of DeGiorgi classes. While originating from quasi-linear elliptic equations, these classes have a life of their own. The investigation of the local and boundary behavior of functions in these classes, involves a combination of methods from PDEs, measure theory, and harmonic analysis. We start by tracing them back to quasi-linear elliptic equations, and then present in detail some of these methods. In particular, we establish that functions in these classes are locally bounded and locally Hölder continuous, and we give conditions for the regularity to extend up to the boundary. Finally, we prove that non-negative functions on the DeGiorgi classes satisfy the Harnack inequality. This, on the one hand, is a surprising fact, since these classes require only some sort of Caccioppoli-type energy bounds. On the other hand, this raises the question of understanding their structure, which to date is still not fully understood. While some facts about these classes are scattered in the literature, this is perhaps the first systematic presentation of DeGiorgi classes in their own right. Some of the material is as recent as last year. In this respect, these last two chapters provide a background on a spectrum of techniques in local behavior of solutions of elliptic PDEs, and build toward research topics of current active investigation.

The presentation is more terse and streamlined than in the first edition. Some elementary background material (Weierstrass Theorem, mollifiers, Ascoli–Arzelá Theorem, Jensen's inequality, etc..) has been removed.

I am indebted to many colleagues and students who, over the past fourteen years, have offered critical suggestions and pointed out misprints, imprecise statements, and points that were not clear on a first reading. Among these Giovanni Caruso, Xu Guoyi, Hanna Callender, David Petersen, Mike O'Leary, Changyong Zhong, Justin Fitzpatrick, Abey Lopez and Haichao Wang. Special thanks go to Matt Calef for reading carefully a large portion of the manuscript and providing suggestions and some simplifying arguments. The help of U. Gianazza has been greatly appreciated. He has read the entire manuscript with extreme care and dedication, picking up points that needed to be clarified. I am very much indebted to Ugo.

I would like to thank Avner Friedman, James Serrin, Constantine Dafermos, Bob Glassey, Giorgio Talenti, Luigi Ambrosio, Juan Manfredi, John Lewis, Vincenzo Vespri, and Gui Qiang Chen for examining the manuscript in detail and for providing valuable comments. Special thanks to David Kinderlehrer for his suggestion to include material on weak formulations and direct methods. Without his input and critical reading, the last two chapters probably would not have been written. Finally, I would like to thank Ann Kostant and the entire team at Birkhäuser for their patience in coping with my delays.

Vanderbilt University
June 2009

Emmanuele DiBenedetto

Preface to the First Edition

These notes are meant to be a self contained, elementary introduction to partial differential equations (PDEs). They assume only advanced differential calculus and some basic L^p theory. Although the basic equations treated in this book, given its scope, are linear, I have made an attempt to approach then from a non-linear perspective.

Chapter I is focused on the Cauchy–Kowalewski theorem. We discuss the notion of characteristic surfaces and use it to classify partial differential equations. The discussion grows from equations of second-order in two variables to equations of second-order in N variables to PDEs of any order in N variables.

In Chapters 2 and 3 we study the Laplace equation and connected elliptic theory. The existence of solutions for the Dirichlet problem is proven by the Perron method. This method clarifies the structure of the sub(super)-harmonic functions, and it is closely related to the modern notion of *viscosity solution*. The elliptic theory is complemented by the Harnack and Liouville theorems, the simplest version of Schauder's estimates, and basic L^p-potential estimates. Then, in Chapter 3 the Dirichlet and Neumann problems, as well as eigenvalue problems for the Laplacian, are cast in terms of integral equations. This requires some basic facts concerning double-layer potentials and the notion of compact subsets of L^p, which we present.

In Chapter 4 we present the Fredholm theory of integral equations and derive necessary and sufficient conditions for solving the Neumann problem. We solve eigenvalue problems for the Laplacian, generate orthonormal systems in L^2, and discuss questions of completeness of such systems in L^2. This provides a theoretical basis for the method of separation of variables.

Chapter 5 treats the heat equation and related parabolic theory. We introduce the representation formulas, and discuss various comparison principles. Some focus has been placed on the uniqueness of solutions to the Cauchy problem and their behavior as $|x| \to \infty$. We discuss Widder's theorem and the structure of the non-negative solutions. To prove the parabolic Harnack estimate we have used an idea introduced by Krylov and Safonov in the context of fully non-linear equations.

The wave equation is treated in Chapter 6 in its basic aspects. We derive representation formulas and discuss the role of the characteristics, propagation of signals, and questions of regularity. For general linear second-order hyperbolic equations in two variables, we introduce the Riemann function and prove its symmetry properties. The sections on Goursat problems represent a concrete application of integral equations of Volterra type.

Chapter 7 is an introduction to conservation laws. The main points of the theory are taken from the original papers of Hopf and Lax from the 1950s. Space is given to the minimization process and the meaning of taking the initial data in the sense of L^1. The uniqueness theorem we present is due to Kruzhkov (1970). We discuss the meaning of *viscosity solution* vis-à-vis the notion of sub-solutions and maximum principle for parabolic equations. The theory is complemented by an analysis of the asymptotic behavior, again following Hopf and Lax.

Even though the layout is theoretical, I have indicated some of the physical origins of PDEs. Reference is made to potential theory, similarity solutions for the porous medium equation, generalized Riemann problems, etc.

I have also attempted to convey the notion of *ill-posed* problems, mainly via some examples of Hadamard.

Most of the background material, arising along the presentation, has been stated and proved in the complements. Examples include the Ascoli–Arzelà theorem, Jensen's inequality, the characterization of compactness in L^p, mollifiers, basic facts on convex functions, and the Weierstrass theorem. A book of this kind is bound to leave out a number of topics, and this book is no exception. Perhaps the most noticeable omission here is some treatment of numerical methods.

These notes have grown out of courses in PDEs I taught over the years at Indiana University, Northwestern University and the University of Rome II, Italy. My thanks go to the numerous students who have pointed out misprints and imprecise statements. Of these, special thanks go to M. O'Leary, D. Diller, R. Czech, and A. Grillo. I am indebted to A. Devinatz for reading a large portion of the manuscript and for providing valuable critical comments. I have also benefited from the critical input of M. Herrero, V. Vespri, and J. Manfredi, who have examined parts of the manuscript. I am grateful to E. Giusti for his help with some of the historical notes. The input of L. Chierchia has been crucial. He has read a large part of the manuscript and made critical remarks and suggestions. He has also worked out in detail a large number of the problems and supplied some of his own. In particular, he wrote the first draft of problems **2.7–2.13** of Chapter 5 and **6.10–6.11** of Chapter 6. Finally I like to thank M. Cangelli and H. Howard for their help with the graphics.

0

Preliminaries

1 Green's Theorem

Let E be an open set in $\mathbb{R}^N$, and let k be a non-negative integer. Denote by $C^k(E)$ the collection of all real-valued, k-times continuously differentiable functions in E. A function f is in $C_o^k(E)$ if $f \in C^k(E)$, and its support is contained in E. A function $f : \bar{E} \to \mathbb{R}$ is in $C^k(\bar{E})$, if $f \in C^k(E)$ and all partial derivatives $\partial^\ell f / \partial x_i^\ell$ for all $i = 1, \dots, N$ and $\ell = 0, \dots, k$, admit continuous extensions up to ∂E. The boundary ∂E is of class C^1 if for all $y \in \partial \Omega$, there exists $\varepsilon > 0$ such that within the ball $B_\varepsilon(y)$ centered at y and radius ε, ∂E can be implicitly represented, in a local system of coordinates, as a level set of a function $\Phi \in C^1(B_\varepsilon(y))$ such that $|\nabla \Phi| \neq 0$ in $B_\varepsilon(y)$.

If ∂E is of class C^1, let $\mathbf{n}(x) = \big(n_1(x), \dots, n_N(x)\big)$ denote the unit normal exterior to E at $x \in \partial E$. Each of the components $n_j(\cdot)$ is well defined as a continuous function on ∂E. A real vector-valued function

$$\bar{E} \ni x \to \mathbf{f}(x) = \big(f_1(x), \dots, f_N(x)\big) \in \mathbb{R}^N$$

is of class $C^k(E)$, $C^k(\bar{E})$, or $C_o^k(E)$ if all components f_j belong to these classes.

Theorem 1.1 *Let E be a bounded domain of $\mathbb{R}^N$ with boundary ∂E of class C^1. Then for every $\mathbf{f} \in C^1(\bar{E})$*

$$\int_E \operatorname{div} \mathbf{f} \, dx = \int_{\partial E} \mathbf{f} \cdot \mathbf{n} \, d\sigma$$

where dx is the Lebesgue measure in E and $d\sigma$ denotes the Lebesgue surface measure on ∂E.

This is also referred to as the *divergence theorem*, or as the formula of *integration by parts*. It continues to hold if $\mathbf{n}$ is only $d\sigma$-a.e. defined in ∂E. For example, ∂E could be a cube in $\mathbb{R}^N$. More generally, ∂E could be the finite

E. DiBenedetto, *Partial Differential Equations: Second Edition*,
Cornerstones, DOI 10.1007/978-0-8176-4552-6_1,

union of portions of surfaces of class C^1. The domain E need not be bounded, provided $|\mathbf{f}|$ and $|\nabla\mathbf{f}|$ decay sufficiently fast as $|x| \to \infty$.[1]

1.1 Differential Operators and Adjoints

Given a symmetric matrix $(a_{ij}) \in \mathbb{R}^N \times \mathbb{R}^N$, a vector $\mathbf{b} \in \mathbb{R}^N$, and $c \in \mathbb{R}$, consider the formal expression

$$\mathcal{L}(\cdot) = a_{ij}\frac{\partial^2}{\partial x_i \partial x_j} + b_i \frac{\partial}{\partial x_i} + c \tag{1.1}$$

where we have adopted the Einstein summation convention, i.e., repeated indices in a monomial expression mean summation over those indices. The formal adjoint of $\mathcal{L}(\cdot)$ is

$$\mathcal{L}^*(\cdot) = a_{ij}\frac{\partial^2}{\partial x_i \partial x_j} - b_i \frac{\partial}{\partial x_i} + c.$$

Thus $\mathcal{L} = \mathcal{L}^*$ if $\mathbf{b} = 0$. If $u, v \in C^2(\bar{E})$ for a bounded open set $E \subset \mathbb{R}^N$ with boundary ∂E of class C^1, the divergence theorem yields the Green's formula

$$\int_E [v\mathcal{L}(u) - u\mathcal{L}^*(v)]dx = \int_{\partial E} [(va_{ij}u_{x_j}n_i - ua_{ij}v_{x_i}n_j) + uv\mathbf{b}\cdot\mathbf{n}]d\sigma. \tag{1.2}$$

If $u, v \in C_o^2(E)$, then

$$\int_E [v\mathcal{L}(u) - u\mathcal{L}^*(v)]dx = 0. \tag{1.2$_o$}$$

More generally, the entries of the matrix (a_{ij}) as well as $\mathbf{b}$ and c might be smooth functions of x. In such a case, for $v \in C^2(\bar{E})$, define

$$\mathcal{L}^*(v) = \frac{\partial^2 (a_{ij}v)}{\partial x_i \partial x_j} - \frac{\partial (b_i v)}{\partial x_i} + cv.$$

The Green's formula $(1.2)_o$ continues to hold for every pair of functions u, $v \in C_o^2(E)$. If u and v do not vanish near ∂E, a version of (1.2) continues to hold, where the right-hand side contains the extra boundary integral

$$\int_{\partial E} uv\frac{\partial a_{ij}}{\partial x_i} n_j \, d\sigma.$$

---

[1] Identifying precise conditions on ∂E and $\mathbf{f}$ for which one can *integrate by parts* is part of *geometric measure theory* ([56]).

2 The Continuity Equation

Let $t \to E(t)$ be a set-valued function that associates to each t in some open interval $I \subset \mathbb{R}$ a bounded open set $E(t) \subset \mathbb{R}^N$, for some $N \geq 2$. Assume that the boundaries $\partial E(t)$ are uniformly of class C^1, and that there exists a bounded open set $E \subset \mathbb{R}^N$ such that $E(t) \subset E$, for all $t \in I$. Our aim is to compute the derivative

$$\frac{d}{dt}\int_{E(t)} \rho(x,t)dx \qquad \text{for a given} \quad \rho \in C^1(E \times I).$$

Regard points $x \in E(t)$ as moving along the trajectories $t \to x(t)$ with velocities $\dot{x} = \mathbf{v}(x,t)$. Assume that the motion, or deformation, of $E(\cdot)$ is smooth in the sense that $(x,t) \to \mathbf{v}(x,t)$ is continuous in a neighborhood of $E \times I$. Forming the difference quotient gives

$$\begin{aligned}
&\frac{d}{dt}\int_{E(t)} \rho(x,t)dx \\
&\quad = \lim_{\Delta t \to 0} \frac{1}{\Delta t}\left(\int_{E(t+\Delta t)} \rho(x,t+\Delta t)dx - \int_{E(t)} \rho(x,t)dx\right) \\
&\quad = \lim_{\Delta t \to 0} \int_{E(t)} \frac{\rho(x,t+\Delta t) - \rho(x,t)}{\Delta t}dx \\
&\qquad + \lim_{\Delta t \to 0} \frac{1}{\Delta t}\left(\int_{E(t+\Delta t)-E(t)} \rho(t)dx - \int_{E(t)-E(t+\Delta t)} \rho(t)dx\right).
\end{aligned} \tag{2.1}$$

The first limit is computed by carrying the limit under the integral, yielding

$$\lim_{\Delta t \to 0} \int_{E(t)} \frac{\rho(x,t+\Delta t) - \rho(x,t)}{\Delta t}dx = \int_{E(t)} \rho_t \, dx.$$

As for the second, first compute the difference of the last two volume integrals by means of Riemann sums as follows. Fix a number $0 < \Delta\sigma \ll 1$, and approximate $\partial E(t)$ by means of a polyhedron with faces of area not exceeding $\Delta\sigma$ and tangent to $\partial E(t)$ at some of their interior points. Let $\{F_1, \dots, F_n\}$ for some $n \in \mathbb{N}$ be a finite collection of faces making up the approximating polyhedron, and let x_i for $i = 1, \dots, n$, be a selection of their tangency points with $\partial E(t)$. Then approximate the set

$$\big[E(t+\Delta t) - E(t)\big] \bigcup \big[E(t) - E(t+\Delta t)\big]$$

by the union of the cylinders of basis F_i and height $\mathbf{v}(x_i,t) \cdot \mathbf{n}\Delta t$, built with their axes parallel to the outward normal to $\partial E(t)$ at x_i. Therefore, for Δt fixed

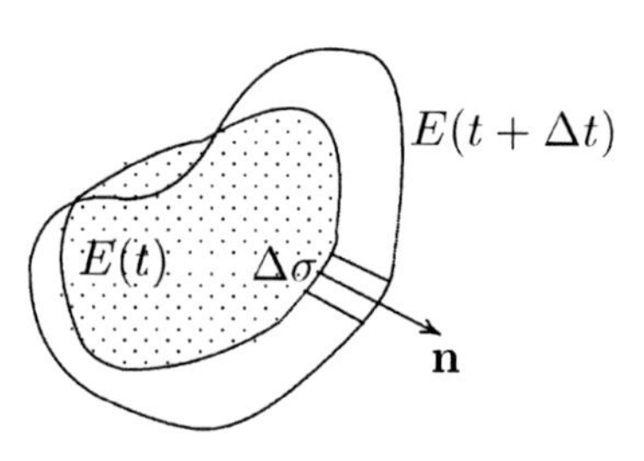

Fig. 2.1.

$$\frac{1}{\Delta t}\left(\int_{E(t+\Delta t)-E(t)} \rho(t)dx - \int_{E(t)-E(t+\Delta t)} \rho(t)dx\right)$$
$$= \lim_{\Delta\sigma\to 0} \sum_{i=1}^{n} \rho(x_i,t)\mathbf{v}(x_i,t)\cdot\mathbf{n}\Delta\sigma + O(\Delta t) = \int_{\partial E(t)} \rho\mathbf{v}\cdot\mathbf{n}d\sigma + O(\Delta t).$$

Letting now $\Delta t \to 0$ in (2.1) yields

$$\frac{d}{dt}\int_{E(t)} \rho\,dx = \int_{E(t)} \rho_t\,dx + \int_{\partial E(t)} \rho\mathbf{v}\cdot\mathbf{n}\,d\sigma. \tag{2.2}$$

By Green's theorem

$$\int_{\partial E(t)} \rho\mathbf{v}\cdot\mathbf{n}\,d\sigma = \int_{E(t)} \operatorname{div}(\rho\mathbf{v})\,dx.$$

Therefore (2.2) can be equivalently written as

$$\frac{d}{dt}\int_{E(t)} \rho\,dx = \int_{E(t)} [\rho_t + \operatorname{div}(\rho\mathbf{v})]\,dx. \tag{2.3}$$

Consider now an ideal fluid filling a region $E \subset \mathbb{R}^3$. Assume that the fluid is compressible (say a gas) and let $(x,t) \to \rho(x,t)$ denote its density. At some instant t, cut a region $E(t)$ out of E and follow the motion of $E(t)$ as if each of its points were identified with the moving particles. Whatever the sub-region $E(t)$, during the motion the mass is conserved. Therefore

$$\frac{d}{dt}\int_{E(t)} \rho\,dx = 0.$$

By the previous calculations and the arbitrariness of $E(t) \subset E$

$$\rho_t + \operatorname{div}(\rho\mathbf{v}) = 0 \qquad \text{in} \quad E\times\mathbb{R}. \tag{2.4}$$

This is referred to as the equation of continuity or conservation of mass.

3 The Heat Equation and the Laplace Equation

Any quantity that is conserved as it *moves* within an open set E with velocity $\mathbf{v}$ satisfies the *conservation law* (2.4). Let u be the temperature of a material homogeneous body occupying the region E. If c is the heat capacity, the thermal energy stored at $x \in E$ at time t is $cu(x,t)$. By Fourier's law the energy "moves" following gradients of temperature, i.e.,

$$cu\mathbf{v} = -k\nabla u \tag{3.1}$$

where k is the conductivity ([42, 17]). Putting this in (2.4) yields the *heat equation*

$$u_t - \frac{k}{c}\Delta u = 0. \tag{3.2}$$

Now let u be the pressure of a fluid moving with velocity $\mathbf{v}$ through a region E of $\mathbb{R}^N$ occupied by a porous medium. The porosity p_o of the medium is the relative infinitesimal fraction of space occupied by the pores and available to the fluid. Let μ, k, and ρ denote respectively kinematic viscosity, permeability, and density. By Darcy's law ([137])

$$\mathbf{v} = -\frac{kp_o}{\mu}\nabla u. \tag{3.3}$$

Assume that k and μ are constant. If the fluid is incompressible, then $\rho =$ const, and it follows from (2.4) that $\operatorname{div}\mathbf{v} = 0$. Therefore the pressure u satisfies

$$\operatorname{div}\nabla u = \Delta u = u_{x_i x_i} = 0 \quad \text{in } E. \tag{3.4}$$

The latter is the Laplace equation for the function u. A fluid whose velocity is given as the gradient of a scalar function is a *potential fluid* ([160]).

3.1 Variable Coefficients

Consider now the same physical phenomena taking place in non-homogeneous, anisotropic media. For heat conduction in such media, temperature gradients might generate heat propagation in preferred directions, which themselves might depend on $x \in E$. As an example one might consider the heat diffusion in a solid of given conductivity, in which is embedded a bundle of curvilinear material fibers of different conductivity. Thus in general, the conductivity of the composite medium is a tensor dependent on the location $x \in E$ and time t, represented formally by an $N \times N$ matrix $k = \big(k_{ij}(x,t)\big)$. For such a tensor, the product on the right-hand side of (3.1) is the row-by-column product of the matrix (k_{ij}) and the column vector ∇u. Enforcing the same conservation of energy (2.4) yields a non-homogeneous, anisotropic version of the heat equation (3.2), in the form

$$u_t - \big(a_{ij}(x,t)u_{x_i}\big)_{x_j} = 0 \quad \text{in } E, \text{ where } a_{ij} = \frac{k_{ij}}{c}. \tag{3.5}$$

Similarly, the permeability of a non-homogeneous, anisotropic porous medium is a position-independent tensor $(k_{ij}(x))$. Then, analogous considerations applied to (3.3), imply that the velocity potential u of the flow of a fluid in a heterogeneous, anisotropic porous medium satisfies the partial differential equation

$$\left(a_{ij}(x)u_{x_i}\right)_{x_j} = 0 \quad \text{in } E, \ \text{ where } a_{ij} = \frac{p_o k_{ij}}{\mu}. \tag{3.6}$$

The physical, tensorial nature of either heat conductivity or permeability of a medium implies that (a_{ij}) is symmetric, bounded, and positive definite in E. However, no further information is available on these coefficients, since they reflect interior properties of physical domains, not accessible without altering the physical phenomenon we are modeling. This raises the question of the meaning of (3.5)–(3.6). Indeed, even if $u \in C^2(E)$, the indicated operations are not well defined for $a_{ij} \in L^\infty(E)$. A notion of solution will be given in Chapter 9, along with solvability methods.

Equations (3.5)–(3.6) are said to be in divergence form. Equations in non-divergence form are of the type

$$a_{ij}(x)u_{x_i x_j} = 0 \quad \text{in } E \tag{3.7}$$

and arise in the theory of stochastic control ([89]).

4 A Model for the Vibrating String

Consider a material string of constant linear density ρ whose end points are fixed, say at 0 and 1. Assume that the string is vibrating in the plane (x, y), set the interval $(0, 1)$ on the x-axis, and let $(x, t) \to u(x, t)$ be the y-coordinate of the string at the point $x \in (0, 1)$ at the instant $t \in \mathbb{R}$. The basic physical assumptions are:

(i) The dimensions of the cross sections are negligible with respect to the length, so that the string can be identified, for all t, with the graph of $x \to u(x, t)$.
(ii) Let $(x, t) \to \mathbf{T}(x, t)$ denote the *tension*, i.e., the sum of the internal forces per unit length, generated by the displacement of the string. Assume that $\mathbf{T}$ at each point $(x, u(x, t))$ is tangent to the string. Letting $T = |\mathbf{T}|$, assume that $(x, t) \to T(x, t)$ is t-independent.
(iii) Resistance of the material to flexure is negligible with respect to the tension.
(iv) Vibrations are small in the sense that $|u|^\theta$ and $|u_x|^\theta$ for $\theta > 1$ are negligible when compared with $|u|$ and $|u_x|$.

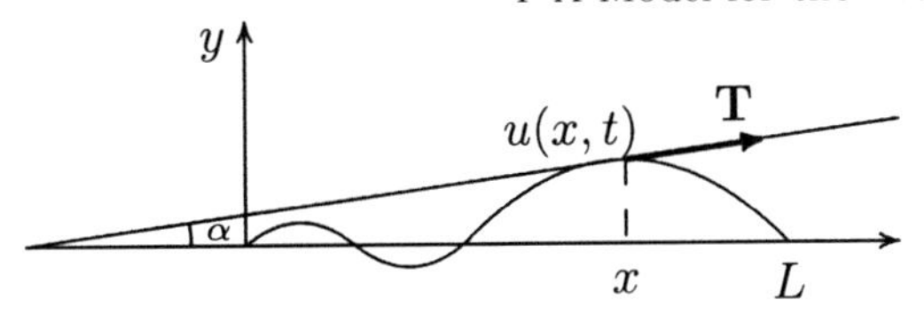

Fig. 4.1.

The tangent line to the graph of $u(\cdot, t)$ at $(x, u(x,t))$ forms with the x-axis an angle $\alpha \in (0, \pi/2)$ given by

$$\sin \alpha = \frac{u_x}{\sqrt{1+u_x^2}}.$$

Therefore the vertical component of the tension $\mathbf{T}$ at $(x, u(x,t))$ is

$$T \sin \alpha = T \frac{u_x}{\sqrt{1+u_x^2}}.$$

Consider next, for t fixed, a small interval $(x_1, x_2) \subset (0,1)$ and the corresponding portion of the string of extremities $\big(x_1, u(x_1,t)\big)$ and $\big(x_2, u(x_2,t)\big)$. Such a portion is in instantaneous equilibrium if both the x and y components of the sum of all forces acting on it are zero. The components in the y-direction are:

1. The difference of the y-components of $\mathbf{T}$ at the two extremities, i.e.,

$$\left(T\frac{u_x}{\sqrt{1+u_x^2}}\right)(x_2,t) - \left(T\frac{u_x}{\sqrt{1+u_x^2}}\right)(x_1,t) = \int_{x_1}^{x_2} \frac{\partial}{\partial x}\left(T\frac{u_x}{\sqrt{1+u_x^2}}\right) dx.$$

2. The total load acting on the portion, i.e.,

$$-\int_{x_1}^{x_2} p(x,t)dx, \qquad \text{where } p(\cdot,t) = \{\text{load per unit length}\}.$$

3. The inertial forces due to the vertical acceleration $u_{tt}(x,t)$, i.e.,

$$\int_{x_1}^{x_2} \rho \frac{\partial^2}{\partial t^2} u(x,t)dx.$$

Therefore the portion of the string is instantaneously in equilibrium if

$$\int_{x_1}^{x_2} \rho \frac{\partial^2}{\partial t^2} u(x,t)dx = \int_{x_1}^{x_2} \left[\frac{\partial}{\partial x}\left(T\frac{u_x}{\sqrt{1+u_x^2}}\right)(x,t) + p(x,t)\right] dx.$$

Dividing by $\Delta x = x_2 - x_1$ and letting $\Delta x \to 0$ gives

$$\rho \frac{\partial^2}{\partial t^2} u - \frac{\partial}{\partial x}\left(T\frac{u_x}{\sqrt{1+u_x^2}}\right) = p \quad \text{in } (0,1)\times\mathbb{R}. \tag{4.1}$$

The balance of forces along the x-direction involves only the tension and gives

$$(T\cos\alpha)(x_1,t) = (T\cos\alpha)(x_2,t)$$

or equivalently

$$\left(\frac{T}{\sqrt{1+u_x^2}}\right)(x_1,t) = \left(\frac{T}{\sqrt{1+u_x^2}}\right)(x_2,t).$$

From this

$$\int_{x_1}^{x_2} \frac{\partial}{\partial x}\left(\frac{T}{\sqrt{1+u_x^2}}\right)dx = 0 \quad \text{for all } (x_1,x_2)\subset(0,1).$$

Therefore

$$\frac{\partial}{\partial x}\left(\frac{T}{\sqrt{1+u_x^2}}\right) = 0 \quad \text{and} \quad x \to \left(\frac{T}{\sqrt{1+u_x^2}}\right)(x,t) = T_o$$

for some $T_o > 0$ independent of x. In view of the physical assumptions (ii) and (iv), may take T_o also independent of t. These remarks in (4.1) yield the partial differential equation

$$\frac{\partial^2}{\partial t^2}u - c^2\frac{\partial^2 u}{\partial x^2} = f \quad \text{in } (0,1)\times\mathbb{R} \tag{4.2}$$

where

$$c^2 = \frac{T_o}{\rho} \quad \text{and} \quad f(x,t) = \frac{p(x,t)}{\rho}.$$

This is the *wave equation* in one space variable.

Remark 4.1 The assumption that ρ is constant is a "linear" assumption in the sense that leads to the linear wave equation (4.2). Non-linear effects due to variable density were already observed by D. Bernoulli ([11]), and by S.D. Poisson ([120]).

5 Small Vibrations of a Membrane

A *membrane* is a rigid thin body of constant density ρ, whose thickness is negligible with respect to its extension. Assume that, at rest, the membrane occupies a bounded open set $E \subset \mathbb{R}^2$, and that it begins to vibrate under the action of a vertical load, say $(x,t) \to p(x,t)$. Identify the membrane with the graph of a smooth function $(x,t) \to u(x,t)$ defined in $E\times\mathbb{R}$ and denote by $\nabla u = (u_{x_1}, u_{x_2})$ the spatial gradient of u. The relevant physical assumptions are:

(i) Forces due to flexure are negligible.
(ii) Vibrations occur only in the direction $\mathbf{u}$ normal to the position of rest of the membrane. Moreover, vibrations are small, in the sense that $u_{x_i}u_{x_j}$ and uu_{x_i} for $i,j=1,2$ are negligible when compared to u and $|\nabla u|$.
(iii) The tension $\mathbf{T}$ has constant modulus, say $|\mathbf{T}| = T_o > 0$.

Cut a small ideal region $G_o \subset E$ with boundary ∂G_o of class C^1, and let G be the corresponding portion of the membrane. Thus G is the graph of $u(\cdot,t)$ restricted to G_o, or equivalently, G_o is the projection on the plane $u=0$ of the portion G of the membrane. Analogously, introduce the curve Γ limiting G and its projection $\Gamma_o = \partial G_o$. The tension $\mathbf{T}$ acts at points $P \in \Gamma$ and is tangent to G at P and normal to Γ. If $\boldsymbol{\tau}$ is the unit vector of $\mathbf{T}$ and $\mathbf{n}$ is the exterior unit normal to G at P, let $\mathbf{e}$ be the unit tangent to Γ at P oriented so that the triple $\{\boldsymbol{\tau}, \mathbf{e}, \mathbf{n}\}$ is positive and $\boldsymbol{\tau} = \mathbf{e} \wedge \mathbf{n}$. Our aim is to compute the vertical component of $\mathbf{T}$ at $P \in \Gamma$. If $\{\mathbf{i}, \mathbf{j}, \mathbf{k}\}$ is the positive unit triple along the coordinate axes, we will compute the quantity $\mathbf{T}\cdot\mathbf{k} = T_o \boldsymbol{\tau}\cdot\mathbf{k}$.

Consider a parametrization of Γ_o, say

$$s \to P_o(s) = \big(x_1(s), x_2(s)\big) \quad \text{for } s \in \{\text{some interval of } \mathbb{R}\}.$$

The unit exterior normal to ∂G_o is given by

$$\boldsymbol{\nu} = \frac{(x_2', -x_1')}{\sqrt{x_1'^2 + x_2'^2}}.$$

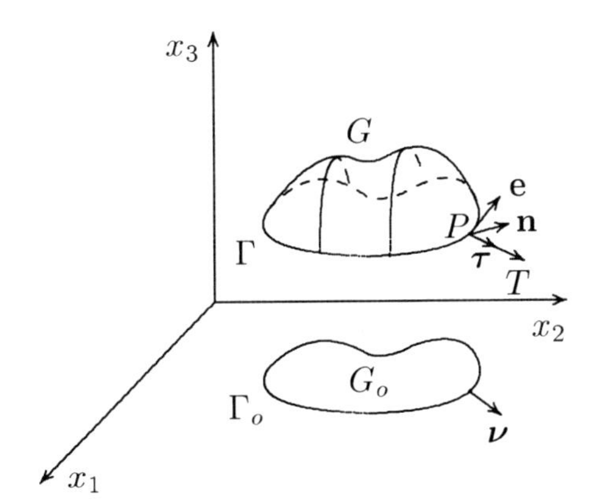

Fig. 5.1.

Consider also the corresponding parametrization of Γ

$$s \to P(s) = \big(P_o(s), u(x_1(s), x_2(s), t)\big).$$

The unit tangent $\mathbf{e}$ to Γ is

$$\mathbf{e} = \frac{(x_1', x_2', x_3')}{\sqrt{x_1'^2 + x_2'^2 + x_3'^2}} = \frac{(\dot{P}_o, \dot{P}_o \cdot \nabla u)}{\sqrt{|\dot{P}_o|^2 + |\dot{P}_o \cdot \nabla u|^2}}$$

and the exterior unit normal to G at Γ is

$$\mathbf{n} = \frac{(-\nabla u, 1)}{\sqrt{1 + |\nabla u|^2}}.$$

Therefore

$$\boldsymbol{\tau} = \mathbf{e} \wedge \mathbf{n} = \det \begin{pmatrix} \mathbf{i} & \mathbf{j} & \mathbf{k} \\ x_1' & x_2' & x_3' \\ -u_{x_1} & -u_{x_2} & 1 \end{pmatrix} \frac{1}{J}$$

where

$$J = \sqrt{1 + |\nabla u|^2} \sqrt{|\dot{P}_o|^2 + |\dot{P}_o \cdot \nabla u|^2}.$$

From this

$$J \boldsymbol{\tau} \cdot \mathbf{k} = (x_2', -x_1') \cdot \nabla u = |\dot{P}_o| \, \nabla u \cdot \boldsymbol{\nu}.$$

If β is the cosine of the angle between the vectors ∇u and $\dot{P}_o$

$$\boldsymbol{\tau} \cdot \mathbf{k} = \frac{\nabla u \cdot \boldsymbol{\nu}}{J_\beta}, \qquad J_\beta = \sqrt{1 + |\nabla u|^2} \sqrt{1 + \beta^2 |\nabla u|^2}.$$

Since

$$(1 + \beta^2 |\nabla u|^2) \le J_\beta \le (1 + |\nabla u|^2)$$

by virtue of the physical assumption (ii)

$$\mathbf{T} \cdot \mathbf{k} \approx T_o \nabla u \cdot \boldsymbol{\nu}.$$

Next, write down the equation of instantaneous equilibrium of the portion G of the membrane. The vertical loads on G, the vertical contribution of the tension $\mathbf{T}$, and the inertial force due to acceleration u_{tt} are respectively

$$\int_{G_o} p(x,t) \, dx, \qquad \int_{\partial G_o} T_o \nabla u \cdot \boldsymbol{\nu} \, d\sigma, \qquad \int_{G_o} \rho u_{tt} \, dx$$

where $d\sigma$ is the measure along the curve ∂G_o. Instantaneous equilibrium of every portion of the membrane implies that

$$\int_{G_o} \rho u_{tt} \, dx = \int_{\partial G_o} T_o \nabla u \cdot \boldsymbol{\nu} \, d\sigma + \int_{G_o} p(x,t) \, dx.$$

By Green's theorem

$$\int_{\partial G_o} T_o \nabla u \cdot \boldsymbol{\nu} \, d\sigma = \int_{G_o} T_o \operatorname{div}(\nabla u u) \, dx.$$

Therefore

$$\int_{G_o} [\rho u_{tt} - T_o \operatorname{div} \nabla u - p]\, dx = 0$$

for all $t \in \mathbb{R}$ and all $G_o \subset E$. Thus

$$u_{tt} - c^2 \Delta u = f \qquad \text{in } E \times \mathbb{R} \tag{5.1}$$

where

$$c^2 = \frac{T_o}{\rho}, \qquad f = \frac{p}{\rho}, \qquad \Delta u = \operatorname{div}(\nabla u).$$

Equation (5.1), modeling small vibrations of a stretched membrane, is the two-dimensional wave equation.

6 Transmission of Sound Waves

An ideal *compressible* fluid is moving within a region $E \subset \mathbb{R}^3$. Let $\rho(x,t)$ and $\mathbf{v}(x,t)$ denote its density and velocity at $x \in E$ at the instant t. Each x can be regarded as being in motion along the trajectory $t \to x(t)$ with velocity $x'(t)$. Therefore, denoting by $v_i(x,t)$ the components of $\mathbf{v}$ along the x_i-axes, then

$$\dot{x}_i(t) = v_i(x(t),t), \qquad i = 1,2,3.$$

The acceleration has components

$$\ddot{x}_i = \frac{\partial v_i}{\partial t} + \frac{\partial v_i}{\partial x_j}\dot{x}_j = \frac{\partial v_i}{\partial t} + (\mathbf{v} \cdot \nabla) v_i$$

where ∇ denotes the gradient with respect to the space variables only. Cut any region $G_o \subset E$ with boundary ∂G_o of class C^1. Since G_o is instantaneously in equilibrium, the balance of forces acting on G_o must be zero. These are:

(i) The inertial forces due to acceleration

$$\int_{G_o} \rho \big[\mathbf{v}_t + (\mathbf{v} \cdot \nabla)\mathbf{v}\big]\, dx.$$

(ii) The Kelvin forces due to pressure. Let $p(x,t)$ be the pressure at $x \in E$ at time t. The forces due to pressure on G are

$$\int_{\partial G_0} p \boldsymbol{\nu}\, d\sigma, \qquad \boldsymbol{\nu} = \{\text{outward unit normal to } \partial G_o\}.$$

(iii) The sum of the external forces, and the internal forces due to friction

$$-\int_{G_o} \mathbf{f}\, dx.$$

Therefore

$$\int_{G_o} \rho[\mathbf{v}_t + (\mathbf{v}\cdot\nabla)\mathbf{v}]\, dx = -\int_{\partial G_o} p\boldsymbol{\nu}\, d\sigma + \int_{G_o} \mathbf{f}\, dx.$$

By Green's theorem

$$\int_{\partial G_o} p\boldsymbol{\nu}\, d\sigma = \int_{G_o} \nabla p\, dx.$$

Therefore by the arbitrariness of $G_o \subset E$

$$\rho\,[\mathbf{v}_t + (\mathbf{v}\cdot\nabla)\mathbf{v}] = -\nabla p + \mathbf{f} \quad \text{in } E\times\mathbb{R}. \tag{6.1}$$

Assume the following physical, modeling assumptions:

(a) The fluid moves with small relative velocity and small time variations of density. Therefore second-order terms of the type $v_i v_{j,x_h}$ and $\rho_t v_i$ are negligible with respect to first order terms.
(b) Heat transfer is slower than pressure drops, i.e., the process is adiabatic and $\rho = h(p)$ for some $h \in C^2(\mathbb{R})$.

Expanding $h(\cdot)$ about the equilibrium pressure p_o, renormalized to be zero, gives

$$\rho = a_o p + a_1 p^2 + \cdots.$$

Assume further that the pressure is close to the equilibrium pressure, so that all terms of order higher than one are negligible when compared to $a_o p$. These assumptions in (6.1) yield

$$\frac{\partial}{\partial t}(\rho\mathbf{v}) = -\nabla p + \mathbf{f} \quad \text{in } E\times\mathbb{R}.$$

Now take the divergence of both sides to obtain

$$\frac{\partial}{\partial t}\operatorname{div}(\rho\mathbf{v}) = -\Delta p + \operatorname{div}\mathbf{f} \quad \text{in } E\times\mathbb{R}.$$

From the continuity equation

$$\operatorname{div}(\rho\mathbf{v}) = -\rho_t = -a_o p_t.$$

Combining these remarks gives the equation of the pressure in the propagation of sound waves in a fluid, in the form

$$\frac{\partial^2 p}{\partial t^2} - c^2\Delta p = f \quad \text{in } E\times\mathbb{R} \tag{6.2}$$

where

$$c^2 = \frac{1}{a_o}, \qquad f = -\frac{\operatorname{div}\mathbf{f}}{a_o}.$$

Equation (6.2) is the wave equation in three space dimensions ([119]).

7 The Navier–Stokes System

The system (6.1) is rather general and holds for any ideal fluid. If the fluid is incompressible, then $\rho = \text{const}$, and the continuity equation (2.4) gives

$$\operatorname{div} \mathbf{v} = 0. \tag{7.1}$$

If in addition the fluid is viscous, the internal forces due to friction can be represented by $\mu\rho\Delta\mathbf{v}$, where $\mu > 0$ is the *kinematic viscosity* ([160]). Therefore (6.1) yields the system

$$\frac{\partial}{\partial t}\mathbf{v} - \mu\Delta\mathbf{v} + (\mathbf{v}\cdot\nabla)\mathbf{v} + \frac{1}{\rho}\nabla p = \mathbf{f}_e \tag{7.2}$$

where $\mathbf{f}_e = \mathbf{f}/\rho$ are the external forces acting on the system. The unknowns are the three components of the velocity and the pressure p, to be determined from the system of four equations (7.1) and (7.2).

8 The Euler Equations

Let S denote the entropy function of a gas undergoing an adiabatic process. The pressure p and the density ρ are linked by the equation of state

$$p = f(S)\rho^{1+\alpha}, \qquad \alpha > 0 \tag{8.1}$$

for some smooth function $f(\cdot)$. The entropy $S\big(x(t), t\big)$ of an infinitesimal portion of the gas moving along the Lagrangian path $t \to x(t)$ is conserved. Therefore ([160])

$$\frac{d}{dt}S = 0$$

where formally

$$\frac{d}{dt} = \frac{\partial}{\partial t} + \mathbf{v}\cdot\nabla$$

is the *total derivative*. The system of the Euler equations of the process is

$$\rho\big[\mathbf{v}_t + (\mathbf{v}\cdot\nabla)\mathbf{v}\big] = -\nabla p + \mathbf{f} \tag{8.2}$$

$$\rho_t + \operatorname{div}(\mathbf{v}\rho) = 0 \tag{8.3}$$

$$\frac{\partial}{\partial t}\frac{p}{\rho^{1+\alpha}} + \mathbf{v}\cdot\nabla\frac{p}{\rho^{1+\alpha}} = 0. \tag{8.4}$$

The first is the pointwise balance of forces following Newton's law along the Lagrangian paths of the motion. The second is the conservation of mass, and the last is the conservation of entropy.

9 Isentropic Potential Flows

A flow is isentropic if $S = \text{const}$. In this case, the equation of state (8.1) permits one to define the pressure as a function of the density alone. Let $u \in C^2(\mathbb{R}^3 \times \mathbb{R})$ be the velocity potential, so that $\mathbf{v} = \nabla u$. Assume that $\mathbf{f} = 0$, and rewrite (8.2) as

$$\frac{\partial}{\partial t} u_{x_i} + u_{x_j} u_{x_i x_j} = -\frac{p_{x_i}}{\rho}, \qquad i = 1, 2, 3. \tag{9.1}$$

From this

$$\frac{\partial}{\partial x_i} \left(u_t + \frac{1}{2} |\nabla u|^2 + \int_0^p \frac{ds}{\rho(s)} \right) = 0, \qquad i = 1, 2, 3.$$

From the equation of state

$$\int_0^p \frac{ds}{\rho(s)} = \frac{1+\alpha}{\alpha} \frac{p}{\rho}.$$

Combining these calculations, gives the Bernoulli law for an isentropic potential flow[2]

$$u_t + \frac{1}{2} |\nabla u|^2 + \frac{1+\alpha}{\alpha} \frac{p}{\rho} = g \tag{9.2}$$

where $g(\cdot)$ is a function of t only. The positive quantity

$$c^2 = \frac{dp}{d\rho} = (1+\alpha) \frac{p}{\rho}$$

has the dimension of the square of a velocity, and c represents the local speed of sound. Notice that c need not be constant. Next multiply the ith equation in (9.1) by u_{x_i} and add for $i = 1, 2, 3$ to obtain

$$\frac{1}{2} \frac{\partial}{\partial t} |\nabla u|^2 + \frac{1}{2} \nabla u \cdot \nabla |\nabla u|^2 = -\frac{1}{\rho} \nabla p \cdot \nabla u. \tag{9.3}$$

Using the continuity equation

$$\begin{aligned} -\frac{1}{\rho} \nabla p \cdot \nabla u &= -\nabla \frac{p}{\rho} \nabla u - \frac{p}{\rho} \frac{1}{\rho} \nabla \rho \cdot \nabla u \\ &= -\nabla \frac{p}{\rho} \cdot \nabla u + \frac{p}{\rho} \frac{1}{\rho} \rho_t + \frac{p}{\rho} \Delta u. \end{aligned}$$

---

[2]Daniel Bernoulli, 1700–1782, botanist and physiologist, made relevant discoveries in hydrodynamics. His father, Johann B. 1667–1748, and his uncle Jakob 1654–1705, brother of Johann, were both mathematicians. Jakob and Johann are known for their contributions to the calculus of variations.

From the equation of state

$$\frac{d}{dt}f(S) = \frac{d}{dt}\frac{p}{\rho^{1+\alpha}} = \frac{1}{\rho^\alpha}\frac{d}{dt}\frac{p}{\rho} + \frac{p}{\rho}\frac{d}{dt}\frac{1}{\rho^\alpha} = 0.$$

From this, expanding the total derivative

$$\frac{\partial}{\partial t}\frac{p}{\rho} + \nabla u \cdot \nabla\frac{p}{\rho} - \alpha\frac{p}{\rho^2}[\rho_t + \nabla u \cdot \nabla\rho] = 0.$$

Again, by the equation of continuity

$$\rho_t + \nabla u \cdot \nabla\rho = \rho\Delta u.$$

Therefore

$$-\nabla\frac{p}{\rho} \cdot \nabla u = \frac{\partial}{\partial t}\frac{p}{\rho} + \alpha\frac{p}{\rho}\Delta u.$$

Combining these calculations in (9.3) gives

$$c^2\Delta u - \frac{1}{2}\nabla u \cdot \nabla|\nabla u|^2 = \frac{1}{2}\frac{\partial}{\partial t}|\nabla u|^2 - \frac{1}{\rho}\frac{\partial}{\partial t}p. \tag{9.4}$$

9.1 Steady Potential Isentropic Flows

For steady flows, rewrite (9.4) in the form

$$(c^2\delta_{ij} - u_{x_i}u_{x_j})u_{x_ix_j} = 0. \tag{9.5}$$

The matrix of the coefficients of the second derivatives $u_{x_ix_j}$ is

$$\left(\delta_{ij} - \frac{u_{x_i}u_{x_j}}{c^2}\right)$$

and its eigenvalues are

$$\lambda_1 = 1 - \frac{|\nabla u|^2}{c^2} \qquad \text{and} \qquad \lambda_2 = 1.$$

Using the steady-state version of the Bernoulli law (9.2) gives the first eigenvalue in terms only of the pressure p and the density ρ. The ratio $M = |\nabla u|/c$ of the speed of a body to the speed of sound in the surrounding medium is called the Mach number.[3]

[3]Ernst Mach, 1838–1916. Mach one is the speed of sound; Mach two is twice the speed of sound;...

10 Partial Differential Equations

The equations and systems of the previous sections are examples of PDEs. Let $u \in C^m(E)$ for some $m \in \mathbb{N}$, and for $j = 1, 2, \dots, m$, let $D^j u$ denote the vector of all the derivatives of u of order j. For example, if $N = m = 2$, denoting (x, y) the coordinates in $\mathbb{R}^2$

$$D^1 u = (u_x, u_y) \qquad \text{and} \qquad D^2 u = (u_{xx}, u_{xy}, u_{yy}).$$

A partial differential equation (PDE) is a functional link among the variables

$$x,\ u,\ D^1 u,\ D^2 u,\ \dots,\ D^m u$$

that is

$$F(x, u, D^1 u, D^2 u, \dots, D^m u) = 0.$$

The PDE is of order m if the gradient of F with respect to $D^m u$ is not identically zero. It is linear if for all $u, v \in C^m(E)$ and all $\alpha, \beta \in \mathbb{R}$

$$\begin{aligned}
&F\big(x, (\alpha u + \beta v), D^1(\alpha u + \beta v), D^2(\alpha u + \beta v), \dots, D^m(\alpha u + \beta v)\big) \\
&= \alpha F\big(x, u, D^1 u, D^2 u, \dots, D^m u\big) + \beta F\big(x, v, D^1 v, D^2 v, \dots, D^m v\big).
\end{aligned}$$

It is *quasi-linear* if it is linear with respect to the highest order derivatives. Typically a quasi-linear PDE takes the form

$$\sum_{m_1 + \cdots + m_N = m} a_{m_1, \dots, m_N} \frac{\partial^m u}{\partial^{m_1} x_1 \cdots \partial^{m_N} x_N} + F_o = 0$$

where m_j are non-negative integers and the coefficients $a_{m_1, \dots, m_N}$, and the forcing term F_o, are given smooth functions of $(x, u, D^1 u, D^2 u, \dots, D^{m-1} u)$. If the PDE is quasi-linear, the sum of the terms involving the derivatives of highest order, is the principal part of the PDE.

1

Quasi-Linear Equations and the Cauchy–Kowalewski Theorem

1 Quasi-Linear Second-Order Equations in Two Variables

Let (x, y) denote the variables in $\mathbb{R}^2$, and consider the quasi-linear equation

$$Au_{xx} + 2Bu_{xy} + Cu_{yy} = D \tag{1.1}$$

where $(x, y, u_x, u_y) \to A, B, C, D(x, y, u_x, u_y)$ are given smooth functions of their arguments. The equation is of order two if at least one of the coefficients A, B, C is not identically zero. Let Γ be a curve in $\mathbb{R}^2$ of parametric representation

$$\Gamma = \begin{cases} x = \xi(t) \\ y = \eta(t) \end{cases} \in C^2(-\delta, \delta) \text{ for some } \delta > 0.$$

On Γ, prescribe the Cauchy data

$$u\big|_\Gamma = v, \quad u_x\big|_\Gamma = \varphi, \quad u_y\big|_\Gamma = \psi \tag{1.2}$$

where $t \to v(t), \varphi(t), \psi(t)$ are given functions in $C^2(-\delta, \delta)$. Since

$$\frac{d}{dt}u\big(\xi(t), \eta(t)\big) = u_x\xi' + u_y\eta' = \varphi\xi' + \psi\eta' = v'$$

of the three functions v, φ, ψ, only two can be assigned independently. The Cauchy problem for (1.1) and Γ, consists in finding $u \in C^2(\mathbb{R}^2)$ satisfying the PDE and the Cauchy data (1.2). Let u be a solution of the Cauchy problem (1.1)–(1.2), and compute its second derivatives on Γ. By (1.1) and the Cauchy data

$$\begin{aligned} Au_{xx} + 2Bu_{xy} + Cu_{yy} &= D \\ \xi' u_{xx} + \eta' u_{xy} \qquad\quad &= \varphi' \\ \xi' u_{xy} + \eta' u_{yy} &= \psi'. \end{aligned}$$

E. DiBenedetto, *Partial Differential Equations: Second Edition*,
Cornerstones, DOI 10.1007/978-0-8176-4552-6_2,

Here A, B, C are known, since they are computed on Γ. Precisely

$$A, B, C\big|_\Gamma = A, B, C(\xi, \eta, v, \varphi, \psi).$$

Therefore, u_{xx}, u_{xy}, and u_{yy} can be computed on Γ, provided

$$\det \begin{pmatrix} A & 2B & C \\ \xi' & \eta' & 0 \\ 0 & \xi' & \eta' \end{pmatrix} \neq 0. \tag{1.3}$$

We say that Γ is a characteristic curve if (1.3) does not hold, i.e., if

$$A\eta'^2 - 2B\xi'\eta' + C\xi'^2 = 0. \tag{1.4}$$

In general, the property of Γ being a characteristic depends on the Cauchy data assigned on it. If Γ admits a local representation of the type

$$y = y(x) \quad \text{in a neighborhood of some } x_o \in \mathbb{R}, \tag{1.5}$$

the characteristics are the graphs of the possible solutions of the differential equation

$$y' = \frac{B \pm \sqrt{B^2 - AC}}{A}.$$

Associate with (1.1) the matrix of the coefficients

$$M = \begin{pmatrix} A & B \\ B & C \end{pmatrix}.$$

Using (1.4) and the matrix M, we classify, locally, the family of quasi-linear equations (1.1) as **elliptic** if $\det M > 0$, i.e., if there exists no real characteristic; **parabolic** if $\det M = 0$, i.e., if there exists one family of real characteristics; **hyperbolic** if $\det M < 0$, i.e., if there exist two families of real characteristics. The elliptic, parabolic, or hyperbolic nature of (1.1) may be different in different regions of $\mathbb{R}^2$. For example, the Tricomi equation ([152])

$$yu_{xx} - u_{yy} = 0$$

is elliptic in the region $[y < 0]$, parabolic on the x-axis and hyperbolic in the upper half-plane $[y > 0]$. The characteristics are solutions of $\sqrt{y}y' = \pm 1$ in the upper half-plane $[y > 0]$.

The elliptic, parabolic, or hyperbolic nature of the PDE may also depend upon the solution itself. As an example, consider the equation of steady compressible fluid flow of a gas of density u and velocity $\nabla u = (u_x, u_y)$ in $\mathbb{R}^2$, introduced in (9.5) of the Preliminaries

$$(c^2 - u_x^2)u_{xx} - 2u_x u_y u_{xy} + (c^2 - u_y^2)u_{yy} = 0$$

where $c > 0$ is the speed of sound. Compute

$$\det M = \det \begin{pmatrix} c^2 - u_x^2 & -u_x u_y \\ -u_x u_y & c^2 - u_y^2 \end{pmatrix} = c^2(c^2 - |\nabla u|^2).$$

Therefore the equation is elliptic for sub-sonic flow ($|\nabla u| < c$), parabolic for sonic flow ($|\nabla u| = c$), and hyperbolic for super-sonic flow ($|\nabla u| > c$). The Laplace equation

$$\Delta u = u_{xx} + u_{yy} = 0$$

is elliptic. The heat equation

$$H(u) = u_y - u_{xx} = 0$$

is parabolic with characteristic lines $y =$ const. The wave equation

$$\Box u = u_{yy} - c^2 u_{xx} = 0 \quad c \in \mathbb{R}$$

is hyperbolic with characteristic lines $x \pm cy =$ const.

2 Characteristics and Singularities

If Γ is a characteristic, the Cauchy problem (1.1)–(1.2) is in general not solvable, since the second derivatives of u cannot be computed on Γ. We may attempt to solve the PDE (1.1) on each side of Γ and then piece together the functions so obtained. Assume that Γ divides $\mathbb{R}^2$ into two regions E_1 and E_2 and let $u_i \in C^2(\bar{E}_i)$, for $i = 1, 2$, be possible solutions of (1.1) in E_i satisfying the Cauchy data (1.2). These are taken in the sense of

$$\lim_{\substack{(x,y)\to(\xi(t),\eta(t)) \\ (x,y)\in E_i}} u_i(x,y),\ u_{i,x}(x,y),\ u_{i,y}(x,y) = v(t), \varphi(t), \psi(t).$$

Setting

$$u = \begin{cases} u_1 & \text{in } E_1 \\ u_2 & \text{in } E_2 \end{cases}$$

the function u is of class C^1 across Γ. If $f_i \in C(\bar{E}_i)$, for $i = 1, 2$, and

$$f = \begin{cases} f_1 & \text{in } E_1 \\ f_2 & \text{in } E_2 \end{cases}$$

let $[f]$ denote the jump of f across Γ, i.e.,

$$[f](t) = \lim_{\substack{(x,y)\to(\xi(t),\eta(t)) \\ (x,y)\in E_1}} f_1(x,y) - \lim_{\substack{(x,y)\to(\xi(t),\eta(t)) \\ (x,y)\in E_2}} f_2(x,y).$$

From the assumptions on u,

$$[u] = [u_x] = [u_y] = 0. \tag{2.1}$$

From (1.1),

$$A[u_{xx}] + 2B[u_{xy}] + C[u_{yy}] = 0. \tag{2.2}$$

Assume that Γ has the local representation (1.5). Then using (2.1), compute

$$[u_{xx}] + [u_{xy}]y' = 0, \qquad [u_{xy}] + [u_{yy}]y' = 0.$$

Therefore

$$[u_{xx}] = [u_{yy}]y'^2, \qquad [u_{xy}] = -[u_{yy}]y'. \tag{2.3}$$

Let $J = [u_{yy}]$ denote the jump across Γ of the second y-derivative of u. From (2.2) and (2.3)

$$J(Ay'^2 - 2By' + C) = 0.$$

If Γ is not a characteristic, then $(Ay'^2 - 2By' + C) \neq 0$. Therefore $J = 0$, and u is of class C^2 across Γ. If $J \neq 0$, then Γ must be a characteristic. Thus if a solution of (1.1) in a region $E \subset \mathbb{R}^2$ suffers discontinuities in the second derivatives across a smooth curve, these must occur across a characteristic.

2.1 Coefficients Independent of u_x and u_y

Assume that the coefficients A, B, C and the term D are independent of u_x and u_y, and that $u \in C^3(\bar{E}_i)$, $i = 1, 2$. Differentiate (1.1) with respect to y in E_i, form differences, and take the limit as $(x, y) \to \Gamma$ to obtain

$$A[u_{xxy}] + 2B[u_{xyy}] + C[u_{yyy}] = 0. \tag{2.4}$$

Differentiating the jump J of u_{yy} across Γ gives

$$J' = [u_{xyy}] + [u_{yyy}]y'. \tag{2.5}$$

From the second jump condition in (2.3), by differentiation

$$-y'J' - y''J = [u_{xxy}] + [u_{xyy}]y'. \tag{2.6}$$

We eliminate $[u_{xxy}]$ from (2.4) and (2.6) and use (2.5) to obtain

$$\begin{aligned} A(y'J' + y''J) &= (2B - Ay')[u_{xyy}] + C[u_{yyy}] \\ &= (2B - Ay')J' + (Ay'^2 - 2By' + C)[u_{yyy}]. \end{aligned}$$

Therefore, if Γ is a characteristic

$$2(B - Ay')J' = Ay''J.$$

This equation describes how the jump J of u_{yy} at some point $P \in \Gamma$ propagates along Γ. In particular, either J vanishes identically, or it is never zero on Γ.

3 Quasi-Linear Second-Order Equations

Let E be a region in $\mathbb{R}^N$, and let $u \in C^2(E)$. A quasi-linear equation in E takes the form

$$A_{ij} u_{x_i x_j} = F \tag{3.1}$$

where we have adopted the summation notation and

$$(x, u, \nabla u) \to A_{ij}, F(x, u, \nabla u) \quad \text{for} \quad i, j = 1, 2, \dots, N$$

are given smooth functions of their arguments. The equation is of order two if not all the coefficients A_{ij} are identically zero. By possibly replacing A_{ij} with

$$\frac{A_{ij} + A_{ji}}{2}$$

we may assume that the matrix (A_{ij}) of the coefficients is symmetric. Let Γ be a hypersurface of class C^2 in $\mathbb{R}^N$, given as a level set of $\Phi \in C^2(E)$; say for example, $\Gamma = [\Phi = 0]$. Assume that $\nabla\Phi \neq 0$ and let $\boldsymbol{\nu} = \nabla\Phi/|\nabla\Phi|$ be the unit normal to Γ oriented in the direction of increasing Φ. For $x \in \Gamma$, introduce a local system of $N-1$ mutually orthogonal unit vectors $\{\boldsymbol{\tau}_1, \dots, \boldsymbol{\tau}_{N-1}\}$ chosen so that the n-tuple $\{\boldsymbol{\tau}_1, \dots, \boldsymbol{\tau}_{N-1}, \boldsymbol{\nu}\}$ is congruent to the orthonormal Cartesian system $\{\mathbf{e}_1, \dots, \mathbf{e}_N\}$. Given $f \in C^1(E)$, compute the derivatives of f, normal and tangential to Γ from

$$D_{\boldsymbol{\nu}} f = \nabla f \cdot \boldsymbol{\nu}, \quad D_{\boldsymbol{\tau}_j} f = \nabla f \cdot \boldsymbol{\tau}_j, \quad j = 1, \dots, N-1.$$

If $\tau_{i,j} = \boldsymbol{\tau}_i \cdot \mathbf{e}_j$, and $\nu_j = \boldsymbol{\nu} \cdot \mathbf{e}_j$, are the projections of $\boldsymbol{\tau}_i$ and $\boldsymbol{\nu}$ on the coordinate axes

$$D_{\boldsymbol{\tau}_j} f = (\tau_{j,1}, \dots, \tau_{j,N}) \cdot \nabla f \qquad \text{and} \qquad D_{\boldsymbol{\nu}} f = (\nu_1, \dots, \nu_N) \cdot \nabla f.$$

Introduce the unitary matrix

$$T = \begin{pmatrix} \tau_{1,1} & \tau_{1,2} & \dots & \tau_{1,N} \\ \tau_{2,1} & \tau_{2,2} & \dots & \tau_{2,N} \\ \vdots & \vdots & \ddots & \vdots \\ \tau_{N-1,1} & \tau_{N-1,2} & \dots & \tau_{N-1,N} \\ \nu_1 & \nu_2 & \dots & \nu_N \end{pmatrix} \tag{3.2}$$

and write

$$\nabla f = T^{-1} \begin{pmatrix} D_{\boldsymbol{\tau}} f \\ D_{\boldsymbol{\nu}} f \end{pmatrix} \qquad \text{where } T^{-1} = T^t. \tag{3.3}$$

The Cauchy data of u on Γ are

$$u\big|_\Gamma = v, \quad D_{\boldsymbol{\tau}_j} u\big|_\Gamma = \varphi_j,\ j = 1, \dots, N-1, \quad D_{\boldsymbol{\nu}} u\big|_\Gamma = \psi \tag{3.4}$$

regarded as restrictions to Γ of smooth functions defined on the whole of E. These must satisfy the compatibility conditions

$$D_{\boldsymbol{T}_j} v = \varphi_j, \qquad j = 1, \ldots, N-1. \tag{3.5}$$

Therefore only v and ψ can be given independently. The Cauchy problem for (3.1) and Γ consists in finding a function $u \in C^2(E)$ satisfying the PDE and the Cauchy data (3.4). If u is a solution to the Cauchy problem (3.1)–(3.4), compute the second derivatives of u on Γ

$$\begin{aligned} u_{x_i} &= \tau_{k,i} u_{\tau_k} + \nu_i u_\nu \\ u_{x_i x_j} &= \tau_{k,i} \frac{\partial}{\partial x_j} u_{\tau_k} + \nu_i \frac{\partial}{\partial x_j} u_\nu \\ &= \tau_{k,i}\tau_{l,j} u_{\tau_k \tau_l} + \tau_{k,i}\nu_j u_{\tau_k \nu} + \tau_{k,j}\nu_i u_{\tau_k \nu} + \nu_i \nu_j u_{\nu\nu}. \end{aligned} \tag{3.6}$$

From the compatibility conditions (3.4) and (3.5)

$$D_{\boldsymbol{T}_i}(D_{\boldsymbol{T}_j} u) = D_{\boldsymbol{T}_i} \varphi_j, \qquad D_{\boldsymbol{T}_i}(D_{\boldsymbol{\nu}} u) = D_{\boldsymbol{T}_i} \psi.$$

Therefore, of the terms on the right-hand side of (3.6), all but the last are known on Γ. Using the PDE, one computes

$$A_{ij} \nu_i \nu_j u_{\nu\nu} = \tilde{F} \quad \text{on } \Gamma \tag{3.7}$$

where $\tilde{F}$ is a known function of Γ and the Cauchy data on it. We conclude that $u_{\nu\nu}$, and hence all the derivatives $u_{x_i x_j}$, can be computed on Γ provided

$$A_{ij} \Phi_{x_i} \Phi_{x_j} \neq 0. \tag{3.8}$$

Both (3.7) and (3.8) are computed at fixed points $P \in \Gamma$. We say that Γ is a characteristic at P if (3.8) is violated, i.e., if

$$(\nabla \Phi)^t (A_{ij}) (\nabla \Phi) = 0 \qquad \text{at } P.$$

Since (A_{ij}) is symmetric, its eigenvalues are real and there is a unitary matrix U such that

$$U^{-1}(A_{ij})U = \begin{pmatrix} \lambda_1 & 0 & \ldots & 0 \\ 0 & \lambda_2 & \ldots & 0 \\ \vdots & \vdots & \ddots & \vdots \\ 0 & 0 & \ldots & \lambda_N \end{pmatrix}.$$

Let $\xi = Ux^t$ denote the coordinates obtained from x by the rotation induced by U. Then

$$(\nabla \Phi)^t (A_{ij}) (\nabla \Phi) = [U^{-1}(\nabla \Phi)]^t U^{-1}(A_{ij}) U [U^{-1}(\nabla \Phi)] = \lambda_i \Phi_{\xi_i}^2.$$

Therefore Γ is a characteristic at P if

$$\lambda_i \Phi_{\xi_i}^2 = 0. \tag{3.9}$$

Writing this for all $P \in E$ gives a first-order PDE in Φ. Its solutions permit us to find the characteristic surfaces as the level sets $[\Phi = \text{const}]$.

3.1 Constant Coefficients

If the coefficients A_{ij} are constant, (3.9) is a first-order PDE with constant coefficients. The PDE in (3.1) is classified according to the number of positive and negative eigenvalues of (A_{ij}). Let p and n denote the number of positive and negative eigenvalues of (A_{ij}), and consider the pair (p, n). The equation (3.1) is classified as **elliptic** if either $(p, n) = (N, 0)$ or $(p, n) = (0, N)$. In either of these cases, it follows from (3.9) that

$$0 = |\lambda_i \Phi_{\xi_i}^2| \geq \min_{1 \leq i \leq N} |\lambda_i| |\nabla \Phi|^2.$$

Therefore there exist no characteristic surfaces.

Equation (3.1) is classified as **hyperbolic** if $p + n = N$ and $p, n \geq 1$. Without loss of generality, we may assume that eigenvalues are ordered so that $\lambda_1, \dots, \lambda_p$ are positive and that $\lambda_{p+1}, \dots, \lambda_N$ are negative. In such a case, (3.9) takes the form

$$\sum_{i=1}^{p} \lambda_i \Phi_{\xi_i}^2 = \sum_{j=p+1}^{N} |\lambda_j| \Phi_{\xi_j}^2.$$

This is solved by

$$\Phi^{\pm}(\xi) = \sum_{i=1}^{p} \sqrt{\lambda_i} \xi_i \pm k \sum_{j=p+1}^{N} \sqrt{|\lambda_j|} \xi_j$$

where

$$k^2 = \Big(\sum_{i=1}^{p} \lambda_i^2 \Big) \Big/ \Big(\sum_{j=p+1}^{N} \lambda_j^2 \Big).$$

The hyperplanes $[\Phi^{\pm} = \text{const}]$ are two families of characteristic surfaces for (3.1). In the literature these PDE are further classified according to the values of p and n. Namely they are called *hyperbolic* if either $p = 1$ or $n = 1$. Otherwise they are called *ultra-hyperbolic*.

Equation (3.1) is classified as **parabolic** if $p + n < N$. Then at least one of the eigenvalues is zero. If, say, $\lambda_1 = 0$, then (3.9) is solved by any function of ξ_1 only, and the hyperplane $\xi_1 = \text{const}$ is a characteristic surface.

3.2 Variable Coefficients

In analogy with the case of constant coefficients we classify the PDE in (3.1) at each point $P \in E$ as elliptic, hyperbolic, or parabolic according to the number of positive and negative eigenvalues of (A_{ij}) at P. The classification is local, and for coefficients depending on the solution and its gradient, the nature of the equation may depend on its own solutions.

4 Quasi-Linear Equations of Order $m \geq 1$

An N-dimensional multi-index α, of size $|\alpha|$, is an N-tuple of non-negative integers whose sum is $|\alpha|$, i.e.,

$$\alpha = (\alpha_1, \ldots, \alpha_N), \quad \alpha_i \in \mathbb{N} \cup \{0\},\ i = 1, \ldots, N, \quad |\alpha| = \sum_{i=1}^{N} \alpha_i.$$

If $f \in C^m(E)$ for some $m \in \mathbb{N}$, and α is a multi-index of size m, let

$$D^\alpha f = \frac{\partial^{\alpha_1}}{\partial x_1^{\alpha_1}} \frac{\partial^{\alpha_2}}{\partial x_2^{\alpha_2}} \cdots \frac{\partial^{\alpha_N}}{\partial x_N^{\alpha_N}} f.$$

If $|\alpha| = 0$ let $D^\alpha f = f$. By $D^{m-1} f$ denote the vector of all the derivatives $D^\alpha f$ for $0 \leq |\alpha| \leq m-1$. Consider the quasi-linear equation

$$\sum_{|\alpha|=m} A_\alpha D^\alpha u = F \tag{4.1}$$

where $(x, D^{m-1}u) \to A_\alpha, F(x, D^{m-1}u)$ are given smooth functions of their arguments. The equation is of order m if not all the coefficients A_α are identically zero. If $\mathbf{v} = (v_1, \ldots, v_N)$ is a vector in $\mathbb{R}^N$ and α is an N-dimensional multi-index, let

$$\mathbf{v}^\alpha = v_1^{\alpha_1} v_2^{\alpha_2} \cdots v_N^{\alpha_N}.$$

Prescribe a surface Γ as in the previous section and introduce the matrix T as in (3.2), so that the differentiation formula (3.3) holds. Denoting by $\beta = (\beta_1, \ldots, \beta_{N-1})$ an $(N-1)$-dimensional multi-index of size $|\beta| \leq m$, set

$$D_{\boldsymbol{T}}^\beta f = D_{\boldsymbol{T}_1}^{\beta_1} D_{\boldsymbol{T}_2}^{\beta_2} \cdots D_{\boldsymbol{T}_{N-1}}^{\beta_{N-1}} f.$$

Write N-dimensional multi-indices as $\alpha = (\beta, s)$, where s is a non-negative integer, and for $|\alpha| \leq m$, set

$$D_{\boldsymbol{T},\boldsymbol{\nu}}^\alpha f = D_{\boldsymbol{T}}^\beta D_{\boldsymbol{\nu}}^s f.$$

The Cauchy data of u on Γ are

$$D_{\boldsymbol{T},\boldsymbol{\nu}}^\alpha u\big|_\Gamma = f_\alpha \in C^m(E) \quad \text{for all } |\alpha| < m. \tag{4.2}$$

Among these we single out the Dirichlet data

$$u\big|_\Gamma = f_o, \tag*{(4.2)$_{\rm D}$}$$

the normal derivatives

$$D_{\boldsymbol{\nu}}^s u\big|_\Gamma = f_s, \qquad |\alpha| = s \leq m-1, \tag*{(4.2)$_{\boldsymbol{\nu}}$}$$

and the tangential derivatives

$$D_{\boldsymbol{T}}^{\beta} u\big|_{\Gamma} = f_{\beta}, \qquad |\beta| < m. \tag{4.2$_{\boldsymbol{T}}$}$$

Of these, only $(4.2)_{\mathrm{D}}$ and $(4.2)_{\boldsymbol{\nu}}$ can be given independently. The remaining ones must be assigned to satisfy the compatibility conditions

$$D_{\boldsymbol{\nu}}^{s} f_{\beta} = D_{\boldsymbol{T}}^{\beta} f_{s} \quad \text{for all } |\beta| \ge 0, \quad |\beta| + s \le m - 1. \tag{4.3}$$

The Cauchy problem for (4.1) consists in finding a function $u \in C^m(E)$ satisfying (4.1) in E and the Cauchy data (4.2) on Γ.

4.1 Characteristic Surfaces

If u is a solution of the Cauchy problem, compute its derivatives of order m on Γ, by using (4.1), the Cauchy data (4.2) and the compatibility conditions (4.3). Proceeding as in formula (3.6), for a multi-index α of size $|\alpha| = m$

$$D^{\alpha} u = \nu_1^{\alpha_1} \cdots \nu_N^{\alpha_N} D_{\boldsymbol{\nu}}^{m} u + g \qquad \text{on } \Gamma$$

where g is a known function that can be computed a-priori in terms of Γ, the Cauchy data (4.2), and the compatibility conditions (4.3). Putting this in (4.1) gives

$$\sum_{|\alpha|=m} A_{\alpha} \boldsymbol{\nu}^{\alpha} D_{\boldsymbol{\nu}}^{m} u = \tilde{F}, \qquad \boldsymbol{\nu}^{\alpha} = \nu_1^{\alpha_1} \cdots \nu_N^{\alpha_N}$$

where $\tilde{F}$ is known in terms of Γ and the data. Therefore all the derivatives, normal and tangential, up to order m can be computed on Γ, provided

$$\sum_{|\alpha|=m} A_{\alpha} \boldsymbol{\nu}^{\alpha} \ne 0. \tag{4.4}$$

We say that Γ is a characteristic surface if (4.4) is violated, i.e.,

$$\sum_{|\alpha|=m} A_{\alpha} (D\Phi)^{\alpha} = 0. \tag{4.5}$$

In general, the property of Γ being a characteristic depends on the Cauchy data assigned on it, unless the coefficients A_{α} are independent of $D^{m-1}u$.

Condition (4.5) was derived at a fixed point of Γ. Writing it at all points of E gives a first-order non-linear PDE in Φ whose solutions permit one to find the characteristics associated with (4.1) as the level sets of Φ. To (4.1) associate the *characteristic form*

$$\mathcal{L}(\xi) = A_{\alpha} \xi^{\alpha}.$$

If $\mathcal{L}(\xi) \ne 0$ for all $\xi \in \mathbb{R}^N - \{0\}$, then there are no characteristic hypersurfaces, and (4.1) is said to be *elliptic*.

5 Analytic Data and the Cauchy–Kowalewski Theorem

A real-valued function f defined in an open set $G \subset \mathbb{R}^k$, for some $k \in \mathbb{N}$, is analytic at $\eta \in G$, if in a neighborhood of η, $f(y)$ can be represented as a convergent power series of $y - \eta$. The function f is analytic in G if it is analytic at every $\eta \in G$.

Consider the Cauchy problem for (4.1) with analytic data. Precisely, assume that Γ is non-characteristic and analytic about one of its points x_o; the Cauchy data (4.2) satisfy the compatibility conditions (4.3) and are analytic at x_o. Finally, the coefficients A_α and the free term F are analytic about the point $(x_o, u(x_o), D^{m-1}u(x_o))$.

The Cauchy–Kowalewski Theorem asserts that under these circumstances, the Cauchy problem (4.1)–(4.2) has a solution u, analytic at x_o. Moreover, the solution is unique within the class of analytic solutions at x_o.

5.1 Reduction to Normal Form ([19])

Up to an affine transformation of the coordinates, we may assume that x_o coincides with the origin and that Γ is represented by the graph of $x_N = \Phi(\bar{x})$, with $\bar{x} = (x_1, \dots, x_{N-1})$, where $\bar{x} \to \Phi(\bar{x})$ is analytic at the origin of $\mathbb{R}^{N-1}$. Flatten Γ about the origin by introducing new coordinates $(\bar{x}, t)$ where $t = x_N - \Phi(\bar{x})$. In this way Γ becomes a $(N-1)$-dimensional open neighborhood of the origin lying on the hyperplane $t = 0$. Continue to denote by u, A_α, and F the transformed functions and rewrite (4.1) as

$$A_{(0,\dots,m)} \frac{\partial^m}{\partial t^m} u = \sum_{\substack{|\beta|+s=m \\ 0<s<m}} A_{(\beta,s)} \frac{\partial^s}{\partial t^s} D^\beta u + \sum_{|\beta|=m} A_{(\beta,0)} D^\beta u + F. \tag{5.1}$$

Here D^β operates only on the variables $\bar{x}$. Since $[t = 0]$ is not a characteristic surface, (4.4) implies

$$A_{(0,\dots,m)} \big(\bar{x}, 0, D^{m-1}u(\bar{x}, 0)\big) \neq 0$$

and this continues to hold in a neighborhood of $\big(0, 0, D^{m-1}u(0,0)\big)$, since the functions A_α are analytic near such a point. Divide (5.1) by the coefficient of $D_t^m u$ and continue to denote by the same letters the transformed terms on the right-hand side. Next, introduce the vector $\mathbf{u} = D^{m-1}u$, and let $u_\alpha = D^\alpha u$ for $|\alpha| \le m-1$, be one component of this vector. If $\alpha = (0, \dots, m-1)$, then the derivative

$$\frac{\partial}{\partial t} u_\alpha = \frac{\partial^m}{\partial t^m} u$$

satisfies (5.1). If $\alpha = (\beta, s)$ and $s \le (m-2)$, then

$$\frac{\partial}{\partial t} u_\alpha = \frac{\partial}{\partial x_i} \tilde{u}_\alpha$$

for some $i = 1, \dots, N-1$ and some component $\tilde{u}_\alpha$ of the vector $\mathbf{u}$. Therefore, the PDE of order m in (5.1) can be rewritten as a first-order system in the normal form

$$\frac{\partial}{\partial t} u_i = A_{ijk} \frac{\partial}{\partial x_j} u_k + F_i(\bar{x}, t, \mathbf{u}).$$

The Cauchy data on Γ reduce to $\mathbf{u}(\bar{x}, 0) = \mathbf{f}(\bar{x})$. Therefore setting

$$\mathbf{v}(\bar{x}, t) = \mathbf{u}(\bar{x}, t) - \mathbf{f}(0)$$

and transforming the coefficients A_{ijk} and the function F accordingly, reduces the problem to one with Cauchy data vanishing at the origin.

The coefficients A_{ijk} as well as the free term F, can be considered independent of the variables $(\bar{x}, t)$. Indeed these can be introduced as auxiliary dependent variables by setting

$$u_j = x_j, \qquad \text{satisfying} \quad \frac{\partial}{\partial t} u_j = 0 \text{ and } u_j(\bar{x}, 0) = x_j$$

for $j = 1, \dots, N-1$, and

$$u_N = t, \qquad \text{satisfying} \quad \frac{\partial}{\partial t} u_N = 1 \text{ and } u_N(\bar{x}, 0) = 0.$$

These remarks permit one to recast the Cauchy problem (4.1)–(4.2), in the *normal form*

$$\begin{gathered} \mathbf{u} = (u_1, u_2, \dots, u_\ell), \quad \ell \in \mathbb{N} \\ \frac{\partial}{\partial t} \mathbf{u} = \mathbf{A}_j(\mathbf{u}) \frac{\partial}{\partial x_j} \mathbf{u} + \mathbf{F}(\mathbf{u}) \\ \mathbf{u}(x, 0) = \mathbf{u}_o(x), \qquad \mathbf{u}_o(0) = 0 \end{gathered} \tag{5.2}$$

where $\mathbf{A}_j = (A_{ik})_j$ are $\ell \times \ell$ matrices and $\mathbf{F} = (F_1, \dots, F_\ell)$ are known functions of their arguments. We have also renamed and indexed the *space* variables, on the hyperplane $t = 0$, as $x = (x_1, \dots, x_N)$.

Theorem 5.1 (Cauchy–Kowalewski) *Assume that $\mathbf{u} \to A_{ikj}(\mathbf{u}), F_i(\mathbf{u})$ and $x \to u_{o,i}(x)$ for $i, j, k = 1, \dots, \ell$ are analytic in a neighborhood of the origin. Then there exists a unique analytic solution of the Cauchy problem (5.2) in a neighborhood of the origin.*

For linear systems, the theorem was first proved by Cauchy ([20]). It was generalized to non-linear systems by Sonja Kowalewskaja ([84]). A generalization is also due to G. Darboux ([24]).

6 Proof of the Cauchy–Kowalewski Theorem

First, use the system in (5.2) to compute all the derivatives, at the origin, of a possible solution. Then using these numbers, construct the formal Taylor's series of an anticipated solution, say

$$\mathbf{u}(x,t) = \sum_{|\beta|+s\ge 0} \frac{D^\beta D_t^s \mathbf{u}(0,0)}{\beta! s!} x^\beta t^s \tag{6.1}$$

where $x^\beta = x_1^{\beta_1} \cdots x_N^{\beta_N}$, and $\beta! = \beta_1! \cdots \beta_N!$. If this series converges in a neighborhood of the origin, then (6.1) defines a function $\mathbf{u}$, analytic near $(0,0)$. Such a $\mathbf{u}$ is a solution to (5.2). Indeed, substituting the power series (6.1) on the left- and right-hand sides of the system in (5.2), gives two analytic functions whose derivatives of any order coincide at $(0,0)$. Thus they must coincide in a neighborhood of the origin. Uniqueness within the class of analytic solutions follows by the same unique continuation principle. Therefore the proof of the theorem reduces to showing that the series in (6.1), with all the coefficients $D^\beta D^s \mathbf{u}(0,0)$ computed from (5.2), converges about the origin.

The convergence of the series could be established, indirectly, by the method of the majorant ([77], 73–78). This was the original approach of A. Cauchy, followed also by S. Kowalewskaja and G. Darboux. The convergence of the series, can also be established by a direct estimation of all the derivatives of $\mathbf{u}$. This is the method we present here. This approach, originally due to Lax ([97]), has been further elaborated and extended by A. Friedman [49]. It has also been extended to an infinite dimensional setting by M. Shimbrot and R.E. Welland ([141]).

Let α and β be N-dimensional multi-indices, denote by m a non-negative integer and set $\alpha^m = \alpha_1^m \cdots \alpha_N^m$, and

$$\alpha + \beta = (\alpha_1 + \beta_1, \dots, \alpha_N + \beta_N), \qquad \binom{\alpha+\beta}{\alpha} = \frac{(\alpha+\beta)!}{\alpha!\beta!}.$$

Denote by ι the multi-index $\iota = (1, \dots, 1)$, so that

$$\beta + \iota = (\beta_1 + 1, \beta_2 + 1, \dots, \beta_N + 1).$$

The convergence of the series in (6.1) is a consequence of the following

Lemma 6.1 *There exist constants C_o and C such that,*

$$|D^\beta D_t^s \mathbf{u}(0,0)| \le C_o C^{|\beta|+s-1} \frac{\beta! s!}{(\beta+\iota)^2 (s+1)^2} \tag{6.2}$$

for all N-dimensional multi-indices β, and all non-negative integers s.

6.1 Estimating the Derivatives of u at the Origin

We will establish first the weaker inequality

Lemma 6.2 *There exist constants C_o and C such that*

$$|D^\beta D_t^s \mathbf{u}(0,0)| \le C_o C^{|\beta|+s-1} \frac{(|\beta|+s)!}{(\beta+\iota)^2 (s+1)^2} \tag{6.3}$$

for all N-dimensional multi-indices β and all non-negative integers s.

This inequality holds for $s = 0$, since $D^\beta \mathbf{u}(0,0) = D^\beta \mathbf{u}_o(0)$, and $\mathbf{u}_o$ is analytic. We will show that if it does hold for s it continues to hold for $s+1$.

Let $\gamma = (\gamma_1, \ldots, \gamma_\ell)$ be an ℓ-dimensional multi-index. Since $\mathbf{A}_j(\cdot)$ and $\mathbf{F}(\cdot)$ are analytic at the origin

$$A_{ikj}(\mathbf{u}) = \sum_{|\gamma|\ge 0} A_\gamma^{ikj} \mathbf{u}^\gamma, \qquad F_i(\mathbf{u}) = \sum_{|\gamma|\ge 0} F_\gamma^i \mathbf{u}^\gamma \tag{6.4}$$

where $\mathbf{u}^\gamma = u_1^{\gamma_1} \cdots u_\ell^{\gamma_\ell}$, and A_γ^{ikj} and F_γ^i are constants satisfying

$$|F_\gamma^i| + |A_\gamma^{ikj}| \le M_o M^{|\gamma|} \tag{6.5}$$

for all $i, k = 1, \ldots, \ell$ and all $j = 1, \ldots, N$, for two given positive constants M_o and M. From (5.2) and (6.4), compute

$$D^\beta D_t^{s+1} u_i = \sum_{|\gamma|\ge 1} A_\gamma^{ikj} D^\beta D_t^s \left(\mathbf{u}^\gamma \frac{\partial u_k}{\partial x_j} \right) + \sum_{|\gamma|\ge 1} F_\gamma^i D^\beta D_t^s \mathbf{u}^\gamma. \tag{6.6}$$

The induction argument requires some preliminary estimates.

7 Auxiliary Inequalities

Lemma 7.1 *Let m, n, i, j be non-negative integers such that $n \ge i$ and $m \ge j$. Then*

$$\sum_{i=0}^{n} \frac{1}{(n-i+1)^2 (i+1)^2} \le \frac{16}{(n+1)^2} \tag{7.1}$$

and

$$\text{and} \qquad \binom{n}{i}\binom{m}{j} \le \binom{n+m}{i+j}. \tag{7.2}$$

Proof If $1 \le i \le \frac{1}{2}n$, the argument of the sum is majorized by

$$\frac{4}{(n+1)^2}\frac{1}{(i+1)^2} \qquad \text{for all integers} \quad 1 \le i \le \tfrac{1}{2}n.$$

If $\frac{1}{2}n < i \le n$, it is majorized by

$$\frac{4}{(n+1)^2}\frac{1}{(n-i+1)^2} \qquad \text{for all integers} \quad \tfrac{1}{2}n < i \le n.$$

Thus in either case

$$\sum_{i=0}^{n} \frac{1}{(n-i+1)^2 (i+1)^2} \le \frac{8}{(n+1)^2} \sum_{i=0}^{\infty} \frac{1}{(i+1)^2}.$$

Inequality (7.2) is proved by induction on m, by making use of the identity

$$\binom{m+1}{j} = \binom{m}{j} + \binom{m}{j-1}. \qquad \blacksquare$$

Next, we establish a multi-index version of Lemma 7.1. If β and σ are N-dimensional multi-indices, we say that $\sigma \le \beta$, if and only if $\sigma_i \le \beta_i$ for all $i = 1, \dots, N$.

Lemma 7.2 *Let α and β be N-dimensional multi-indices. Then*

$$\sum_{\sigma\le\beta} \frac{1}{(\beta-\sigma+\iota)^2(\sigma+\iota)^2} \le \frac{16^N}{(\beta+\iota)^2}. \tag{7.3}$$

and

$$\binom{\alpha+\beta}{\alpha} \le \binom{|\alpha+\beta|}{|\alpha|}. \tag{7.4}$$

Proof (of (7.3)) For 1-dimensional multi-indices, (7.3) is precisely (7.1). Assuming that (7.3) holds true for multi-indices of dimension $k \in \mathbb{N}$, will show that it continues to hold for multi-indices of dimension $k+1$. Let β and σ be k-dimensional multi-indices and let

$$\tilde\beta = (\beta_1, \dots, \beta_k, \beta_{k+1}), \qquad \tilde\sigma = (\sigma_1, \dots, \sigma_k, \sigma_{k+1})$$

denote multi-indices of dimension $k+1$. Then by the induction hypothesis and (7.1)

$$\begin{aligned}\sum_{\tilde\sigma\le\tilde\beta} \frac{1}{(\tilde\beta-\tilde\sigma+\iota)^2(\tilde\sigma+\iota)^2} &= \sum_{\sigma\le\beta} \frac{1}{(\beta-\sigma+\iota)^2(\sigma+\iota)^2} \\ &\quad\times \sum_{\sigma_{k+1}=0}^{\beta_{k+1}} \frac{1}{(\beta_{k+1}-\sigma_{k+1}+1)^2(\sigma_{k+1}+1)^2} \\ &\le \frac{16^k}{(\beta+\iota)^2}\frac{16}{(\beta_{k+1}+1)^2} = \frac{16^{k+1}}{(\tilde\beta+\iota)^2}.\end{aligned}$$

■

Proof (of (7.4)) By (7.2), the inequality holds for 2-dimensional multi-indices. Assuming that it holds for k-dimensional multi-indices, we show that it continues to hold for multi-indices of dimension $k+1$. Let α and β be $(k+1)$-dimensional multi-indices. Then by the induction hypothesis

$$\begin{aligned}\binom{\alpha+\beta}{\alpha} &= \frac{(\alpha+\beta)!}{\alpha!\beta!} = \frac{\prod_{j=1}^k(\alpha_j+\beta_j)}{\prod_{j=1}^k \alpha_j!\beta_j!}\frac{(\alpha_{k+1}+\beta_{k+1})!}{\alpha_{k+1}!\beta_{k+1}!} \\ &\le \frac{\left[\sum_{j=1}^k(\alpha_j+\beta_j)\right]!}{\left[\sum_{j=1}^k\alpha_j\right]!\left[\sum_{j=0}^k\beta_j\right]!}\frac{(\alpha_{k+1}+\beta_{k+1})!}{\alpha_{k+1}!\beta_{k+1}!} \\ &\le \binom{\sum_{j=1}^k(\alpha_j+\beta_j)}{\sum_{j=1}^k\alpha_j}\binom{\alpha_{k+1}+\beta_{k+1}}{\alpha_{k+1}} \le \binom{\sum_{j=1}^{k+1}(\alpha_j+\beta_j)}{\sum_{j=1}^{k+1}\alpha_j} = \binom{|\alpha+\beta|}{|\alpha|}.\end{aligned}$$

■

8 Auxiliary Estimations at the Origin

Lemma 8.1 *Let* $\mathbf{u} = (u_1, \ldots, b_\ell)$ *satisfy (6.3). Then for every* $1 \le p, q \le \ell$ *and every* N*-dimensional multi-index* β

$$|D^\beta D_t^s (u_p u_q)(0,0)| \le \left(\frac{cC_o}{C}\right)^2 C^{|\beta|+s} \frac{(|\beta|+s)!}{(\beta+\iota)^2(s+1)^2} \tag{8.1}$$

where $c = 16^{N+1}$*. For every* ℓ*-dimensional multi-index* γ

$$|D^\beta D_t^s \mathbf{u}^\gamma(0,0)| \le \left(\frac{cC_o}{C}\right)^{|\gamma|} C^{|\beta|+s} \frac{(|\beta|+s)!}{(\beta+\iota)^2(s+1)^2}. \tag{8.2}$$

Moreover, for all indices $h = 1, \ldots, N$ *and* $k = 1, \ldots, \ell$

$$|D^\beta D_t^s (\mathbf{u}^\gamma D_{x_h} u_k)(0,0)| \le C_o \left(\frac{cC_o}{C}\right)^{|\gamma|} C^{|\beta|+s} \frac{(|\beta|+s+1)!}{(\beta+\iota)^2(s+1)^2}. \tag{8.3}$$

Proof (of (8.1)) By the generalized Leibniz rule

$$D^\beta D_t^s u_p u_q = \sum_{j=0}^{s} \binom{s}{j} \sum_{\sigma \le \beta} \binom{\beta}{\sigma} (D^{\beta-\sigma} D_t^{s-j} u_p)(D^\sigma D_t^j u_q)$$

where σ is an N-dimensional multi-index of size $|\sigma| \le |\beta|$. Compute this at the origin and estimate the right-hand side by using (6.3), which is assumed to hold for all the t-derivatives of order up to s. By (7.4)

$$\binom{\beta}{\sigma}\binom{s}{j} \le \binom{|\beta|}{|\sigma|}\binom{s}{j} \le \binom{|\beta|+s}{|\sigma|+j}.$$

Therefore

$$\begin{aligned}|D^\beta D_t^s (u_p u_q)(0,0)| \le & C_o^2 C^{|\beta|+s-2} (|\beta|+s)! \\ & \times \sum_{j=0}^{s} \frac{1}{(s-j+1)^2 (j+1)^2} \sum_{\sigma \le \beta} \frac{1}{(\beta-\sigma+\iota)^2(\sigma+\iota)^2}.\end{aligned}$$

To prove (8.1) estimate the two sums on the right-hand side with the aid of (7.1) of Lemma 7.1 and (7.3) of Lemma 7.2. ■

Proof (of (8.2)) The proof is by induction. If γ is a ℓ-dimensional multi-index of either form

$$(\ldots, 1, \ldots, 1, \ldots), \qquad (\ldots, 2, \ldots)$$

then (8.2) is precisely (8.1) for such a multi-index. Therefore (8.2) holds for multi-indices of size $|\gamma| = 2$. Assuming it does hold for multi-indices of size $|\gamma|$, we will show that it continues to hold for all ℓ-dimensional multi-indices

$$\tilde{\gamma} = (\gamma_1, \ldots, \gamma_p + 1, \ldots, \gamma_\ell), \qquad 1 \le p \le \ell$$

of size $|\gamma| + 1$. By the Leibniz rule

$$D^\beta D_t^s \mathbf{u}^{\tilde{\gamma}} = \sum_{j=0}^{s} \binom{s}{j} \sum_{\sigma \le \beta} \binom{\beta}{\sigma} (D^{\beta-\sigma} D_t^{s-j} \mathbf{u}^\gamma)(D^\sigma D_t^j u_p).$$

First compute this at the origin. Then estimate $D^{\beta-\sigma} D_t^{s-j} \mathbf{u}^\gamma(0,0)$ by the induction hypothesis, and the terms $D^\sigma D_t^j u_p(0,0)$ by (6.3).

The estimation is concluded by proceeding as in the proof of (8.1). ■

Proof (of (8.3)) By the generalized Leibniz rule

$$D^\beta D_t^s (\mathbf{u}^\gamma D_{x_h} u_k) = \sum_{j=0}^{s} \binom{s}{j} \sum_{\sigma \le \beta} \binom{\beta}{\sigma} (D^{\beta-\sigma} D_t^{s-j} \mathbf{u}^\gamma)(D^\sigma D_t^j D_{x_h} u_k).$$

First compute this at the origin. Then majorize the terms involving $\mathbf{u}^\gamma$, by means of (8.2), and the terms $D^\sigma D_{x_h} D_t^j u_k(0,0)$ by (6.3). This gives

$$\begin{aligned} |D^\beta D_t^s (\mathbf{u}^\gamma D_{x_h} u_k)(0,0)| &\le C_o \left(\frac{cC_o}{C}\right)^{|\gamma|} C^{|\beta|+s} \\ &\times \sum_{j=0}^{s} \sum_{\sigma \le \beta} \binom{|\beta|+s}{|\sigma|+j} \frac{(|\beta-\sigma|+s-j)!(|\sigma|+j+1)!}{(\beta-\sigma+\iota)^2(\tilde{\sigma}+\iota)^2(s-j+1)^2(j+1)^2} \end{aligned} \tag{8.4}$$

where $\tilde{\sigma}$ is the N-dimensional multi-index $\tilde{\sigma} = (\sigma_1, \dots, \sigma_h + 1, \dots, \sigma_N)$. Estimate

$$(|\sigma| + j + 1)! \le (|\sigma| + j)!(|\beta| + s + 1) \quad \text{and} \quad (\tilde{\sigma} + \iota)^2 \ge (\sigma + \iota)^2.$$

These estimates in (8.4) yield

$$\begin{aligned} |D^\beta D_t^s (\mathbf{u}^\gamma D_{x_h} u_k)(0,0)| &\le C_o \left(\frac{cC_o}{C}\right)^{|\gamma|} C^{|\beta|+s} (|\beta|+s)!(|\beta|+s+1) \\ &\quad \times \sum_{j=0}^{s} \frac{1}{(s-j+1)^2(j+1)^2} \sum_{\sigma \le \beta} \frac{1}{(\beta-\sigma+\iota)^2(\sigma+\iota)^2} \\ &\le cC_o \left(\frac{cC_o}{C}\right)^{|\gamma|} C^{|\beta|+s} (|\beta|+s+1)!. \end{aligned}$$

■

9 Proof of the Cauchy–Kowalewski Theorem (Concluded)

To prove (6.3) return to (6.6) and estimate

$$\begin{aligned} |D^\beta D_t^{s+1} \mathbf{u}(0,0)| \le& 2(cC_o + 1) M_o N \ell^2 C^{|\beta|+s} \\ &\times \frac{(|\beta|+s+1)!}{(\beta+\iota)^2(s+2)^2} \sum_{|\gamma| \ge 1} \left(\frac{cC_o M}{C}\right)^{|\gamma|} \end{aligned}$$

where M and M_o are the constants appearing in (6.5). It remains to choose C so large that

$$2(C_o+1)M_oN\ell^2 \sum_{|\gamma|\ge 1}\left(\frac{cC_oM}{C}\right)^{|\gamma|} \le C_o. \tag{9.1}$$

Remark 9.1 The choice of C gives a lower estimate of the radius of convergence of the series (6.1) and (6.4).

9.1 Proof of Lemma 6.1

Given the inequality (6.3), the proof of the Cauchy–Kowalewski theorem is a consequence of the following algebraic lemma

Lemma 9.1 *Let α be a N-dimensional multi-index. Then $|\alpha|! \le N^{|\alpha|}\alpha!$.*

Proof Let x_i for $i = 1,\dots,N$ be given real numbers, and let k be a positive integer. If α denotes an N-dimensional multi-index of size k, by the Leibniz version of Newton's formula (**9.2** of the Complements)

$$\left(\sum_{i=1}^{N} x_i\right)^k = \sum_{|\alpha|=k}\frac{k!}{\alpha!}\prod_{i=1}^{N} x_i^{\alpha_i}. \tag{9.2}$$

From this, taking $x_i = 1$ for all $i = 1,\dots,N$

$$N^{|\alpha|} = \sum_{|\alpha|=k}\frac{|\alpha|!}{\alpha!}.$$

■

Problems and Complements

1c Quasi-Linear Second-Order Equations in Two Variables

1.1. Assume that the functions A, B, C, D in (1.1) are of class C^∞. Assume also that Γ is of class C^∞. Prove that all the derivatives

$$k, h, \ell \in \mathbb{N}, \qquad \frac{\partial^k u}{\partial x^h \partial y^\ell}, \qquad h+\ell = k$$

can be computed on Γ provided (1.4) holds.

1.2. Assume that in (1.1), A, B, C are constants and $D = 0$. Introduce an affine transformation of the coordinate variables that transforms (1.1) into either the Laplace equation, the heat equation, or the wave equation.

1.3. Prove the last statement of §2.1. Discuss the case of constant coefficients.

5c Analytic Data and the Cauchy–Kowalewski Theorem

5.1. Denote points in $\mathbb{R}^{N+1}$ by (x,t) where $x \in \mathbb{R}^N$ and $t \in \mathbb{R}$. Let φ and ψ be analytic in $\mathbb{R}^N$. Find an analytic solution, about $t = 0$ of

$$\Delta u = 0, \quad u(x,0) = \varphi(x), \quad u_t(x,0) = \psi(x).$$

5.2. Let f_1 and f_2 be analytic and periodic of period 2π in $\mathbb{R}$. Solve the problem

$$\Delta u = 0 \quad \text{in} \quad 1 - \varepsilon < |x| < 1 + \varepsilon$$

$$u \Big|_{|x|=1} = f_1(\theta), \qquad \frac{\partial u}{\partial |x|} \Big|_{|x|=1} = f_2(\theta)$$

for some $\varepsilon \in (0,1)$. Compare with the Poisson integral (3.11) of Chapter 2.

6c Proof of the Cauchy–Kowalewski Theorem

6.1. Prove that (6.2) ensures the convergence of the series (6.1) and give an estimate of the radius of convergence.

6.2. Let β be a N-dimensional multi-index of size $|\beta|$. Prove that the number of derivatives D^β of order $|\beta|$ does not exceed $|\beta|^N$.

6.3. Let $\mathbf{u} : \mathbb{R}^N \to \mathbb{R}^\ell$ be analytic at some point $x_o \in \mathbb{R}^N$. Prove that there exist constants C_o and C such that for all N-dimensional multi-indices β

$$|D^\beta \mathbf{u}(x_o)| \le C_o C^{|\beta|-1} \frac{|\beta|!}{(\beta + \iota)^2}.$$

6.4. Prove (6.4)–(6.4).

8c The Generalized Leibniz Rule

8.1. Let $u, v \in C^\infty(\mathbb{R})$ be real-valued. The Leibniz rule states that for every $n \in \mathbb{N}$

$$D^n(uv) = \sum_{i=0}^{n} \binom{n}{i} D^{n-i} u D^i v.$$

In particular if $u, v \in C^\infty(\mathbb{R}^N)$, then

$$D^n_{x_r}(uv) = \sum_{i=0}^{n} \binom{n}{i} D^{n-i}_{x_r} u D^i_{x_r} v.$$

Prove, by induction, the generalized Leibniz rule

$$D^\beta(uv) = \sum_{\sigma \le \beta} \binom{\beta}{\sigma} D^{\beta-\sigma} u D^\sigma v.$$

9c Proof of the Cauchy–Kowalewski Theorem (Concluded)

9.1. Prove that C can be chosen such that (9.1) holds.

9.2. Prove (9.2) by induction, starting from the binomial formula

$$(x_1 + x_2)^k = \sum_{j=0}^{k} \binom{k}{k-j} x_1^j x_2^{k-j}.$$

2

The Laplace Equation

1 Preliminaries

Let E be a domain in $\mathbb{R}^N$ for some $N \geq 2$, with boundary ∂E of class C^1. Points in E are denoted by $x = (x_1, \dots, x_N)$. A function $u \in C^2(E)$ is *harmonic* in E if

$$\Delta u = \operatorname{div} \nabla u = \sum_{i=1}^{N} \frac{\partial^2}{\partial x_i^2} u = 0 \quad \text{in} \quad E. \tag{1.1}$$

The formal operator Δ is called the *Laplacian*.[1] The interest in this equation stems from its connection to physical phenomena such as

1. Steady state heat conduction in a homogeneous body with constant heat capacity and constant conductivity.
2. Steady state potential flow of an incompressible fluid in a porous medium with constant permeability.
3. Gravitational potential in $\mathbb{R}^N$ generated by a uniform distribution of masses.

The interest is also of pure mathematical nature in view of the rich structure exhibited by (1.1). The formal operator in (1.1) is invariant under rotations or translations of the coordinate axes. Precisely, if A is a (unitary, orthonormal) rotation matrix and $y = A(x - \xi)$ for some fixed $\xi \in \mathbb{R}^N$, then formally

$$\Delta_x = \sum_{i=1}^{N} \frac{\partial^2}{\partial x_i^2} = \sum_{i=1}^{N} \frac{\partial^2}{\partial y_i^2} = \Delta_y.$$

This property is also called *spherical symmetry* of the Laplacian in $\mathbb{R}^N$.

---

[1]Pierre Simon, Marquis de Laplace, 1749–1827. Author of *Traité de Mécanique Céleste* (1799–1825). Also known for the frequent use of the phrase *il est aisé de voir* which has unfortunately become all too popular in modern mathematical writings. The same equation had been introduced, in the context of potential fluids, by Joseph Louis, Compte de Lagrange, 1736–1813, author of *Traité de Mécanique Analytique* (1788).

E. DiBenedetto, *Partial Differential Equations: Second Edition*,
Cornerstones, DOI 10.1007/978-0-8176-4552-6_3,

1.1 The Dirichlet and Neumann Problems

Given $\varphi \in C(\partial E)$, the Dirichlet problem for the operator Δ in E consists in finding a function $u \in C^2(E) \cap C(\bar{E})$ satisfying

$$\Delta u = 0 \text{ in } E, \qquad \text{and} \qquad u\big|_{\partial E} = \varphi. \tag{1.2}$$

Given $\psi \in C(\partial E)$, the Neumann problem consists in finding a function $u \in C^2(E) \cap C^1(\bar{E})$ satisfying

$$\Delta u = 0 \text{ in } E, \qquad \text{and} \qquad \frac{\partial}{\partial \mathbf{n}} u = \nabla u \cdot \mathbf{n} = \psi \text{ on } \partial E \tag{1.3}$$

where $\mathbf{n}$ denotes the outward unit normal to ∂E. The Neumann datum ψ is also called *variational*.

We will prove that if E is bounded, the Dirichlet problem is always uniquely solvable. The Neumann problem, on the other hand, is not always solvable. Indeed, integrating the first of (1.3) in E, we arrive at the necessary condition

$$\int_{\partial E} \psi \, d\sigma = 0 \tag{1.4}$$

where $d\sigma$ denotes the surface measure on ∂E. Thus ψ cannot be assigned arbitrarily.

Lemma 1.1 *Let E be a bounded open set with boundary ∂E of class C^1 and assume that (1.2) and (1.3) can both be solved within the class $C^2(\bar{E})$. Then the solution of (1.2) is uniquely determined by φ, and the solution of (1.3) is uniquely determined by ψ up to a constant.*

Proof We prove only the statement regarding the Dirichlet problem. If u_i for $i = 1, 2$, are two solutions of (1.2), the difference $w = u_1 - u_2$ is a solution of the Dirichlet problem with homogeneous data

$$\Delta w = 0 \quad \text{in } E, \qquad w\big|_{\partial E} = 0.$$

Multiplying the first of these by w and integrating over E gives

$$\int_E |\nabla w|^2 dx = 0. \qquad \blacksquare$$

Remark 1.1 Arguments of this kind are referred to as *energy methods*. The assumption $w \in C^2(\bar{E})$ is used to justify the various calculations in the integration by parts. The lemma continues to hold for solutions in the class $C^2(E) \cap C^1(\bar{E})$. Indeed, one might first carry the integration over an open, proper subset $E' \subset E$, with boundary $\partial E'$ of class C^1, and then let E' expand to E. We will show later that uniqueness for the Dirichlet problem holds within the class $C^2(E) \cap C(\bar{E})$, required by the formulation (1.2).

Remark 1.2 A consequence of the lemma is that the problem

$$\Delta u = 0 \text{ in } E \qquad \text{and } u\big|_{\partial E} = \varphi, \;\; \nabla \cdot \mathbf{n} = \psi$$

in general is not solvable.

1.2 The Cauchy Problem

Let Γ be an $(N-1)$-dimensional surface of class C^1 contained in E and prescribe $N+1$ functions $\psi_i \in C^2(\Gamma)$, for $i = 0, 1, \dots, N$. The Cauchy problem consists in finding $u \in C^2(E)$ satisfying

$$\Delta u = 0 \text{ in } E \quad \text{and} \quad u = \psi_o,\ u_{x_i} = \psi_i,\ i = 1, \dots, N \text{ on } \Gamma. \tag{1.5}$$

The Cauchy problem is not always solvable. First the data ψ_i, must be compatible, i.e., derivatives of u along Γ computed using ψ_o and computed using ψ_i must coincide. Even so, in general, the solution, if any, can only be found near Γ. The Cauchy–Kowalewski theorem gives some sufficient conditions to ensure *local* solvabilty of (1.5).

1.3 Well-Posedness and a Counterexample of Hadamard

A boundary value problem for the Laplacian, say the Dirichlet, Neumann or Cauchy problem, is *well-posed* in the sense of Hadamard if one can identify a class of boundary data, say $\mathcal{C}$, such that each datum in $\mathcal{C}$ yields a *unique* solution, and small variations of the data within $\mathcal{C}$ yield small variations on the corresponding solutions. The meaning of *small variation* is made precise in terms of the topology suggested by the problem. This is referred to as the problem of *stability*. A problem that does not meet any one of these criteria is called *ill-posed.*

Consider the problem of finding a harmonic function in E taking either Dirichlet data or Neumann data on a portion Σ_1 of ∂E and both Dirichlet and variational data on the remaining part $\Sigma_2 = \partial E - \Sigma_1$. Such a problem is ill-posed. Even if a solution exists, in general it is not stable in any reasonable topology, as shown by the following example due to Hadamard ([62]).

The boundary value problem

$$\begin{aligned} &u_{xx} + u_{yy} = 0 && \text{in } \left(-\tfrac{\pi}{2} < x < \tfrac{\pi}{2}\right) \times (y > 0) \\ &u(\pm\tfrac{\pi}{2}, y) = 0 && \text{for } y > 0 \\ &u(x, 0) = 0 && \text{for } -\tfrac{\pi}{2} < x < \tfrac{\pi}{2} \\ &u_y(x, 0) = e^{-\sqrt{n}} \cos nx && \text{for } -\tfrac{\pi}{2} < x < \tfrac{\pi}{2} \end{aligned}$$

admits the family of solutions

$$u_n(x, y) = \frac{1}{n} e^{-\sqrt{n}} \cos nx \sinh ny, \qquad \text{where } n \text{ is an odd integer.}$$

One verifies that

$$\|u_{n,y}(\cdot, 0)\|_{\infty,(-\frac{\pi}{2},\frac{\pi}{2})} \to 0 \text{ as } n \to \infty$$

and that for all $y > 0$

$$\|u_n(\cdot, y)\|_{\infty,(-\frac{\pi}{2},\frac{\pi}{2})}, \quad \|u_n(\cdot, y)\|_{2,(-\frac{\pi}{2},\frac{\pi}{2})} \to \infty \text{ as } n \to \infty.$$

1.4 Radial Solutions

The invariance of Δ under orthonormal linear transformations suggests that we look for solutions of (1.1) in $\mathbb{R}^N$ depending only on $\rho = |x-y|$, for any fixed $y \in \mathbb{R}^N$. Any such solution $\rho \to V(\rho; y)$ must satisfy

$$V'' + \frac{N-1}{\rho} V' = 0, \qquad y \in \mathbb{R}^N \text{ fixed}$$

where the derivatives are meant with respect to ρ. By integration this gives, up to additive and multiplicative constants

$$\mathbb{R}^N - \{y\} \ni x \to \begin{cases} \dfrac{1}{|x-y|^{N-2}} & \text{if } N \geq 3 \\ \ln|x-y| & \text{if } N = 2. \end{cases} \tag{1.6}$$

These are the *potentials* of the Laplacian in $\mathbb{R}^N$ with a *pole* at y. Consider a finite distribution $\{(e_i, y_i)\}$ for $i = 1, \dots, n$, of electrical charges e_i, concentrated at the points y_i. The function

$$\mathbb{R}^N - \{y_1, \dots, y_n\} \ni x \to \sum_{i=1}^{n} \frac{e_i}{|x-y_i|^{N-2}}, \qquad x \neq y_i, \ i = 1, 2, \dots, n$$

is harmonic, and it represents the potential generated by the charges (e_i, y_i) outside them.

Let E be a bounded, Lebesgue measurable set in $\mathbb{R}^N$, and let $\mu \in C(\bar{E})$. The function

$$\mathbb{R}^N - \bar{E} \ni x \to \int_E \frac{\mu(y)}{|x-y|^{N-2}} dy, \qquad N \geq 3$$

is harmonic in $\mathbb{R}^N - \bar{E}$, and it represents the *Newtonian potential* generated outside $\bar{E}$, by the distribution of *masses* (or charges) $\mu(y)dy$ in E. Let Σ be an $(N-1)$-dimensional bounded surface of class C^1 in $\mathbb{R}^N$, for some $N \geq 3$, and let $\mathbf{n}(y)$ denote the unit normal at $y \in \Sigma$. The orientation of $\mathbf{n}(y)$ is arbitrary but fixed, so that $y \to \mathbf{n}(y)$ is continuous on Σ. Given $\varphi, \psi \in C(\Sigma)$ the two functions

$$\mathbb{R}^N - \bar{\Sigma} \ni x \to \begin{cases} \displaystyle\int_\Sigma \frac{\varphi(y)}{|x-y|^{N-2}} d\sigma \\ \displaystyle\int_\Sigma \frac{\psi(y)}{|x-y|^{N}} (x-y) \cdot \mathbf{n}(y) d\sigma \end{cases}$$

are harmonic in $\mathbb{R}^N - \bar{\Sigma}$. The first is called *single-layer potential*, and it gives the potential generated, outside $\bar{\Sigma}$, by a distribution of charges (or masses) on Σ, of density $\varphi(\cdot)$. The second is called *double-layer potential* and it represents the electrical potential generated, outside Σ, by a distribution of dipoles on Σ, with density $\psi(\cdot)$.

Analogous harmonic functions can be constructed for $N = 2$, by using the second of (1.6). These would be called logarithmic potentials.

2 The Green and Stokes Identities

Let E be a bounded open set in $\mathbb{R}^N$ with boundary ∂E of class C^1, and let $u, v \in C^2(\bar{E})$. By the divergence theorem we obtain the Green's identities

$$\int_E v \Delta u \, dx = - \int_E \nabla v \cdot \nabla u \, dx + \int_{\partial E} v \frac{\partial u}{\partial \mathbf{n}} \, d\sigma \tag{2.1}$$

$$\int_E (v\Delta u - u \Delta v) \, dx = \int_{\partial E} \left(v \frac{\partial u}{\partial \mathbf{n}} - u \frac{\partial v}{\partial \mathbf{n}} \right) d\sigma. \tag{2.2}$$

Remark 2.1 By approximation, (2.1)–(2.2) continue to hold for functions $u, v \in C^2(E) \cap C^1(\bar{E})$ such that Δu and Δv are essentially bounded in E.

Remark 2.2 If u is harmonic in E, then

$$\int_{\partial E} \frac{\partial u}{\partial \mathbf{n}} d\sigma = 0 \quad \text{and} \quad \int_E |\nabla u|^2 dx = \int_{\partial E} u \frac{\partial u}{\partial \mathbf{n}} d\sigma.$$

2.1 The Stokes Identities

Let $u \in C^2(\bar{E})$ and let ω_N denote the area of the unit sphere in $\mathbb{R}^N$ for $N \geq 3$. Then for all $x \in E$

$$\begin{aligned} u(x) =& \frac{1}{\omega_N(N-2)} \int_{\partial E} \left(|x-y|^{2-N} \frac{\partial u}{\partial \mathbf{n}} - u(y) \frac{\partial |x-y|^{2-N}}{\partial \mathbf{n}} \right) d\sigma \\ &- \frac{1}{\omega_N(N-2)} \int_E |x-y|^{2-N} \Delta u \, dy. \end{aligned} \tag{2.3}$$

If $N = 2$

$$\begin{aligned} u(x) =& \frac{1}{2\pi} \int_{\partial E} \left(u \frac{\partial \ln |x-y|}{\partial \mathbf{n}} - \ln |x-y| \frac{\partial u}{\partial \mathbf{n}} \right) d\sigma \\ &+ \frac{1}{2\pi} \int_E \ln |x-y| \Delta u \, dy. \end{aligned} \tag{2.4}$$

Remark 2.3 These are implicit representation formulas of smooth functions in $\bar{E}$.

Proof We prove only (2.3). Fix $x \in E$ and let $B_\varepsilon(x)$ be the ball of radius ε centered at x. Assume that ε is so small that $B_\varepsilon(x) \subset E$, and apply (2.2) in $E - B_\varepsilon(x)$ for $y \to v(y)$ equal to the potential with pole at x introduced in (1.6). Since $V(\cdot; x)$ is harmonic in $E - B_\varepsilon(x)$, (2.2) yields

$$\begin{aligned} \frac{N-2}{\varepsilon^{N-1}} \int_{|x-y|=\varepsilon} u(y) \, d\sigma =& \int_{\partial E} \left(|x-y|^{2-N} \frac{\partial u}{\partial \mathbf{n}} - u \frac{\partial |x-y|^{2-N}}{\partial \mathbf{n}} \right) d\sigma \\ &+ \varepsilon^{2-n} \int_{|x-y|=\varepsilon} \nabla u \cdot \frac{x-y}{|x-y|} \, d\sigma - \int_{E - B_\varepsilon(x)} |x-y|^{2-N} \Delta u \, dy. \end{aligned} \tag{2.5}$$

As $\varepsilon \to 0$

$$\int_{E-B_\varepsilon(x)} |x-y|^{2-N} \Delta u \, dy \longrightarrow \int_E |x-y|^{2-N} \Delta u \, dx$$

and

$$\varepsilon^{2-N} \int_{|x-y|=\varepsilon} \nabla u \cdot \frac{x-y}{|x-y|} d\sigma \longrightarrow 0.$$

As for the left-hand side of (2.5)

$$\frac{1}{\varepsilon^{N-1}} \int_{|x-y|=\varepsilon} u(y) d\sigma = \omega_N u(x) + \frac{1}{\varepsilon^{N-1}} \int_{|x-y|=\varepsilon} [u(y) - u(x)] d\sigma.$$

The last integral tends to zero as $\varepsilon \to 0$, since

$$\begin{aligned} \frac{1}{\varepsilon^{N-1}} \int_{|x-y|=\varepsilon} |u(y) - u(x)| d\sigma &\le \frac{\|\nabla u\|_{\infty,E}}{\varepsilon^{N-1}} \int_{|x-y|=\varepsilon} |x-y| d\sigma \\ &\le \varepsilon \omega_N \|\nabla u\|_{\infty,E}. \end{aligned}$$

These remarks in (2.5) prove (2.3) after we let $\varepsilon \to 0$. ■

Motivated by the Stokes identities, set

$$F(x;y) = \begin{cases} \dfrac{1}{\omega_N(N-2)} \dfrac{1}{|x-y|^{N-2}} & \text{if } N \ge 3 \\ \\ \dfrac{-1}{2\pi} \ln|x-y| & \text{if } N = 2. \end{cases} \tag{2.6}$$

The function $F(\cdot;y)$ is called the fundamental solution of the Laplacian with pole at y.

Corollary 2.1 *Let E be a bounded open set in $\mathbb{R}^N$ with boundary ∂E of class C^1 and let $u \in C^2(\bar{E})$ be harmonic in E. Then for all $x \in E$*

$$u(x) = \int_{\partial E} \left(F(x;\cdot) \frac{\partial u}{\partial \mathbf{n}} - u \frac{\partial F(x;\cdot)}{\partial \mathbf{n}} \right) d\sigma. \tag{2.7}$$

A consequence of this corollary, and the structure of the fundamental solution $F(\cdot, y)$ is the following

Proposition 2.1 *Let E be an open set in $\mathbb{R}^N$ and let $u \in C^2(E)$ be harmonic in E. Then $u \in C^\infty(E)$, and for every multi-index α, the function $D^\alpha u$ is harmonic in E.*

Proof If E is bounded, ∂E is of class C^1, and $u \in C^2(\bar{E})$, the statement follows from the representation (2.7). Otherwise, apply (2.7) to any bounded open subset $E' \subset E$ with boundary of class C^1. ■

Corollary 2.2 $u \in C_o^2(E) \Longrightarrow u(x) = -\int_E F(x;y) \Delta u \, dy$ *for all* $x \in E$.

3 Green's Function and the Dirichlet Problem for a Ball

Given a bounded open set $E \subset \mathbb{R}^N$ with boundary ∂E of class C^1, consider the problem of finding, for each fixed $x \in E$, a function $y \to \Phi(x;y) \in C^2(\bar{E})$ satisfying

$$\Delta_y \Phi(x;\cdot) = 0 \text{ in } E, \qquad \text{and} \qquad \Phi(x;\cdot)\big|_{\partial E} = F(x;\cdot) \tag{3.1}$$

where $F(x;y)$ is the fundamental solution of the Laplacian introduced in (2.6). Assume for the moment that (3.1) has a solution. Assume also that the Dirichlet problem (1.2) has a solution $u \in C^2(\bar{E})$. Then the second Green's identity (2.2) written for $y \to u(y)$ and $y \to \Phi(x;y)$ gives

$$0 = \int_{\partial E} \left(\Phi(x;\cdot)\frac{\partial u}{\partial \mathbf{n}} - u \frac{\partial \Phi(x;\cdot)}{\partial \mathbf{n}} \right) d\sigma \qquad \text{for all fixed } x \in E.$$

Subtract this from the implicit representation (2.7), to obtain

$$u(x) = -\int_{\partial E} \varphi \frac{\partial G(x;\cdot)}{\partial \mathbf{n}} d\sigma \tag{3.2}$$

where

$$(x,y) \to G(x;y) = F(x;y) - \Phi(x;y). \tag{3.3}$$

The function $G(\cdot;\cdot)$ is the Green function for the Laplacian in E. Its relevance is in that *every* solution $u \in C^2(\bar{E})$ of the Dirichlet problem (1.2) admits the explicit representation (3.2), through the Dirichlet data φ and $G(\cdot;\cdot)$. Its relevance is also in that it permits a pointwise representation of a smooth function u defined in E and vanishing near ∂E.

Corollary 3.1 $u \in C_o^2(E) \Longrightarrow u(x) = -\int_E G(x;y)\Delta u\, dy$ *for all* $x \in E$.

Lemma 3.1 *The Green function is symmetric, i.e.,* $G(x;y) = G(y;x)$.

Proof Fix $x_1, x_2 \in E$ and let $\varepsilon > 0$ be small enough that

$$B_\varepsilon(x_i) \subset E \text{ for } i = 1,2, \text{ and } B_\varepsilon(x_1) \cap B_\varepsilon(x_2) = \emptyset.$$

Apply Green's identity (2.2) to the pair of functions $G(x_i;\cdot)$ for $i = 1,2$ in the domain $E - [B_\varepsilon(x_1) \cup B_\varepsilon(x_2)]$. Since $G(x_i;\cdot)$ are harmonic in such a domain, and vanish on ∂E

$$\begin{aligned} &-\int_{\partial B_\varepsilon(x_1)} \left[G(x_1;y)\frac{\partial}{\partial \mathbf{n}(y)} G(x_2;y) - G(x_2;y)\frac{\partial}{\partial \mathbf{n}(y)} G(x_1;y) \right] d\sigma \\ &= \int_{\partial B_\varepsilon(x_2)} \left[G(x_1;y)\frac{\partial}{\partial \mathbf{n}(y)} G(x_2;y) - G(x_2;y)\frac{\partial}{\partial \mathbf{n}(y)} G(x_1;y) \right] d\sigma. \end{aligned}$$

Let $\varepsilon \to 0$ and observe that

$$\lim_{\varepsilon\to 0}\int_{\partial B_\varepsilon(x_1)} \frac{\partial}{\partial \mathbf{n}(y)} G(x_2;y)d\sigma = 0$$
$$\lim_{\varepsilon\to 0}\int_{\partial B_\varepsilon(x_2)} \frac{\partial}{\partial \mathbf{n}(y)} G(x_1;y)d\sigma = 0.$$

Therefore

$$\lim_{\varepsilon\to 0}\int_{\partial B_\varepsilon(x_1)} G(x_2;y)\frac{\partial G(x_1;y)}{\partial \mathbf{n}(y)}d\sigma = \lim_{\varepsilon\to 0}\int_{\partial B_\varepsilon(x_2)} G(x_1;y)\frac{\partial G(x_2;y)}{\partial \mathbf{n}(y)}d\sigma.$$

From the definition of $G(\cdot;\cdot)$

$$\frac{\partial G(x_i;y)}{\partial \mathbf{n}(y)} = \frac{\partial F(x_i;y)}{\partial \mathbf{n}(y)} - \frac{\partial \Phi(x_i;y)}{\partial \mathbf{n}(y)},$$

and observe that

$$\lim_{\varepsilon\to 0}\int_{\partial B_\varepsilon(x_1)} G(x_1;y)\frac{\partial}{\partial \mathbf{n}(y)}\Phi(x_2;y)d\sigma = 0$$
$$\lim_{\varepsilon\to 0}\int_{\partial B_\varepsilon(x_2)} G(x_2;y)\frac{\partial}{\partial \mathbf{n}(y)}\Phi(x_1;y)d\sigma = 0$$

since $y \to \Phi(x_i;y)$ are regular. This implies that

$$\lim_{\varepsilon\to 0}\int_{\partial B_\varepsilon(x_1)} G(x_2;y)\frac{\partial F(x_1;y)}{\partial \mathbf{n}(y)}d\sigma = \lim_{\varepsilon\to 0}\int_{\partial B_\varepsilon(x_2)} G(x_1;y)\frac{\partial F(x_2;y)}{\partial \mathbf{n}(y)}d\sigma.$$

Computing the limits as in the the proof of the Stokes identity gives

$$\lim_{\varepsilon\to 0}\int_{\partial B_\varepsilon(x_j)} G(x_i;y)\frac{\partial F(x_j;y)}{\partial \mathbf{n}(y)}d\sigma = G(x_i;x_j) \quad \text{for } x_i \neq x_j.$$

Thus $G(x_1;x_2) = G(x_2;x_1)$. ■

Corollary 3.2 *The functions $G(\cdot;y)$ and $\partial G(\cdot;y)/\partial \mathbf{n}(y)$, for fixed $y \in \partial E$, are harmonic in E.*

To solve the Dirichlet problem (1.2) we may find $G(\cdot;\cdot)$, and write down (3.2). This would be a candidate for a solution. By Corollary 3.2 it is harmonic. It would remain to show that

$$\lim_{x\to x_*} u(x) = \varphi(x_*) \qquad \text{for all } x_* \in \partial E. \tag{3.4}$$

We will show that this is indeed the case if (3.1) has a solution, i.e., if the Green's function for Δ in E can be determined. Thus solving the Dirichlet problem (1.2) reduces to solving the family of Dirichlet problems (3.1). The advantage in dealing with the latter is that the boundary datum $F(x;\cdot)$ is specific and given by (2.6). Nevertheless, (3.1) can be solved explicitly only for domains E exhibiting a simple geometry.

3.1 Green's Function for a Ball

Let B_R be the ball of radius R centered at the origin of $\mathbb{R}^N$. The map[2]

$$\xi = \frac{R^2}{|x|^2}x, \qquad x \neq 0 \tag{3.5}$$

transforms $B_R - \{0\}$ into $\mathbb{R}^N - B_R$, and ∂B_R into itself. Referring to Figure 3.1, the two triangles $\Delta(y, 0, \xi)$ and $\Delta(x, 0, y)$, are similar whenever $y \in \partial B_R$. Indeed, they have in common the angle θ, and in view of (3.5), the ratios $|x|/|y|$ and $|y|/|\xi|$ are equal, provided $|y| = R$. Therefore

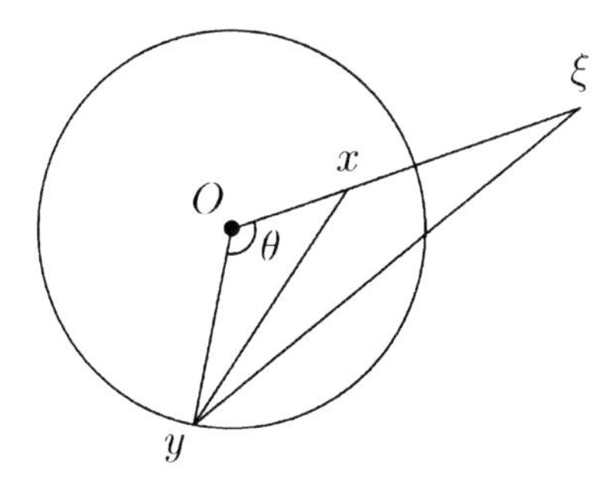

Fig. 3.1.

$$|x-y| = |\xi - y|\frac{|x|}{R} = |\xi - y|\frac{R}{|\xi|}, \qquad x \in B_R \text{ and } y \in \partial B_R. \tag{3.6}$$

Then for each fixed $x \in B_R - \{0\}$, the solution of (3.1) is given by

$$\Phi(x;y) = \begin{cases} \dfrac{1}{\omega_N(N-2)}\left(\dfrac{R}{|x|}\right)^{N-2}\dfrac{1}{|\xi - y|^{N-2}} & \text{if } N \geq 3 \\[2ex] \dfrac{-1}{2\pi}\ln|\xi - y|\dfrac{|x|}{R} & \text{if } N = 2. \end{cases} \tag{3.7}$$

While constructed for $x \neq 0$, the function $\Phi(x;\cdot)$ is well defined also for $x = 0$, modulo taking the limit as $|x| \to 0$. For all $x \in B_R$, the function $\Phi(x;\cdot)$ is harmonic in B_R since its pole ξ lies outside B_R. Moreover, by virtue of (3.6), the boundary conditions in (3.1) are satisfied. Thus the Green function for the ball B_R is

$$G(x;y) = \frac{1}{\omega_N(N-2)}\left[\frac{1}{|x-y|^{N-2}} - \left(\frac{R}{|x|}\right)^{N-2}\frac{1}{|\xi - y|^{N-2}}\right] \tag{3.8$_{N\geq 3}$}$$

[2] Called also the Kelvin transform ([151]).

for $N \geq 3$, and

$$G(x;y) = \frac{1}{2\pi}\left[\ln|\xi - y|\frac{|x|}{R} - \ln|x-y|\right] \quad \text{for } N = 2. \qquad (3.8)_{N=2}$$

The derivative of $G(x;\cdot)$, normal to the sphere $|y| = R$, is computed from

$$\frac{\partial}{\partial|y|}G(x;y)\big|_{y\in\partial B_R}.$$

From Figure 3.1, by elementary trigonometry

$$|x-y|^2 = |y|^2 + |x|^2 - 2|y||x|\cos\theta$$
$$|\xi - y|^2 = |y|^2 + \frac{R^4}{|x|^2} - 2|y|\frac{R^2}{|x|}\cos\theta.$$

Therefore for θ fixed and $y \in \partial B_R$

$$\frac{\partial|x-y|}{\partial|y|} = \frac{|y| - |x|\cos\theta}{|x-y|}, \qquad \frac{\partial|\xi - y|}{\partial|y|} = \frac{|y| - \dfrac{R^2}{|x|}\cos\theta}{|\xi - y|}.$$

First let $N \geq 3$. Computing from $(3.8)_{N\geq3}$ and $(3.8)_{N=2}$, and using (3.6), gives

$$-\frac{\partial}{\partial|y|}G(x;y)\big|_{y\in\partial B_R} = \frac{1}{R\omega_N}\frac{R^2 - |x|^2}{|x-y|^N}.$$

Such a formula also holds for $N = 2$ with $\omega_2 = 2\pi$. Put this in (3.3) to derive the following Poisson representation.

Lemma 3.2 *Let $u \in C^2(\bar{E})$ be a solution of the Dirichlet problem (1.2) in the ball B_R. Then*

$$u(x) = \frac{1}{\omega_N R}\int_{\partial B_R}\varphi(y)\frac{R^2 - |x|^2}{|x-y|^N}d\sigma, \qquad N \geq 2. \qquad (3.9)$$

Setting $u \equiv 1$ in (3.9) gives

$$\frac{1}{\omega_N R}\int_{\partial B_R}\frac{R^2 - |x|^2}{|x-y|^N}d\sigma = 1 \quad \text{for all } x \in B_R. \qquad (3.10)$$

Even though the representation (3.9) has been derived for solutions of (1.2) of class $C^2(\bar{B}_R)$, it actually gives the unique solution of the Dirichlet problem for the sphere, as shown by the following existence theorem.

Theorem 3.1 *The Dirichlet problem (1.2) for $E = B_R$ has a unique solution given by (3.9).*

Proof (existence) By Corollary 3.2, the function u given by (3.9) is harmonic in B_R. To prove (3.4), fix $x_* \in \partial B_R$, choose an arbitrarily small positive number ε, and let $\delta \in (0,1)$ be so small that

$$|\varphi(y) - \varphi(x_*)| < \varepsilon \quad \text{for all } \; y \in \Sigma_\delta = \big\{y \in \partial B_R \;\big|\; |y - x_*| < \delta\big\}. \tag{3.11}$$

By (3.10)

$$\varphi(x_*) = \frac{1}{R\omega_N} \int_{\partial B_R} \varphi(x_*) \frac{R^2 - |x|^2}{|x-y|^N} d\sigma.$$

Therefore

$$\begin{aligned} u(x) - \varphi(x_*) &= \frac{1}{R\omega_N} \int_{\partial B_R} [\varphi(y) - \varphi(x_*)] \frac{R^2 - |x|^2}{|x-y|^N} d\sigma \\ &= \frac{1}{R\omega_N} \int_{\Sigma_\delta} [\varphi(y) - \varphi(x_*)] \frac{R^2 - |x|^2}{|x-y|^N} d\sigma \\ &\quad + \frac{1}{R\omega_N} \int_{\partial B_R - \Sigma_\delta} [\varphi(y) - \varphi(x_*)] \frac{R^2 - |x|^2}{|x-y|^N} d\sigma \\ &= I_\delta^{(1)}(x, x_*) + I_\delta^{(2)}(x, x_*). \end{aligned}$$

For δ fixed, $I_\delta^{(2)}(x, x_*) \to 0$ as $x \to x_*$. Moreover, in view of (3.10) and (3.11), $|I_\delta^{(1)}(x, x_*)| < \varepsilon$. Therefore $\lim_{x \to x_*} |u(x) - \varphi(x_*)| \le \varepsilon$ for all $\varepsilon > 0$. ∎

Remark 3.1 By Lemma 1.1 and the Remark 1.1, such a solution is unique in the class $C^2(B_R) \cap C^1(\bar{B}_R)$. It will be shown in the next section that uniqueness holds for solutions $u \in C^2(B_R) \cap C(\bar{B}_R)$.

4 Sub-Harmonic Functions and the Mean Value Property

Let E be a bounded open set in $\mathbb{R}^N$ with boundary ∂E of class C^1. Let $u \in C^2(\bar{E})$, and assume that the solution $\Phi(x;\cdot)$ of (3.1) exists for all $x \in E$. Subtracting the second Green's identity (2.2) for u and $\Phi(x;\cdot)$ from the Stokes identity (2.3) gives

$$u(x) = -\int_{\partial E} u \frac{\partial G(x;y)}{\partial \mathbf{n}} d\sigma - \int_E G(x;y) \Delta u(y) dy \quad \text{for all } \; x \in E.$$

In particular, if E is a ball $B_R(x_o)$ of radius R centered at x_o, by setting $x = x_o$, we obtain

$$u(x_o) = \fint_{\partial B_R(x_o)} u d\sigma - \int_{B_R(x_o)} G(0; y - x_o) \Delta u(y) dy \tag{4.1}$$

where for a measurable set $D \subset \mathbb{R}^N$ of finite measure and $f \in L^1(D)$

$$\fint_D f\,dy = \frac{1}{|D|}\int_D f\,dy, \qquad |D| = \text{meas}\,(D).$$

Let $E \subset \mathbb{R}^N$ be open. A function $u \in C(E)$ is *sub-harmonic* in E if

$$u(x_o) \le \fint_{\partial B_R(x_o)} u\,d\sigma \qquad \text{for all } \; B_R(x_o) \subset E. \tag{4.2}$$

This implies that if $u \in C(E)$ is sub-harmonic in E, then (**4.1** of the Complements)

$$u(x_o) \le \fint_{B_R(x_o)} u\,dy \qquad \text{for all } \; B_R(x_o) \subset E. \tag{4.3}$$

A function $u \in C(E)$ is *super-harmonic* if $-u$ is sub-harmonic in E.

The Green function $G(\cdot;\cdot)$ for a ball, as defined in $(3.8)_{N\ge3}$ and $(3.8)_{N=2}$, is non-negative. A consequence is that if $u \in C^2(E)$ is such that $\Delta u \ge (\le)0$ in E, then, by (4.1), u is sub(super)-harmonic in E. Conversely if $u \in C^2(E)$ is sub(super)-harmonic, then $\Delta u \ge (\le)0$ in E. Indeed, (4.1) implies

$$\int_{B_R(x_o)} G(0; y - x_o)\Delta u(y)dy \ge 0 \qquad \text{for all } \; B_R(x_o) \subset E.$$

From this

$$\Delta u(x_o)\int_{B_R(x_o)} G(0; y - x_o)dy \ge \int_{B_R(x_o)} G(0; y - x_o)[\Delta u(x_o) - \Delta u(y)]dy$$

and the assertion follows upon dividing by the coefficient of $\Delta u(x_o)$, and letting $R \to 0$.

Lemma 4.1 *Let E be a bounded, connected, open set in $\mathbb{R}^N$. If $u \in C(\bar{E})$ is sub-harmonic in E, then either u is constant, or*

$$u(x) < \sup_{\partial E} u \qquad \textit{for all } \; x \in E.$$

Proof Let $x_o \in \bar{E}$ be a point where $u(x_o) = \sup_{\bar{E}} u$, and assume that u is not identically equal to $u(x_o)$. If $x_o \in E$, for every ball $B_R(x_o) \subset E$

$$u(y) \le u(x_o) \qquad \text{for all } y \in B_R(x_o)$$

which implies

$$\fint_{\partial B_R(x_o)} [u(y) - u(x_o)]d\sigma \le 0.$$

On the other hand, since u is sub-harmonic, this same integral must be non-negative. Therefore

$$\fint_{\partial B_R(x_o)} [u(y) - u(x_o)]d\sigma = 0 \qquad \text{and} \qquad u(y) \le u(x_o).$$

Thus there exists a ball $B_r(x_o)$ for some $r > 0$ such that $u(y) = u(x_o)$ for all $y \in B_r(x_o)$. Consider the set $\mathcal{E}_o = \{y \in E | u(y) = u(x_o)\}$. The previous remarks prove that $\mathcal{E}_o$ is open. By the continuity of u, it is closed in the relative topology of E. Therefore, since E is connected, $\mathcal{E}_o = E$ and $u \equiv u(x_o)$. The contradiction implies that $x_o \in \partial E$. ∎

A function $u \in C(E)$ satisfies the *mean value property* in E if

$$u(x_o) = \fint_{\partial B_R(x_o)} u\, d\sigma \qquad \text{for all } B_R(x_o) \subset E, \tag{4.4}$$

equivalently if (**4.2** of the Complements)

$$u(x_o) = \fint_{B_R(x_o)} u\, dy \qquad \text{for all } B_R(x_o) \subset E. \tag{4.5}$$

Functions satisfying such a property are both sub- and super-harmonic. By (4.1), harmonic functions in E satisfy the mean value property.

Lemma 4.2 *Let E be a bounded, connected, open set in $\mathbb{R}^N$. If $u \in C(\bar{E})$ satisfies the mean value property in E, then either it is constant or*

$$\sup_E |u| = \sup_{\partial E} |u|.$$

Proof (Theorem 3.1, uniqueness) If u, v are two solutions of the Dirichlet problem (1.2) for $E = B_R$, the difference $w = u - v$ is harmonic in B_R and vanishes on ∂B_R. Thus $w \equiv 0$ by Lemma 4.2. ∎

Lemma 4.3 *The following are equivalent:*

$$u \in C(E) \ \textit{satisfies the mean value property} \tag{i}$$

$$u \in C^2(E) \textit{ and } \quad \Delta u = 0. \tag{ii}$$

Proof We have only to prove (i)$\Longrightarrow$(ii). Having fixed $B_R(x_o) \subset E$, let $v \in C^2(B_R(x_o)) \cap C(\bar{B}_R(x_o)$ be the unique solution of the Dirichlet problem

$$\Delta v = 0 \ \text{ in } \ B_R(x_o) \qquad \text{and} \qquad v\big|_{\partial B_R(x_o)} = u.$$

Such a solution is given by the Poisson formula (3.9) up to a change of variables that maps x_o into the origin. The difference $w = u - v$ satisfies the mean value property in $B_R(x_o)$, and by Lemma 4.2, $w \equiv 0$. ∎

4.1 The Maximum Principle

We restate some of these properties in a commonly used form. Let E be a bounded, connected, open set in $\mathbb{R}^N$ with boundary ∂E of class C^1, and let $u \in C^2(E) \cap C(\bar{E})$ be non-constant in E. Then

$$\begin{array}{lll} \Delta u \ge 0 \text{ in } E \Longrightarrow u(x) < \sup_{\partial E} u & \forall x \in E \\ \Delta u \le 0 \text{ in } E \Longrightarrow u(x) > \inf_{\partial E} u & \forall x \in E \\ \Delta u = 0 \text{ in } E \Longrightarrow |u(x)| < \sup_{\partial E} |u| & \forall x \in E. \end{array}$$

Remark 4.1 The assumption that ∂E is of class C^1 can be removed by applying the maximum principle to a family of expanding connected open sets with smooth boundary, exhausting E.

Remark 4.2 The assumption of E being bounded cannot be removed, as shown by the following counterexample. Let E be the sector $x_2 > |x_1|$ in $\mathbb{R}^2$. The function $u(x) = x_2^2 - x_1^2$ is harmonic in E, vanishes on ∂E, and takes arbitrarily large values in E.

4.2 Structure of Sub-Harmonic Functions

Set

$$\begin{aligned} \sigma(E) &= \{v \in C(E) \mid v \text{ is sub-harmonic in } E\} \\ \Sigma(E) &= \{v \in C(E) \mid v \text{ is super-harmonic in } E\}. \end{aligned} \tag{4.6}$$

Proposition 4.1 *Let $v, v_i \in \sigma(E)$ and $c_i \in \mathbb{R}^+$ for $i = 1, \dots, n$. Then*

$$v \in \sigma(E') \quad \textit{for every open subset } E' \subset E \tag{i}$$

$$\sum_{i=1}^n c_i v_i \in \sigma(E) \tag{ii}$$

$$\max\{v_1, v_2, \dots, v_n\} \in \sigma(E) \tag{iii}$$

For every non-decreasing convex function $f(\cdot)$ in $\mathbb{R}$ (iv)

$$v \in \sigma(E) \Longrightarrow f(v) \in \sigma(E).$$

Proof The statements (i)–(ii) are obvious. To prove (iii), observe that having fixed $B_R(x_o) \subset E$, for some $1 \le i \le n$

$$\begin{aligned} \max\{v_1(x_o), \dots, v_n(x_o)\} = v_i(x_o) &\le \fint_{\partial B_R(x_o)} v_i \, d\sigma \\ &\le \fint_{\partial B_R(x_o)} \max\{v_1, \dots, v_n\} d\sigma. \end{aligned}$$

To prove (iv), write (4.2) for v and apply $f(\cdot)$ to both sides. By Jensen's inequality

$$f(v(x_o)) \le f\Big(\fint_{\partial B_R(x_o)} v d\sigma\Big) \le \fint_{\partial B_R(x_o)} f(v) d\sigma. \qquad \blacksquare$$

Remark 4.3 For simplicity, (iii) and (iv) have been stated separately. In fact (iv) implies (iii).

An important subclass of $\sigma(E)$ is that of the sub-harmonic functions in E that actually are harmonic in some sphere contained in E. Given $v \in \sigma(E)$, fix $B_\rho(\xi) \subset E$ and solve the Dirichlet problem

$$\Delta H_v = 0 \ \text{ in } \ B_\rho(\xi) \qquad \text{and} \qquad H_v \big|_{\partial B_\rho(\xi)} = v.$$

The unique solution H_v is the harmonic extension of $v\big|_{\partial B_\rho(\xi)}$ into $B_\rho(\xi)$. The function that coincides with v in $E - B_\rho(\xi)$ and that equals H_v in $B_\rho(\xi)$ is denoted by $v_{\xi,\rho}$, i.e.,

$$v_{\xi,\rho}(x) = \begin{cases} v(x) & \text{if } x \in E - B_\rho(\xi) \\ H_v(x) & \text{if } x \in B_\rho(\xi). \end{cases} \tag{4.7}$$

Since $v \in \sigma(B_\rho(\xi))$ and H_v is harmonic in $B_\rho(\xi)$, we have $v - H_v \in \sigma(B_\rho(\xi))$. Therefore $v \le H_v$ in $B_\rho(\xi)$. The definition of $v_{\xi,\rho}$ then implies

$$v \le v_{\xi,\rho} \qquad \text{in } \ E. \tag{4.8}$$

Proposition 4.2 *Let $v \in \sigma(E)$. Then $v_{\xi,\rho} \in \sigma(E)$.*

Proof One needs to verify that $v_{\xi,\rho}$ satisfies (4.2) for all $B_R(x_o) \subset E$. This is obvious for $x_o \in E - B_\rho(\xi)$ in view of (4.8). Fix $x_o \in B_\rho(\xi)$ and assume, by contradiction, that there is a ball $B_R(x_o) \subset E$ such that

$$v_{\xi,\rho}(x_o) > \fint_{\partial B_R(x_o)} v_{\xi,\rho}\, d\sigma.$$

Construct the function

$$w = (v_{\xi,\rho})_{x_o,R} = \begin{cases} v_{\xi,\rho} & \text{in } E - B_R(x_o) \\ H_{v_{\xi,\rho}} & \text{in } B_R(x_o). \end{cases}$$

Since $v_{\xi,\rho} \ge v$, by the maximum principle $w \ge v_{x_o,R}$. Since w satisfies the mean value property in $B_R(x_o)$, the contradiction assumption implies that

$$v_{\xi,\rho}(x_o) - w(x_o) > 0. \tag{4.9}$$

The difference $v_{\xi,\rho} - w$ is harmonic in $B_R(x_o) \cap B_\rho(\xi)$. The boundary of such a set is the union of ∂_1 and ∂_2, where

$$\partial_1 = \partial B_R(x_o) \cap \bar{B}_\rho(\xi) \qquad \text{and} \qquad \partial_2 = \partial B_\rho(\xi) \cap \bar{B}_R(x_o).$$

Because of (4.9), the function $x \to (v_{\xi,\rho} - w)(x)$, restricted to $B_R(x_o) \cap B_\rho(\xi)$, must take its positive maximum at some point $x_* \in \partial_1 \cup \partial_2$. Since it vanishes on ∂_1, there exists some $x_* \in \partial_2$ such that $v_{\xi,\rho}(x_*) > w(x_*)$. By construction, $v_{\xi,\rho} = v$ on ∂_2. Therefore $v(x_*) > w(x_*)$. Since $w \ge v_{x_o,R}$, we conclude that $v(x_*) > v_{x_o,R}(x_*)$. This contradicts (4.8) and proves the proposition. ∎

Remark 4.4 Analogous facts hold for super-harmonic functions.

5 Estimating Harmonic Functions and Their Derivatives

We will prove that if u is harmonic and is non-negative in E, then in any compact subset $K \subset E$, its maximum and minimum value are comparable. We also establish sharp estimates for the derivatives of u in the interior of E.

5.1 The Harnack Inequality and the Liouville Theorem

Theorem 5.1 (Harnack ([68])) *Let u be a non-negative harmonic function in E. Then for all $x \in B_\rho(x_o) \subset B_R(x_o) \subset E$*

$$\left(\frac{R}{R+\rho}\right)^{N-2}\frac{R-\rho}{R+\rho}u(x_o) \le u(x) \le \left(\frac{R}{R-\rho}\right)^{N-2}\frac{R+\rho}{R-\rho}u(x_o). \tag{5.1}$$

Proof Modulo a translation, we may assume that $x_o = 0$. By the Poisson formula (3.9) and the mean value property (4.7), for all $x \in B_R$

$$\begin{aligned} u(x) &= \frac{R^2-|x|^2}{R\omega_N}\int_{\partial B_R}\frac{u(y)}{|x-y|^N}d\sigma \le \frac{R^2-|x|^2}{R\omega_N}\int_{\partial B_R}\frac{u(y)}{(|y|-|x|)^N}d\sigma \\ &= \frac{R^2-|x|^2}{(R-|x|)^N}R^{N-2}\fint_{\partial B_R} u\,d\sigma = \left(\frac{R}{R-|x|}\right)^{N-2}\frac{R+|x|}{R-|x|}u(0). \end{aligned}$$

This proves the estimate above in (5.1). For the estimate below, observe that

$$u(x) = \frac{R^2-|x|^2}{R\omega_N}\int_{\partial B_R}\frac{u(y)}{|x-y|^N}d\sigma \ge \frac{R^2-|x|^2}{R\omega_N}\int_{\partial B_R}\frac{u(y)}{(|y|+|x|)^N}d\sigma$$

and conclude as above. ■

Corollary 5.1 (Harnack Inequality ([68])) *For every compact, connected subset $K \subset E$, there exists a constant C depending only on N and* $\text{dist}(K;\partial E)$, *such that*

$$C\min_K u \ge \max_K u. \tag{5.2}$$

Proof Let $x_1, x_2 \in K$ be such that $\min_K u = u(x_1)$ and $\max_K u = u(x_2)$. Fix a path Γ in K connecting x_1 and x_2, and cover Γ with finitely many spheres for each of which (5.1) holds. ■

Corollary 5.2 (Liouville Theorem) *A non-negative harmonic function in $\mathbb{R}^N$ is constant.*

Proof In (5.1) fix $x_o \in \mathbb{R}^N$ and $\rho > 0$. Letting $R \to \infty$ gives $u(x) = u(x_o)$ for all $x \in B_\rho(x_o)$. Since x_o and $\rho > 0$ are arbitrary, $u = \text{const}$ in $\mathbb{R}^N$. ■

Corollary 5.3 *Let u be harmonic in $\mathbb{R}^N$ and such that $u \ge k$ for some constant k. Then u is constant.*

Proof The function $u - k$ is harmonic and non-negative in $\mathbb{R}^N$. ∎

Remark 5.1 The proof of Theorem 5.1 shows that in (5.1), for $x \in B_R(x_o)$ fixed, the number ρ can be taken to be $|x|$. This permits us to estimate from below the normal derivative of any harmonic function u in $B_R(x_o)$ at points $x_* \in \partial B_R(x_o)$ where u attains its minimum.

Proposition 5.1 *Let $u \in C^2(B_R(x_o)) \cap C(\bar{B}_R(x_o))$ be harmonic in $B_R(x_o)$, let $x_* \in \partial B_R(x_o)$ be a minimum point of u in $\bar{B}_R(x_o)$, and set*

$$u(x_*) = \min_{\bar{B}_R(x_o)} u \quad \textit{and} \quad \mathbf{n} = \frac{x_* - x_o}{|x_* - x_o|}.$$

Then

$$-\frac{\partial u}{\partial \mathbf{n}}(x_*) \geq 2^{1-N} \frac{u(x_o) - u(x_*)}{R}. \tag{5.3}$$

Proof The function $u - u(x_*)$ is harmonic and non-negative in $B_R(x_o)$. Apply (5.1) to such a function, with $\rho = |x|$, to get

$$\frac{u(x) - u(x_*)}{R - |x|} \geq 2^{1-N} \frac{u(x_o) - u(x_*)}{R}.$$

Letting now $x \to x_*$ along $\mathbf{n}$ proves (5.3). ∎

5.2 Analyticity of Harmonic Functions

If u is harmonic in E, by Proposition 2.1, $D^\alpha u$ is also harmonic in E, for every multi-index α. Therefore $D^\alpha u$ satisfies the mean value property (4.5) for all multi-indices α. In particular, for all $i = 1, \ldots, N$ and all $B_R(x_o) \subset E$

$$\frac{\partial u}{\partial x_i}(x_o) = \fint_{B_R(x_o)} u_{x_i}(y) dy = \frac{N}{\omega_N R^N} \int_{\partial B_R(x_o)} u \frac{(y - x_o)_i}{|y - x_o|} d\sigma.$$

From this

$$\left|\frac{\partial u}{\partial x_i}(x_o)\right| \leq \frac{N}{R} \sup_{B_R(x_o)} |u|. \tag{5.4}$$

This estimate is a particular case of the following

Theorem 5.2 *Let u be harmonic in E. Then for all $B_R(x_o) \subset E$, and for all multi-indices α*

$$|D^\alpha u(x_o)| \leq \left(\frac{Ne}{R}\right)^{|\alpha|} \frac{|\alpha|!}{e} \sup_{B_R(x_o)} |u|. \tag{5.5}$$

Proof By (5.4) the estimate holds for multi-indices of size 1. It will be shown by induction that if (5.5) holds for multi-indices of size $|\alpha|$, it continues to hold for multi-indices β of size $|\beta| = |\alpha| + 1$. For any such β

$$D^\beta u = \frac{\partial}{\partial x_i} D^\alpha u \qquad \text{for some } 1 \le i \le N.$$

Fix $\tau \in (0,1)$ and apply (4.5) to $D^\beta u$ in the ball $B_{\tau R}(x_o)$. This gives

$$\begin{aligned} D^\beta u(x_o) &= \fint_{B_{\tau R}(x_o)} \frac{\partial}{\partial x_i} D^\alpha u \, dy \\ &= \frac{N}{\omega_N \tau^N R^N} \int_{\partial B_{\tau R}(x_o)} D^\alpha u(y) \frac{(y - x_o)_i}{|y - x_o|} \, d\sigma. \end{aligned}$$

By (5.5) applied over balls centered at $y \in \partial B_{\tau R}(x_o)$ and radius $(1-\tau)R$

$$|D^\alpha u(y)| \le \left(\frac{Ne}{(1-\tau)R} \right)^{|\alpha|} \frac{|\alpha|!}{e} \sup_{B_R(x_o)} |u| \qquad \text{for all } y \in \partial B_{\tau R}(x_o).$$

Therefore

$$|D^\beta u(x_o)| \le \left(\frac{Ne}{R} \right)^{|\alpha|+1} \frac{1}{(1-\tau)^{|\alpha|}\tau} \frac{|\alpha|!}{e^2} \sup_{B_R(x_o)} |u|.$$

To prove the theorem, choose

$$\tau = \frac{1}{|\alpha|+1} = \frac{1}{|\beta|} \qquad \text{so that} \qquad (1-\tau)^{-|\alpha|} \le \left(1 - \frac{1}{|\beta|} \right)^{-|\beta|} \le e. \qquad \blacksquare$$

Corollary 5.4 *Let u be harmonic in E. Then u is locally analytic in E.*

Proof Let k be a positive number to be chosen, and having fixed $x_o \in E$, let R be so small that $B_{(k+1)R}(x_o) \subset E$. The Taylor expansion of u in $B_R(x_o)$ about x_o is

$$u(x) = \sum_{|\alpha| \le n} \frac{D^\alpha u(x_o)}{\alpha!} (x - x_o)^\alpha + \sum_{|\beta| = n+1} \frac{D^\beta u(\xi)}{\beta!} (x - x_o)^\beta$$

for some $\xi \in B_R(x_o)$. Estimate the terms of the remainder by applying (5.5) to the ball centered at ξ and radius kR. This gives

$$\begin{aligned} \frac{|D^\beta u(\xi)|}{\beta!} |(x - x_o)^\beta| &\le \left(\frac{Ne}{kR} \right)^{|\beta|} \frac{|\beta|!}{\beta!} \frac{R^{|\beta|}}{e} \sup_{B_{(k+1)R}(x_o)} |u| \\ &\le \left(\frac{Ne^{N+1}}{k} \right)^{|\beta|} \sup_{B_{(k+1)R}(x_o)} |u| \end{aligned}$$

where we have also used the inequality $|\beta|! \le e^{N|\beta|} \beta!$. Set

$$\frac{Ne^{N+1}}{k} = \theta \qquad \text{and} \qquad \sup_{B_{(k+1)R}(x_o)} |u| = M.$$

Choose k so that $\theta < 1$, and majorize the remainder of the Taylor series by

$$\left| \sum_{|\beta|=n+1} \frac{D^\beta u(\xi)}{\beta!}(x - x_o)^\beta \right| \le M \sum_{|\beta|=n+1} \theta^{|\beta|} \le M\,|\beta|^N \theta^{|\beta|}.$$

Since this tends to zero as $|\beta| \to \infty$, the Taylor series of u about x_o converges to u uniformly in $B_R(x_o)$. ∎

6 The Dirichlet Problem

We will establish that the boundary value problem (1.2) has a unique solution for any given $\varphi \in C(\partial E)$. In the statement of the Dirichlet problem (1.2), the boundary ∂E was assumed to be of class C^1. In particular ∂E satisfies the *exterior sphere condition*, i.e.,

$$\begin{gathered} \text{for all } x_* \in \partial E \text{ there exists an exterior ball} \\ B_R(x_o) \subset \mathbb{R}^N - \bar{E} \text{ such that } \partial B_R(x_o) \cap \partial E = x_*. \end{gathered} \tag{6.1}$$

The ball $B_R(x_o)$ is exterior to E, and its boundary $\partial B_R(x_o)$ touches ∂E only at x_*. Such a property is shared by domains whose boundary could be irregular. For example, it is satisfied if ∂E exhibits corners or even spikes pointing outside E.

Theorem 6.1 *Let E be a bounded domain in $\mathbb{R}^N$ whose boundary ∂E satisfies the exterior sphere condition (6.1). Then for every $\varphi \in C(\partial E)$ there exists a unique solution to the Dirichlet problem*

$$u \in C^2(E) \cap C(\bar{E}), \quad \Delta u = 0 \ \text{ in } E, \quad \text{and} \quad u\big|_{\partial E} = \varphi. \tag{6.2}$$

Proof (Perron ([117])) Recall the definition (4.6) of the classes $\sigma(E)$ and $\Sigma(E)$ and for a fixed $\varphi \in C(\partial E)$, consider the two classes

$$\begin{aligned} \sigma(\varphi; E) &= \left\{ v \in \sigma(E) \cap C(\bar{E}) \ \text{ and } \ v\,\big|_{\partial E} \le \varphi \right\} \\ \Sigma(\varphi; E) &= \left\{ v \in \Sigma(E) \cap C(\bar{E}) \ \text{ and } \ v\,\big|_{\partial E} \ge \varphi \right\}. \end{aligned}$$

Any constant $k \le \min_{\partial E} \varphi$ is in $\sigma(\varphi; E)$, and any constant $h \ge \max_{\partial E} \varphi$ is in $\Sigma(\varphi; E)$. Therefore $\sigma(\varphi; E)$ and $\Sigma(\varphi; E)$ are not empty. If a solution u to (6.2) exists, it must satisfy

$$v \le u \le w \quad \text{for all } v \in \sigma(\varphi; E) \ \text{ and } \ \text{for all } w \in \Sigma(\varphi; E).$$

This suggests to look for u as the unique element of separation of the two classes $\sigma(\varphi; E)$ and $\Sigma(\varphi; E)$, i.e.,

$$\sup_{v \in \sigma(\varphi; E)} v(x) \stackrel{\text{def}}{=} u(x) \stackrel{\text{def}}{=} \inf_{w \in \Sigma(\varphi; E)} w(x), \qquad \forall x \in \bar{E}. \tag{6.3}$$

To prove the theorem we have to prove the following two facts.

Lemma 6.1 *The function u defined by (6.3) is harmonic in E.*

Lemma 6.2 $u \in C(\bar{E})$ *and* $u|_{\partial E} = \varphi$. ■

Proof (Lemma 6.1) Fix $x_o \in E$ and select a sequence $\{v_n\} \subset \sigma(\varphi; E)$ such that $v_n(x_o) \to u(x_o)$. The functions

$$V_n = \max\{v_1, v_2, \dots, v_n\} \tag{6.4}$$

belong to $\sigma(\varphi; E)$, and the sequence $\{V_n\}$ satisfies

$$V_n \le V_{n+1} \quad \text{and} \quad \lim_{n\to\infty} V_n(x_o) = u(x_o).$$

Let $B_\rho(\xi) \subset E$ be a ball containing x_o, and construct the functions $V_{n;\xi,\rho}$ as described in (4.7). By Proposition 4.2, $V_{n;\xi,\rho} \in \sigma(\varphi; E)$, and by the previous remarks

$$V_{n;\xi,\rho} \le V_{n+1;\xi,\rho} \quad \text{and} \quad V_{n;\xi,\rho}(x_o) \to u(x_o).$$

Thus $\{V_{n;\xi,\rho}\}$ converges monotonically to some function $z(\cdot)$, which we claim is harmonic in $B_\rho(\xi)$. Indeed, $V_{n;\xi,\rho} - V_{1;\xi,\rho}$ are all harmonic and non-negative in $B_\rho(\xi)$, and the sequence $\{V_{n;\xi,\rho}(x_o) - V_{1;\xi,\rho}(x_o)\}$ is equi-bounded. Therefore by the Harnack inequality (5.1), $\{V_{n;\xi,\rho} - V_{1;\xi,\rho}\}$ is equi-bounded on compact subsets of $B_\rho(\xi)$. By Theorem 5.2, also all the derivatives $D^\alpha(V_{n;\xi,\rho} - V_{1;\xi,\rho})$ are equi-bounded on compact subsets of $B_\rho(\xi)$. Therefore, by possibly passing to a subsequence, $\{D^\alpha(V_{n;\xi,\rho} - V_{1;\xi,\rho})\}$ converge uniformly on compact subsets of $B_\rho(\xi)$, for all multi-indices α. Thus $z(\cdot)$ is infinitely differentiable in $B_\rho(\xi)$ and $\{D^\alpha V_{n;\xi,\rho}\} \to D^\alpha z$ uniformly on compact subsets of $B_\rho(\xi)$, for all multi-indices α. Since all $V_{n;\xi,\rho}$ are harmonic in $B_\rho(\xi)$, also z is harmonic in $B_\rho(\xi)$.

By construction, $z(x_o) = u(x_o)$. To prove that $z(x) = u(x)$ for all $x \in B_\rho(\xi)$, fix $\tilde{x} \in B_\rho(\xi)$ and construct sequences $\{\tilde{v}_n\}$ and $\{\tilde{V}_n\}$ as follows:

$$\tilde{v}_n \in \sigma(\varphi; E) \quad \text{and} \quad \tilde{v}_n(\tilde{x}) \to u(\tilde{x})$$

$$\tilde{V}_n(x) = \max\{V_n(x); \tilde{v}_1(x), \tilde{v}_2(x), \dots, \tilde{v}_n(x)\} \quad \forall x \in E$$

where V_n are defined in (6.4). Starting from $\tilde{V}_n$, construct the corresponding functions $\tilde{V}_{n;\xi,\rho}$ as indicated in (4.7). Arguing as before, these satisfy

$$\tilde{V}_n \le \tilde{V}_{n;\xi,\rho}, \qquad \tilde{V}_{n;\xi,\rho} \le \tilde{V}_{n+1;\xi,\rho}, \qquad \tilde{V}_{n;\xi,\rho}(\tilde{x}) \to u(\tilde{x}).$$

Moreover, $\{\tilde{V}_{n;\xi,\rho}\}$ converges monotonically in $B_\rho(\xi)$ to a harmonic function $\tilde{z}(\cdot)$ satisfying

$$\tilde{z}(x) \ge z(x) \quad \text{for all } x \in B_\rho(\xi) \quad \text{and} \quad \tilde{z}(\tilde{x}) = u(\tilde{x}).$$

By the construction (6.3) of u

$$u(x_o) = z(x_o) \le \tilde{z}(x_o) = u(x_o).$$

Thus the function $\tilde{z} - z$ is non-negative and harmonic in $B_\rho(\xi)$, and it vanishes in an interior point x_o of $B_\rho(\xi)$. This is impossible unless $\tilde{z}(x) = z(x)$ for all $x \in B_\rho(\xi)$. In particular

$$\tilde{z}(\tilde{x}) = z(\tilde{x}) = u(\tilde{x}).$$

Since $\tilde{x} \in B_\rho(\xi)$ is arbitrary, we conclude that u is harmonic in a neighborhood of x_o and hence in the whole of E, since x_o is an arbitrary point of E. ∎

Proof (Lemma 6.2) Fix $x_* \in \partial E$ and let $B_R(x_o)$ be the ball exterior to E and touching ∂E only at x_* claimed by (6.1). The function

$$H(x) = \begin{cases} \dfrac{1}{R^{N-2}} - \dfrac{1}{|x - x_o|^{N-2}} & \text{if } N \ge 3 \\[2ex] \ln \dfrac{|x - x_o|}{R} & \text{if } N = 2 \end{cases} \tag{6.5}$$

is harmonic in a neighborhood of E and positive on ∂E except at x_*, where it vanishes. Fix an arbitrarily small positive number ε and determine $\delta = \delta(\varepsilon) \in (0, 1)$ so that

$$|\varphi(x) - \varphi(x_*)| \le \varepsilon \qquad \forall [|x - x_*| \le \delta] \cap \partial E.$$

We claim that for all $\varepsilon > 0$ there exists a constant C_ε, depending only on $\|\varphi\|_{\infty,\partial E}$, R, N, and $\delta(\varepsilon)$, such that

$$|\varphi(x) - \varphi(x_*)| < \varepsilon + C_\varepsilon H(x) \qquad \forall x \in \partial E. \tag{6.6}$$

This is obvious if $|x - x_*| \le \delta$. If $x \in \partial E$ and $|x - x_*| > \delta$

$$|\varphi(x) - \varphi(x_*)| \le 2\|\varphi\|_{\infty,\partial E} \frac{H(x)}{H_\delta}, \qquad \text{where} \qquad H_\delta = \min_{[|x - x_*| \ge \delta] \cap \partial E} H(x).$$

To prove (6.6), we have only to observe that $H_\delta > 0$. It follows from (6.6) that for all $x \in \partial E$

$$\varphi(x_*) - \varepsilon - C_\varepsilon H(x) \le \varphi(x) \le \varphi(x_*) + \varepsilon + C_\varepsilon H(x).$$

This implies that

$$\varphi(x_*) - \varepsilon - C_\varepsilon H \in \sigma(\varphi; E)$$

and

$$\varphi(x_*) + \varepsilon + C_\varepsilon H \in \Sigma(\varphi; E).$$

Therefore for all $x \in \bar{E}$

$$\varphi(x_*) - \varepsilon - C_\varepsilon H(x) \le u(x) \le \varphi(x_*) + \varepsilon + C_\varepsilon H(x).$$

This in turn implies

$$|u(x) - \varphi(x_*)| \le \varepsilon + C_\varepsilon H(x) \qquad \forall x \in \bar{E}.$$

We now let $x \to x_*$ for $\varepsilon \in (0,1)$ fixed. Since $H \in C(\bar{E})$ and $H(x_*) = 0$

$$\limsup_{x \to x_*} |u(x) - \varphi(x_*)| \le \varepsilon \qquad \forall \varepsilon \in (0,1). \qquad \blacksquare$$

7 About the Exterior Sphere Condition

The existence theorem is based on an *interior* statement (Lemma 6.1) and a *boundary* statement concerning the behavior of u near ∂E (Lemma 6.2). The first can be established regardless of the structure of ∂E. The second relies on the construction of the function $H(\cdot)$ in (6.5). Such a construction is made possible by the exterior sphere condition (6.1). Indeed, this is the only role played by (6.1). Keeping this in mind, we might impose on ∂E the

Barrier Postulate: $\forall x_* \in \partial E$, $\exists H(x_*; \cdot) \in C(\bar{E})$ satisfying

$$\begin{aligned} &H(x_*; \cdot) \text{ is super-harmonic in a neighborhood of } E \\ &H(x_*; x) > 0 \quad \forall x \in \bar{E} - \{x_*\}, \text{ and } H(x_*; x_*) = 0. \end{aligned} \tag{7.1}$$

Any such function $H(x_*; \cdot)$ is a *barrier* for the Dirichlet problem (6.2) at x_*.

Assume that ∂E satisfies the barrier postulate. Arguing as in the proof of Lemma 6.2, having fixed $x_* \in \partial E$, for all $\varepsilon > 0$ there exists a constant

$$C_\varepsilon = C_\varepsilon(\|\varphi\|_{\infty,\partial E}, N, H(x_*; \cdot), \varepsilon)$$

such that

$$|\varphi(x) - \varphi(x_*)| \le \varepsilon + C_\varepsilon H(x_*; x) \qquad \forall x \in \partial E.$$

Therefore, for all $x \in \partial E$

$$\varphi(x_*) - \varepsilon - C_\varepsilon H(x_*; x) \le \varphi(x) \le \varphi(x_*) + \varepsilon + C_\varepsilon H(x_*; x).$$

Since $H(x_*; \cdot)$ is super-harmonic

$$\varphi(x_*) - \varepsilon - C_\varepsilon H(x_*; x) \in \sigma(\varphi; E)$$

and

$$\varphi(x_*) + \varepsilon + C_\varepsilon H(x_*; x) \in \Sigma(\varphi; E).$$

Therefore for all $x \in \bar{E}$

$$\varphi(x_*) - \varepsilon - C_\varepsilon H(x_*; x) \le u(x) \le \varphi(x_*) + \varepsilon + C_\varepsilon H(x_*; x)$$

and

$$|u(x) - \varphi(x_*)| < \varepsilon + C_\varepsilon H(x_*; x) \qquad \forall x \in \bar{E}.$$

This proves Lemma 6.2 if the exterior sphere condition (6.1) is replaced by the barrier postulate (7.1). We conclude that the Dirichlet problem (6.2) is uniquely solvable for every domain E satisfying the barrier postulate.

7.1 The Case $N = 2$ and ∂E Piecewise Smooth

Let E be a bounded domain in $\mathbb{R}^2$ whose boundary ∂E is the finite union of portions of curves of class C^1. Domains of this kind permit corners and even spikes pointing outside or inside E. Fix $x_* \in \partial E$ and assume, modulo a translation, that x_* coincides with the origin. We may also assume, up to a homothetic transformation, that E is contained in the unit disc about the origin. Identifying $\mathbb{R}^2$ with the complex plane $\mathbb{C}$, points $z = \rho e^{i\theta}$ of E, are determined by a unique value of the argument $\theta \in (-\pi, \pi)$. Therefore $\ln z$ is uniquely defined in E. A barrier at the origin is

$$H(x) = -\mathrm{Re}\left(\frac{1}{\ln z}\right) = -\frac{\ln \rho}{\ln^2 \rho + \theta^2}.$$

7.2 A Counterexample of Lebesgue for $N = 3$ ([101])

If $N \geq 3$, spikes pointing outside E are permitted, since any such point would satisfy the exterior sphere condition (6.1). Spikes pointing inside E are, in general, not permitted as shown by the following example of Lebesgue.

Denote points in $\mathbb{R}^3$ by (x, z), where $x = (x_1, x_2)$ and $z \in \mathbb{R}$. The function

$$v(x, z) = \int_0^1 \frac{s\,ds}{\sqrt{|x|^2 + (s - z)^2}} \tag{7.2}$$

is harmonic outside $[|x| = 0] \cap [0 \leq z \leq 1]$. By integration by parts one computes

$$\begin{aligned} v(x, z) = & \sqrt{|x|^2 + (1 - z)^2} - \sqrt{|x|^2 + z^2} \\ & + z \ln \left| \left[(1 - z) + \sqrt{|x|^2 + (1 - z)^2}\right] \left[z + \sqrt{|x|^2 + z^2}\right] \right| \\ & - 2z \ln |x|. \end{aligned}$$

As $(x, z) \to 0$, the sum of the first three terms on the right-hand tends to 1, whereas the last term is discontinuous at zero. It tends to zero if $(x, z) \to 0$ along the curve $|z|^\beta = |x|$ for all $\beta > 0$. However, if $(x, z) \to 0$ along $|x| = e^{-\gamma/2z}$, for $z > 0$ and $\gamma > 0$, it converges to γ. We conclude that

$$\lim_{\substack{(x,z)\to 0 \\ \text{along } |x| = e^{-\gamma/2z}}} v(x, z) = 1 + \gamma.$$

Therefore, all the level surfaces $[v = 1 + \gamma]$ for all $\gamma > 0$ go through the origin, and as a consequence, v is not continuous at the origin.

Fix $c > 0$ and consider the domain

$$E = [v < 1 + c] \cap [|x, z| < 1].$$

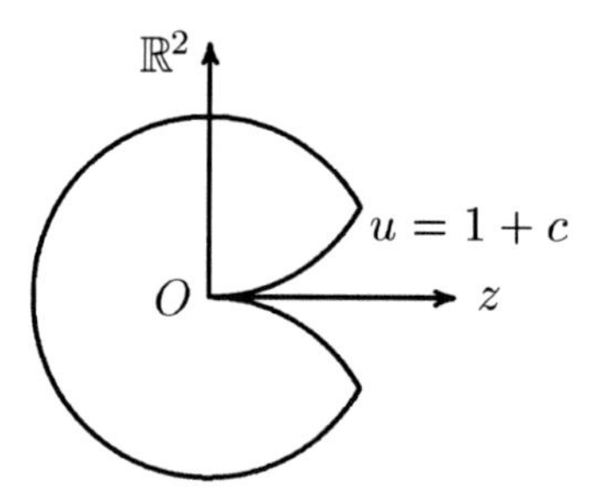

Fig. 7.1.

There exists no solution to the Dirichlet problem

$$u \in C^2(E) \cap C(\bar{E}), \quad \Delta u = 0 \text{ in } E, \quad \text{and} \quad u\big|_{\partial E} = v\big|_{\partial E}. \tag{7.3}$$

Notice that even though v is not continuous in $\bar{E}$, the restriction $v|_{\partial E}$ is continuous on ∂E. The idea of the counterexample is based on showing that any solution of (7.3) must coincide with v, which itself is not a solution.

Fix any $\varepsilon > 0$, and consider the domain $E_\varepsilon = E \cap [|x, z| > \varepsilon]$. Assume that u is a solution of (7.3) and let C be a constant such that $|u - v| < C$ in E. The functions

$$w_\varepsilon = C \frac{\varepsilon}{|x, z|} \pm (u - v)$$

are harmonic in E_ε and non-negative on ∂E_ε. Thus by the maximum principle

$$|u(x, z) - v(x, z)| \le C \frac{\varepsilon}{|x, z|} \quad \text{in} \quad E_\varepsilon.$$

8 The Poisson Integral for the Half-Space

Denote points in $\mathbb{R}^{N+1}$ by (x, t), where $x \in \mathbb{R}^N$ and $t \in \mathbb{R}$. Consider the Dirichlet problem

$$\begin{cases} u \in C^2(\mathbb{R}^N \times \mathbb{R}^+) \cap C(\overline{\mathbb{R}^N \times \mathbb{R}^+}) \\ \Delta u = 0 \quad \text{in } \mathbb{R}^N \times \mathbb{R}^+ \\ u(x, 0) = \varphi(x) \in C(\mathbb{R}^N) \cap L^\infty(\mathbb{R}^N). \end{cases} \tag{8.1}$$

A solution to (8.1) is called the harmonic extension of φ in the upper half-space $\mathbb{R}^N \times \mathbb{R}^+$. Consider the fundamental solution (2.6) of the Laplacian in $\mathbb{R}^{N+1}$ with pole at $(y, 0)$

$$F(x, t; y) = \begin{cases} \dfrac{1}{(N-1)\omega_{N+1}} \dfrac{1}{[|x - y|^2 + t^2]^{\frac{N-1}{2}}} & \text{if } N \ge 2 \\ \\ -\dfrac{1}{2\pi} \ln \left[|x - y|^2 + t^2\right]^{1/2} & \text{if } N = 1. \end{cases}$$

The Poisson kernel for the half-space is defined for all $N \ge 1$ by

$$K(x;y) = -2\frac{\partial F(x,t;y)}{\partial t} = \frac{2t}{\omega_{N+1}\left[|x-y|^2+t^2\right]^{\frac{N+1}{2}}}. \tag{8.2}$$

Theorem 8.1 *Every $\varphi \in C(\mathbb{R}^N) \cap L^\infty(\mathbb{R}^N)$ has a unique bounded harmonic extension H_φ in $\mathbb{R}^N \times \mathbb{R}^+$, given by*

$$H_\varphi(x,t) = \frac{2t}{\omega_{N+1}}\int_{\mathbb{R}^N}\frac{\varphi(y)}{\left[|x-y|^2+t^2\right]^{\frac{N+1}{2}}}dy. \tag{8.3}$$

Proof (Uniqueness) If u and v are both bounded solutions of (8.1), the difference $w = u - v$ is harmonic in $\mathbb{R}^N \times \mathbb{R}^+$ and vanishes for $t = 0$. By reflection about the hyperplane $t = 0$, the function

$$\tilde{w}(x,t) = \begin{cases} w(x,t) & \text{if } t > 0 \\ -w(x,-t) & \text{if } t \le 0 \end{cases}$$

is bounded and harmonic in $\mathbb{R}^{N+1}$. Therefore, by Liouville's theorem (Corollary 5.3), it is constant. Since $w(\cdot,0) = 0$, it vanishes identically. ∎

Remark 8.1 The statement of uniqueness in Theorem 8.1 holds only within the class of bounded solutions. Indeed, the two functions $u = 0$ and $v = t$ are both harmonic extensions of $\varphi = 0$.

Proof (Existence) The function H_φ defined in (8.3) is harmonic in $\mathbb{R}^N \times \mathbb{R}^+$. The boundedness of H_φ follows from the boundedness of φ and the following lemma.

Lemma 8.1 *For all $\varepsilon > 0$ and all $x \in \mathbb{R}^N$*

$$\frac{2\varepsilon}{\omega_{N+1}}\int_{\mathbb{R}^N}\frac{dy}{\left[|x-y|^2+\varepsilon^2\right]^{\frac{N+1}{2}}} = 1. \tag{8.4}$$

Proof Assume $N \ge 2$. The change of variables $y - x = \varepsilon\xi$ transforms the integral in (8.4) into

$$\frac{2}{\omega_{N+1}}\int_{\mathbb{R}^N}\frac{d\xi}{(1+|\xi|^2)^{\frac{N+1}{2}}} = 2\frac{\omega_N}{\omega_{N+1}}\int_0^\infty \frac{\rho^{N-1}}{(1+\rho^2)^{\frac{N+1}{2}}}d\rho = 1. \tag{8.5}$$

The case $N = 1$ is treated analogously (**8.1** of the Complements).

To conclude the proof of Theorem 8.1, it remains to show that for all $x_* \in \mathbb{R}^N$

$$\lim_{(x,t)\to x_*} H_\varphi(x,t) = \varphi(x_*).$$

This is established as in Theorem 3.1 by making use of (8.4). ∎

9 Schauder Estimates of Newtonian Potentials

Let E be a bounded open set in $\mathbb{R}^N$ and continue to denote by $F(\cdot;\cdot)$ the fundamental solution of the Laplacian, introduced in (2.6). The Newtonian potential generated in $\mathbb{R}^N$ by a density distribution $f \in L^p(E)$ for some $p > 1$ is defined by

$$A_F f = \int_E F(\cdot;y)f(y)dy \tag{9.1}$$

provided the right-hand side is finite. If $f \in L^\infty(E)$

$$\|A_F f\|_{\infty,\mathbb{R}^N} + \|\nabla A_F f\|_{\infty,\mathbb{R}^N} \le \gamma \|f\|_{\infty,E} \tag{9.2}$$

where γ is a constant depending only on N and $\operatorname{diam}(E)$. Further regularity of $A_F f$ can be established if f is Hölder continuous and compactly supported in E. For $m \in \mathbb{N} \cup \{0\}$, $\eta \in (0,1)$, and $\varphi \in C^\infty(E)$, set

$$\|\varphi\|_{m,\eta;E} \stackrel{\text{def}}{=} \sum_{|\alpha|\le m} \|D^\alpha \varphi\|_{\infty,E} + \sum_{|\alpha|=m} \sup_{x,y\in E} \frac{|D^\alpha \varphi(x) - D^\alpha \varphi(y)|}{|x-y|^\eta}. \tag{9.3}$$

Denote by $C^{m,\eta}(E)$ the space of functions $\varphi \in C^m(E)$ with finite norm $\|\varphi\|_{m,\eta;E}$ and by $C_o^{m,\eta}(E)$ the space of functions $\varphi \in C^{m,\eta}(E)$ compactly supported in E. If $m=0$, we let $C^{0,\eta}(E) = C^\eta(E)$ and $\|\varphi\|_{0,\eta;E} = \|\varphi\|_{\eta;E}$.

Proposition 9.1 *Let $f \in C_o^{m,\eta}(E)$. Then $A_F f \in C^{m+2,\eta}(E)$, and there exists a constant γ depending upon N, m, η, and* $\operatorname{diam}(E)$, *such that*

$$\|A_F f\|_{m+2,\eta;E} \le \gamma \|f\|_{m,\eta;E}. \tag{9.4}$$

Proof It suffices to prove (9.4) for $m=0$ and for $f \in C_o^\infty(E)$. Assume $N \ge 3$, the proof for $N=2$ being analogous, and rewrite (9.1) as

$$\omega_N(N-2)A_F f = v(\cdot) = \int_{\mathbb{R}^N} |\xi|^{2-N} f(\cdot + \xi)d\xi$$

and compute

$$\begin{aligned} v_{x_i x_j}(x) &= \int_{\mathbb{R}^N} |\xi|^{2-N} f_{\xi_i\xi_j}(x+\xi)d\xi = -\int_{\mathbb{R}^N} (|\xi|^{2-N})_{\xi_j} f_{\xi_i}(x+\xi)d\xi \\ &= -\int_{|\xi|>r} (|\xi|^{2-N})_{\xi_j} f_{\xi_i}(x+\xi)d\xi \\ &\quad -\int_{|\xi|<r} (|\xi|^{2-N})_{\xi_j} [f(x+\xi) - f(x)]_{\xi_i} d\xi \\ &= \int_{|\xi|>r} (|\xi|^{2-N})_{\xi_i\xi_j} f(x+\xi)d\xi \\ &\quad + \int_{|\xi|<r} (|\xi|^{2-N})_{\xi_i\xi_j} (f(x+\xi) - f(x))d\xi \\ &\quad + f(x)\int_{|\xi|=r} (|\xi|^{2-N})_{\xi_j} \frac{\xi_i}{|\xi|} d\sigma. \end{aligned} \tag{9.5}$$

In this representation, r is any positive number, $d\sigma$ is the surface measure over the sphere ∂B_r, and the integral extended over the ball $|\xi| < r$ is meant in the sense of the limit

$$\int_{|\xi|<r} (|\xi|^{2-N})_{\xi_j\xi_j}(f(x+\xi) - f(x))d\xi$$
$$\stackrel{\text{def}}{=} \lim_{\varepsilon\to 0} \int_{\varepsilon<|\xi|<r} (|\xi|^{2-N})_{\xi_j\xi_j}(f(x+\xi) - f(x))d\xi.$$

Such a limit exists, since f is Hölder continuous. From (9.5), by taking $r = \operatorname{diam}(E)$, we estimate

$$\begin{aligned} \sum_{|\alpha|=2} \|D^\alpha v\|_{\infty,\mathbb{R}^N} &\le \gamma \|f\|_{\infty,E} \left(1 + \int_{r<|\xi|<2r} |\xi|^{-N} d\xi\right) \\ &\quad + \gamma \int_{|\xi|<r} |\xi|^{-N+\eta} \frac{|f(x+\xi) - f(x)|}{|\xi|^\eta} d\xi \\ &\le \gamma\,(1 + \operatorname{diam}(E))\, \|\!| f |\!\|_{\eta;E}. \end{aligned} \tag{9.6}$$

Next we fix $y \in \mathbb{R}^N$ and represent $v_{x_ix_j}(y)$. By calculations analogous to those leading to (9.5)

$$\begin{aligned} v_{x_ix_j}(y) &= \int_{\mathbb{R}^N} |\xi|^{2-N} f_{\xi_i\xi_j}(y+\xi)d\xi \\ &= \int_{\mathbb{R}^N} |(x-y)+\xi|^{2-N} f_{\xi_i\xi_j}(x+\xi)d\xi \\ &= \int_{|\xi|>r} \left(|(x-y)+\xi|^{2-N}\right)_{\xi_i\xi_j} f(x+\xi)d\xi \\ &\quad + \int_{|\xi|<r} \left(|(x-y)+\xi|^{2-N}\right)_{\xi_i\xi_j} [f(x+\xi) - f(y)]d\xi \\ &\quad + f(y) \int_{|\xi|=r} \left(|(x-y)+\xi|^{2-N}\right)_{\xi_j} \frac{\xi_i}{|\xi|} d\sigma. \end{aligned} \tag{9.7}$$

From the representations (9.5) and (9.7), we obtain by difference

$$\begin{aligned} &v_{x_ix_j}(x) - v_{x_ix_j}(y) \\ &\quad = \int_{|\xi|>r} \left(|\xi|^{2-N} - |(x-y)+\xi|^{2-N}\right)_{\xi_i\xi_j} [f(x+\xi) - f(y)]d\xi \\ &\quad + \int_{|\xi|<r} (|\xi|^{2-N})_{\xi_j\xi_j} [f(x+\xi) - f(x)]d\xi \\ &\quad - \int_{|\xi|<r} \left(|(x-y)+\xi|^{2-N}\right)_{\xi_i\xi_j} [f(x+\xi) - f(y)]d\xi \\ &\quad + [f(x) - f(y)] \int_{|\xi|=r} (|\xi|^{2-N})_{\xi_j} \frac{\xi_i}{|\xi|} d\sigma \end{aligned}$$

$$+ f(y) \int_{|\xi|=r} \left(|\xi|^{2-N} - |(x-y)+\xi|^{2-N} \right)_{\xi_j} \frac{\xi_i}{|\xi|} d\sigma$$

$$+ f(y) \int_{|\xi|>r} \left(|\xi|^{2-N} - |(x-y)+\xi|^{2-N} \right)_{\xi_i \xi_j} d\xi.$$

The sum of the last two integrals is zero, and the integral extended over the shell $|\xi| = r$ is majorized by

$$\gamma \| f \|_{\eta;E} |x-y|^\eta$$

where γ is a constant depending only upon the dimension N. In the estimates below we denote by γ a constant that can be different in different contexts, and it can be computed quantitatively a priori in terms of N alone. Estimating the first integral extended over the ball $|\xi| < r$ we have

$$\left| \int_{|\xi|<r} (|\xi|^{2-N})_{\xi_j \xi_j} [f(x+\xi) - f(x)] d\xi \right| \le \gamma(N) \| f \|_{\eta;E} \int_{|\xi|<r} |\xi|^{-N+\eta} d\xi$$
$$\le \gamma \| f \|_{\eta;E} r^\eta.$$

Analogously

$$\left| \int_{|\xi|<r} \left(|(x-y)+\xi|^{2-N} \right)_{\xi_i \xi_j} [f(x+\xi) - f(y)] d\xi \right|$$
$$\le \gamma \| f \|_{\eta;E} \int_{|\xi|<r} |(x-y)+\xi|^{-N+\eta} d\xi$$
$$\le \gamma \| f \|_{\eta;E} \int_{|z|<r+|x-y|} |z|^{-N+\eta} dz$$
$$\le \gamma \| f \|_{\eta;E} (r^\eta + |x-y|^\eta).$$

Combining these estimates, we conclude that there is a constant γ depending only upon N, such that for every $r > 0$

$$|v_{x_i x_j}(x) - v_{x_i x_j}(y)| \le \gamma \| f \|_{\eta;E} (r^\eta + |x-y|^\eta)$$
$$+ \gamma \left| \int_{|\xi|>r} \left(|\xi|^{2-N} - |(x-y)+\xi|^{2-N} \right)_{\xi_i \xi_j} [f(x+\xi) - f(y)] d\xi \right|. \tag{9.8}$$

Choose $r = 2|x-y|$ so that over the set $|\xi| > r$

$$2|\xi| \ge |(x-y)+\xi| \ge \frac{1}{2}|\xi|.$$

Then by direct calculation and the mean value theorem

$$\left| \left(|\xi|^{2-N} - |(x-y)+\xi|^{2-N} \right)_{\xi_i \xi_j} \right| \le \left| \frac{\delta_{ij}}{|\xi|^N} - \frac{\delta_{ij}}{|(x-y)+\xi|^N} \right|$$
$$+ \left| \frac{\xi_i \xi_j}{|\xi|^{N+2}} - \frac{[(x-y)+\xi]_i [(x-y)+\xi]_j}{|(x-y)+\xi|^{N+2}} \right|$$
$$\le \gamma \frac{|x-y|}{|\xi|^{N+1}}.$$

Therefore the last integral in (9.8) can be majorized by

$$\gamma|x-y|\int_{|\xi|>r}\frac{|\xi|^\eta}{|\xi|^{N+1}}\frac{|f(x+\xi)-f(y)|}{|(x-y)+\xi|^\eta}d\xi$$
$$\leq \gamma\|f\|_{\eta;E}\, r\int_{|\xi|>r}|\xi|^{-(N+1)+\eta}d\xi = \gamma\|f\|_{\eta;E}\, r^\eta.$$

Combining this with (9.2) and (9.6) proves the proposition. ∎

Remark 9.1 The proof shows that

$$\sup_{x,y\in\mathbb{R}^N}\frac{|v_{x_ix_j}(x)-v_{x_ix_j}(y)|}{|x-y|^\eta}\leq\gamma\sup_{x,y\in E}\frac{|f(x)-f(y)|}{|x-y|^\eta}$$

where the constant γ depends only on N and η and is independent of $|E|$.

Remark 9.2 The dependence on $\mathrm{diam}(E)$ on the right-hand side of (9.3) enters only through (9.2) and (9.6). Therefore the constant γ in (9.4) depends on $\mathrm{diam}(E)$ as

$$\gamma=\gamma_o(1+\mathrm{diam}(E))\qquad\text{for some }\ \gamma_o=\gamma_o(N,m,\eta).$$

10 Potential Estimates in $L^p(E)$

The estimate (9.2) implies that A, as defined by (9.1), is a map from $L^\infty(E)$ into $L^\infty(E)$. More precisely, it maps $L^\infty(E)$ into the subspace of the Lipschitz continuous functions defined in E. It is natural to ask whether $f\in L^p(E)$ for some $p\geq 1$ would imply that $A_F f\in L^q(E)$ for some $q\geq 1$, and what is the relation between p and q.

If $f\in C_o^\eta(E)$, then $A_F f$, as defined by (9.1), is differentiable and

$$\nabla A_F f(x)=\frac{1}{\omega_N}\int_E\frac{(x-y)}{|x-y|^N}f(y)dy.$$

If however $f\in L^p(E)$ for some $p\geq 1$, the symbol $\nabla A_F f$ does not have the classical meaning of derivative, and it is simply defined by its right-hand side. If this is finite a.e. in E, we say that $\nabla A_F f$ is the weak gradient of the potential $A_F f$. We will give sufficient integrability conditions on f to ensure that the weak gradient $\nabla A_F f$ is in $L^q(E)$ for some $q\geq 1$.

Both issues are addressed by investigating the integrability of the *Riesz potential*

$$E\ni x\to w_\alpha(x)=\int_E\frac{f(y)}{|x-y|^{N-\alpha}}dy\qquad\text{for some }\ \alpha>0.\tag{10.1}$$

Proposition 10.1 *Let $f \in L^p(E)$ for some $p \ge 1$. Then*

$$|w_\alpha| \in L^q(E) \quad \textit{where} \quad q \in \begin{cases} \left[1, \dfrac{Np}{N-\alpha p}\right) & \textit{if } p < \dfrac{N}{\alpha} \\ [1,\infty) & \textit{if } p = \dfrac{N}{\alpha} \\ [1,\infty] & \textit{if } p > \dfrac{N}{\alpha}. \end{cases} \tag{10.2}$$

Moreover, there exists a constant γ that can be determined a priori only in terms of N, p, q, α, and diam(E), *such that*

$$\|w_\alpha\|_{q,E} \le \gamma \|f\|_{p,E}. \tag{10.3}$$

The constant $\gamma \to \infty$ as either $p \to N/\alpha$ or as diam$(E) \to \infty$.

Proof Assume first $p < N/\alpha$, and choose $s > 1$ from

$$\frac{1}{s} = 1 + \frac{1}{q} - \frac{1}{p}, \qquad 1 < s < \frac{N}{N-\alpha}.$$

By Hölder's inequality

$$\begin{aligned} |w_\alpha(x)| &= \int_E \left(|x-y|^{(\alpha-N)s}|f|^p\right)^{\frac{1}{q}} |f|^{1-\frac{p}{q}} |x-y|^{(\alpha-N)(1-\frac{s}{q})} dy \\ &\le \left(\int_E |x-y|^{(\alpha-N)s}|f|^p dy\right)^{\frac{1}{q}} \left(\int_E |x-y|^{(\alpha-N)\frac{q-s}{q-1}} |f|^{\frac{q-p}{q-1}} dy\right)^{\frac{q-1}{q}} \\ &\le \left(\int_E |x-y|^{(\alpha-N)s}|f|^p dy\right)^{\frac{1}{q}} \|f\|_{p,E}^{p(1-\frac{1}{s})} \left(\int_E |x-y|^{(\alpha-N)s} dy\right)^{\frac{1}{s}-\frac{1}{q}}. \end{aligned}$$

The last integral involving $|x-y|^{(\alpha-N)s}$ is estimated above by extending the domain of integration to the ball of center x and radius diam(E). It gives

$$\begin{aligned} \int_E |x-y|^{(\alpha-N)s} dy &\le \omega_N \int_0^{\mathrm{diam}(E)} \rho^{(N-1)-(N-\alpha)s} d\rho \\ &= \frac{\omega_N \,\mathrm{diam}(E)^{N(1-s\frac{N-\alpha}{N})}}{N\left(1-s\frac{N-\alpha}{N}\right)} = \gamma(N,s) \end{aligned}$$

provided

$$1 < s < \frac{N}{N-\alpha} \qquad \text{i.e.,} \qquad \frac{1}{p} - \frac{1}{q} < \frac{\alpha}{N}.$$

This determines the range of q in (10.1). In the estimates below, denote by γ a generic positive constant that can be determined a priori only in terms of N, p, q, and diam(E). To proceed, carry this estimate into the right-hand side of the estimation of $|w_\alpha(x)|$, take the qth power of both sides and integrate over E. Interchanging the order of integration with the aid of Fubini's theorem

$$\begin{aligned}\int_E |w_\alpha|^q dx &\le \gamma\Big(\int_E\int_E |x-y|^{(\alpha-N)s}|f(y)|^p dydx\Big)\|f\|_{p,E}^{pq(1-\frac{1}{s})}\\ &\le \gamma\Big[\int_E |f(y)|^p\Big(\int_E |x-y|^{(\alpha-N)s}dx\Big)dy\Big]\|f\|_{p,E}^{pq(1-\frac{1}{s})}\\ &\le \gamma\|f\|_{p,E}^q.\end{aligned}$$

Let now $p > N/\alpha$. Then by Hölder's inequality

$$|w_\alpha(x)| \le \|f\|_{p,E}\Big(\int_E |x-y|^{(\alpha-N)\frac{p}{p-1}}dy\Big)^{\frac{p-1}{p}}. \qquad \blacksquare$$

Corollary 10.1 *Let $f \in L^p(E)$ for some $p > N$. There exists a constant γ depending only upon N, p and diam(E), such that*

$$\|v\|_{\infty,E} + \|\nabla v\|_{\infty,E} \le \gamma\|f\|_{p,E}.$$

The constant $\gamma \to \infty$ as $p \to N$.

Let E a bounded open set in $\mathbb{R}^N$ with boundary ∂E of class C^1. For $f \in L^p(\partial E)$ set

$$v_\alpha(x) = \int_{\partial E}\frac{f(y)d\sigma(y)}{|x-y|^{N-1-\alpha}} \qquad \text{for } \alpha > 0 \tag{10.4}$$

where $d\sigma$ is the Lebesgue surface measure on ∂E.

Corollary 10.2 *Let $f \in L^p(\partial E)$ for some $p \ge 1$. Then*

$$|v_\alpha| \in L^q(E), \qquad \textit{where} \qquad q \in \begin{cases}\Big[1, \dfrac{(N-1)p}{(N-1)-\alpha p}\Big) & \textit{if } p < \dfrac{N-1}{\alpha}\\[2ex] [1,\infty) & \textit{if } p = \dfrac{N-1}{\alpha}\\[2ex] [1,\infty] & \textit{if } p > \dfrac{N-1}{\alpha}.\end{cases} \tag{10.5}$$

Moreover, there exists a constant γ that can be determined a priori in terms of N, p, q, α, and $|\partial E|$ only, such that

$$\|v_\alpha\|_{q,\partial E} \le \gamma\|f\|_{p,\partial E}. \tag{10.6}$$

The constant $\gamma \to \infty$ as either $p \to (N-1)/\alpha$ or as $|\partial E| \to \infty$.

11 Local Solutions

Consider formally local solutions of the Poisson equation

$$\Delta u = f, \quad \text{in } E \tag{11.1}$$

with no reference to possible boundary data on ∂E. Thus we assume that $u \in C^2(E)$ and $f \in C^{m+\eta}_{\text{loc}}(E)$ for some non-negative integer m and some $\eta \in [0,1)$. This means that $|\!|\!| f |\!|\!|_{m,\eta;K} < \infty$ for every compact set $K \subset E$. Let $K \subset K' \subset E$ be such that $\text{dist}(K;\partial K') > 0$. We will derive estimates of the norm $C^{m+\eta}(K)$ for u in terms of the norms $\|u\|_{\infty,K'}$ and $|\!|\!| f |\!|\!|_{m+\eta;K'}$.

Proposition 11.1 *There exists a constant γ depending only upon N, m, and* $\text{dist}(K;\partial K')$ *such that*

$$|\!|\!| u |\!|\!|_{m+2,\eta;K} \le \gamma \left(|\!|\!| f |\!|\!|_{m,\eta;K'} + \|u\|_{\infty,K'} \right).$$

Proof It suffices to prove the proposition for $m = 0$, and for $K \subset K'$ two concentric balls $B_{\sigma\rho}(x_o) \subset B_\rho(x_o) \subset E$, for some $\sigma \in (0,1)$. In such a case the statement takes the following form.

Lemma 11.1 *There exists a constant γ depending only upon N such that for all $B_{\sigma\rho}(x_o) \subset B_\rho(x_o) \subset E$*

$$\begin{aligned} |\!|\!| u |\!|\!|_{2+\eta;B_{\sigma\rho}(x_o)} \le & \gamma\left(1 + \frac{1}{(1-\sigma)\rho^\eta}\right) |\!|\!| f |\!|\!|_{\eta;B_\rho(x_o)} \\ & + \gamma\left(1 + \frac{1}{(1-\sigma)^{N+2}\rho^{2+\eta}}\right) \|u\|_{\infty,B_\rho(x_o)}. \end{aligned}$$

Proof (Lemma 11.1) The point $x_o \in E$ being fixed, we may assume after a translation that it coincides with the origin and write $B_\rho(0) = B_\rho$. Construct a smooth non-negative cutoff function $\zeta \in C_o^\infty(E)$ such that

$$\zeta = 1 \text{ in } B_{\frac{(1+\sigma)}{2}\rho} \quad \text{and} \quad |D^\alpha \zeta| \le \frac{C^{|\alpha|}}{[(1-\sigma)\rho]^{|\alpha|}} \tag{11.2}$$

for all multi-indices α of size $|\alpha| \le 2$ and for some constant C. Multiplying (11.1) by ζ and setting $v = u\zeta$, we find that v satisfies

$$\Delta v = f\zeta + u\Delta\zeta + 2\nabla u \cdot \nabla\zeta, \qquad v \in C_o^2(B_\rho).$$

Let $N \ge 3$, the case $N = 2$ being similar. By the Stokes identity (2.3), we may represent v as the superposition of the two Newtonian potentials

$$\begin{aligned} V_1(x) &= -\frac{1}{\omega_N(N-2)} \int_{\mathbb{R}^N} |x-y|^{2-N} (f\zeta)(y)dy \\ V_2(x) &= -\frac{1}{\omega_N(N-2)} \int_{\mathbb{R}^N} |x-y|^{2-N} (u\Delta\zeta + 2\nabla u \cdot \nabla\zeta)(y)dy. \end{aligned}$$

By virtue of Proposition 9.1 and Remark 9.2

$$\|V_1\|_{2,\eta;\mathbb{R}^N} \le \gamma(1+\rho)\|f\zeta\|_{\eta;B_\rho} \le \gamma(1+\rho+\rho^{-\eta})\|f\|_{\eta;B_\rho}.$$

We estimate $V_2(\cdot)$ within the ball $B_{\sigma\rho}$ by rewriting it as

$$\begin{aligned} V_2(x) = & \frac{1}{\omega_N(N-2)} \int_{\mathbb{R}^N} u(y)|x-y|^{2-N} \Delta\zeta \, dy \\ & + \frac{2}{\omega_N(N-2)} \int_{\mathbb{R}^N} u(y) \nabla |x-y|^{2-N} \cdot \nabla\zeta \, dy. \end{aligned}$$

Since $\nabla\zeta = 0$ within the ball of radius $\frac{(1+\sigma)}{2}\rho$, these integrals are not singular for $x \in B_{\sigma\rho}$ and we estimate

$$\|V_2\|_{2,\eta;B_{\sigma\rho}} \le \gamma\left(1 + \frac{1}{(1-\sigma)^{N+2}\rho^{2+\eta}}\right)\|u\|_{\infty,B_\rho}. \qquad \blacksquare$$

11.1 Local Weak Solutions

The previous remarks imply that if $f \in C^\eta_{\text{loc}}(E)$, the classical local solution of the Poisson equation (11.1) can be implicitly represented about any point $x_o \in E$ as

$$\begin{aligned} u(x) = & -\int_{B_\rho(x_o)} F(x;y) f\zeta dy \\ & + \int_{B_\rho(x_o)} u(y) \left(F(x;y)\Delta\zeta + 2\nabla F(x;y) \cdot \nabla\zeta\right) dy \end{aligned} \tag{11.3}$$

where $F(\cdot;\cdot)$ is the fundamental solution of the Laplace equation, introduced in (2.6), and ζ satisfies (11.2). Consider now the various integrals in (11.3), regardless of their derivation. The second is well defined for all $x \in B_{\sigma\rho}(x_o)$ if $u \in L^1_{\text{loc}}(E)$. The first defines a function

$$x \to \int_{B_\rho(x_o)} F(x;y) f\zeta \, dy \in L^1_{\text{loc}}(E)$$

provided $f \in L^p_{\text{loc}}(E)$ for some $p \ge 1$. Since $B_\rho(x_o) \subset E$ is arbitrary, this suggests the following

Definition Let $f \in L^p_{\text{loc}}(E)$ for some $p \ge 1$. A function $u \in L^1_{\text{loc}}(E)$ is a *weak solution* to the Poisson equation (11.1) in E if it satisfies (11.3).

The estimates of Section 10 imply:

Proposition 11.2 *Let $f \in L^p_{\text{loc}}(E)$ for $p > 1$ and let $u \in L^1_{\text{loc}}(E)$ be a local weak solution of (11.1) in E. There exists a constant γ depending only on N and p such that for all $B_{\sigma\rho}(x_o) \subset B_\rho(x_o) \subset E$*

$$\|\nabla u\|_{q,B_{\sigma\rho}(x_o)} \le \gamma \|f\|_{p,B_\rho(x_o)} + \gamma[(1-\sigma)\rho]^{-(N+1)}\rho^{\frac{N}{q}}\|u\|_{1,B_\rho(x_o)}$$

where

$$q \in \begin{cases} \left[1, \dfrac{N}{N-p}\right) & \text{if } \ 1 \le p < N \\ [1,\infty) & \text{if } \quad p = N \\ [1,\infty] & \text{if } \quad p > N. \end{cases}$$

If $p \in [1, N]$, *the constant* $\gamma \to \infty$ *as* $q \to Np/(N-p)$. *Moreover, if* $p > N/2$

$$\|u\|_{\infty,B_{\sigma\rho}(x_o)} \le \gamma\rho^{2-\frac{N}{p}}\|f\|_{p,B_\rho(x_o)} + \gamma[(1-\sigma)\rho]^{-N}\|u\|_{1,B_\rho(x_o)}.$$

The constant $\gamma \to \infty$ *as* $p \to N/2$.

12 Inhomogeneous Problems

12.1 On the Notion of Green's Function

Let E be a bounded domain in $\mathbb{R}^N$ with boundary ∂E of class C^1. The construction of the Green's function for E, introduced in (3.3), hinges on solving the family of Dirichlet problems (3.1). These solutions $y \to \Phi(x; y)$ were required to be of class $C^2(\bar{E})$. Such a regularity has been used to justify intermediate calculations, and it appears naturally in the explicit construction of Green's function for a ball.

However for each fixed $x \in E$, the Dirichlet problem (3.1) has a unique solution if one merely requires that ∂E satisfies the barrier postulate and that

$$\Phi(x; \cdot) \in C^2(E) \cap C(\bar{E}) \quad \text{for all fixed } \ x \in E.$$

This is the content of Theorem 6.1 and the remarks of Section 7.

Proposition 12.1 *Every bounded open set* $E \subset \mathbb{R}^N$ *with boundary* ∂E *satisfying the barrier postulate admits a Green's function* $G(\cdot;\cdot)$. *Moreover, for all* $(x; y) \in E \times E$,

$$\begin{aligned} &0 \le G(x; y) \le F(x; y) && \text{for } N \ge 3 \\ &0 \le G(x; y) \le \frac{1}{2\pi}\ln\frac{\operatorname{diam}(E)}{|x-y|} && \text{for } N = 2. \end{aligned} \tag{12.1}$$

Proof For fixed $x \in E$, let $\varepsilon > 0$ be so small that $B_\varepsilon(x) \subset E$. The function $G(x; \cdot)$ is harmonic in $E - B_\varepsilon(x)$, and it vanishes on ∂E. The number ε can be chosen to be so small that $G(x; \cdot) > 0$ on $\partial B_\varepsilon(x)$. Therefore, by the maximum principle, $G(x; \cdot) \ge 0$ in $E - B_\varepsilon(x)$ and hence in E since ε is arbitrary. The function $\Phi(x; \cdot)$ is harmonic in E, and by the maximum principle, it takes its maximum and minimum values on ∂E. If $N \ge 3$ it is positive on ∂E and thus $\Phi(\cdot;\cdot) > 0$ in E. This proves the first of (12.1). If $N = 2$ rewrite the Green's function as

$$G(x;y) = \frac{1}{2\pi} \ln \frac{\mathrm{diam}(E)}{|x-y|} - \left[\Phi(x;y) + \frac{1}{2\pi} \ln \mathrm{diam}(E) \right].$$

For fixed $x \in E$, the function of y in $[\cdots]$ is harmonic in E and non-negative on ∂E. ∎

Remark 12.1 The estimate roughly asserts that the singularity of $G(\cdot;\cdot)$ is of the same nature as the singularity of the fundamental solution $F(\cdot;\cdot)$.

Corollary 12.1 *$G(x;\cdot) \in L^p(E)$ uniformly in x, for all $p \in [1, \frac{N}{N-2})$.*

12.2 Inhomogeneous Problems

Given $f \in C^\eta(\bar{E})$ for some $\eta \in (0,1)$, consider the boundary value problem

$$u \in C^2(E) \cap C(\bar{E}), \qquad \Delta u = f \ \text{ in } E, \qquad \text{and} \qquad u\big|_{\partial E} = 0. \tag{12.2}$$

Theorem 12.1 *The boundary value problem (12.2) has a unique solution.*

Proof (Uniqueness) If u_i for $i = 1, 2$ solve (12.2), their difference is harmonic in E and vanishes on ∂E. Thus it vanishes identically in E, by the maximum principle. ∎

Proof (Existence) Assume momentarily that (12.2) has a solution $u \in C^2(\bar{E})$ and that the Green's function $G(x;\cdot)$ for E is of class $C^2(\bar{E})$. Then subtracting Green's identity (2.2) written for the pair of functions u and $\Phi(x;\cdot)$ from the Stokes identity (2.3)–(2.3) gives

$$\begin{aligned} u(x) &= -\int_E G(x;y) f(y) dy \\ &= -\int_E F(x;y) f(y) dy + \int_E \Phi(x;y) f(y) dy. \end{aligned} \tag{12.3}$$

This is a candidate for a solution of (12.2). To show that it is indeed a solution, we have to show that it takes zero boundary data in the sense of continuous functions in $\bar{E}$, it is of class $C^2(E)$, and it satisfies the PDE. Fix $x_* \in \partial E$ and write

$$\begin{aligned} \lim_{x \to x_*} u(x) &= -\lim_{x \to x_*} \int_E G(x;y) f(y) dy \\ &= -\lim_{x \to x_*} \int_{E \cap B_\varepsilon(x_*)} G(x;y) f(y) dy \\ &\quad - \lim_{x \to x_*} \int_{E - B_\varepsilon(x_*)} G(x;y) f(y) dy. \end{aligned}$$

The second integral tends to zero by the property of the Green's function, since y is away from the singularity $x = x_*$. The first integral is estimated by means of (12.1), and it yields

$$\int_{E\cap B_\varepsilon(x_*)} G(x;y)|f(y)|dy \le 2\sup_E |f| \int_{E\cap B_\varepsilon(x_*)} |F(x;y)|dy \le O(\varepsilon)$$

uniformly in x. ∎

To establish in what sense the PDE is satisfied, we assume $N \ge 3$, the arguments for $N = 2$ being similar.

12.3 The Case $f \in C_o^\infty(E)$

The set $K = \text{supp}(f)$ is a compact proper subset of E, and $\Phi(x;\cdot)$ is in $C^\infty(E)$. Therefore by symmetry

$$\Delta \int_E \Phi(x;y)f(y)dy = \int_E \Delta_y \Phi(x;y)f(y)dy = 0.$$

Calculating the Laplacian of the first term on the right-hand side of (12.3) gives

$$\begin{aligned}\Delta \int_E |x-y|^{2-N} f(y)dy &= \int_{\mathbb{R}^N} |\xi|^{2-N} \Delta_\xi f(x-\xi)d\xi \\ &= \lim_{\varepsilon\to 0} \int_{|\xi|>\varepsilon} |\xi|^{2-N} \Delta_\xi f(x\xi)d\xi.\end{aligned}$$

Perform a double integration by parts on the last integral using that f is compactly supported in $\mathbb{R}^N$. Taking into account that $|\xi|^{2-N}$ is harmonic in $|\xi| > \varepsilon$, and proceeding as in the proof of the Stokes identity (2.3) gives

$$\frac{1}{\omega_N(N-2)} \Delta \int_E |x-y|^{2-N} f(y)dy = f(x).$$

Combining these calculations shows that u defined by (12.3) satisfies the Poisson equation (12.2) in the classical sense.

12.4 The Case $f \in C^\eta(\bar{E})$

Let $\{K_j\}$ be a family of nested compact subsets of E exhausting E. Construct a sequence of functions $\{f_j\} \subset C_o^\infty(E)$ satisfying

$$\|f_j\|_{\eta;K_j} \le \|f\|_{\eta;E} \quad \text{and} \quad \lim_{j\to\infty} \|f_j - f\|_{\eta;K} = 0$$

for every compact subset $K \subset E$. Let u_j be the unique classical solution of

$$u_j \in C^2(E) \cap C(\bar{E}), \quad \Delta u_j = f_j \text{ in } E, \quad \text{and} \quad u_j\big|_{\partial E} = 0. \tag{12.4}$$

From the representation formula (12.3)

$$\|u_j\|_{\infty,E} \le \gamma \|f_j\|_{\infty;E} \le \gamma |\!|\!| f |\!|\!|_{\eta;E}, \qquad \forall j \in \mathbb{N}.$$

Combining this with Proposition 11.1 gives

$$|\!|\!| u_j |\!|\!|_{2,\eta;K_1} \le \gamma |\!|\!| f_j |\!|\!|_{\eta;K_1} \le \gamma |\!|\!| f |\!|\!|_{\eta;E} \qquad \forall j \in \mathbb{N}$$

where γ depends on N and $\operatorname{dist}\{K_1; \partial E\}$ and is independent of j. By the Ascoli–Arzelà theorem, we may select a subsequence $\{u_{j_1}\}$ out of $\{u_j\}$ converging in $C^{2,\eta}(K_1)$ to a function $u_1 \in C^{2,\eta}(K_1)$. By the same process, we may select a subsequence $\{u_{j_2}\}$ out of $\{u_{j_1}\}$ converging in $C^{2,\eta}(K_2)$ to a function $u_2 \in C^{2,\eta}(K_2)$, which coincides with u_1 within K_1. Continuing this diagonalization process, we obtain a function $u \in C^{2,\eta}_{\text{loc}}(E)$ and a subsequence $\{u_{j'}\}$ out of the original sequence $\{u_j\}$ such that $\{u_{j'}\} \to u$ in $C^{2,\eta}_{\text{loc}}(E)$. Letting $j \to \infty$ in (12.4) along such a subsequence proves that $u \in C^{2,\eta}(E)$ is a classical solution of the PDE. To verify that $u \in C(\bar{E})$ and that it vanishes on ∂E in the sense of continuous functions, we have only to observe that u satisfies (12.2) by the same limiting process. ∎

Problems and Complements

1c Preliminaries

1.1c Newtonian Potentials on Ellipsoids

Compute the Newtonian potential generated by a uniform distribution of masses, or charges, on the surface of an ellipsoid. Verify that such a potential is constant inside the ellipsoid ([80], pages 22 and 193).

Theorem 1.1c *Let E be a bounded domain in $\mathbb{R}^N$ of boundary ∂E of class C^2. The Newtonian potential $V(\cdot)$, generated by a uniform distribution of masses on ∂E, is constant in E if and only if E is an ellipsoid.*

The sufficient part of the theorem is due to Newton. The necessary part in dimension $N = 2$ was established in 1931 by Dive [37]. The necessary part for all $N \ge 2$ has been recently established is in [28]. The assumption of uniform distribution cannot be removed as shown in [140].

1.2c Invariance Properties

1.4. Prove that the Laplacian is invariant under a unitary affine transformation of coordinates in $\mathbb{R}^N$.

1.5. Find all second-order rotation invariant operators of the type

$$L(u) = \sum_{i,j,h,k=1}^{N} a_{ijhk} u_{x_i x_j} u_{x_h x_k}.$$

1.6. Prove that Δ is the only second-order, linear operator invariant under orthogonal linear transformation of the coordinates axes.

1.7. Find all homogeneous harmonic polynomials of degree n in two and three variables ([70]).
Hint: For $N = 2$ attempt z^n and $\bar{z}^n$, where $z = x_1 + ix_2$ and $\bar{z} = x_1 - ix_2$. For $N = 3$ attempt polynomials of the type $z^j P_{n-j}(|z|^2, x_3)$, where P_{n-j} is a polynomial of degree $n - j$ in the variables $|z|^2 = x_1^2 + x_2^2$ and x_3.

1.8. Let $N = 2$, and identify E with a portion of the complex plane $\mathbb{C}$. Then the real and imaginary part of a holomorphic function in E are harmonic in E ([18], pages 124–125).

2c The Green and Stokes Identities

2.1. Prove that if $u \in C^2(E) \cap C^1(\bar{E})$ is harmonic in E, then it is locally analytic in E. It will be a consequence of the estimates in Section 5 that the hypothesis $u \in C^1(\bar{E})$, can be removed.

2.2. It follows from the Stokes identity (2.3)–(2.3) that if $u \in C^2(E) \cap C^1(\bar{E})$ is harmonic in E, it can be represented as the sum of a single-layer, and a double-layer potential.

2.3. Let ω_N denote the surface area of the unit sphere in $\mathbb{R}^N$. Prove that for $N = 2, 3, \ldots,$

$$\omega_{N+1} = 2\omega_N \int_0^{\pi/2} (\sin t)^{N-1} dt.$$

3c Green's Function and the Dirichlet Problem for the Ball

3.1. Prove that the Green's function in (3.3) is non-negative.

3.2. Verify by direct calculation that the kernel in (3.9) is harmonic.
Hint: One needs to prove that for all $y \in \partial B_R$

$$\Delta_x \frac{R^2 - |x|^2}{|x - y|^N} = 0 \text{ in } B_R.$$

From (3.8)

$$\frac{2|x|}{2-N}\frac{\partial}{\partial|x|}|x-y|^{2-N} = \frac{2|x|}{|x-y|^N}(|x|-|y|\cos\theta)$$
$$= \frac{|x|^2-|y|^2+|x-y|^2}{|x-y|^N}$$

and for $y \in \partial B_R$

$$\frac{R^2-|x|^2}{|x-y|^N} = |x-y|^{2-N} + \frac{2|x|}{N-2}\frac{\partial}{\partial|x|}|x-y|^{2-N}.$$

Therefore it suffices to show that the second term on the right-hand side is harmonic.

3.3. Study the Dirichlet problem

$$\Delta u = 0 \ \text{ in } \ B_R, \qquad \text{and } u|_{\partial B_R} = \begin{cases} 0 & \text{if } x_N < 0 \\ 1 & \text{if } x_N \ge 0. \end{cases}$$

Examine the behavior of the solution for $x_N = 0$ near ∂B_R.

3.4. Find Green's function for the half-space $x_N > 0$.
Hint: Set $\bar{x} = (x_1, \ldots, x_{N-1})$ and consider the reflection map analogous to (3.5), i.e., $\xi(x) = (\bar{x}, -x_N)$.

3.5. Using the results of **3.4**, discuss the solvability of the Dirichlet problem

$$\Delta u = 0 \ \text{ in } \ \mathbb{R}^{N-1} \times [x_N > 0]$$
$$u(\bar{x}, 0) = \varphi(\bar{x}) \in C(\mathbb{R}^{N-1}) \cap L^\infty(\mathbb{R}^{N-1}).$$

3.6. Construct Green's function for the quadrant $[x > 0] \cap [y > 0]$ in $\mathbb{R}^2$.

3.7. Find Green's function for the half-ball in $\mathbb{R}^N$.

3.1c Separation of Variables

3.8. Solve the Dirichlet problem for the rectangle $[0 < x < a] \times [0 < y < b]$ in $\mathbb{R}^2$ by looking for "separated" solutions of the form $u(x,y) = X(x)Y(y)$. Enforcing the PDE, derive the ODEs

$$X'' = (\text{const})X, \qquad Y'' = -(\text{const})Y.$$

Superpose the families of solutions X_n, Y_n of these ODEs, by writing

$$u = \sum A_n X_n(x) Y_n(y), \qquad A_n \in \mathbb{R},\ n = 0, 1, 2, \ldots,$$

Finally, determine the coefficients A_n from the prescribed boundary data. In the actual calculations, it is convenient to split the problem into the sum of Dirichlet problems for each of which the data are zero on three sides of the rectangle. For example, the problem

$$\Delta u = 0 \text{ in } R = [0 < x < 1] \times [0 < y < 1]$$
$$u(0, y) = \cos\frac{\pi}{2}y, \quad u(x, 0) = 1 - x, \quad u(1, y) = u(x, 1) = 0$$

can be solved by superposing the solutions of the two problems

$$\Delta u_1 = 0 \text{ in } R, \quad \text{and} \quad u_1(0, y) = \cos\frac{\pi}{2}y$$
$$u_1(x, 0) = u_1(1, y) = u_1(x, 1) = 0,$$

and

$$\Delta u_2 = 0 \text{ in } R, \quad \text{and} \quad u_2(x, 0) = 1 - x$$
$$u_2(1, y) = u_2(x, 1) = u_2(0, y) = 0.$$

Even though the boundary data are not continuous on ∂R, one might solve formally for u_i, $i = 1, 2$ and verify that $u = u_1 + u_2$ is indeed the unique solution of the given problem.

3.9. Solve the Dirichlet problem for the disc $x^2 + y^2 < 1$ by separation of variables, and show that this produces the same solution as that obtained by the Poisson formula (3.9).
Hint: Write Δu in terms of polar coordinates (ρ, θ) to arrive at

$$u_{\rho\rho} + \frac{1}{\rho}u_\rho + \frac{1}{\rho^2}u_{\theta\theta} = 0 \text{ in } [0 < \rho < 1] \times [0 \le \theta < 2\pi].$$

To this equation apply the method of separation of variables.

3.10. Use a modification of this technique to solve the Dirichlet problem for the annulus $r < |x| < R$ in $\mathbb{R}^2$.

3.11. Solve the Dirichlet problem for the rectangle with vertices $A = (1, 0)$, $B = (2, 1)$, $C = (1, 2)$, $D = (0, 1)$.

4c Sub-Harmonic Functions and the Mean Value Property

4.1. Prove that (4.2) implies (4.3). *Hint*: (4.2) implies

$$\omega_N r^{N-1} u(x_o) \le \int_{\partial B_r(x_o)} u(y)\, d\sigma \quad \text{for all } B_r(x_o) \subset E.$$

Integrate both sides in dr for $r \in (0, R)$.

4.2. Prove that (4.4) and (4.5) are equivalent. *Hint*: Write (4.5) in the form

$$\frac{\omega_N}{N} r^N u(x_o) = \int_{B_r(x_o)} u\, dy$$

and take the derivative of both sides with respect to r.

4.3. Let $u \in C^2(E)$ satisfy $\Delta u = u$ in E. Prove that u has neither a positive maximum nor a negative minimum in E.

4.4. Let u be harmonic in E. Prove that $|\nabla u|^2$ is sub-harmonic in E.

4.1c Reflection and Harmonic Extension

4.5. Denote points in $\mathbb{R}^N$ by $x = (\bar{x}, x_N)$ where $\bar{x} = (x_1, \dots, x_{N-1})$, and let u be the unique solution of

$$\Delta u = 0 \text{ in } B_1, \qquad \text{and} \qquad u\big|_{\partial B_1} = \varphi \in C(\partial B_1).$$

Prove that $\varphi(\bar{x}, x_N) = -\varphi(\bar{x}, -x_N)$ implies $u(\bar{x}, x_N) = -u(\bar{x}, -x_N)$.

4.6. Let u be harmonic in $B_1^+ = B_1 \cap [x_N > 0]$, and vanishing for $x_N = 0$. Extend it with a harmonic function in the whole of B_1.

4.7. Solve explicitly the Dirichlet problem

$$\Delta u = 0 \text{ in } B_1^+, \qquad \text{and} \qquad u\big|_{\partial B_1^+} = \begin{cases} x_N^3 & \text{if } x_N > 0 \\ 0 & \text{if } x_N = 0. \end{cases}$$

4.2c The Weak Maximum Principle

Consider the formal differential operator

$$\mathcal{L}_o = a_{ij}(x) \frac{\partial^2}{\partial x_i x_j} + b_i(x) \frac{\partial}{\partial x_i} \tag{4.1c}$$

where $a_{ij}, b_i \in C(\bar{E})$ and the matrix (a_{ij}) is symmetric and positive definite in E.

Theorem 4.1c *Let $u \in C^2(E) \cap C(\bar{E})$ satisfy $\mathcal{L}_o(u) \ge 0$ in E. Then*

$$u(x) \le \max_{\partial E} u \qquad \textit{for all } x \in E.$$

Proof Fix $y \in \mathbb{R}^N - \bar{E}$, let γ be a constant to be chosen later, and consider the function

$$v = u + \varepsilon e^{\gamma |x-y|^2} \qquad \text{for } \varepsilon > 0.$$

It satisfies

$$\begin{aligned} \mathcal{L}_o(v) &\ge 2\gamma\varepsilon \left[a_{ij}\delta_{ij} + 2\gamma a_{ij}(x-y)_i(x-y)_j \right] e^{\gamma|x-y|^2} \\ &\quad + 2\gamma\varepsilon \left[b_i(x-y)_i \right] e^{\gamma|x-y|^2} \\ &\ge 2\gamma\varepsilon \left[N\lambda + 2\gamma\lambda|x-y|^2 - B|x-y| \right] e^{\gamma|x-y|^2} \end{aligned}$$

where $B = \max_{1 \le i \le n} \max_{\bar{E}} |b_i|$. By the Cauchy inequality

$$N\lambda + 2\gamma\lambda|x-y|^2 - B|x-y| \ge N\lambda + \left(2\gamma\lambda - \frac{B^2}{4N\lambda} \right) |x-y|^2 - N\lambda.$$

Therefore γ can be chosen a priori dependent only upon B, N, and λ, such that $\mathcal{L}_o(v) > 0$ in E. If x_o is an interior maximum for v, $b_i(x_o) v_{x_i}(x_o) = 0$ and

$$a_{ij}v_{x_ix_j}\Big|_{x=x_o} > 0 \quad \text{and} \quad v_{x_ix_j}\Big|_{x=x_o} \le 0. \tag{$*$}$$

Next observe that $a_{ij}v_{x_ix_j} = \text{trace}\big((a_{ij})(v_{x_ix_j})\big)$. Using that (a_{ij}) and $(v_{x_ix_j})$ are symmetric, and that the trace is invariant under orthogonal linear transformations, prove that

$$a_{ij}(x_o)v_{x_ix_j}(x_o) \le 0.$$

This and $(*)$ give a contradiction. Therefore

$$\max_{\bar{E}} v \le \max_{\partial E} v$$

and the theorem follows on letting $\varepsilon \to 0$. ∎

4.8. Let $u \in C^2(E) \cap C(\bar{E})$ satisfy

$$\mathcal{L}(u) \equiv \mathcal{L}_o(u) + c(x)u \ge 0 \quad \text{in } E.$$

Prove that if $c \le 0$, then u cannot have a non-negative maximum in the interior of E. Give a counterexample to show that the assumption $c \le 0$ cannot be removed.

4.3c Sub-Harmonic Functions

4.9. Prove that $v \in \sigma(E)$ if and only if for every open set $E' \subset E$ and every harmonic function u such that $u\big|_{\partial E'} = v$, then $v \le u$ in E'.

4.10. Prove that $x \to \ln|x|$ is sub-harmonic in $\mathbb{R}^N - \{0\}$.

4.11. Give examples of non-differentiable sub-harmonic functions.

4.12. Let $u \in C^2(E) \cap C(\bar{E})$ be a solution of

$$\Delta u = -1 \text{ in } E, \qquad \text{and} \qquad u\big|_{\partial E} = 0.$$

Prove that for all $x_o \in E$

$$u(x_o) \ge \frac{1}{2N} \inf_{x\in\partial E} |x - x_o|^2.$$

4.3.1c A More General Notion of Sub-Harmonic Functions

Certain questions in potential theory require a notion of sub-harmonic functions that does not assume continuity. First, by a real valued function in E is meant $u : E \to [-\infty, +\infty)$. Thus u is defined everywhere in E and is permitted to take the "value" $-\infty$. A real-valued function $u : E \to [-\infty, +\infty)$ is *upper semi-continuous* if $[u < s]$ is open for all $s \in \mathbb{R}$.

Definition of F. Riesz ([126], see also [122]) Let E be a connected, open subset of $\mathbb{R}^N$. A real valued function $u : E \to [-\infty, +\infty)$, is sub-harmonic in E, if it is upper semi-continuous, and if for every compact subset $K \subset E$ and for every function $H \in C(K)$ and harmonic in the interior of K

$$u\,\big|_{\partial K} \le H\,\big|_{\partial K} \quad \Longrightarrow \quad u \le H \text{ in } K.$$

4.13. Prove that this notion of upper semi-continuity is equivalent to

$$\limsup_{x \to x_o} u(x) \le u(x_o) \qquad \text{for all } x_o \in E.$$

4.14. Prove that the function

$$u(x) = \begin{cases} \ln|x| & \text{if } x \neq 0 \\ -\infty & \text{if } x = 0 \end{cases}$$

is upper semi-continuous. Such a function is sub-harmonic in the sense of F. Riesz, and it is not sub-harmonic in the sense of Section 4.

4.15. Prove that apart from the continuity requirement, Riesz notion of a sub-harmonic function is equivalent to the notion of Section 4.

4.16. Let $\{u_n\}$ be a decreasing sequence of sub-harmonic functions, in the sense of F. Riesz and let

$$u(x) = \lim_{n \to \infty} u_n(x) \qquad \forall x \in E.$$

Prove that either $u \equiv -\infty$ in E, or u is sub-harmonic in E, in the sense of F. Riesz ([75], page 16).

5c Estimating Harmonic Functions

5.1. Let u be harmonic in E. Estimate the radius of convergence of its Taylor's series about $x_o \in E$.

5.2. (Unique Continuation) Let u be harmonic in E and vanishing in an open subset of E. Prove that if E is connected, u vanishes identically in E. As a consequence, if u and v are harmonic in a connected domain E and coincide in a open subset of E, then $u \equiv v$ in E.

5.3. Find two harmonic functions in the unit ball that coincide on the set $[|x| < \frac{1}{2}] \cap [x_N = 0]$.

5.4. (Phragmen–Lindelöf-Type Theorems) Let $\mathbb{R}^N_+ = \mathbb{R}^N \cap [x_N > 0]$, and let u be a non-negative harmonic function in $\mathbb{R}^N_+$. Prove that if u is bounded and vanishes on the hyperplane $x_N = 0$, then it is identically zero.

Remark 5.1c The function $u = x_N$ shows that the assumption of u being bounded cannot be removed. However, this is in some sense the only counterexample as shown by the following theorem of Serrin ([133]).

Theorem 5.1c *Let u be a non-negative harmonic function in $\mathbb{R}^N_+$ vanishing for $x_N = 0$. There exists a constant C depending only on N, such that*

$$\limsup_{|x| \to \infty} \frac{u(x)}{|x|} \le C.$$

5.1c Harnack-Type Estimates

5.5. Let u be harmonic in $\mathbb{R}^N$, and let $\mathcal{G}$ be its graph. If $P \in \mathcal{G}$, denote by π_P be the tangent plane to $\mathcal{G}$ at P. Prove that $[\pi_P \cap \mathcal{G}] - P \neq \emptyset$.

5.6. Prove that a non-negative harmonic function in a connected open set E is either identically zero or strictly positive in E.

5.7. Let Q be the rectangle with vertices $(0,0)$, $(nr,0)$, $(nr,2r)$, $(0,2r)$, for some $r > 0$ and $n \in \mathbb{N}$. Let $P_o = (r,r)$ and $P_* = \big((n-1)r, r\big)$. Prove that a non-negative harmonic function u in Q satisfies

$$2^{-2n} u(P_o) \le u(P_*) \le 2^{2n} u(P_o).$$

5.8. Assume that the mixed boundary problem

$$\begin{gathered} u \in C^2(B_1) \cap C^1(\bar{B}_1), \qquad \Delta u = -1 \text{ in } B_1 \\ u = 0 \text{ in } \partial B_1 \cap [x_N > 0], \quad \nabla u \cdot \mathbf{n} = -u \text{ on } \partial B_1 \cap [x_N < 0] \end{gathered}$$

has a unique solution. Prove that $u \ge 0$ in $\bar{B}_1$, and that $u \,\big|_{\partial B_1 \cap [x_N > 0]} > 0$.

5.2c Ill-Posed Problems: An Example of Hadamard

The following problem is in general ill-posed.

$$\begin{gathered} \Delta u = 0 \text{ in } E = [0 < x < 1] \times [0 < y < 1] \\ u(\cdot, 0) = \varphi, \quad u_y(\cdot, 0) = \psi \end{gathered} \tag{$*$}$$

Proposition 5.1c *A solution of $(*)$ exists if and only if the function*

$$(0,1) \ni x \to \varphi(x) - \frac{1}{\pi} \int_0^1 \psi(s) \ln |xs| ds$$

is analytic.

Proof If u solves $(*)$, write $u = v + w$, where

$$v(x,y) = \frac{1}{2\pi} \int_0^1 \psi(s) \ln[(x-s)^2 + y^2] ds.$$

Then $\Delta w = 0$ in E and $w_y(\cdot, 0) = 0$. Therefore, by the reflection principle $(x,y) \to w(x,|y|)$ is harmonic in $[0 < x < 1] \times [-1 < y < 1]$, and $x \to w(x,0)$ is analytic. ∎

5.9. Prove that the following problem is ill-posed.

$$\begin{aligned} &\Delta u = 0 \text{ in } [|x| < 1] \times [0 < y < 1] \\ &u(-1, \cdot) = u(\cdot, 1) = u(1, \cdot) = 0 \\ &u_y(\cdot, 0) = 0, \quad u(\cdot, 0) = 1 - |x|. \end{aligned}$$

5.3c Removable Singularities

Let $x_o \in E$ and let u be harmonic in $E - \{x_o\}$. The function u is analytic in $E - \{x_o\}$, and it might be singular at x_o. An example is the fundamental solution $F(\cdot; x_o)$ of the Laplace equation with pole at x_o. A point $x_o \in E$ is a *removable singularity* for u if u can be extended continuously in x_o so that the resulting function is harmonic in the whole of E.

The pole x_o is not a removable singularity for $F(\cdot; x_o)$. This suggests that for a singularity at x_o to be removable, the behavior of u near x_o should be better than that of $F(\cdot; x_o)$.

Theorem 5.2c *Assume that*

$$\lim_{x\to x_o} \frac{u(x)}{F(x; x_o)} = 0. \tag{$**$}$$

Then x_o is a removable singularity.

Proof Let v be the harmonic extension in the ball $B_\rho(x_o)$ of $u \big|_{\partial B_\rho(x_o)}$. Such an extension can be constructed by the Poisson formula (3.9) and Theorem 3.1. The proof consists in showing that $u = v$ in $B_\rho(x_o)$. Assume $N \geq 3$, the proof for $N = 2$ being similar. Consider the ball $B_\varepsilon(x_o) \subset B_\rho(x_o)$ and set

$$M_\varepsilon = \|u - v\|_{\infty, \partial B_\varepsilon(x_o)}.$$

By $(**)$, for every fixed $\eta > 0$, there exists $\varepsilon_o \in (0, \rho)$ such that $M_\varepsilon \leq \varepsilon^{2-N}\eta$ for all $\varepsilon \leq \varepsilon_o$. The two functions

$$w^\pm = M_\varepsilon \left(\frac{\varepsilon}{|x - x_o|}\right)^{N-2} \pm (u - v)$$

are harmonic in the annulus $\varepsilon < |x - x_o| < \rho$ and non-negative for $|x - x_o| = \rho$. Moreover, on the sphere $|x - x_o| = \varepsilon$

$$w^\pm \Big|_{|x-x_o|=\varepsilon} = M_\varepsilon \pm (u - v) \Big|_{|x-x_o|=\varepsilon} \geq 0.$$

Therefore, by the maximum principle, for all $\varepsilon < |x - x_o| < \rho$

$$|u - v|(x) \leq \frac{M_\varepsilon \varepsilon^{N-2}}{|x - x_o|^{N-2}} \leq \frac{\eta}{|x - x_o|^{N-2}}.$$ ■

Theorem 5.3c *Assume that*

$$\lim_{x\to x_o} |x - x_o|^{N-1} \nabla u \cdot \frac{x - x_o}{|x - x_o|} = 0.$$

Then x_o is a removable singularity.

Proof (Hint) Let v be the harmonic extension of $u|_{\partial B_\rho(x_o)}$ into $B_\rho(x_o)$, and for $\varepsilon \in (0, \rho)$ set

$$D_\varepsilon = \sup_{\partial B_\rho(x_o)} \left| \nabla(u - v) \cdot \frac{x - x_o}{|x - x_o|} \right|.$$

Introduce the two functions

$$w^\pm = \frac{\varepsilon^{N-1} D_\varepsilon}{(N-2)|x - x_o|^{N-2}} \pm (u - v)$$

and prove that a minimum for $w^\pm$ cannot occur on $\partial B_\varepsilon(x_o)$. ∎

5.10. Prove that if

$$\lim_{x \to x_o} \frac{u(x)}{F(x; x_o)} = c \qquad \text{for some } c \in \mathbb{R}$$

then $u = cF(\cdot; x_o) + v$, where v is harmonic in E.

5.11. The previous statements assume that u has a limit as $x \to x_o$. Prove that if u is a non-negative harmonic function in the punctured ball $B_1 - \{0\}$, then the limit of $u(x)$ as $|x| \to 0$ exists, finite or infinite ([55]).

7c About the Exterior Sphere Condition

The counterexample of Lebesgue leads to a question that can be roughly formulated as follows. How wide should the cusp in Figure 7.1 be, to ensure the existence of solutions? The correct way of measuring "how wide" a cusp should be is by means of the concept of *capacity* introduced by Wiener ([161]). The capacity of a compact set $K \subset \mathbb{R}^N$ is defined by

$$\text{cap}(K) = \inf_{\substack{v \in C_o^\infty(\mathbb{R}^N) \\ v \geq 1 \text{ on } K}} \int_{\mathbb{R}^N} |\nabla v|^2 dx.$$

Such a definition can be extended to Borel sets. Now consider a domain E whose boundary has a cusp pointing inside E as in Figure 7.1. Let x_* be the "vertex" of the cusp, and consider the compact sets

$$K_n = (\mathbb{R}^N - E) \cap \bar{B}_{2^{-n}}(x_*) \qquad n \in \mathbb{N}$$

obtained by intersecting the region enclosed by the cusp, outside E, with balls centered at x_* and radius 2^{-n}.

Theorem 7.1c (Wiener [162]) *The following are equivalent:*

(i). There exists a barrier $H(x_*; \cdot)$ for the Dirichlet problem (6.2) at x_*
(ii). The series $\sum$ cap(K_n) is divergent.

For a theory of capacity and *capacitable* sets, see [80, 104, 94].

8c Problems in Unbounded Domains

8.1. Compute (8.5) for $N = 2, 3$ first. Then proceed by induction for all N. Use **2.3** and

$$\int_0^\infty \frac{\rho^{N-2}}{(1+\rho^2)^{N/2}} d\rho = \int_0^{\pi/2} (\sin t)^{N-2} dt.$$

8.2. Having in mind **3.7** and the representation formula (3.2), justify the definition (8.2) of the Poisson kernel for the half-space.

8.3. Give a solution formula for the Neumann problem in the half-space. Discuss uniqueness.

8.4. Let $E \subset \mathbb{R}^N$ be bounded, connected, and with boundary ∂E of class C^1. Prove that there exists at most one solution to the boundary value problem in the exterior of E

$$u \in C^2(\mathbb{R}^N - E) \cap C(\overline{\mathbb{R}^N - E}), \quad \Delta u = 0 \text{ in } \mathbb{R}^N - E$$
$$u\big|_{\partial E} = \varphi \in C(\partial E), \quad \lim_{|x|\to\infty} u(x) = \gamma$$

for a given constant γ.

8.1c The Dirichlet Problem Exterior to a Ball

Let $N \geq 3$, set $E = |x| > R$ for some $R > 0$, and consider the exterior problem

$$u \in C^2(E) \cap C(\bar{E}), \quad \Delta u = 0 \text{ in } E \tag{8.1c}$$
$$u\big|_{|x|=R} = \varphi \in C(\partial E), \quad \lim_{|x|\to\infty} u(x) = \gamma \tag{8.2c}$$

for a given constant γ.

Step 1. First apply the Kelvin transform to map E into $B_R - \{0\}$. Then introduce the new unknown function

$$v(y) = |y|^{2-N} u\left(\frac{R^2}{|y|^2} y\right) \qquad y \neq 0.$$

With the aid of Theorem 5.2c, verify that the singularity $y = 0$ is removable. Then, in terms of the new coordinates, (8.1c) becomes

$$v \in C^2(B_R) \cap C(\bar{B}_R), \quad \Delta v = 0 \text{ in } B_R \tag{8.3c}$$
$$v\big|_{\partial B_R} = R^{2-N} \varphi \in C(\partial B_R). \tag{8.4c}$$

Step 2. Solve (8.3c) by means of Poisson formula (3.9). Return to the original coordinates x by inverting the Kelvin transform. In this process use formula (3.6). To the function so obtained add a radial harmonic function vanishing for $|x| = R$ and satisfying the last of (8.1c). The solution is

$$u(x) = \gamma\left[1 - \left(\frac{R}{|x|}\right)^{N-2}\right] - \frac{1}{R\omega_N} \int_{\partial B_R} \varphi(y) \frac{R^2 - |x|^2}{|x-y|^N} d\sigma.$$

9c Schauder Estimates up to the Boundary ([135, 136])

Let $u \in C^2(E) \cap C(\bar{E})$ be the unique solution of the Dirichlet problem (1.2). If ∂E and the boundary datum φ are of class $C^{2,\eta}$, it is natural to expect that $u \in C^{2,\eta}(\bar{E})$. Prove the following[3].

Proposition 9.1c *Let $u \in C^2(\mathbb{R}^N \times \mathbb{R}^+) \cap C(\overline{\mathbb{R}^N \times \mathbb{R}^+})$ be the unique bounded solution of the Dirichlet problem (8.1). If $\varphi(x) \in C_o^{2,\eta}(\mathbb{R}^N)$ then $u \in C^{2,\eta}(\overline{\mathbb{R}^N \times \mathbb{R}^+})$ and there exists a constant γ depending only upon N, η and the diameter of the support of φ such that*

$$\|u\|_{2,\eta;\overline{\mathbb{R}^N \times \mathbb{R}^+}} \le \gamma \|\varphi\|_{2,\eta;\mathbb{R}^N}.$$

Moreover there exist a constant γ depending only upon N and η and independent of the support of φ such that

$$\sup_{\substack{x,y\in\mathbb{R}^N \\ x_{N+1},y_{N+1}\in\mathbb{R}^+}} \frac{|u_{x_ix_j}(x,x_{N+1}) - u_{x_ix_j}(y,y_{N+1})|}{[|x-y|^2 + (x_{N+1}-y_{N+1})^2]^{\eta/2}} \le \gamma \sup_{x,y\in\mathbb{R}^N} \frac{|\varphi(x)-\varphi(y)|}{|x-y|^\eta}.$$

Proof (Hint:) Apply the same technique of proof of Theorem 9.1 to the Poisson integral (8.1). ∎

The result of Theorem 9.1c is the key step in deriving $C^{2,\eta}$ estimates up to the boundary for solutions of the Dirichlet problem (1.2). The technique consists of performing a local flattening of ∂E.

10c Potential Estimates in $L^p(E)$

10.1. Compute the last integral in the proof of Proposition 10.1 by introducing polar coordinates. Prove that for $\alpha = 1$, the constant γ in (10.3) is

$$\gamma = \frac{1}{\omega_N^{1/p}(N-2)} \left(\frac{p-1}{2p-N}\right)^{(p-1)/p} \operatorname{diam}(E)^{2-N/p}.$$

10.2. Verify that for $N = 2$ and $\alpha = 1$, the constant γ in (10.3) is

$$\gamma = \frac{1}{(2\pi)^{1/p}} \begin{cases} \dfrac{1}{e}\dfrac{p}{p-1} & \text{if } \operatorname{diam}(E) \le 1 \\ \operatorname{diam}(E)^{2\frac{p-1}{p}} \ln \operatorname{diam}(E) & \text{if } \operatorname{diam}(E) > 1. \end{cases}$$

10.3. Prove the following

[3] A version of these estimates is in [54, 91]. Their parabolic counterparts are in [45, 92].

Corollary 10.1c *Let $E = B_R$ and $N \ge 3$. Then*

$$\|A_F f\|_{\infty,E} \le \gamma(N,p,R)\left(\fint_{B_R} |f(y)|^p dy\right)^{1/p}$$

where

$$\gamma(N,p,R) = 2R^2 \frac{N^{1/p}2^{2-N/p}}{N-2}\left(\frac{p-1}{2p-N}\right)^{\frac{p-1}{p}}.$$

State and prove a similar corollary for the case $N = 2$.

10.4. In Proposition 10.1 the boundedness of E is essential. For $\alpha = 2$, give an example of $f \in L^p(E)$ for some $p > N/2$ and E unbounded for which $A_F f$ is unbounded.

10.5. If E is unbounded, the boundedness of $A_F f$ can be recovered by imposing on f a fast decay as $|x| \to \infty$. Prove the following

Lemma 10.1c *Assume $f \in L^p(E)$ for some $p > N/2$, and*

$$|f(x)| \le C|x|^{-(2+\varepsilon)} \qquad \textit{for } |x| > R_o$$

for given positive constants C, R_o, ε. Then $A_F f \in L^\infty(E)$, and there exists a constant γ depending only on N, C, R_o, ε such that

$$\|A_F f\|_{\infty,E} \le \gamma(1 + \|f\|_{p,E}).$$

10.6. Give an example of $A_F f$ such that $|\nabla A_F f| \in L^2(E)$ and $w_2 \notin L^\infty(E)$.

10.1c Integrability of Riesz Potentials

The proof of Proposition 10.1 shows that the constant γ in (10.3) deteriorates as $q \to Np/(N-\alpha p)_+$. However, (10.3) continues to hold also for the limiting case

$$q = \frac{Np}{(N-\alpha p)_+}, \qquad \text{provided} \quad \alpha p < N.$$

The proof for such a limiting case, however, is rather delicate and is based on Hardy's inequality ([65], also in [31], Chapter VIII, §18).

10.2c Second Derivatives of Potentials

If $f \in C_o^\eta(E)$, then $A_F f$ is twice continuously differentiable, and formally

$$(A_F f)_{x_i x_j}(x) = \int_E F_{x_i x_j}(x;y) f(y) dy. \tag{10.1c}$$

The integral is meant in the sense of the improper integral

$$\int_E F_{x_i x_j}(x;y) f(y) dy = \lim_{\varepsilon \to} \int_{E \cap [|x-y| > \varepsilon]} F_{x_i x_j}(x;y) f(y) dy.$$

The limit exists, since $f \in C_o^\eta(E)$. However, if $f \in L^p(E)$ for some $p > 1$, such a representation loses its classical meaning. It is natural to ask, in analogy with Proposition 10.1, whether one may use (10.1c) to define $(A_F f)_{x_i x_j}$ in a weak sense and whether such weak second derivatives are in $L^q(E)$ for some $q \ge 1$. It turns out that

$$f \in L^p(E) \Longrightarrow (A_F f)_{x_i x_j} \in L^p(E) \quad \text{for } 1 < p < \infty.$$

The proof of this fact cannot be constructed from (10.1) for $\alpha = 0$, since the latter would be a divergent integral even if $f \in C_o^\infty(E)$. One has to rely instead on cancellation properties of the kernel in (10.1c). These estimates are due to Calderón and Zygmund ([16], see also [144]).

3

Boundary Value Problems by Double-Layer Potentials

1 The Double-Layer Potential

Let Σ be an $(N-1)$-dimensional bounded surface of class C^1 in $\mathbb{R}^N$ for $N \ge 2$, whose boundary $\partial\Sigma$ is an $(N-2)$-dimensional surface of class C^1. Fix $x_o \in \mathbb{R}^N - \bar{\Sigma}$ and consider the cone $C(\Sigma, x_o)$ generated by half-lines originating at x_o and passing through points of $\partial\Sigma$. Let $\alpha(x_o)$ denote the solid angle spanned by $C(\Sigma, x_o)$, that is, the area of the portion of the unit sphere centered at x_o, cut by the cone. The double-layer potential generated in x_o by a distribution of dipoles identically equal to 1 on Σ is defined by

$$W(\Sigma, x_o) = -\int_\Sigma \frac{\partial F(x_o; y)}{\partial \mathbf{n}(y)} d\sigma = \frac{-1}{\omega_N} \int_\Sigma \frac{(x_o - y) \cdot \mathbf{n}(y)}{|x_o - y|^N} d\sigma. \tag{1.1}$$

Here $\mathbf{n}(\cdot)$ is the unit normal to Σ exterior to the cone $C(\Sigma, x_o)$, and $F(\cdot;\cdot)$ is the fundamental solution of the Laplacian, introduced in (2.6) of Chapter 2.

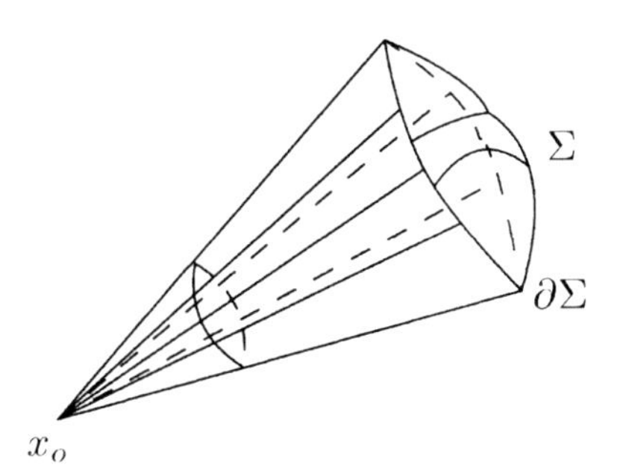

Fig. 1.1.

The same cone is generated by infinitely many surfaces; however, the double-layer potential depends only on x_o and the solid angle $\alpha(x_o)$. This is the content of the next proposition.

E. DiBenedetto, *Partial Differential Equations: Second Edition*,
Cornerstones, DOI 10.1007/978-0-8176-4552-6_4,

Proposition 1.1 *Let Σ_1 and Σ_2 be any two surfaces generating the same cone $C(\Sigma, x_o)$. Then*

$$W(\Sigma_1, x_o) = W(\Sigma_2, x_o) = W(\Sigma, x_o) = \frac{\alpha(x_o)}{\omega_N}.$$

Proof Let E be the portion of the cone $C(\Sigma_1, x_o) = C(\Sigma_2, x_o)$ included by the surfaces Σ_1 and Σ_2. Since x_o is outside $\bar{E}$

$$\begin{aligned} 0 = \int_E \Delta_y F(x_o; y) dy &= \frac{1}{\omega_N} \int_{\partial E - (\Sigma_1 \cup \Sigma_2)} \frac{(x_o - y) \cdot \mathbf{n}(y)}{|x_o - y|^N} d\sigma \\ &+ \int_{\Sigma_1} \frac{\partial F(x_o; y)}{\partial \mathbf{n}(y)} d\sigma - \int_{\Sigma_2} \frac{\partial F(x_o; y)}{\partial \mathbf{n}(y)} d\sigma. \end{aligned}$$

The first integral on the right-hand side vanishes since $(x_o - y)$ is tangent to the cone and thus normal $\mathbf{n}(y)$. Therefore $W(\Sigma_1, x_o) = W(\Sigma_2, x_o)$.

Next, since $W(\Sigma, x_o)$ is independent of Σ, we replace Σ with the portion of the sphere $\partial B_R(x_o)$ cut by the cone, that is

$$\Sigma_o = \partial B_R(x_o) \cap C(\Sigma, x_o) \qquad \text{for some } R > 0.$$

The normal to Σ_o exterior to the cone is

$$\mathbf{n}(y) = \frac{y - x_o}{|y - x_o|}.$$

This in (1.1) gives

$$W(\Sigma, x_o) = \frac{1}{\omega_N R^{N-1}} \int_{\Sigma_o} d\sigma = \frac{\alpha(x_o)}{\omega_N}. \qquad \blacksquare$$

In what follows E, is a bounded open set in $\mathbb{R}^N$ with boundary ∂E of class $C^{1,\alpha}$ for some $\alpha \in (0, 1]$. The double-layer potential generated at a point $x_o \in \mathbb{R}^N - \partial E$ by a continuous distribution of dipoles $y \to v(y)$ on ∂E is

$$\begin{aligned} W(\partial E, x_o; v) &= -\int_{\partial E} v(y) \frac{\partial F(x_o; y)}{\partial \mathbf{n}(y)} d\sigma \\ &= \frac{-1}{\omega_N} \int_{\partial E} v(y) \frac{(x_o - y) \cdot \mathbf{n}(y)}{|x_o - y|^N} d\sigma. \end{aligned} \tag{1.2}$$

Proposition 1.2 *In (1.2) let $v = 1$. Then*

$$W(\partial E, x_o; 1) = \begin{cases} 1 & \text{for } x_o \in E \\ 0 & \text{for } x_o \in \mathbb{R}^N - \bar{E}. \end{cases} \tag{1.3}$$

Proof The first follows from the Stokes identity (2.3)–(2.4) of Chapter 2, written for $u = 1$. If x_o is outside $\bar{E}$, the function $y \to F(x_o; y)$ is harmonic in E. Therefore

$$\int_E \Delta_y F(x_o; y) dy = \int_{\partial E} \frac{\partial F(x_o; y)}{\partial \mathbf{n}(y)} d\sigma = 0. \qquad \blacksquare$$

2 On the Integral Defining the Double-Layer Potential

As x_o tends to a point $x \in \partial E$ the integrand in (1.2), becomes singular. Such a singularity however is integrable. This is the content of the next lemma.

Lemma 2.1 *There exists a constant C depending only on N, α, and the structure of ∂E such that*

$$\frac{|(x-y)\cdot \mathbf{n}(x)|}{|x-y|^N} \le C\frac{1}{|x-y|^{N-1-\alpha}} \qquad \textit{for all } x,y \in \partial E.$$

Proof It will suffice to prove the lemma for all $|x-y| < \eta$ for some $\eta > 0$. Fix $x \in \partial E$ and assume, after a translation, that it coincides with the origin. Since ∂E is of class $C^{1,\alpha}$, there exists $\eta > 0$ such that the portion of ∂E within the ball B_η, centered at the origin, can be represented, in a local system of coordinates, as the graph of a function φ satisfying

$$\begin{gathered} \xi_N = \varphi(\xi), \quad \xi = (\xi_1, \dots, \xi_{N-1}), \quad |\xi| < \eta \\ \varphi \in C^{1,\alpha}(|\xi| < \eta), \quad \varphi(0) = 0, \quad |\nabla\varphi(0)| = 0 \\ |\nabla\varphi(\xi)| \le C_\varphi |\xi|^\alpha \text{ for a given positive constant } C_\varphi. \end{gathered} \tag{2.1}$$

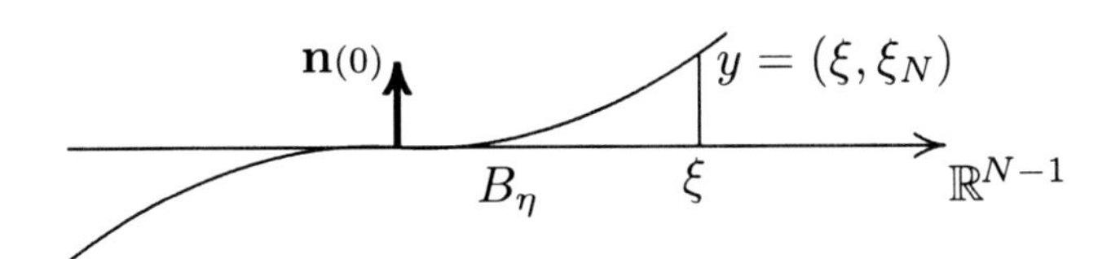

Fig. 2.2.

Then $\mathbf{n}(0)$ is the unit vector of the ξ_N-axis, exterior to E, and in the new coordinates

$$|\xi| \le |y| \le (1 + \|\nabla\varphi\|_{\infty,B_\eta})|\xi|$$

and

$$|-y\cdot\mathbf{n}(0)| = |\varphi(\xi)| \le C_\varphi|\xi|^{1+\alpha}. \qquad \blacksquare$$

Let $x \in \partial E$ and for $\varepsilon > 0$ let $S_\varepsilon(x) = \partial E \cap B_\varepsilon(x)$ denote the portion of ∂E within the ball $B_\varepsilon(x)$ centered at x and radius ε.

Lemma 2.2 *There exist constants C and ε_o, depending only on N, α, and the structure of ∂E, such that for every $\varepsilon \le \varepsilon_o$*

$$\int_{S_\varepsilon(x)} \frac{|(z-y)\cdot\mathbf{n}(y)|}{|z-y|^N} d\sigma \le C$$

uniformly for all $z \in B_\varepsilon(x)$.

Proof Fix such a $z \in B_\varepsilon(x)$. Since ∂E is of class $C^{1+\alpha}$, we may choose ε_o so small that z has a unique projection, say z_p, on ∂E. Set $\delta = |z - z_p|$ and compute

$$|z - y|^2 = |z_p - y|^2 + \delta^2 - 2(z_p - y) \cdot (z_p - z).$$

Since $z_p \in \partial E$, by Lemma 2.1

$$\begin{aligned} |(z_p - y) \cdot (z_p - z)| &= \left|(z_p - y) \cdot \frac{z_p - z}{|z_p - z|}\right| \\ &= |(z_p - y) \cdot \mathbf{n}(z_p)|\delta \le C|z_p - y|^{1+\alpha}\delta. \end{aligned}$$

Therefore

$$|z - y|^2 \ge |z_p - y|^2 + \delta^2 - 2C|z_p - y|^{1+\alpha}\delta \ge \frac{1}{4}(|z_p - y| + \delta)^2$$

provided ε_o is chosen sufficiently small. Also

$$|(z - y) \cdot \mathbf{n}(y)| \le |(z_p - y) \cdot \mathbf{n}(y)| + \delta \le C|z_p - y|^{1+\alpha} + \delta.$$

Therefore

$$\frac{|(z - y) \cdot \mathbf{n}(y)|}{|z - y|^N} \le C\frac{|z_p - y|^{1+\alpha} + \delta}{(|z_p - y| + \delta)^N}.$$

From this

$$\begin{aligned} \int_{S_\varepsilon(x)} \frac{|(z - y) \cdot \mathbf{n}(y)|}{|z - y|^N} d\sigma &\le \int_{S_\varepsilon(x)} \frac{1}{|z_p - y|^{N-1-\alpha}} d\sigma \\ &\quad + C\delta \int_{S_\varepsilon(x)} \frac{d\sigma}{(|z_p - y| + \delta)^N}. \end{aligned}$$

The first integral is convergent, since $x \in \partial E$, and it is bounded above by

$$\sup_{x \in \partial E} \int_{\partial E} \frac{1}{|x - y|^{N-1-\alpha}} d\sigma.$$

To estimate the second integral, first extend the integration to the larger set $S_{2\varepsilon}(z_p)$. Then introduce a local system of coordinates with the origin at z_p, so that $S_{2\varepsilon}(z_p)$ is represented as in (2.1). This gives

$$\begin{aligned} \delta \int_{S_\varepsilon(x)} \frac{d\sigma}{(|x - y| + \delta)^N} &\le C_\varphi \delta \int_{|\xi| < 2\varepsilon} \frac{d\xi}{(|\xi| + \delta)^N} \\ &\le C'_\varphi \delta \int_0^{2\varepsilon} \frac{dr}{(r + \delta)^2} \le C''_\varphi. \end{aligned}$$

■

3 The Jump Condition of $W(\partial E, x_o; v)$ Across ∂E

We will compute the limit of $W(\partial E, x_o; v)$ as $x_o \to \partial E$ either from within E or from outside $\bar{E}$. Having fixed $x \in \partial E$, denote by $\{x^i\} \subset E$ a sequence of points approaching x from the interior of E. Likewise, denote by $\{x^e\} \subset \mathbb{R}^N - \bar{E}$ a sequence of points approaching x from the exterior of $\bar{E}$.

Proposition 3.1 *Let $v \in C(\partial E)$. Then for all $x \in \partial E$*

$$\lim_{x^i \to x} W(\partial E, x^i; v) = \frac{1}{2} v(x) - \frac{1}{\omega_N} \int_{\partial E} v(y) \frac{(x-y)\cdot \mathbf{n}(y)}{|x-y|^N} d\sigma$$
$$\lim_{x^e \to x} W(\partial E, x^e; v) = -\frac{1}{2} v(x) - \frac{1}{\omega_N} \int_{\partial E} v(y) \frac{(x-y)\cdot \mathbf{n}(y)}{|x-y|^N} d\sigma.$$

Combining these limits gives the jump condition of the potential $W(\partial E, x_o; v)$ across ∂E.

Corollary 3.1 *Let $v \in C(\partial E)$. Then for all $x \in \partial E$*

$$\lim_{x^i \to x} W(\partial E, x^i; v) - \lim_{x^e \to x} W(\partial E, x^e; v) = v(x).$$

Proof (Proposition 3.1) For $\varepsilon > 0$, let $S_\varepsilon(x) = \partial E \cap B_\varepsilon(x)$ and write

$$\begin{aligned}\int_{\partial E} v(y) \frac{(x^i-y)\cdot \mathbf{n}(y)}{|x^i-y|^N} d\sigma &= v(x) \int_{S_\varepsilon(x)} \frac{(x^i-y)\cdot \mathbf{n}(y)}{|x^i-y|^N} d\sigma \\ &+ \int_{S_\varepsilon(x)} [v(y)-v(x)] \frac{(x^i-y)\cdot \mathbf{n}(y)}{|x^i-y|^N} d\sigma \\ &+ \int_{\partial E - S_\varepsilon(x)} v(y) \frac{(x^i-y)\cdot \mathbf{n}(y)}{|x^i-y|^N} d\sigma.\end{aligned}$$

Choose $\varepsilon \le \varepsilon_o$, where ε_o is the number claimed by Lemma 2.2, and set

$$-\int_{S_\varepsilon(x)} \frac{(x^i-y)\cdot \mathbf{n}(y)}{|x^i-y|^N} d\sigma = \alpha(\varepsilon, x^i)$$

where by Proposition 1.1, $\alpha(\varepsilon, x^i)$ is the solid angle of the cone generated by the lines through x^i and the points of $\partial S_\varepsilon(x)$. By Lemma 2.2

$$\begin{aligned}&\left| \int_{S_\varepsilon(x)} [v(y)-v(x)] \frac{(x^i-y)\cdot \mathbf{n}(y)}{|x^i-y|^N} d\sigma \right| \\ &\qquad \le \sup_{y \in S_\varepsilon(x)} |v(y)-v(x)| \left| \int_{S_\varepsilon(x)} \frac{(x^i-y)\cdot \mathbf{n}(y)}{|x^i-y|^N} d\sigma \right| \\ &\qquad \le C \sup_{y \in S_\varepsilon(x)} |v(y)-v(x)|.\end{aligned}$$

Therefore

$$\int_{\partial E} v(y)\frac{(x^i - y)\cdot \mathbf{n}(y)}{|x^i - y|^N}d\sigma = - v(x)\alpha(\varepsilon, x^i) + O(\varepsilon) + \int_{\partial E - S_\varepsilon(x)} v(y)\frac{(x^i - y)\cdot \mathbf{n}(y)}{|x^i - y|^N}d\sigma.$$

Now let $x^i \to x$ to obtain

$$\lim_{x^i \to x} W(\partial E, x^i; v) = v(x)\frac{\alpha(\varepsilon; x)}{\omega_N} - O(\varepsilon) - \int_{\partial E - S_\varepsilon(x)} v(y)\frac{(x - y)\cdot \mathbf{n}(y)}{|x - y|^N}d\sigma \tag{3.1}$$

where $\alpha(\varepsilon; x)$ is the solid angle of the cone generated by lines through x and points of $\partial S_\varepsilon(x)$. As $\varepsilon \to 0$, $\alpha(\varepsilon; x) \to \frac{1}{2}\omega_N$. To prove the proposition, we let $\varepsilon \to 0$ in (3.1) with the aid of Lemma 2.1. ■

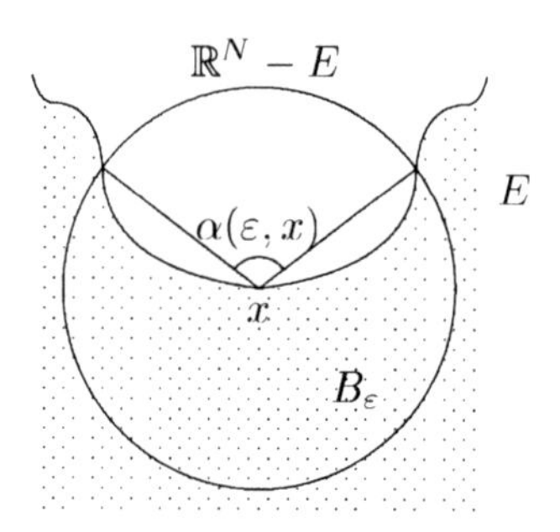

Fig. 3.3.

Corollary 3.2 *For all* $x \in \partial E$

$$-\frac{1}{\omega_N}\int_{\partial E}\frac{(x - y)\cdot \mathbf{n}(y)}{|x - y|^N}d\sigma = \frac{1}{2}.$$

Proof Apply (3.1) with $v = 1$ and use (1.3). ■

Remark 3.1 Combining this with Proposition 1.2, we conclude that the double-layer potential $x \to W(\partial E; x)$ generated by a constant distribution of dipoles on ∂E, at points $x \in \mathbb{R}^N$, is a function that is discontinuous across $\partial\Omega$, and its values are, up to a multiplicative constant

$$W(\partial E; x) = \begin{cases} 1 & \text{for } x \in E \\ \frac{1}{2} & \text{for } x \in \partial E \\ 0 & \text{for } x \in \mathbb{R}^N - \bar{E}. \end{cases} \tag{3.2}$$

4 More on the Jump Condition Across ∂E

Fix $x \in \partial E$ and denote by $\mathbf{n}(x)$ the outward unit normal at x. For $x_o \in \mathbb{R}^N - \partial E$, set

$$\begin{aligned}\tilde{W}(\partial E, x_o; v) &= -\frac{1}{\omega_N}\int_{\partial E} v(y)\frac{(x_o - y)\cdot \mathbf{n}(x)}{|x_o - y|^N}d\sigma \\ &= \nabla_z\left(\int_{\partial E} v(y)F(z;y)d\sigma\right)\cdot \mathbf{n}(x)\Big|_{z=x_o}.\end{aligned} \tag{4.1}$$

As $x_o \to x$ the behavior of $\tilde{W}(\partial E, x_o; v)$ is similar to that of the double-layer potential $W(\partial E, x_o; v)$, provided x_o approaches x along the normal $\mathbf{n}(x)$. Let $\{x^i\}$ and $\{x^e\}$ denote sequences approaching x from the inside and respectively outside of E, say for example

$$x^i = x - \delta_i \mathbf{n}(x), \qquad x^e = x + \delta_e \mathbf{n}(x)$$

where $\{\delta_i\}$ and $\{\delta_e\}$ are sequences of positive numbers decreasing to zero.

Proposition 4.1 *Let $v \in C(\partial E)$. Then for all $x \in \partial E$*

$$\begin{aligned}\lim_{\delta_i \to 0}\tilde{W}(\partial E, x^i; v) &= \frac{1}{2}v(x) - \frac{1}{\omega_N}\int_{\partial E} v(y)\frac{(x-y)\cdot \mathbf{n}(x)}{|x-y|^N}d\sigma \\ \lim_{\delta_e \to 0}\tilde{W}(\partial E, x^e; v) &= -\frac{1}{2}v(x) - \frac{1}{\omega_N}\int_{\partial E} v(y)\frac{(x-y)\cdot \mathbf{n}(x)}{|x-y|^N}d\sigma.\end{aligned}$$

Corollary 4.1 *Let $v \in C(\partial E)$. Then for all $x \in \partial E$*

$$\lim_{x^i \to x}\tilde{W}(\partial E, x^i; v) - \lim_{x^e \to x}\tilde{W}(\partial E, x^e; v) = v(x).$$

Proof (Proposition 4.1) We prove only the first statement. Write

$$\begin{aligned}\lim_{\delta_i \to 0}\tilde{W}(\partial E, x^i; v) &= \lim_{\delta_i \to 0}\frac{1}{\omega_N}\int_{\partial E} v(y)\frac{(x^i - y)\cdot(\mathbf{n}(y) - \mathbf{n}(x))}{|x^i - y|^N}d\sigma \\ &\quad + \lim_{\delta_i \to 0}\frac{-1}{\omega_N}\int_{\partial E} v(y)\frac{(x^i - y)\cdot \mathbf{n}(y)}{|x^i - y|^N}d\sigma.\end{aligned}$$

The second limit is computed by means of Proposition 3.1, and equals

$$\frac{1}{2}v(x) - \frac{1}{\omega_N}\int_{\partial E} v(y)\frac{(x-y)\cdot \mathbf{n}(y)}{|x-y|^N}d\sigma.$$

To compute the first limit, set $S_{2\varepsilon} = \partial E \cap B_{2\varepsilon}(x)$ for $0 < \varepsilon \ll 1$, and write

$$\begin{aligned}&\int_{\partial E} v(y)\frac{(x^i - y)\cdot(\mathbf{n}(y) - \mathbf{n}(x))}{|x^i - y|^N}d\sigma \\ &\qquad = \int_{\partial E - B_{2\varepsilon}(x)} v(y)\frac{(x^i - y)\cdot(\mathbf{n}(y) - \mathbf{n}(x))}{|x^i - y|^N}d\sigma \\ &\qquad\quad + \int_{S_{2\varepsilon}(x)} v(y)\frac{(x^i - y)\cdot(\mathbf{n}(y) - \mathbf{n}(x))}{|x^i - y|^N}d\sigma.\end{aligned}$$

Without loss of generality, we may assume that the sequence $\{x^i\}$ is contained in $B_\varepsilon(x)$. Then for ε fixed

$$\lim_{\delta_i\to 0}\int_{\partial E - B_{2\varepsilon}(x)} v(y)\frac{(x^i-y)\cdot(\mathbf{n}(y)-\mathbf{n}(x))}{|x^i-y|^N}d\sigma = \int_{\partial E - B_{2\varepsilon}(x)} v(y)\frac{(x-y)\cdot(\mathbf{n}(y)-\mathbf{n}(x))}{|x-y|^N}d\sigma.$$

It remains to prove that

$$\lim_{\delta_i\to 0}\int_{S_{2\varepsilon}(x)} v(y)\frac{(x^i-y)\cdot(\mathbf{n}(y)-\mathbf{n}(x))}{|x^i-y|^N}d\sigma = \int_{S_{2\varepsilon}(x)} v(y)\frac{(x-y)\cdot(\mathbf{n}(y)-\mathbf{n}(x))}{|x-y|^N}d\sigma.$$

Compute

$$|x^i-y|^2 = |x-y|^2 + \delta_i^2 + 2\delta_i(x-y)\cdot\mathbf{n}(x).$$

By Lemma 2.1, since $|x-y|<\varepsilon$

$$|(x-y)\cdot\mathbf{n}(x)| \le \text{const}|x-y|^{1+\alpha} \le \varepsilon^\alpha\text{const}|x-y|.$$

Therefore, if ε is chosen so small that $\varepsilon^\alpha\text{const}\le 1$

$$|x^i-y|^2 \ge |x-y|^2 + \delta_i^2 - 2\varepsilon^\alpha\text{const}|x-y|\delta_i \ge \frac{1}{4}(|x-y|+\delta_i)^2.$$

Next estimate

$$\frac{|(x^i-y)\cdot(\mathbf{n}(y)-\mathbf{n}(x))|}{|x^i-y|^N} \le \gamma\frac{|x-y|^\alpha}{|x^i-y|^{N-1}} \le \gamma\frac{|x-y|^\alpha}{(|x-y|+\delta_i)^{N-1}} \le \frac{\gamma}{|x-y|^{N-1-\alpha}}$$

for a constant γ depending only upon N and the structure of ∂E. Therefore to compute the last limit, it suffices to pass to the limit under the integral with the aid of the Lebesgue dominated convergence theorem. ■

5 The Dirichlet Problem by Integral Equations ([111])

Let E be a bounded domain in $\mathbb{R}^N$ with boundary ∂E of class $C^{1,\alpha}$ for some $\alpha\in(0,1]$, and consider the Dirichlet problem (1.2) of Chapter 2. Seek a solution of such a problem in the form of a double-layer potential

$$u(x) = \frac{-1}{\omega_N}\int_{\partial E} v(y)\frac{(x-y)\cdot\mathbf{n}(y)}{|x-y|^N}d\sigma \tag{5.1}$$

for some unknown density $v \in C(\partial E)$. To impose the boundary data $u = \varphi$ on $\partial\Omega$, let x tend to points of ∂E. Using Proposition 3.1, we conclude that $v(\cdot)$ must satisfy the integral equation

$$\frac{1}{2}v = \int_{\partial E} K_D(\cdot; y)v(y)d\sigma + \varphi \tag{5.2}$$

where $K_D(\cdot;\cdot)$ is the Dirichlet kernel

$$K_D(x; y) = \frac{1}{\omega_N}\frac{(x-y)\cdot\mathbf{n}(y)}{|x-y|^N}. \tag{5.3}$$

Proposition 5.1 *Suppose that (5.2) has a solution $v \in C(\partial E)$. Then (5.1) for such a v defines a solution of the Dirichlet problem (1.2) of Chapter 2.*

Proof The function u defined by (5.1) is harmonic in E, and by Proposition 3.1, it takes the boundary values φ on ∂E. ∎

For $v \in L^\infty(\partial E)$, set

$$A_D v = \int_{\partial E} K_D(\cdot; y)v(y)d\sigma. \tag{5.4}$$

Proposition 5.2 *The function $x \to A_D v(x)$ is continuous in ∂E.*

Proof Let $\{x_n\}$ be a sequence of points in ∂E converging to some $x_o \in \partial E$. First observe that $\{K_D(x_n;\cdot)\} \to K_D(x_o;\cdot)$ a.e. in ∂E. Then write

$$\lim |A_D v(x_n) - A_D v(x_o)| \le \lim \int_{\partial E} |K_D(x_n; y) - K_D(x_o; y)||v(y)|dy.$$

Pass to the limit under integral, with the aid of Lemma 2.1 (**5.2.** of the Complements). ∎

Corollary 5.1 *Let $\{v_n\}$ be equi-bounded in $L^\infty(\partial E)$. Then $\{A_D v_n\}$ is equi-bounded and equi-continuous in ∂E (**5.3.** of the Problems and Complements).*

6 The Neumann Problem by Integral Equations ([111])

Consider the Neumann problem (1.3) of Chapter 2 for a datum $\psi \in C(\partial E)$. For $x \in \partial E$, such a datum is taken in the sense

$$\lim_{\delta\to 0} \nabla u(x - \delta\mathbf{n}(x))\cdot\mathbf{n}(x) = \psi(x). \tag{6.1}$$

Seek solutions of the Neumann problem in the form of a single-layer potential

$$u = \int_{\partial E} F(\cdot; y)v(y)d\sigma \tag{6.2}$$

where $F(\cdot;\cdot)$ is the fundamental solution of the Laplacian, introduced in (2.6) of Chapter 2, and $v(\cdot)$ is an unknown surface density. To compute v, impose the boundary condition in the sense of (6.1). First, for $x^i \in E$ and $x \in \partial E$, compute

$$\nabla u(x^i) \cdot \mathbf{n}(x) = -\frac{1}{\omega_N} \int_{\partial E} v(y) \frac{(x^i - y) \cdot \mathbf{n}(x)}{|x^i - y|^N} d\sigma.$$

Then take $x^i = x - \delta \mathbf{n}(x)$ and take the limit as $\delta \to 0$ of the integral on the right-hand side by making use of Proposition 4.1. We conclude that $v(\cdot)$ must satisfy the integral equation

$$\frac{1}{2} v = \int_{\partial E} K_N(\cdot; y) v(y) d\sigma + \psi \tag{6.3}$$

where $K_N(\cdot;\cdot)$ is the Neumann kernel

$$K_N(x, y) = \frac{1}{\omega_N} \frac{(x - y) \cdot \mathbf{n}(x)}{|x - y|^N}. \tag{6.4}$$

Proposition 6.1 *Suppose that (6.3) has a solution $v \in C(\partial E)$. Then (6.2), for such a v, defines a solution to the Neumann problem (1.3) of Chapter 2.*

Proof The function defined by (6.2) is harmonic in E and by the previous calculations, satisfies the Neumann data on ∂E. ■

Theorem 6.1 *A necessary and sufficient condition of solvability of (6.3) is that ψ be of zero-average over ∂E, i.e., that (1.4) of Chapter 2 holds.*

Proof (Necessity) Integrate (6.3) over ∂E in $d\sigma(x)$. By Corollary 3.1

$$\int_{\partial E} K_N(x; y) d\sigma(x) = -\frac{1}{\omega_N} \int_{\partial E} \frac{(x - y) \cdot \mathbf{n}(x)}{|x - y|^N} d\sigma(x) = \frac{1}{2}.$$ ■

The sufficient part of the theorem will be proved in the next chapter.

For $v \in L^\infty(\partial E)$ set

$$A_N v = \int_{\partial E} K_N(\cdot; y) v(y) d\sigma. \tag{6.5}$$

Proposition 6.2 *The function $x \to A_N v(x)$ is continuous in ∂E.*

Corollary 6.1 *Let $\{v_n\}$ be equi-bounded in $L^\infty(\partial E)$. Then $\{A_N v_n\}$ is equi-bounded and equi-continuous in ∂E.*

7 The Green Function for the Neumann Problem

Consider the family of Neumann problems

$$\begin{aligned} N(x;\cdot)\in C^2(E)\cap C^1(\bar{E}) \text{ for all } x\in E, \quad \Delta_y N(x;y)=k \text{ in } E \\ \frac{\partial}{\partial \mathbf{n}(y)}N(x;y)=\frac{\partial}{\partial \mathbf{n}(y)}F(x;y), \quad y\in\partial E \end{aligned} \tag{7.1}$$

where F is the fundamental solution of the Laplacian, introduced in (2.6) of Chapter 2, and k is a constant. Integrating the equation by parts and using (1.3) gives

$$k\,|E|=\int_{\partial E}\frac{\partial F(x;y)}{\partial \mathbf{n}(y)}d\sigma=-1.$$

Therefore, a necessary condition of solvability is $k=-|E|^{-1}$. Assuming for the moment that (7.1) has a solution set

$$(x,y)\to \mathcal{G}(x;y)=F(x;y)-N(x;y) \quad \text{for } x\neq y.$$

This is called Green's function for the Neumann problem, and it satisfies

$$\Delta_y\mathcal{G}=-k \text{ in } E, \quad \frac{\partial}{\partial \mathbf{n}(y)}\mathcal{G}(x;y)=0 \text{ on } \partial E; \quad x\neq y. \tag{7.2}$$

Green's function is not unique; indeed if $\mathcal{G}(x;\cdot)$, is a Green's function, then

$$\mathcal{G}(x;\cdot)+v(x) \quad \text{for all } v\in C^2(E)$$

is still a Green's function for the Neumann problem. Having determined one such a function, say for example $\mathcal{G}_1(x;y)$, we let

$$v=-\fint_E \mathcal{G}_1(\cdot;y)dy \quad \text{and} \quad \mathcal{G}(x;y)\stackrel{\text{def}}{=}\mathcal{G}_1(x;y)+v(x).$$

In this way, among all the possible Green's functions for the Neumann problem, we have selected the one with zero-average for all $x\in E$. Such a selection implies that $\mathcal{G}(\cdot;\cdot)$ is symmetric. This is a particular case of the following

Lemma 7.1 *Let $\mathcal{G}(\cdot;\cdot)$ be Green's function for the Neumann problem satisfying*

$$x\to\int_E \mathcal{G}(x;y)dy=const.$$

Then $\mathcal{G}(x;y)=\mathcal{G}(y;x)$.

The proof is the same as in Lemma 3.1 of Chapter 2. From now on, we will select $\mathcal{G}$ satisfying the zero-average property. Therefore, by symmetry

$$\Delta_y\mathcal{G}=\Delta_x\mathcal{G}=-k \quad \text{in } E \text{ for } x\neq y.$$

Let u be a solution of the Neumann problem (1.3) of Chapter 2. From the Stokes identity (2.3)–(2.4) of Chapter 2, subtract the Green's identity (2.2), of the same Chapter, written for u and $N(x;\cdot)$. This gives

$$u = \int_{\partial E} \psi \mathcal{G}(\cdot;y) d\sigma - k \int_E u\, dy.$$

If u is a solution of the Neumann problem, then $u + C$ are also solutions of the same problem, for all constants C. Choosing

$$C = -\fint_E u dy$$

we select, among all solutions of the Neumann problem, the one with the zero-average property and satisfying the representation

$$u = \int_{\partial E} \psi \mathcal{G}(\cdot;y) d\sigma. \tag{7.3}$$

This representation is a candidate for a solution of the Neumann problem. By the symmetry of $\mathcal{G}(\cdot;\cdot)$

$$\Delta_x u = \int_{\partial E} \psi \Delta_x \mathcal{G}(\cdot;y)\, d\sigma = -k \int_{\partial E} \psi\, d\sigma = 0.$$

Thus the condition that ψ be of zero-average over ∂E is necessary for (7.3) to define a harmonic function. It would remain to establish that the boundary datum is taken in the sense of (6.1). This verification could be carried out if one had an explicit expression for $\mathcal{G}(\cdot;\cdot)$. This would be analogous to the Dirichlet problem for the ball, where a verification of the boundary data was possible via the explicit Poisson representation of Theorem 3.1 of Chapter 2.

Even though the method is elegant, the actual calculation of the Green's function $\mathcal{G}(\cdot;\cdot)$ can be effected explicitly only for domains with a simple geometry such as balls or cubes (see Section 7c of the Complements).

7.1 Finding $\mathcal{G}(\cdot;\cdot)$

One might look for $\mathcal{G}(\cdot;\cdot)$ of the form

$$\mathcal{G}(x;y) = F(x;y) - \gamma_o |y|^2 + h(x;y)$$

up to the addition of a function $x \to v(x)$. Here γ_o is a constant to be determined, and

$$h(x;\cdot) \in C^2(E) \cap C^1(\bar{E}) \text{ is harmonic for all } x \in E.$$

Such a $\mathcal{G}(\cdot;\cdot)$ satisfies the first of (7.2) for the choice $2N\gamma_o = k$. Imposing the boundary conditions on $\mathcal{G}(x;\cdot)$ implies that $h(x;\cdot)$ must satisfy

$$\Delta h(x;\cdot) = 0 \text{ in } E, \quad \frac{\partial h(x;y)}{\partial \mathbf{n}(y)} = 2\gamma_o y \cdot \mathbf{n}(y) - \frac{\partial F(x;y)}{\partial \mathbf{n}(y)} \text{ on } \partial E.$$

This family of Neumann problems can be solved by the method of integral equations outlined in the previous section. Specifically, one looks for $h(x;\cdot)$ in the form of a single-layer potential

$$h(x;y) = \int_{\partial E} v(x;\eta) F(y;\eta)\, d\sigma(\eta), \quad x,y \in E$$

where the unknown x-dependent density distribution $v(x;\cdot)$ satisfies the integral equation

$$\frac{1}{2} v(x;y) = 2\gamma_o y \cdot \mathbf{n}(y) - \frac{\partial F(x;y)}{\partial \mathbf{n}(y)} + \frac{1}{\omega_N} \int_{\partial E} v(x;\eta) \frac{(y-\eta)\cdot \mathbf{n}(y)}{|y-\eta|^N}\, d\sigma(\eta).$$

This integral equation is solvable if and only if

$$\int_{\partial E} \frac{\partial h(x;y)}{\partial \mathbf{n}(y)}\, d\sigma = 0 \quad \text{for all } x \in E.$$

This is part of the existence theory for such integral equations that will be developed in the next chapter. To verify the zero-average condition, compute

$$\int_{\partial E} \frac{\partial F(x;y)}{\partial \mathbf{n}(y)}\, d\sigma = -W(\partial E, x; 1) = -1$$

by Proposition 1.2. On the other hand, by the divergence theorem and the indicated choices of γ_o and k

$$2\gamma_o \int_{\partial E} y \cdot \mathbf{n}(y) d\sigma = \gamma_o \int_E \operatorname{div} \nabla |y|^2 dx = 2N\gamma_o |E| = -1.$$

8 Eigenvalue Problems for the Laplacian

Consider the problem of finding $\lambda \in \mathbb{R} - \{0\}$ and a non-trivial $u \in C^2(E) \cap C^\eta(\bar{E})$, for some $\eta \in (0,1)$ satisfying

$$\Delta u = -\lambda u \text{ in } E, \quad \text{and} \quad u = 0 \text{ on } \partial E. \tag{8.1}$$

This is the eigenvalue problem for the Laplacian with homogeneous Dirichlet data on ∂E. If (8.1) has a non-trivial solution, by the results of Section 12 of Chapter 2, such a solution u satisfies

$$u = \lambda \int_E G(\cdot;y) u\, dy \tag{8.2}$$

where $G(\cdot;\cdot)$ is the Green's function for the Laplacian in E. The non-trivial pair (λ, u) represents an eigenvalue and an eigenfunction for the integral equation (8.2). Conversely, if (8.2) has a non-trivial solution pair (λ, u), such that $u \in C^\eta(\bar{E})$, for some $\eta \in (0,1)$, then by the same procedure of Section 12 of Chapter 2, such a u is also a solution of (8.1). We summarize

Lemma 8.1 *A non-trivial pair* (λ, u) *is a solution of (8.1) if and only if it solves (8.2).*

8.1 Compact Kernels Generated by Green's Function

Let E be a bounded open set in $\mathbb{R}^N$ with boundary ∂E of class C^1 and let $G(\cdot;\cdot)$ be the Green's function for the Laplacian in E. Set

$$L^p(E) \ni f \to A_G f = \int_E G(\cdot; y) f(y) dy \quad \text{for some } p \geq 1 \tag{8.3}$$

provided the right-hand side defines a function in $L^q(E)$ for some $q \geq 1$.

Theorem 8.1 A_G *is a compact mapping in* $L^p(E)$ *for all* $1 \leq p \leq \infty$, *i.e., it maps bounded sets in* $L^p(E)$, *into pre-compact sets in* $L^p(E)$.

Green's function G is defined in (3.3) of Chapter 2, where $F(\cdot;\cdot)$ is the fundamental solution of the Laplace equation in $\mathbb{R}^N$, for $N \geq 2$, defined in (2.6) of Chapter 2, and Φ is introduced in (3.1) of the same chapter. Setting

$$L^p(E) \ni f \to \begin{cases} A_F f = \displaystyle\int_E F(\cdot; y) f(y) dy \\ \\ A_\Phi f = \displaystyle\int_E \Phi(\cdot; y) f(y) dy \end{cases} \quad \text{for some } p \geq 1 \tag{8.4}$$

the proof reduces to showing that both A_F and A_Φ are compact in $L^p(E)$.

9 Compactness of A_F in $L^p(E)$ for $1 \leq p \leq \infty$

The operator A_F maps bounded sets in $L^p(E)$ into bounded sets of $L^p(E)$, for all $1 \leq p \leq \infty$. This is content of Proposition 10.1 of Chapter 2. For $p > N$, the compactness of A_F follows from Corollary 10.1 of Chapter 2. Indeed, in such a case, A_F maps bounded sequences $\{f_n\} \subset L^p(E)$ into sequences $\{A_F f_n\}$ of equi-Lipschitz continuous functions in $\bar{E}$. Therefore compactness follows from the Ascoli–Arzelà theorem.

To establish compactness in $L^p(E)$ for $1 \leq p \leq N$, for a fixed vector $h \in \mathbb{R}^N$, introduce the translation operator

$$L^p(E) \ni v \to T_h v = \begin{cases} v(\cdot + h) & \text{if } \cdot + h \in E \\ 0 & \text{otherwise.} \end{cases}$$

Also for $\delta > 0$ set

$$E_\delta = \{x \in E \mid \text{dist}(x, \partial E) > \delta\}. \tag{9.1}$$

The proof uses the following characterization of pre-compact subsets of $L^p(E)$ ([31] Chapter 5, Section 22).

Theorem 9.1 *A bounded subset $K \subset L^p(E)$, for $1 \le p < \infty$ is pre-compact in $L^p(E)$ if and only if for every $\varepsilon > 0$ there exists $\delta > 0$ such that for all vectors $h \in \mathbb{R}^N$ of length $|h| < \delta$, and for all $v \in K$*

$$\|T_h v - v\|_{p,E} < \varepsilon \qquad \textit{and} \qquad \|v\|_{p,E-E_\delta} < \varepsilon. \tag{9.2}$$

To verify the assumptions of the theorem, set $v = A_F f$ and use Proposition 10.1 of Chapter 2 and (10.3) of the same Chapter, for $\alpha = 1$. Fix $\delta > 0$ so small that $E - E_\delta$ is not empty and let $h \in \mathbb{R}^N$ be such that $|h| < \delta$. Then

$$\begin{aligned}\int_{E_\delta} |v(x+h) - v(x)| dx &\le \int_{E_\delta} \int_0^1 \left| \frac{d}{dt} v(x+th) \right| dt dx \\ &\le |h| \int_0^1 \int_{E_\delta} |\nabla v(x+th)| dx dt \\ &\le |h| |E|^{1-\frac{1}{q}} \|\nabla A_F f\|_{q,E} \le \gamma |h| \|f\|_{p,E}\end{aligned}$$

where p and q are as in the indicated proposition, including possibly the limiting cases. Therefore for all $\sigma \in (0, \frac{1}{p})$

$$\begin{aligned}\int_{E_\delta} |T_h v - v|^p dx &= \int_{E_\delta} |T_h v - v|^{p\sigma + p(1-\sigma)} dx \\ &\le \left(\int_{E_\delta} |T_h v - v| dx \right)^{p\sigma} \left(\int_{E_\delta} |T_h v - v|^{\frac{p(1-\sigma)}{1-p\sigma}} dx \right)^{1-p\sigma}.\end{aligned}$$

Choose σ from

$$\frac{p(1-\sigma)}{1-p\sigma} = q \in \left(p, \frac{Np}{(N-p)_+} \right), \qquad \text{i.e.,} \qquad \sigma = \frac{q-p}{p(q-1)}.$$

One verifies that such a choice is possible if q is in the range (10.2) of Chapter 2, for $\alpha = 1$. Therefore

$$\int_{E_\delta} |T_h v - v|^p dx \le \gamma \delta^{p\sigma} \|f\|_{p,E}^{p(1-\sigma)}.$$

On the other hand

$$\begin{aligned}\int_{E-E_\delta} |T_h v - v|^p dx &\le 2^p |E - E_\delta|^{\frac{q-p}{q}} \left(\int_E |v|^q dx \right)^{p/q} \\ &\le \gamma |E - E_\delta|^{\frac{q-p}{q}} \|f\|_{p,E}^p.\end{aligned}$$

Since ∂E is of class C^1, there exists a constant γ depending on N and the structure of ∂E such that $|E - E_\delta| \le \gamma\delta$. Combining these estimates

$$\|T_h v - v\|_{p,E} = \|T_h A_F f - A_F f\|_{p,E} \le \gamma \delta^{\frac{q-p}{pq}} (1 + \|f\|_{p,E}). \qquad \blacksquare$$

10 Compactness of A_Φ in $L^p(E)$ for $1 \le p < \infty$

Since $\Phi(x;\cdot)$ is harmonic in E and equals $F(x;\cdot)$ on ∂E, possible singularities of $\Phi(x;y)$ occur for $x \in \partial E$. Therefore, for $\delta > 0$ fixed, $\Phi(\cdot;y) \in C^\infty(E_\delta)$ uniformly in $y \in \bar{E}$. If $\{f_n\}$ is a bounded sequence in $L^p(E)$, then the sequence $\{A_\Phi f_n\}$ is equi-bounded and equi-continuous in $\bar{E}_\delta$. By the Ascoli–Arzelà theorem, a subsequence can be selected, and relabeled with n, such that for every $\varepsilon > 0$, there exists a positive integer $n(\varepsilon)$ such that

$$\|A_\Phi f_n - A_\Phi f_m\|_{\infty,E_\delta} \le \varepsilon \qquad \text{for } \; n,m \ge n(\varepsilon). \tag{10.1}$$

The selection of the subsequence depends on δ. Let $\{\delta_j\}$ be a sequence of positive numbers decreasing to zero, and let $\{A_\Phi f_{n_j}\}$ be the subsequence, out of $\{A_\Phi f_n\}$, for which (10.1) holds within E_{δ_j}. By diagonalization, one may select a subsequence, and relabel it with n, such that $\{A_\Phi f_n\}$ is a Cauchy sequence in $C(\bar{E}_{\delta_j})$ for each fixed $j = 1, 2, \dots$. We claim that $\{A_\Phi f_n\}$ is a Cauchy sequence in $L^p(E)$. Fix $\varepsilon > 0$ arbitrarily small and $j \in \mathbb{N}$ arbitrarily large. There exists a positive integer $m(\varepsilon, j)$ such that

$$\|A_\Phi f_n - A_\Phi f_m\|_{\infty,E_{\delta_j}} \le \varepsilon \qquad \text{for } \; n,\, m \ge m(\varepsilon, j).$$

Next, for $n, m \ge m(\varepsilon, j)$

$$\begin{aligned}\|A_\Phi f_n - A_\Phi f_m\|_{p,E}^p &= \int_{E_{\delta_j}} |A_\Phi f_n - A_\Phi f_m|^p dx + \int_{E-E_{\delta_j}} |A_\Phi f_n - A_\Phi f_m|^p dx \\ &\le \varepsilon^p |E| + \int_{E-E_{\delta_j}} \Big| \int_E \Phi(x;y)|f_n(y) - f_m(y)|dy\Big|^p dx.\end{aligned}$$

Let us assume that $N \ge 3$, the proof for $N = 2$ being similar. By Proposition 12.1 of Chapter 2 the last integral is majorized by

$$\Big(\int_E \Big| \int_E F(x;y)|f_n(y) - f_m(y)|dy\Big|^q dx\Big)^{p/q} |E - E_{\delta_j}|^{1-p/q}$$

for some $q > p$. ∎

11 Compactness of A_Φ in $L^\infty(E)$

Lemma 11.1 *Let $f \in L^\infty(E)$. Then $A_\Phi f$ is Hölder continuous in $\bar{E}$. Namely, there exist constants $\gamma > 1$ and $0 < \eta \le 1$, that can be determined a priori in terms of N and the structure of ∂E only such that*

$$|A_\Phi f(x_1) - A_\Phi f(x_2)| \le \gamma \|f\|_{\infty,E} |x_1 - x_2|^\eta \quad \textit{for all } \; x_1, x_2 \in \bar{E}.$$

Proof The points $x_i \in \bar{E}$ being fixed, set $\delta = |x_1 - x_2|^\alpha$, where $\alpha > 0$ is to be chosen, and denote by $B_\delta(\bar{x})$ the ball of radius δ centered at $\bar{x} = \frac{1}{2}(x_1 + x_2)$. For such a δ, let E_δ be defined as in (9.1), where without loss of generality we may assume that δ is so small that $E_{4\delta} \neq \emptyset$. Assume first that $B_\delta(\bar{x}) \subset E_{2\delta}$. Then

$$\begin{aligned}
A_\Phi f(x_1) - A_\Phi f(x_2) &= \int_E [\Phi(x_1; y) - \Phi(x_2; y)] f(y) dy \\
&= \int_E \left(\int_0^1 \frac{d}{ds} \Phi(s x_1 + (1-s) x_2; y) ds \right) f(y) dy \\
&= (x_1 - x_2) \cdot \int_0^1 \nabla_x \left(\int_E \Phi(s x_1 + (1-s) x_2; y) f(y) dy \right) ds.
\end{aligned}$$

The function $A_\Phi f$ is harmonic in $B_{2\delta}(\bar{x})$, and $s x_1 + (1-s) x_2 \in B_\delta(\bar{x})$ for all $s \in [0,1]$. Therefore by Theorem 5.2 of Chapter 2

$$\begin{aligned}
\left| \nabla_x \int_E \Phi(x; y) f(y) dy \right| &\le \frac{\gamma}{\delta} \sup_{x \in B_{2\delta}(\bar{x})} \left| \int_E \Phi(x; y) f(y) dy \right| \\
&\le \frac{\gamma}{\delta} \|f\|_{\infty, E} \sup_{x \in E} \int_E F(x; y) dy \le \frac{\gamma}{\delta} \|f\|_{\infty, E}.
\end{aligned}$$

Combining these estimates

$$|A_\Phi f(x_1) - A_\Phi f(x_2)| \le \tilde{\gamma} \|f\|_{\infty, E} \frac{|x_1 - x_2|}{\delta}$$

provided $B_\delta(\bar{x}) \subset E_{2\delta}$. Assume now that $B_\delta(\bar{x}) \cap (E - E_{2\delta}) \neq \emptyset$, and write

$$\begin{aligned}
A_\Phi f(x_1) - A_\Phi f(x_2) &= \int_E [\Phi(x_1; y) - \Phi(x_2; y)] f(y) dy \\
&= \int_{E_{4\delta}} [\Phi(x_1; y) - \Phi(x_2; y)] f(y) dy \\
&\quad + \int_{E - E_{4\delta}} [\Phi(x_1; y) - \Phi(x_2; y)] f(y) dy = I_1 + I_2.
\end{aligned}$$

$$\begin{aligned}
I_1 &= (x_1 - x_2) \cdot \int_0^1 \nabla_x \left(\int_{E_{4\delta}} \Phi(s x_1 + (1-s) x_2; y) f(y) dy \right) ds \\
&= (x_1 - x_2) \cdot \int_0^1 \left(\int_{E_{4\delta}} \nabla_y \Phi(s x_1 + (1-s) x_2; y) f(y) dy \right) ds
\end{aligned}$$

by symmetry of $\Phi(\cdot;\cdot)$. Since $y \to \Phi(s x_1 + (1-s) x_2; y)$ is harmonic in $E_{4\delta}$, by Theorem 5.2 of Chapter 2

$$\begin{aligned}
|\nabla \Phi(s x_1 + (1-s) x_2; y)| &\le \frac{\gamma}{\delta} \sup_{z \in B_\delta(y)} \Phi(s x_1 + (1-s) x_2; z) \\
&\le \frac{\gamma}{\delta} \sup_{z \in B_\delta(y)} F(s x_1 + (1-s) x_2; z) \le \frac{\hat{\gamma}}{\delta^{N-1}}.
\end{aligned}$$

Therefore

$$|I_1| \le \tilde{\gamma} \|f\|_{\infty,E} \frac{|x_1 - x_2|}{\delta^{N-1}}.$$

Next

$$\begin{aligned}
|I_2| &\le \int_{E-E_{4\delta}} [\Phi(x_1;y) + \Phi(x_2;y)]|f(y)|dy \\
&\le \int_{E-E_{4\delta}} [F(x_1;y) + F(x_2;y)]|f(y)|dy \\
&\le \gamma \|f\|_{\infty,E} |E - E_{4\delta}|^{1/N} \le \gamma \|f\|_{\infty,E} \delta^{1/N}.
\end{aligned}$$

Therefore if $B_\delta(\bar{x}) \cap (E - E_{2\delta}) \neq \emptyset$

$$|A_\Phi f(x_1) - A_\Phi f(x_2)| \le \tilde{\gamma} \Big(\frac{|x_1 - x_2|}{\delta^{N-1}} + \delta^{1/N} \Big).$$

The remaining cases are treated by inserting between x_1 and x_2, finitely many points $\{\xi_1, \dots, \xi_n\}$, so that each of the pairs (x_1, ξ_1), (ξ_j, ξ_{j+1}), and (ξ_n, x_2) falls in one of the previous cases. The number n will depend on the structure of ∂E. ■

Corollary 11.1 *A_G is compact in $L^\infty(E)$.*

Corollary 11.2 *Let $u \in L^2(E)$ be a solution of (8.2). Then $u \in C^\eta(\bar{E})$ for some $\eta > 0$.*

Proof Applying Proposition 10.1 of Chapter 2 a finite number of times implies that $u \in L^\infty(E)$. Then the conclusion follows from Lemma 11.1 and Corollary 10.1 of Chapter 2. ■

Remark 11.1 The corollary provides the necessary regularity for the eigenvalue problems (8.1) and (8.2) to be equivalent.

Problems and Complements

2c On the Integral Defining the Double-Layer Potential

2.1. Prove a sharper version of Lemma 2.1. In particular, find the optimal conditions on ∂E to ensure that

$$\frac{(x - y) \cdot \mathbf{n}(y)}{|x - y|^N} \in L^1(\partial E).$$

2.2. Consider the integral in (2.1). As $x_o \to x \in \partial E$, the integrand tends to

$$\frac{(x-y)\cdot\mathbf{n}(y)}{|x-y|^N}v(y) \qquad \text{for a.e. } y \in \partial E.$$

Moreover, such a function is in $L^{1+\varepsilon}(\partial E)$ for some $\varepsilon > 0$. This follows from Lemma 2.1. However, the limit cannot be carried under the integral. Explain.

5c The Dirichlet Problem by Integral Equations

5.1. Let $E \subset \mathbb{R}^N$ be open and bounded and with boundary ∂E of class C^1. Formulate the following exterior Dirichlet problem in terms of an integral equation:

$$\begin{gathered} u \in C^2(\mathbb{R}^N - \bar{E}) \cap C(\overline{\mathbb{R}^N - E}), \qquad \Delta u = 0 \text{ in } \mathbb{R}^N - E \\ u\big|_{\partial E} = \varphi \in C(\partial E), \qquad \lim_{|x|\to\infty} u(x) = 0. \end{gathered}$$

Compare with Section 8.1c of the Problems and Complements of Chapter 2.

5.2. In the proof of Proposition 5.2, justify the passage of the limit under the integral. *Hint*:

$$\begin{aligned} \int_{\partial E} &|K_D(x_n;y) - K_D(x_o;y)||v(y)|d\sigma \\ &= \int_{\partial E\cap[|y-x_o|<\varepsilon]} |K_D(x_n;y) - K_D(x_o;y)||v(y)|d\sigma \\ &\quad + \int_{\partial E\cap[|y-x_o|\ge\varepsilon]} |K_D(x_n;y) - K_D(x_o;y)||v(y)|d\sigma. \end{aligned}$$

5.3. Prove a stronger version of Proposition 5.2. Namely, if $v \in L^\infty(\partial E)$, then $\{A_D v\}$ is Hölder continuous with exponent α/N, where α is the constant appearing in Lemma 2.1. *Hint*: For $x_1, x_2 \in \partial E$

$$|A_D v(x_1) - A_D v(x_2)| \le \frac{\|v\|_{\infty,\partial E}}{\omega_N} \int_{\partial E} \left| \frac{(x_1-y)\cdot\mathbf{n}(y)}{|x_1-y|^N} - \frac{(x_2-y)\cdot\mathbf{n}(y)}{|x_2-y|^N} \right| d\sigma.$$

May assume that $|x_1 - x_2| < 1$ and set

$$\partial_i = \{y \in \partial E \big| |x_i - y| < |x_1 - x_2|^{1/N}\}.$$

Divide the integral into one extended over $\partial_1 \cap \partial_2$ and another one extended over the complement.

6c The Neumann Problem by Integral Equations

6.1. Let $E \subset \mathbb{R}^N$ be open and bounded and with boundary ∂E of class C^1. Formulate the following exterior Neumann problem in terms of an integral equation:

$$u \in C^2(\mathbb{R}^N - \bar{E}) \cap C^1(\overline{\mathbb{R}^N - E}), \qquad \Delta u = 0 \text{ in } \mathbb{R}^N - E$$
$$\frac{\partial}{\partial \mathbf{n}} u\big|_{\partial E} = \varphi \in C(\partial E), \qquad \lim_{|x|\to\infty} u(x) = 0.$$

6.2. Prove a stronger version of Proposition 6.2. Namely, if $v \in L^\infty(\partial E)$, then $\{A_N v\}$ is Hölder continuous with exponent α/N, where α is the constant appearing in Lemma 2.1.

7c Green's Function for the Neumann Problem

7.1c Constructing $\mathcal{G}(\cdot;\cdot)$ for a Ball in $\mathbb{R}^2$ and $\mathbb{R}^3$

Attempt finding $\mathcal{G}(\cdot;\cdot)$ of the form

$$\mathcal{G}(x;y) = F(x;y) + \underbrace{\Phi(x;y) + h(x;y) + \gamma|y|^2}_{-N(x;y)}$$

where F and Φ are defined in (2.6) and (3.7) of Chapter 2 respectively, γ is a constant, and $h(x;\cdot)$ is a suitable harmonic function in B_R to be chosen to satisfy the last of (7.2). As in Section 7.1 one computes $2N\gamma_o|E| = -1$. Using the explicit expression of Φ

$$\frac{\partial}{\partial \mathbf{n}(y)}(F + \Phi + \gamma|y|^2) = \frac{1}{\omega_N R|x-y|^{N-2}} + 2\gamma_o R.$$

Therefore the harmonic function $h(x;\cdot)$ has to be chosen to satisfy

$$\frac{\partial}{\partial \mathbf{n}(y)} h(x;y) = \frac{-1}{\omega_N R|x-y|^{N-2}} - 2N\gamma_o R.$$

7.1.1c The Case $N = 2$

Choose $h = 0$ and $\gamma_o = -1/4\pi R^2$ to conclude that Green's function for the Neumann problem for the disc $D_R \subset \mathbb{R}^2$ is

$$\mathcal{G}(x;y) = \frac{-1}{2\pi}\left(\ln|\xi - y|\frac{|x|}{R} + \ln|x-y|\right) - \frac{1}{4\pi R^2}|y|^2, \quad \xi = \frac{R^2}{|x|^2}x$$

up to an arbitrary additive smooth function of x.

7.1.2c The Case $N = 3$

The function $h(x;\cdot)$ is given by

$$h(x;y) = \frac{1}{4\pi}\ln\mathcal{H}, \quad \mathcal{H} = \frac{(\xi - y)\cdot x}{|x|} + |\xi - y|, \quad \xi = \frac{R^2}{|x|^2}x.$$

One computes

$$4\pi\nabla h = \frac{-1}{\mathcal{H}}\left(\frac{x}{|x|} + \frac{\xi - y}{|\xi - y|}\right) \tag{$*$}$$

and

$$\begin{aligned}4\pi\Delta h &= \frac{2}{\mathcal{H}|\xi - y|} - \frac{1}{\mathcal{H}^2}\left(\frac{x}{|x|} + \frac{\xi - y}{|\xi - y|}\right)\cdot\left(\frac{x}{|x|} + \frac{\xi - y}{|\xi - y|}\right)\\ &= \frac{2}{\mathcal{H}|\xi - y|} - \frac{2}{\mathcal{H}^2}\left(1 + \frac{\xi - y}{|\xi - y|}\cdot\frac{x}{|x|}\right) = 0.\end{aligned}$$

From $(*)$, taking into account that $\xi = R^2 x/|x|^2$

$$\begin{aligned}4\pi\frac{\partial h}{\partial\mathbf{n}(y)} &= 4\pi\nabla h\cdot\frac{y}{R}\\ &= \frac{-1}{R\mathcal{H}}\left(\frac{x\cdot y}{|x|} + \frac{(\xi - y)\cdot(y - \xi)}{|\xi - y|} + \frac{(\xi - y)\cdot\xi}{|\xi - y|}\right)\\ &= \frac{1}{R\mathcal{H}}\left((\xi - y)\cdot\frac{x}{|x|} + |\xi - y| - \frac{x}{|x|}\cdot\xi - \frac{(\xi - y)\cdot\xi}{|\xi - y|}\right)\\ &= \frac{1}{R\mathcal{H}}\left(\mathcal{H} - \frac{\left[(\xi - y)\cdot\frac{x}{|x|} + |\xi - y|\right]}{|\xi - y|}\frac{R^2}{|x|}\right) = \frac{1}{R} - \frac{1}{|x - y|}.\end{aligned}$$

The last equality follows from (3.6) of Chapter 2, since $y \in \partial B_R$. We conclude that Green's function for the Neumann problem for the ball $B_R \subset \mathbb{R}^3$ is

$$\begin{aligned}\mathcal{G}(x;y) =& \frac{1}{4\pi}\left(\frac{1}{|x - y|} + \frac{R}{|x|}\frac{1}{|\xi - y|}\right)\\ &+ \frac{1}{4\pi}\ln\left((\xi - y)\cdot\frac{x}{|x|} + |\xi - y|\right) - \frac{1}{8\pi R^3}|y|^2\end{aligned}$$

up to an arbitrary additive smooth function of x.

8c Eigenvalue Problems

8.1. Formulate the homogeneous Neumann eigenvalue problem

$$\Delta u = -\lambda u \ \text{ in } E, \quad \tfrac{\partial}{\partial\mathbf{n}}u = 0 \ \text{ on } \partial E. \tag{8.1c}$$

8.2. Find eigenvalues and eigenfunctions of (8.1) and (8.1c), when E is a parallelepiped in $\mathbb{R}^3$ of sides a, b, c. Do the same for a disc $D_R \subset \mathbb{R}^2$.

8.3. Let $A_{\mathcal{G}}$ be defined as in (8.3) with G replaced by $\mathcal{G}$. Prove the analogue of Theorem 8.1.

4

Integral Equations and Eigenvalue Problems

1 Kernels in $L^2(E)$

Let E be a bounded open set in $\mathbb{R}^N$ with boundary ∂E of class C^1. For complex-valued f and g in $L^2(E)$, set

$$\langle f, g\rangle = \int_E f\bar{g}\,dx \quad \text{and} \quad \|f\|^2 = \langle f, f\rangle$$

and say that f and g are orthogonal if $\langle f, g\rangle = 0$. A complex-valued $dx \times dx$-measurable function $K(\cdot;\cdot)$ defined in $E \times E$ is a kernel acting in $L^2(E)$ if the two operators

$$Af = \int_E K(\cdot; y)f(y)dy, \qquad A^* f = \int_E \bar{K}(y; \cdot)f(y)dy$$

map bounded subsets of $L^2(E)$ into bounded subsets of $L^2(E)$, equivalently, if there is a constant γ depending only upon N and E such that

$$\|Af\| \le \gamma\|f\| \quad \text{and} \quad \|A^* f\| \le \gamma\|f\|, \quad \text{for all } f \in L^2(E).$$

It would be sufficient to require only one of these, since any of them implies the other. Indeed, assuming that the first holds

$$\begin{aligned}
|\langle A^* f, g\rangle| &= \left| \int_E \left(\int_E \bar{K}(y; x)f(y)dy \right) \bar{g}(x)dx \right| \\
&= \left| \int_E f(y) \overline{\left(\int_E K(y; x)g(x)dx \right)} dy \right| = |\langle f, Ag\rangle| \le \gamma\|f\|\|g\|.
\end{aligned}$$

The operators A and A^* are adjoint in the sense that

$$\langle Af, g\rangle = \langle f, A^* g\rangle \quad \text{for all } f, g \in L^2(E).$$

E. DiBenedetto, *Partial Differential Equations: Second Edition*,
Cornerstones, DOI 10.1007/978-0-8176-4552-6_5,

Their norm is defined by

$$\|A\| = \sup_{\|f\|=1} \|Af\|, \qquad \|A^*\| = \sup_{\|f\|=1} \|A^* f\|.$$

By the characterization of the $L^2(E)$-norm ([31], Chapter V, Section 4)

$$\begin{aligned}
\|A\| &= \sup_{\|f\|=1} \sup_{\|g\|=1} \left| \int_E \left(\int_E K(x;y) f(y) dy \right) \bar{g}(x) dx \right| \\
&= \sup_{\|f\|=1} \sup_{\|g\|=1} \left| \int_E f(y) \overline{\left(\int_E \bar{K}(x;y) g(x) dx \right)} dy \right| \\
&\le \sup_{\|f\|=1} \sup_{\|g\|=1} \|f\| \left(\int_E \left| \int_E \bar{K}(x;y) g(x) dx \right|^2 dy \right)^{1/2} \\
&= \sup_{\|g\|=1} \left(\int_E \left| \int_E \bar{K}(x;y) g(x) dx \right|^2 dy \right)^{1/2} = \|A^*\|.
\end{aligned}$$

A similar calculation gives $\|A^*\| \le \|A\|$. Thus $\|A\| = \|A^*\|$.

A kernel $K(\cdot;\cdot)$ in $L^2(E)$ is *compact* if the resulting operators A and A^* are compact in $L^2(E)$, i.e., if they map bounded subsets of $L^2(E)$ into precompact subsets of $L^2(E)$. If A is compact, A^* is also compact.

A kernel $K(\cdot;\cdot)$ in $L^2(E)$ is *symmetric* if $K(x;y) = K(y,x)$ for a.e. $(x,y) \in E \times E$. If it is symmetric and real-valued, then $A = A^*$.

1.1 Examples of Kernels in $L^2(E)$

Given two n-tuples $\{\varphi_1, \dots, \varphi_n\}$ and $\{\psi_1, \dots, \psi_n\}$ of linearly independent, complex-valued functions in $L^2(E)$, set

$$K(x;y) = \sum_{i=1}^{n} \varphi_i(x) \bar{\psi}_i(y) \quad \text{for a.e. } (x,y) \in E \times E. \tag{1.1}$$

A kernel of this kind is called *separable*, or of *finite rank*, or *degenerate*. For such a kernel for any $f \in L^2(E)$

$$\int_E K(\cdot;y) f(y) dy = \sum_{i=1}^{n} \varphi_i \langle f, \psi_i \rangle.$$

Thus separable kernels are compact, but need not be symmetric. Green's function $G(\cdot;\cdot)$ for the Laplacian with homogeneous Dirichlet data is a real-valued, symmetric, compact kernel in $L^2(E)$ (see Theorem 8.1 of Chapter 3). This last example shows that a kernel $K(\cdot;\cdot)$ in $L^2(E)$ need not be in $L^2(E \times E)$.

1.1.1 Kernels in $L^2(\partial E)$

If ∂E is of class $C^{1,\alpha}$ for some $\alpha > 0$, one might consider complex-valued kernels defined and measurable in $\partial E \times \partial E$ and introduce in a similar manner integral operators A and A^* for such kernels, where the integrals are over ∂E for the Lebesgue surface measure on it. Examples of such kernels are the Dirichlet kernel K_D introduced in (5.3) and the Neumann kernel K_N introduced in (6.4) of Chapter 3. The corresponding operators A_D and A_N are introduced in (5.4) and (6.5) of the same chapter. By Corollary 5.1 and Corollary 6.1 they map $L^2(\partial E)$ into $L^2(\partial E)$. The kernels K_D and K_N are real-valued but not symmetric.

2 Integral Equations in $L^2(E)$

A Fredholm integral equation of the second kind in $L^2(E)$ is an expression of the form ([44])

$$u = \lambda \int_E K(\cdot; y)u(y)dy + f \tag{2.1}$$

where λ is a complex parameter, f is a complex-valued function in $L^2(E)$, and $K(\cdot;\cdot)$ is a complex-valued kernel in $L^2(E)$. A solution of (2.1) is a complex-valued function $u \in L^2(E)$ for which (2.1) holds a.e. in E.

To the integral equation (2.1) associate the homogeneous, and the adjoint homogeneous, equations

$$\mathcal{U} = \lambda \int_E K(x; y)\mathcal{U}(y)dy, \qquad \mathcal{V} = \bar{\lambda} \int_E \bar{K}(y; x)\mathcal{V}(y)dy. \tag{2.2}$$

The general solution of (2.1) is the sum of a particular solution and a solution of the associated homogeneous equation.

Lemma 2.1 *The integral equation (2.1) has at most one solution if and only if $\mathcal{U} = 0$ is the only solution of the associated homogeneous equation (2.2).*

Denoting by $\mathbb{I}$ the identity operator, (2.1) can be rewritten in the operator form

$$(\mathbb{I} - \lambda A)u = f. \tag{2.3}$$

2.1 Existence of Solutions for Small $|\lambda|$

Theorem 2.1 *Let λ and $K(\cdot;\cdot)$ satisfy $|\lambda|\|A\| < 1$. Then for every $f \in L^2(E)$, there exists a solution to the integral equation (2.1). The solution is unique if the associated homogeneous equation (2.1) admits only the trivial solution.*

Proof If $|\lambda|$ is small, a first approximation to a possible solution u is $u_o = f$. Then progressively improve the approximation by setting

$$u_n = \lambda A u_{n-1} + f \qquad n = 1, 2, \dots. \tag{2.4}$$

Set $A^o = \mathbb{I}$ and $A^n = AA^{n-1}$ for $n \in \mathbb{N}$, and estimate

$$\begin{aligned} \|A^n\| &= \sup_{\|f\|=1} \|A^n f\| = \sup_{\|f\|=1} \|A(A^{n-1}f)\| \\ &= \sup_{\|f\|=1} \frac{\|A(A^{n-1}f)\|}{\|A^{n-1}f\|} \|A^{n-1}f\| \le \|A\| \|A^{n-1}\|. \end{aligned}$$

By iteration, $\|A^n\| \le \|A\|^n$ for all $n \in \mathbb{N}$. With this symbolism, the approximating solutions u_n take the explicit form

$$u_n = \sum_{i=0}^{n} \lambda^i A^i f. \tag{2.5}$$

From this, for every pair of positive integers $n > m$

$$\|u_n - u_m\| \le \|f\| \sum_{i=m+1}^{n} |\lambda|^i \|A\|^i \to 0 \qquad \text{as } m, n \to \infty.$$

Therefore $\{u_n\}$ is a Cauchy sequence in $L^2(E)$ and we let u denote its limit. Also

$$\|Au_n - Au\| \le \|A\| \|u_n - u\| \to 0 \qquad \text{as } n \to \infty.$$

Therefore $\{Au_n\} \to Au$ in $L^2(E)$. To prove the theorem, we let $n \to \infty$ in (2.4), in the sense of $L^2(E)$. ∎

Motivated by the convergence of the series $\sum |\lambda|^i \|A\|^i$ and by the formal symbolism of (2.5), write

$$u = \sum \lambda^i A^i f \overset{\text{def}}{=} (\mathbb{I} - \lambda A)^{-1} f. \tag{2.6}$$

The operator $(\mathbb{I} - \lambda A)^{-1} : L^2(E) \to L^2(E)$ is called the *resolvent*, and it satisfies

$$(\mathbb{I} - \lambda A)(\mathbb{I} - \lambda A)^{-1} = (\mathbb{I} - \lambda A)^{-1}(\mathbb{I} - \lambda A) = \mathbb{I}.$$

Since $\langle A^n f, g\rangle = \langle f, A^{*n} g\rangle$ for all $f, g \in L^2(E)$, there also hold

$$\langle (\mathbb{I} - \lambda A)^{-1} f, g\rangle = \langle f, (\mathbb{I} - \bar{\lambda} A^*)^{-1} g\rangle. \tag{2.7}$$

3 Separable Kernels

If the kernel $K(\cdot;\cdot)$ is of finite rank, as in (1.1), rewrite (2.1) in the form

$$u - f = \lambda \sum_{i=1}^{n} \varphi_i \int_E (u - f)\bar{\psi}_i \, dy + \lambda \sum_{i=1}^{n} \varphi_i \int_E f \bar{\psi}_i \, dy. \tag{3.1}$$

The associated homogeneous and adjoint homogeneous equations are

$$\mathcal{U} = \lambda \sum_{i=1}^{n} \varphi_i \int_E \mathcal{U} \bar{\psi}_i \, dy, \qquad \mathcal{V} = \bar{\lambda} \sum_{i=1}^{n} \psi_i \int_E \mathcal{V} \bar{\varphi}_i \, dy. \tag{3.2}$$

3.1 Solving the Homogeneous Equations

Solutions of (3.2) are of the form $\mathcal{U} = \sum_{i=1}^{n} w_i \varphi_i$ and $\mathcal{V} = \sum_{i=1}^{n} \tilde{w}_i \psi_i$ where the numbers $w_i = \langle \mathcal{U}, \psi_i \rangle$ and $\tilde{w}_i = \langle \mathcal{V}, \varphi_i \rangle$ are to be determined. Putting this form of $\mathcal{U}$ into the first of (3.2) gives

$$\begin{aligned}\sum_{i=1}^{n} w_i \varphi_i &= \lambda \sum_{i=1}^{n} \varphi_i \int_E \Big(\sum_{j=1}^{n} w_j \varphi_j(y) \Big) \bar{\psi}_i(y) dy \\ &= \lambda \sum_{i=1}^{n} \sum_{j=1}^{n} w_j \varphi_i \int_E \varphi_j \bar{\psi}_i dy = \lambda \sum_{i=1}^{n} \Big(\sum_{j=1}^{n} a_{ij} w_j \Big) \varphi_i\end{aligned}$$

where $a_{ij} = \langle \varphi_j, \psi_i \rangle$. Since the set of functions $\{\varphi_i\}_1^n$ is linearly independent, this leads to the linear system

$$[\mathbb{I} - \lambda(a_{ij})]\mathbf{w} = 0, \qquad \mathbf{w} = (w_1, \dots, w_n)^t. \tag{3.3}$$

Analogously, putting the form of $\mathcal{V}$ into the second of (3.2) and taking into account that the set $\{\psi_i\}_1^n$ is linearly independent leads to the linear system

$$[\mathbb{I} - \bar{\lambda}(\bar{a}_{ji})]\tilde{\mathbf{w}} = 0, \qquad \tilde{\mathbf{w}} = (\tilde{w}_1, \dots, \tilde{w}_n)^t. \tag{3.3*}$$

Let r be the rank of $[\mathbb{I} - \lambda(a_{ij})]$. If $r = n$ then $\det[\mathbb{I} - \lambda(a_{ij})] \neq 0$ and (3.3)–(3.3)* have only the trivial solution. Otherwise, the systems have $n-r$ linearly independent solutions, say $\mathbf{w}_j$ and $\tilde{\mathbf{w}}_j$ for $j = 1, \dots, n-r$, and (3.2) have, respectively, the $n-r$ solutions

$$\mathcal{U}_j = \sum_{i=1}^{n} w_{ij} \varphi_i, \qquad \mathcal{V}_j = \sum_{i=1}^{n} \tilde{w}_{ij} \psi_i, \qquad j = 1, \dots, n-r.$$

3.2 Solving the Inhomogeneous Equation

Solutions to (3.1) are of the form $u - f = \sum_{i=1}^{n} v_i \varphi_i$ where the complex numbers v_i are to be determined from (3.1). Putting this in (3.1), and setting $f_i = \langle f, \psi_i \rangle$, yields the linear system

$$[\mathbb{I} - \lambda(a_{ij})]\mathbf{v} = \lambda \mathbf{f}.$$

If $\det[\mathbb{I} - \lambda(a_{ij})] \neq 0$, then for every $\mathbf{f} \in \mathbb{R}^N$, this system admits a unique solution. Otherwise, the system is solvable if and only if $\mathbf{f}$ is orthogonal to the $(n-r)$-dimensional subspace spanned by the solutions of (3.3)*, that is, if and only if

$$\begin{aligned}0 = \mathbf{f} \cdot \tilde{\mathbf{w}}_j &= \sum_{i=1}^{n} \int_E f \overline{\tilde{w}}_{ij} \bar{\psi}_i \, dy = \int_E f \overline{\sum_{i=1}^{n} \tilde{w}_{ij} \psi_i} \, dy \\ &= \int_E f \bar{\mathcal{V}}_j dy = \langle f, \mathcal{V}_j \rangle, \qquad j = 1, \dots, n-r.\end{aligned}$$

Theorem 3.1 *Let $K(\cdot;\cdot)$ be separable. Then the integral equation (2.1) is solvable if and only if f is orthogonal to all the solutions of the adjoint homogeneous equation (2.2). In particular, if λ is not an eigenvalue of the matrix (a_{ij}), then (2.1) is uniquely solvable for every $f \in L^2(E)$.*

More generally, one might consider separable kernels of the type

$$K_{\mathrm{sep}}(\cdot;\cdot) = \sum_{i=1}^{n} \varphi_i(\cdot;\lambda)\overline{\psi}_i(\cdot;\lambda) \tag{3.4}$$

where

$$\lambda \to a_{ij}(\lambda) = \int_E \varphi_i(y;\lambda)\bar{\psi}_j(y;\lambda)dy \tag{3.5}$$

are analytic functions of λ in the complex plane $\mathbb{C}$. Then $\det[\mathbb{I} - \lambda\big(a_{ij}(\lambda)\big)]$ can vanish only at isolated points of $\mathbb{C}$. Therefore, for such kernels, (2.1) is uniquely solvable for every $f \in L^2(E)$ except for isolated values of λ in $\mathbb{C}$.

Remark 3.1 The theorem, due to Fredholm, discriminates between those values of λ that are eigenvalues of (a_{ij}) and the remaining ones. For this reason it is referred to as the Fredholm alternative ([43, 44], see also Mikhlin [106] and Tricomi [153]).

4 Small Perturbations of Separable Kernels

Consider the integral equation (2.1) for kernels of the form

$$K(\cdot;\cdot) = K_{\mathrm{sep}}(\cdot;\cdot) + K_o(\cdot;\cdot) \tag{4.1}$$

where $K_{\mathrm{sep}}(\cdot;\cdot)$ is separable and $K_o(\cdot;\cdot)$ is a kernel in $L^2(E)$. Setting

$$A_o f = \int_E K_o(\cdot;y)f(y)dy, \qquad A_o^* f = \int_E \bar{K}_o(y;\cdot)f(y)dy$$

the *perturbation* $K_o(\cdot;\cdot)$ is said to be small if

$$|\lambda|\|A_o\| < 1 \quad \text{and} \quad |\lambda|\|A_o^*\| < 1. \tag{4.2}$$

This implies that the resolvent $(\mathbb{I} - \lambda A_o)^{-1}$ is well defined in the sense of (2.6) and permits one to rewrite (2.1) in the form

$$\begin{aligned}(\mathbb{I} - \lambda A_o)u &= \lambda \int_E K_{\mathrm{sep}}(\cdot;y)u\,dy + f \\ &= \lambda \sum_{i=1}^{n} \varphi_i \int_E \bar{\psi}_i(\mathbb{I} - \lambda A_o)^{-1}(\mathbb{I} - \lambda A_o)u\,dy + f.\end{aligned}$$

Set $z = (\mathbb{I} - \lambda A_o)u$ and observe that by (2.7)

$$\int_E \bar{\psi}_i (\mathbb{I} - \lambda A_o)^{-1} (\mathbb{I} - \lambda A_o) u \, dy = \int_E \overline{(\mathbb{I} - \bar{\lambda} A_o^*)^{-1} \psi_i} \, z \, dy.$$

Therefore, solving the integral equation is equivalent to solving

$$z = \lambda \sum_{i=1}^n \varphi_i \int_E \overline{(\mathbb{I} - \bar{\lambda} A_o^*)^{-1} \psi_i} \, z \, dy + f. \tag{4.3}$$

This, in turn, has the associated adjoint homogeneous equation

$$\mathcal{V} = \bar{\lambda} \sum_{i=1}^n (\mathbb{I} - \bar{\lambda} A_o^*)^{-1} \psi_i \int_E \bar{\varphi}_i \mathcal{V} \, dy \tag{4.4}$$

which can be rewritten as

$$(\mathbb{I} - \bar{\lambda} A_o^*) \mathcal{V} = \bar{\lambda} \sum_{i=1}^n \psi_i \int_E \bar{\varphi}_i \mathcal{V} \, dy$$

or equivalently

$$\mathcal{V} = \bar{\lambda} \sum_{i=1}^n \psi_i \int_E \bar{\varphi}_i \mathcal{V} \, dy + \bar{\lambda} \int_E \bar{K}_o(y; \cdot) \mathcal{V} \, dy = \bar{\lambda} \int_E \bar{K}(y; \cdot) \mathcal{V} \, dy.$$

This is precisely the adjoint homogeneous equation in (2.2) associated with the original kernel $K(\cdot; \cdot)$. Thus $\mathcal{V} \in L^2(E)$ is a solution of (4.4) if and only if it is a solution of the adjoint homogeneous equation

$$\mathcal{V} = \bar{\lambda} A^* \mathcal{V}, \qquad Af = \int_E K(\cdot; y) f \, dy \tag{4.5}$$

associated with the original kernel.

4.1 Existence and Uniqueness of Solutions

By Theorem 3.1, the integral equation (4.3) is solvable if and only if f is orthogonal to all the solutions of the adjoint homogeneous equation (4.4), that is, if and only if f is orthogonal to all the solutions of the adjoint homogeneous equation (4.5). The solution of (4.3) is unique if the associated homogeneous equation

$$v = \lambda \sum_{i=1}^n \varphi_i \int_E v \bar{\psi}_i \, dy + \lambda A_o v \tag{4.6}$$

admits only the trivial solution. We may regard this as an integral equation with separable kernel and with *forcing term* $f = \lambda A_o v$. By Theorem 3.1, such an equation has at most one solution if λ is not an eigenvalue of the matrix (a_{ij}). In such a case, since $v = 0$ solves (4.6), it must be the only solution. Recalling that λ is restricted by the condition (4.2), we conclude that if λ is not an eigenvalue of (a_{ij}) satisfying $|\lambda| < \|A_o\|^{-1}$, then (4.6) has only the trivial solution, and consequently (4.3) has at most one solution.

Analogously, if λ is not an eigenvalue of (a_{ij}) satisfying $|\lambda| < \|A_o\|^{-1}$, then (4.5) has only the trivial solution. Therefore any $f \in L^2(E)$ would be orthogonal to all the solutions of the adjoint homogeneous equation (4.5), and therefore (4.3) is uniquely solvable for every $f \in L^2(E)$.

Theorem 4.1 *If the kernel $K(\cdot;\cdot)$ is a small perturbation of a separable kernel, in the sense of (4.1)–(4.2), the integral equation (2.1) is solvable if and only if f is orthogonal to all the solutions of the adjoint homogeneous equation (4.5). In particular, if λ is not an eigenvalue of (a_{ij}) such that $|\lambda| < \|A_o\|^{-1}$, then (2.1) is uniquely solvable for every $f \in L^2(E)$.*

More generally, one might consider kernels that are small perturbations of a separable kernel of the type of (3.4), with the corresponding functions in (3.5) analytic in the disc $D = \{|\lambda| < \|A_o\|^{-1}\} \subset \mathbb{C}$. Then $\det[\mathbb{I} - \lambda(a_{ij}(\lambda))]$ can vanish only at isolated points of D. Therefore, for such kernels, (2.1) is uniquely solvable for every $f \in L^2(E)$ except for isolated values of λ in D.

5 Almost Separable Kernels and Compactness

A kernel $K(\cdot;\cdot)$ in $L^2(E)$ is almost separable if for all $\varepsilon > 0$ it can be decomposed as in (4.1) with $K_{\text{sep}}(\cdot;\cdot)$ separable and a *small* perturbation $K_o(\cdot;\cdot)$ such that

$$\|A_o\| = \sup_{\|f\|=1} \left\| \int_E K_o(\cdot;y) f\, dy \right\| \le \varepsilon. \tag{5.1}$$

The perturbation kernel $K_o(\cdot;\cdot)$ is not required to be compact. It turns out however that the original kernel $K(\cdot;\cdot)$ is compact.

Proposition 5.1 (F. Riesz [125]) *An almost separable kernel $K(\cdot;\cdot)$ in $L^2(E)$ is compact.*

Proof Let A be the operator generated by $K(\cdot;\cdot)$. It will suffice to prove that every bounded sequence $\{f_n\} \subset L^2(E)$ contains a subsequence $\{f_{n'}\} \subset \{f_n\}$ such that $\{Af_{n'}\}$ is Cauchy in $L^2(E)$. For $j \in \mathbb{N}$, let $A_{\text{sep},j}$ and $A_{o,j}$ be the operators corresponding to the decomposition (4.1) and (5.1) for the choice $\varepsilon = 1/j$. Since $A_{\text{sep},j}$ are compact, select by diagonalization, a subsequence $\{f_{n'}\} \subset \{f_n\}$ such that $\{A_{\text{sep},j} f_{n'}\}$ is convergent for all $j \in \mathbb{N}$. Then

$$\|Af_{n'} - Af_{m'}\| < \|A_{\text{sep},j} f_{n'} - A_{\text{sep},j} f_{m'}\| + 2\|A_{o,j}\| \sup_n \|f_n\|. \qquad \blacksquare$$

Corollary 5.1 *Let $A : L^2(E) \to L^2(E)$ be such that for all $\varepsilon > 0$ it can be decomposed into the sum of a compact operator and a "small perturbation" A_ε of norm $\|A_\varepsilon\| < \varepsilon$. Then A is compact.*

5.1 Solving Integral Equations for Almost Separable Kernels

It follows from Theorem 4.1, and the remarks following it, that (2.1) for an almost separable kernel can always be solved for every $f \in L^2(E)$ except at most finitely many values of λ within the disc $[|z| < \varepsilon^{-1}] \subset \mathbb{C}$. Since $\varepsilon > 0$ is arbitrary, we conclude that solvability is ensured for all $f \in L^2(E)$, except for countably many complex numbers $\{\lambda_n\}$. Every disc $|z| < \varepsilon^{-1}$ of the complex plane contains at most finitely many such exceptional values of λ. Therefore if the sequence $\{\lambda_n\}$ is infinite, then $\{|\lambda_n|\} \to \infty$. Equivalently, for all $\varepsilon > 0$, the adjoint homogeneous equation associated to such almost separable kernels has only the trivial solution, except for finitely many values of λ within the disc $[|z| < \varepsilon^{-1}] \subset \mathbb{C}$. These are the *eigenvalues* of the adjoint homogeneous equation. Since $\varepsilon > 0$ is arbitrary, we conclude:

Theorem 5.1 *The integral equation (2.1) with almost separable kernel $K(\cdot;\cdot)$ is solvable if and only if f is orthogonal to all the solutions of the associated adjoint homogeneous equation in (2.2). Moreover, it is always uniquely solvable for every $f \in L^2(E)$ except for countably many values of λ. These are the eigenvalues of the associated adjoint homogeneous equation (2.2). If the sequence $\{\lambda_n\}$ is infinite, then $\{|\lambda_n|\} \to \infty$.*

5.2 Potential Kernels Are Almost Separable

A *potential* kernel is a measurable function $K(\cdot;\cdot) : E \times E \to \mathbb{R}$ such that

$$|K(x;y)| \le C|x-y|^{-N+\alpha} \qquad \text{for almost all } (x,y) \in E \times E \tag{5.2}$$

for some positive constants C and α. For $\delta > 0$ let

$$K_\delta(x;y) = \begin{cases} \delta^{-N+\alpha} & \text{if } K(x,y) \ge \delta^{-N+\alpha} \\ K(x;y) & \text{if } |K(x;y)| < \delta^{-N+\alpha} \\ -\delta^{-N+\alpha} & \text{if } K(x,y) \le -\delta^{-N+\alpha}. \end{cases}$$

Since $K_\delta(\cdot;\cdot)$ is uniformly continuous in $\bar{E}\times\bar{E}$, by the Weierstrass theorem, for each $\varepsilon > 0$ there exists a polynomial $P_n(x;y)$ in the $2N$ variables $(x,y) \in \bar{E}\times\bar{E}$ such that

$$\|K_\delta - P_n\|_{\infty,E\times E} \le \frac{\varepsilon}{2\sqrt{|E|}}.$$

Writing $K = P_n + (K - P_n)$, the perturbation $K_o = K - P_n$ satisfies (5.1). Indeed, for all $f \in L^2(E)$ of norm $\|f\| = 1$

$$\begin{aligned} \int_E |K_o(\cdot;y)f|dy &\le \int_E |K_\delta(\cdot;y) - P_n(\cdot;y)||f|dy + \int_E |K(\cdot;y) - K_\delta(\cdot;y)||f|dy \\ &\le \frac{\varepsilon}{2} + C\int_{|x-y|<C^{\frac{1}{N-\alpha}}\delta} |x-y|^{-N+\alpha}|f|dy. \end{aligned}$$

It remains to choose δ so small that the last integral is less than $\varepsilon/2$. This is possible by virtue of Proposition 10.1 of Chapter 2.

Remark 5.1 Analogous considerations can be carried almost verbatim for integral equations set on ∂E such as (5.2) and (6.3) of Chapter 3. By virtue of Lemma 2.1 of that Chapter, the Dirichlet and Neumann kernels K_D and K_N, introduced in (5.3) and (6.4) of Chapter 3 respectively, are potential, almost separable kernels in $\partial E\times\partial E$. For these integral equations, Theorem 5.1 continues to hold with the proper modifications.

The idea of approximating potential kernels by separable ones, appears in E. Schmidt ([138]) and J. Radon ([123]).

6 Applications to the Neumann Problem

Solving the Neumann problem (1.3) of Chapter 2 is equivalent to solving the integral equation (6.3) of Chapter 3. The latter is in turn solvable if and only if the Neumann datum ψ is orthogonal, in the sense of $L^2(\partial E)$, to all the solutions of the adjoint, homogeneous equation

$$\begin{aligned}\frac{1}{2}\mathcal{V}(x) &= \int_{\partial E} K_N(y;x)\mathcal{V}\,d\sigma = -\frac{1}{\omega_N}\int_{\partial E}\frac{(x-y)\cdot\mathbf{n}(y)}{|x-y|^N}\mathcal{V}\,d\sigma \\ &= -\int_{\partial E} K_D(x;y)\mathcal{V}\,d\sigma = -\int_{\partial E}\frac{\partial F(x;y)}{\partial\mathbf{n}}\mathcal{V}\,d\sigma.\end{aligned} \tag{6.1}$$

By Corollary 3.2 of Chapter 3, this is solved by $\mathcal{V}=$ const. If the constants are the only solutions, then the zero-average condition (1.4) of Chapter 2 on the Neumann datum ψ would imply that such a ψ is orthogonal to all the solutions of the homogeneous, adjoint equation (6.1) and thus would provide the characteristic condition of solvability of the Neumann problem. Therefore the sufficient part of Theorem 6.1 of Chapter 3 will be a consequence of the following

Proposition 6.1 *The integral equation (6.1) admits only the constant solutions.*

Proof Solutions $\mathcal{V}\in L^2(\partial E)$ of (6.1) are continuous in ∂E. Indeed, applying Corollary 10.2 of Chapter 2 a finite number of times, one finds $\mathcal{V}\in L^\infty(\partial E)$. Then $\mathcal{V}\in C(\partial E)$, by Proposition 5.2 of Chapter 3. Let v be the harmonic extension of $\mathcal{V}$ in E and denote by φ the resulting normal derivative of such an extension on ∂E, that is

$$\Delta v = 0 \ \text{ in } \ E,\quad v\big| = \mathcal{V},\quad \text{and}\quad \varphi \stackrel{\text{def}}{=} \frac{\partial v}{\partial\mathbf{n}} \ \text{ on } \partial E.$$

By the Stokes representation formula (2.7) of Chapter 2

$$E\ni x\to v(x) = -\int_{\partial E}\mathcal{V}\frac{\partial F(x;y)}{\partial\mathbf{n}}\,d\sigma + \int_{\partial E}\varphi F(x;y)\,d\sigma.$$

Letting $x \to \partial E$ and using Proposition 3.1 of Chapter 3, gives

$$\mathcal{V} = \frac{1}{2}\mathcal{V} - \int_{\partial E} K_D(\cdot;y)\mathcal{V}\,d\sigma + \int_{\partial E} \varphi F(\cdot;y)\,d\sigma.$$

Therefore if $\mathcal{V}$ is a solution of (6.1), then

$$\int_{\partial E} \varphi F(x;y)\,d\sigma = 0 \qquad \text{for all } x \in \partial E. \tag{6.2}$$

To prove the proposition it suffices to establish that (6.2) implies $\varphi = 0$. Indeed, in such a case, v would be harmonic in E and with zero flux on ∂E; thus $v = \mathcal{V} = \text{const}$. Assume $N \geq 3$ and consider the function

$$H = \int_{\partial E} \varphi F(\cdot;y)d\sigma.$$

It is harmonic in E and it vanishes on ∂E. Therefore it vanishes identically in E. Likewise, it is harmonic in $\mathbb{R}^N - E$, it vanishes on ∂E, and $|H(x)| \to 0$ as $|x| \to \infty$. Therefore $H = 0$ in $\mathbb{R}^N - E$. Fix $x \in \partial E$ and let $\{x^i\}$ and $\{x^e\}$ be sequences of points approaching x respectively from the interior and the exterior of E. Let also $\tilde{W}(\partial E, x_o; \varphi)$ be defined as in (4.1) of Chapter 3. By the previous remarks

$$\tilde{W}(\partial E, x_o; \varphi) = -\nabla H(x_o) \cdot \mathbf{n}(x) = 0 \qquad \text{in } x_o \in \mathbb{R}^N - \partial E.$$

On the other hand by the jump condition of Corollary 4.1 of Chapter 3

$$0 = \lim_{x^i \to x} \tilde{W}(\partial E, x^i; \varphi) - \lim_{x^e \to x} \tilde{W}(\partial E, x^e; \varphi) = \varphi(x).$$

For $N = 2$ see Section 6c of the Problems and Complements. ∎

7 The Eigenvalue Problem

In what follows we will assume that A is generated by a real-valued, compact, symmetric, almost separable kernel $K(\cdot;\cdot)$ in $L^2(E)$, so that $A = A^*$. Consider the problem of finding non-trivial pairs (λ, u), solutions of

$$u = \lambda A u, \qquad \lambda \in \mathbb{C}, \quad u \in L^2(E). \tag{7.1}$$

The numbers λ are called the *eigenvalues* of the operator A, and the functions u are its corresponding *eigenfunctions*.

Proposition 7.1 *Any two distinct eigenfunctions corresponding to two distinct eigenvalues are orthogonal in $L^2(E)$. Moreover, the eigenvalues of A are real and the eigenfunctions of A are real-valued.*

Remark 7.1 If (λ, u) is a solution pair to (7.1), then also $(\lambda, \mu u)$, for all $\mu \in \mathbb{C}$, are solution pairs. Therefore a more precise statement of the proposition would be that the eigenvalues of A are real, and the eigenfunctions can be taken to be real-valued.

Proof (Proposition 7.1) Let (λ_i, u_i), $i = 1, 2$, be distinct solution pairs, say

$$u_1 = \lambda_1 A u_1, \qquad u_2 = \lambda_2 A u_2, \qquad \lambda_1 \neq \lambda_2.$$

Multiply the first by u_2, and integrate over E to obtain

$$\frac{1}{\lambda_1} \int_E u_1 u_2 \, dx = \int_E A u_1 u_2 \, dy = \int_E u_1 A u_2 \, dy = \frac{1}{\lambda_2} \int_E u_1 u_2 \, dy.$$

Therefore $\langle u_1, \bar{u}_2 \rangle = 0$. Let now (λ, u) be a non-trivial solution of (7.1). Then

$$\bar{u} = \overline{\lambda A u} = \bar{\lambda} A \bar{u}.$$

Therefore the pair $(\bar{\lambda}, \bar{u})$ is also a solution of (7.1). If $\lambda \neq \bar{\lambda}$, then $\langle u, u \rangle = \|u\|^2 = 0$. Thus $\lambda = \bar{\lambda}$. Since both u and $\bar{u}$ are eigenfunctions for the same eigenvalue λ, the functions

$$\frac{u + \bar{u}}{2} = \mathrm{Re}(u), \qquad \frac{u - \bar{u}}{2i} = \mathrm{Im}(u)$$

are also eigenfunctions for the same eigenvalue λ. Thus u can be taken to be real. ∎

Proposition 7.2 *The operator A admits at most countably many distinct eigenvalues $\{\lambda_n\}$. If the sequence $\{\lambda_n\}$ is infinite, then $\{|\lambda_n|\} \to \infty$. Moreover, to each eigenvalue λ there correspond finitely many, linearly independent eigenfunctions $\{u_{\lambda,1}, \dots, u_{\lambda,n_\lambda}\}$, for some $n_\lambda \in \mathbb{N}$.*

Proof Regard (7.1) as an integral equation of the type of (2.1) with $f = 0$. According to Theorem 5.1 this is uniquely solvable except for countably many numbers $\{\lambda_n\}$. If $\lambda \neq \lambda_n$, then $u = 0$ is the only solution. Therefore non-trivial pairs (λ, u) occur for at most countably many values of λ. To prove the second statement, let λ be a fixed eigenvalue of (7.1). Since $K(\cdot; \cdot)$ is real-valued and almost separable, it can be written as in (4.1) with $|\lambda| \|A_o\| < 1$, and the integral equation (7.1) can be rewritten as an analogue of (4.3) for $z = (\mathbb{I} - \lambda A_o) u$, that is

$$z = \lambda \sum_{i=1}^{n} \varphi_i \int_E (\mathbb{I} - \lambda A_o^*)^{-1} \psi_i z \, dy. \tag{7.2}$$

Solutions of this are of the form

$$z = \lambda \sum_{i=1}^{n} z_i \varphi_i, \qquad \text{where} \qquad z_i = \int_E (\mathbb{I} - \lambda A_o^*)^{-1} \psi_i z \, dy.$$

Multiply (7.2) by $(\mathbb{I}-\lambda A_o^*)^{-1}\psi_j$ and integrate over E to arrive at the algebraic system

$$z_j = \lambda a_{ij} z_i, \qquad a_{ij} = \int_E \varphi_i(\mathbb{I} - \lambda A_o^*)^{-1}\psi_j dy.$$

This has at most n linearly independent vector solutions $(z_{1,j}, \dots, z_{n,j})$ for $j = 1, \dots, n$. Accordingly, (7.2) has finitely many solutions, and (7.1) has finitely many linearly independent solutions $u = (\mathbb{I} - \lambda A_o)z$ for a given λ. ∎

An eigenvalue λ of A is *simple* if to λ there corresponds only one eigenfunction u up to a multiplicative constant μ. The eigenvalues of (7.1) need not be simple. To an eigenvalue λ there corresponds a maximal set of linearly independent eigenfunctions $\{v_{\lambda,1}, \dots, v_{\lambda,n_\lambda}\}$. Any linear combination of these is an eigenfunction for the same eigenvalue λ. We let

$$\mathcal{E}_\lambda = \{\text{the linear span of the eigenvectors of } \lambda\}.$$

By the Gram–Schmidt orthonormalization procedure we may arrange for $\mathcal{E}_\lambda$ to be spanned by an orthonormal system of eigenvectors $\{u_{\lambda,1}, \dots, u_{\lambda,n_\lambda}\}$.

Corollary 7.1 *The set of eigenfunctions of (7.1) can be chosen to be orthonormal in $L^2(E)$.*

8 Finding a First Eigenvalue and Its Eigenfunctions

Let S_1 be the unit sphere of $L^2(E)$. If (λ, u) is a a non-trivial solution pair of (7.1), by possibly replacing u with $u\|u\|^{-1}$, one may assume that $u \in S_1$. Thus

$$\|A\varphi\|^2 \le \|A\|^2 \text{ for all } \varphi \in S_1 \text{ and } \|Au\|^2 = \lambda^{-2}.$$

This suggests that the eigenvalue λ and the eigenfunction u satisfy

$$\sup_{\varphi\in S_1} \|A\varphi\|^2 = \|Au\|^2 = \lambda^{-2} \tag{8.1}$$

and that they can be found by solving such an extremal problem.[1]

Theorem 8.1 *The eigenvalue problem (7.1) admits a non-trivial solution.*

Proof Select $\varphi_1 \in S_1$. If $\|A\varphi_1\| \ge \|A\varphi\|$ for all $\varphi \in S_1$, then the supremum in (8.1) is achieved at φ_1. If not, there exists $\varphi_2 \in S_1$ such that $\|A\varphi_2\| > \|A\varphi_1\|$. Proceeding in this manner, we generate a *maximizing* sequence $\{\varphi_n\} \subset S_1$. Since A is compact, one may select out of $\{\varphi_n\}$ a subsequence, relabeled with n, such that $\{\varphi_n\} \to u$ weakly in $L^2(E)$ and $\{A\varphi_n\} \to w$, strongly in $L^2(E)$, and in addition

$$\lim_{n\to\infty} \|A\varphi_n\|^2 = \sup_{\varphi\in S_1} \|A\varphi\|^2 \overset{\text{def}}{=} \lambda^{-2}.$$

[1]This idea, due to Hilbert [69], applies to general, linear, symmetric, compact operators in $L^2(E)$; also in F. Reillich [128].

Lemma 8.1 $u = \lambda^2 A^2 u$.

Proof Observe first $A^2\varphi_n \to Aw$ strongly in $L^2(E)$. Indeed

$$\|A^2\varphi_n - Aw\| = \|A[A\varphi_n - w]\| \le \|A\|\|A\varphi_n - w\|.$$

For the assertion, it will suffice to show that

$$\|\varphi_n - \lambda^2 A^2\varphi_n\| \to 0 \quad \text{as } n \to \infty. \tag{*}$$

Indeed, since $\{A^2\varphi_n\}$ converges to Aw strongly in $L^2(E)$, this would imply that $\{\varphi_n\} \to u$ strongly in $L^2(E)$, and as a consequence $w = Au$ and $Aw = A^2u$. To prove (*) write

$$\|\varphi_n - \lambda^2 A^2\varphi_n\|^2 = \|\varphi_n\|^2 - 2\lambda^2\langle A^2\varphi_n, \bar{\varphi}_n\rangle + \lambda^4\|A^2\varphi_n\|^2.$$

Since A is symmetric $\langle A^2\varphi_n, \bar{\varphi}_n\rangle = \|A\varphi_n\|^2$. Moreover $\|A\varphi_n\|^2 \le \lambda^{-2}$, and

$$\|A^2\varphi_n\|^2 = \frac{\|A(A\varphi_n)\|^2}{\|A\varphi_n\|^2}\|A\varphi_n\|^2 \le \sup_{\varphi\in S_1}\|A\varphi\|^4 \le \frac{1}{\lambda^4}.$$

Combining these estimates

$$\|\varphi_n - \lambda^2 A^2\varphi_n\|^2 \le 2 - 2\lambda^2\|A\varphi_n\|^2 \to 0. \qquad \blacksquare$$

To prove the theorem, rewrite the conclusion of Lemma 8.1 as

$$(\mathbb{I} - \lambda^2 A^2)u = (\mathbb{I} + \lambda A)(\mathbb{I} - \lambda A)u = 0.$$

If $(\mathbb{I} - \lambda A)u = 0$, then the pair (λ, u) is a non-trivial solution of the eigenvalue problem (7.1). Otherwise, setting $(\mathbb{I} - \lambda A)u = \psi$, the pair $(-\lambda, \psi)$ solves $(\mathbb{I} - (-\lambda)A)\psi = 0$, and therefore is a non-trivial solution of (7.1). $\blacksquare$

9 The Sequence of Eigenvalues

Let λ_1 be the eigenvalue claimed by Theorem 8.1, denote by $\mathcal{E}_1$ the linear span of all the eigenvectors of λ_1, and let $\mathcal{E}_1^\perp$ be its orthogonal complement. Motivated by the previous maximization procedure, we construct another eigenvalue λ_2 by the formula

$$\sup_{\varphi\in S_1\cap\mathcal{E}_1^\perp}\|A\varphi\|^2 = \lambda_2^{-2}. \tag{9.1}$$

If $A\varphi = 0$ for all $\varphi \in S_1 \cap \mathcal{E}_1^\perp$, then $\lambda_2 = \infty$. Otherwise, we proceed as before and find a non-trivial pair (λ_2, u_2) such that $\lambda_1^2 < \lambda_2^2$ and $u_2 \perp \mathcal{E}_1$. Having constructed the first n eigenvalues $\{\lambda_1, \ldots, \lambda_n\}$ and the corresponding eigenspaces $\{\mathcal{E}_1, \ldots, \mathcal{E}_n\}$, construct λ_{n+1} by the maximization problem

$$\sup_{\varphi \in S_1 \cap [\mathcal{E}_1 \cup \cdots \cup \mathcal{E}_n]^\perp} \|A\varphi\| = \lambda_{n+1}^{-2}. \tag{9.2}$$

If for some $n \in \mathbb{N}$, $A\varphi = 0$ for all $\varphi \in S_1 \cap [\mathcal{E}_1 \cup \cdots \cup \mathcal{E}_n]^\perp$, then $\lambda_{n+1} = \infty$ and the process terminates. Otherwise, proceeding inductively, we construct a sequence $\{\lambda_n\}$ of eigenvalues such that $\lambda_n^2 < \lambda_{n+1}^2$, and a sequence $\{u_m\}$ of eigenfunctions that can be chosen to form an orthonormal sequence in $L^2(E)$. Such a sequence in general need not be complete. Necessary and sufficient conditions on the kernel $K(\cdot;\cdot)$ to ensure completeness will be given in the next section.

9.1 An Alternative Construction Procedure of the Sequence of Eigenvalues

Having determined λ_1, let $\{u_{\lambda_1,1}, \dots, u_{\lambda_1,n_1}\}$ be a set of real-valued linearly independent orthonormal eigenfunctions spanning the eigenspace $\mathcal{E}_1$, and set

$$K_1(x;y) = \frac{1}{\lambda_1} \sum_{i=1}^{n_1} u_{\lambda_1,i}(x) u_{\lambda_1,i}(y).$$

The kernel $K_1(\cdot;\cdot)$ is symmetric, and the corresponding operator

$$L^2(E) \ni f \to A_1 f = \int_E K_1(\cdot;y) f(y) dy = \frac{1}{\lambda_1} \sum_{i=1}^{n_1} u_{\lambda_1,i} \langle u_{\lambda_1,i}, \bar{f} \rangle$$

is compact and symmetric. Then compute an eigenvalue, λ_2, for the problem

$$u = \lambda(A - A_1)u \tag{9.3}$$

by the previous procedure, that is

$$\sup_{\varphi \in S_1} \|(A - A_1)\varphi\|^2 = \lambda_2^{-2}.$$

Unlike the maximization in (9.1) and (9.2), here the supremum is taken over the entire unit sphere S_1 of $L^2(E)$. However, the operator $(A - A_1)$ is, roughly speaking, *inactive* on the eigenspace $\mathcal{E}_1$. In particular, if $A = A_1$, then $\lambda_2 = \infty$, and the process terminates.

Lemma 9.1 *A pair (λ, u) different from $(\lambda_1, u_{\lambda_1,i})$ for all $i = 1, \dots, n_1$, is a solution of the eigenvalue problem (7.1) if and only if it is a solution of the eigenvalue problem (9.3).*

Proof If (λ, u) solves (7.1) and $\lambda \neq \lambda_1$, then $u \perp \mathcal{E}_1$, and therefore it is a solution of (9.3). Now let (λ, u) solve (9.3). Multiply both sides by $u_{\lambda_1,i}$ and integrate over E. Since $A - A_1$ is symmetric

$$\begin{aligned} \langle u, u_{\lambda_1,i} \rangle &= \lambda \langle (A - A_1)u, u_{\lambda_1,i} \rangle = \lambda \langle u, (A - A_1) u_{\lambda_1,i} \rangle \\ &= \lambda \left(\langle u, A u_{\lambda_1,i} \rangle - \frac{\langle u, u_{\lambda_1,i} \rangle}{\lambda_1} \right) = 0 \end{aligned}$$

since $u_{\lambda_1,i}$ is an eigenfunction for A corresponding to λ_1. This implies that

$$u = \lambda(A - A_1)u = \lambda Au - \lambda \sum_{i=1}^{n_1} u_{\lambda_1,i}\langle u_{\lambda_1,i}, u\rangle = \lambda Au. \qquad \blacksquare$$

Proceeding inductively, construct λ_{n+1} from

$$\sup_{\varphi \in S_1} \left\|(A - \sum_{j=1}^{n} A_j)\varphi\right\|^2 = \lambda_{n+1}^{-2} \tag{9.4}$$

where A_j is the compact symmetric operator in $L^2(E)$ corresponding to the kernel

$$K_j(x;y) = \frac{1}{\lambda_j} \sum_{i=1}^{n_j} u_{\lambda_j,i}(x) u_{\lambda_j,i}(y)$$

and $\{u_{\lambda_j,1}, \dots, u_{\lambda_j,n_j}\}$ are real-valued linearly independent orthonormal eigenfunctions corresponding to the eigenvalue λ_j. If, for some $n \in \mathbb{N}$

$$K(x;y) = \sum_{j=1}^{n} K_j(x;y) = \sum_{j=1}^{n} \frac{1}{\lambda_j} \sum_{i=1}^{n_j} u_{\lambda_j,i}(x) u_{\lambda_j,i}(y)$$

set $\lambda_{n+1} = \infty$, and the process terminates. If not, recall that $\{\lambda_n^2\} \to \infty$ and deduce from (9.4) that

$$\limsup_{n\to\infty} \left\|(A - \sum_{j=1}^{n} A_j)f\right\| = 0 \quad \text{for all } f \in L^2(E).$$

Therefore, letting $n \to \infty$ in (9.4) gives a *formal* expansion of the kernel $K(\cdot;\cdot)$, in terms of its eigenvalues and eigenfunctions. Since the right-hand side of (9.4) tends to zero as $n \to \infty$, such an expansion can be rigorously interpreted in the sense of

$$\langle K(x;\cdot), f\rangle = \sum_{j=1}^{\infty} \frac{1}{\lambda_j} \sum_{i=1}^{n_j} u_{\lambda_j,i}(x) \langle u_{\lambda_j,i}, f\rangle \tag{9.5}$$

for all $f \in L^2(E)$. Equivalently

$$\sum_{j=1}^{n} A_j f \to Af \quad \text{in} \quad L^2(E). \tag{9.6}$$

10 Questions of Completeness and the Hilbert–Schmidt Theorem

Let $\{\lambda_n\}$ and $\{u_{\lambda_j,i}\}$ be the sequences of eigenvalues and corresponding eigenfunctions of A. If $\{\lambda_n\}$ is finite, say $\{\lambda_1, \dots, \lambda_m\}$ for some $m \in \mathbb{N}$, set $\lambda_j = \infty$ for $j > m$ and $u_{\lambda_j,i} = 0$ for $i = 1, \dots, n_j$ for all $j > m$. Reorder $\{u_{\lambda_j,i}\}$ into a sequence $\{v_n\}$ of real-valued linearly independent orthonormal eigenfunctions, and rewrite (9.5) in the form

$$\langle K(x;\cdot), f\rangle = \sum_{i=1}^{\infty} \frac{1}{\lambda_i} v_i(x)\langle v_i, f\rangle \tag{10.1}$$

where λ_i remains the same as $\{v_i\}$, for $i = 1, \dots, n_{\lambda_i}$ spans the eigenspace $\mathcal{E}_j$. The system $\{v_n\}$ is *complete* if it spans the whole of $L^2(E)$. Equivalently if $[\mathrm{span}\{v_n\}]^{\perp} = \{0\}$, that is if $\langle f, v_i\rangle = 0$ for all i implies $f = 0$. If $\{v_n\}$ is complete then every $f \in L^2(E)$ can be represented as

$$f = \sum_{i=1}^{\infty} \langle f, v_i\rangle v_i \quad \text{in the sense } \Big\| f - \sum_{i=1}^{n} \langle f, v_i\rangle v_i \Big\| \to 0 \text{ as } n \to \infty.$$

The series on the right-hand side is the Fourier series of f.

Proposition 10.1 $Af = 0 \iff \langle f, v_i\rangle = 0$ *for all* $i \in \mathbb{N}$.

Proof For fixed (λ_i, v_i), $\langle f, v_i\rangle = \lambda_i\langle Af, v_i\rangle$. This proves the implication $\Longrightarrow$. The converse statement follows from (10.1). ∎

Corollary 10.1 *The orthonormal system* $\{v_n\}$ *is complete in* $L^2(E)$ *if and only if* $f \neq 0$ *implies* $Af \neq 0$.

Remark 10.1 If the kernel $K(\cdot;\cdot)$ is of finite rank, then $\{v_n\}$ cannot be complete in $L^2(E)$.

The corollary gives a simple criterion to check the completeness of $\{v_n\}$. We will apply it to the case when $K(\cdot;\cdot)$ is Green's function $G(\cdot;\cdot)$ for Laplacian with homogeneous Dirichlet data on $\partial\Omega$.

Proposition 10.2 $L^2(E) \ni f \neq 0 \Longrightarrow \int_E G(\cdot; y) f dy \neq 0$.

Proof Let $\varphi \in C_o^{\infty}(E)$ and recall the representation of Corollary 3.1 of Chapter 2. For $f \in L^2(E)$

$$\begin{aligned}
\langle Af, \Delta\varphi\rangle &= \int_E \int_E G(x;y) f(y) \Delta\bar{\varphi}(x)\, dy dx \\
&= \int_E f(y) \int_E G(x;y)\Delta\bar{\varphi}(x)\, dx dy \\
&= -\int_E f\bar{\varphi} dy = -\langle f, \varphi\rangle.
\end{aligned}$$

∎

10.1 The Case of $K(x;\cdot) \in L^2(E)$ Uniformly in x

Assume that $K(\cdot;\cdot)$ is a real-valued compact symmetric kernel acting on $L^2(E)$ that generates an orthonormal system $\{v_n\}$ complete in $L^2(E)$. It is natural to ask whether, or under what conditions, the Fourier series of a function $f \in$

$L^2(E)$ converges to f in some stronger topology, for example in the topology of the uniform convergence in $\bar{E}$. This requires more stringent assumptions on f and on the kernel $K(\cdot;\cdot)$.

Assume that $\|K(x;\cdot)\| \le C$ for some positive constant C uniformly in x. In such a case, in (10.1), we may take $f(x;y) = K(x;y)$ to obtain

$$\sum_{i=1}^{\infty} \frac{1}{\lambda_i^2} v_i^2(x) \le C^2 \quad \text{for all } x \in E. \tag{10.2}$$

Theorem 10.1 (Hilbert–Schmidt) *Let $K(\cdot;\cdot)$ be a real-valued compact symmetric kernel in $L^2(E)$, which generates an orthonormal system $\{v_n\}$ complete in $L^2(E)$ and such that $\|K(x;\cdot)\| \le C$ for some $C > 0$, uniformly in x. Then every function $f \in L^2(E)$ that can be represented as*

$$f = \int_E K(\cdot;y) g\, dy \quad \text{for some } g \in L^2(E) \tag{10.3}$$

has a Fourier series $\sum_{i=1}^{\infty} \langle f, v_i\rangle v_i$, absolutely and uniformly convergent to f in E, that is, $\left\| f - \sum_{i=1}^{n} \langle f, v_i\rangle v_i \right\|_\infty \to 0$ as $n \to \infty$.

Proof It suffices to show that $\left\| \sum_{i=m}^{\infty} |\langle f, v_i\rangle| |v_i| \right\|_\infty \to 0$ as $m \to \infty$. This is a consequence of (10.2) and the representation (10.3). Indeed, for all $x \in E$

$$\begin{aligned} \left[\sum_{i=m}^{\infty} |\langle f, v_i\rangle| |v_i(x)| \right]^2 &= \left[\sum_{i=m}^{\infty} |\langle g, v_i\rangle| \frac{|v_i(x)|}{|\lambda_i|} \right]^2 \\ &\le \left[\sum_{i=m}^{\infty} |\langle g, v_i\rangle|^2 \right] \left[\sum_{i=m}^{\infty} \frac{1}{\lambda_i^2} v_i^2(x) \right]. \end{aligned}$$ ■

Remark 10.2 The Green's function $G(\cdot;\cdot)$ satisfies the assumptions of the Hilbert–Schmidt theorem only for $N = 2, 3$.

Corollary 10.2 *Let $N = 2, 3$. Then a function $f \in C_o^1(E) \cap C^2(\bar{E})$ has a Fourier series that converges absolutely and uniformly to f in E.*

Proof It follows from the Hilbert–Schmidt theorem and the representation of Corollary 3.1 of Chapter 2. ■

11 The Eigenvalue Problem for the Laplacean

Eingenvalues and eigenfunctions for the Laplacian with homogeneous Dirichlet data are those related to (8.1)–(8.2) of the previous chapter, which were shown to be equivalent.

Theorem 11.1 *The eigenvalues of the Laplacian on a bounded open set $E \subset \mathbb{R}^N$, with homogeneous Dirichlet data on ∂E, are positive and form a monotone increasing sequence $\{\lambda_n\} \to \infty$ as $n \to \infty$. Moreover, the corresponding orthonormal system of eigenfunctions $\{v_n\}$ is complete in $L^2(E)$. Finally, the first eigenfunction can be taken to be positive, and λ_1 is simple.*

Proof If the pair (λ, u) solves (8.1) of Chapter 3, is non-trivial, and $\lambda < 0$, then u cannot take a positive maximum in E. Indeed, if a positive maximum were taken at some $x_o \in E$

$$-\Delta u(x_o) = \lambda u(x_o) < 0.$$

By the same argument, u cannot attain a negative minimum in E. Therefore $u = 0$. The statement about completeness follows from Proposition 10.2.

The maximization process (8.1) implies that if the supremum is achieved for some u_1 then it is achieved also for $|u_1|$. Indeed, since $G(\cdot;\cdot) \ge 0$

$$\lambda_1^{-2} = \|Au_1\|^2 = \left\| \int_E G(\cdot;y) u_1 dy \right\|^2 \le \left\| \int_E G(\cdot;y)|u_1| dy \right\|^2 = \lambda_1^{-2}.$$

Thus u_1 and $|u_1|$ are both eigenfunctions for the same eigenvalue λ_1. In particular

$$-\Delta |u_1| = \lambda_1 |u_1|.$$

This in turn implies that the function

$$E \times \mathbb{R} \ni (x,t) \to w(x,t) = |u_1(x)| e^{\sqrt{\lambda_1} t}$$

is a non-negative harmonic function in the $(N+1)$-dimensional strip $E \times \mathbb{R}$. By the Harnack estimate of Corollary 5.1 of Chapter 2, $|u_1| > 0$ in E, and therefore $u_1 = |u_1|$. We conclude that all the eigenfunctions corresponding to λ_1 can be taken to be positive. In particular, no two of them can be orthogonal. Thus λ_1 is simple. ∎

11.1 An Expansion of Green's Function

Formula (9.5) provides an expansion of the Green's function $G(\cdot;\cdot)$ in terms of its eigenvalues and eigenfunctions. Namely, for all $f \in L^2(E)$ and for a.e. $x \in E$

$$\langle G(x;\cdot), f \rangle = \frac{1}{\lambda_1} u_{\lambda_1}(x) \langle u_{\lambda_1}, f \rangle + \sum_{j=2}^{\infty} \frac{1}{\lambda_j} \sum_{i=1}^{n_j} u_{\lambda_j,i}(x) \langle u_{\lambda_j,i}, f \rangle.$$

Problems and Complements

2c Integral Equations

2.1c Integral Equations of the First Kind

Equation (2.1) was also called an integral equation of the *second* kind. An integral equation of the *first* kind is of the form

$$\int_E K(\cdot; y)u\,dy = f. \tag{2.1c}$$

Here $f \in L^2(E)$ is given, $K(\cdot;\cdot)$ is a kernel in $L^2(E)$, and u is the unknown function. Below we give an example of an integral equation of the first kind.

2.2c Abel Equations ([2, 3])

A particle constrained on a vertical plane falls from rest under the action of gravity along a trajectory γ. On the vertical plane introduce a Cartesian system originating at ground level, and with $\mathbf{j}$ directed along the ascending vertical. If the particle is initially at level x from the ground, we seek the trajectory γ such that it will hit the ground after a time $t = \bar{f}(x)$, where f is a given function. Parametrize γ by the angle θ that the tangent line at points of γ forms with the horizontal axis, taken counterclockwise starting from the positive direction of the horizontal axis.

The speed of the falling particle at level $y \in [0, x]$ is $\sqrt{2g(x-y)}$, where g is the acceleration of gravity. The velocity along $\mathbf{j}$ is

$$\frac{dy}{dt} = -\sqrt{2g(x-y)}\sin\theta$$

or, separating the variables

$$\frac{dy}{\sqrt{2g(x-y)}\sin\theta} = -dt.$$

Integrate the left-hand side from the initial level x to the final level 0, and the right-hand side from the initial time 0 to the final time $\bar{f}(x)$. This gives the Abel integral equation of the first kind

$$\int_0^x \frac{v(y)dy}{\sqrt{x-y}} = f(x) \quad \text{where } v(y) = \frac{1}{\sin\theta} \quad \text{and } f = -\sqrt{2g}\bar{f}. \tag{2.2c}$$

When $f = \text{const}$, this is the problem of the *tautochrone* trajectory. More generally, an Abel integral equation takes the form

$$\int_0^x \frac{v(y)dy}{(x-y)^\alpha} = f(x) \tag{2.3c}$$

for some $\alpha > 0$ and $f \in C^1[0,\infty)$. This can be recast in the form (2.1c) as follows. First limit x not to exceed some fixed positive number a. Then set

$$K(x;y) = \begin{cases} (x-y)^{-\alpha} & \text{if } 0 \le y < x \\ 0 & \text{if } x \le y \le a. \end{cases}$$

and rewrite (2.3c) as

$$\int_0^a K(\cdot;y)v dy = f.$$

Kernels of this kind are said to be of Volterra type ([158, 159]).

2.3c Solving Abel Integral Equations

In (2.3c) replace x by a running variable η, multiply both sides by $(x-\eta)^{\alpha-1}$ and integrate in $d\eta$ over $(0,x)$. Interchanging the order of integration gives

$$\int_0^x v(y)\left[\int_y^x \frac{d\eta}{(x-\eta)^{1-\alpha}(\eta-y)^\alpha}\right]dy = \int_0^x \frac{f(\eta)}{(x-\eta)^{1-\alpha}}d\eta.$$

Compute the integral in braces by the change of variables

$$\eta = y + \frac{1}{s+1}(x-y), \qquad s \in [0,\infty).$$

This gives

$$\int_y^x \frac{d\eta}{(x-\eta)^{1-\alpha}(\eta-y)^\alpha}d\eta = \int_0^\infty \frac{ds}{s^{1-\alpha}(1+s)} = \frac{\pi}{\sin\alpha\pi}$$

where the last integral has been computed by the method of residues ([18] page 107). Combining these calculations

$$\begin{aligned} \frac{\pi}{\sin\alpha\pi}\int_0^x v(y)dy &= \int_0^x \frac{f(y)}{(x-y)^{1-\alpha}}dy \\ &= \frac{x^\alpha}{\alpha}f(0) + \frac{1}{\alpha}\int_0^x (x-y)^\alpha f'(y)dy. \end{aligned}$$

Taking the derivative gives an explicit representation, of the solution of the Abel integral equation (2.3c), in the form

$$v(x) = \frac{\pi}{\sin\alpha\pi}\left[\frac{f(0)}{x^{1-\alpha}} + \int_0^x \frac{f'(y)}{(x-y)^{1-\alpha}}dy\right]. \tag{2.4c}$$

2.4c The Cycloid ([3])

To find the parametric equations of the tautochrone trajectory, in (2.4c) take $\alpha = \frac{1}{2}$ and $f = C$. Using (2.2c)

$$v(x) = \frac{C}{\pi\sqrt{x}}, \qquad \sin\theta = \frac{\pi\sqrt{x}}{C}.$$

Denoting by $x = x(\theta)$ the vertical component of the parametrization of γ

$$x(\theta) = \frac{C^2}{\pi^2}\sin^2\theta = \frac{C^2}{2\pi^2}(1 - \cos 2\theta). \tag{2.5c}$$

Let $y = y(\theta)$ denote the horizontal component of the parametrization of γ, and let γ have the local representation $y = y(x)$. Then

$$dy = \frac{dx}{\tan\theta} = \frac{C^2}{\pi^2}\frac{2\sin\theta\cos\theta}{\tan\theta}d\theta = \frac{C^2}{\pi^2}(1 + \cos 2\theta)d\theta$$

and by integration

$$y(\theta) = \frac{C^2}{\pi^2}\left(\theta + \frac{1}{2}\cos 2\theta\right) + c_o \tag{2.6c}$$

for a constant c_o. The equations (2.5c) and (2.6c) are the parametric equations of a cycloid.

2.5c Volterra Integral Equations ([158, 159])

Let f be bounded and continuous in $\mathbb{R}^+$, and consider the Volterra equation

$$u(x) = \lambda\int_0^x K(x;y)u(y)dy + f(x)$$

where $K(\cdot;\cdot)$ is bounded and continuous in $\mathbb{R}^+\times\mathbb{R}^+$. Assuming that $K(x;y) = 0$ for $y > x$, rewrite this as

$$u = \lambda\int_0^\infty K(\cdot;y)u(y)dy + f.$$

Prove that for all $x \in \mathbb{R}^+$

$$|(A^n f)| \le \sup_{\mathbb{R}^+}\|f\|\frac{K^n x^n}{n!}, \qquad \text{where } (Af)(x) = \int_0^x K(x;y)f(y)dy$$

and where K is an upper bound for $|K(\cdot;\cdot)|$. Conclude that a solution must be continuous and locally bounded in $\mathbb{R}^+$.

2.1. Say in what sense the Dirichlet and Neumann problems for the Laplacian in a bounded domain are mutually adjoint. *Hint:* See Sections 5–6 of Chapter 3 and the arguments of Section 6 of this chapter.

2.2. Find A^2 if $K(x;y) = e^{|x-y|}$ and $E = (0,1)$.
2.3. Find A^2 and A^3 if $K(x;y) = x - y$ and $E = (0,1)$.

One might ask whether these integral equations set in $\mathbb{R}^+$, have a solution if $K(\cdot;\cdot)$ does not vanish for $y > x$. It turns out that some decay has to imposed on $K(\cdot;\cdot)$. For kernels of the type $K(x;y) = K(x-y)$ and $K(s) \to 0$ exponentially fast as $s \to \infty$ a theory is developed by N. Wiener and E. Hopf ([163]). See also G. Talenti [147].

3c Separable Kernels

3.1c Hammerstein Integral Equations ([64])

Consider the non-linear integral equation of Hammerstein type

$$u = \int_E K(x;y) f(y, u(y)) dy.$$

If the kernel is separable, set

$$\gamma_i = \int_E \psi_i f(y, u(y)) dy \qquad \big(K(\cdot;\cdot) = \sum_{i=1}^n \varphi_i \psi_i\big)$$

where the numbers γ_i are to be determined from

$$u = \sum_{i=1}^n \gamma_i \varphi_i = \sum_{i=1}^n \varphi_i \int_E \psi_i f\big(y, \sum_{i=1}^n \gamma_i \varphi_i(y)\big) dy.$$

Therefore, the numbers γ_i are the possible solutions (real or complex) of the system

$$\gamma_i = \int_E \psi_i f\big(y, \sum_{i=1}^n \gamma_i \varphi_i(y)\big) dy.$$

3.1. Solve the Hammerstein equations

$$u(x) = \lambda \int_0^1 x y u^2(y) dy, \qquad u(x) = \lambda \int_{-1}^1 \frac{|x-y|}{1+u^2(y)} dy.$$

3.2. Let $\varphi \in L^2[0,1]$ be non-negative. Show that if $\|\varphi\|_{2,[0,1]} > 1$, there are no real-valued solutions of the Hammerstein equation

$$u(x) = \frac{1}{2} \int_0^1 \varphi(x)\varphi(y)\big(1 + u^2(y)\big) dy.$$

6c Applications to the Neumann Problem

Prove Proposition 6.1 for $N = 2$ by the following steps.

Step 1: Consider the double-layer potential

$$\mathbb{R}^2 - \partial E \ni x \to W(\partial E, x; \varphi) = \frac{1}{2\pi} \int_{\partial E} \varphi(y) \frac{\partial \ln |x - y|}{\partial \mathbf{n}(y)} d\sigma.$$

By the first part of the proof of Proposition 6.1, such a function is identically zero in E. Prove that it vanishes on ∂E. Thus $W(\partial E, \cdot; \varphi)$ is harmonic in $\mathbb{R}^2 - E$, vanishes as $|x| \to \infty$ and has zero normal derivative on ∂E.

Step 2: Prove that $\forall N \geq 2$, there exists at most one solution to the problem

$$u \in C^2(\mathbb{R}^N - \bar{E}) \cap C^1(\mathbb{R}^N - E), \quad \Delta u = 0 \text{ in } \mathbb{R}^N - \bar{E}$$
$$\frac{\partial u}{\partial \mathbf{n}} = 0 \text{ on } \partial E, \quad \text{and} \quad \lim_{|x| \to \infty} u(x) = 0.$$

Prove that positive maxima or negative minima cannot occur on ∂E.

9c The Sequence of Eigenvalues

9.1. Let A be generated by the Green's function for the Laplacian with homogeneous Dirichlet data. Prove that the maximization process (8.1) is *formally* equivalent to

$$\min_{\varphi \in \mathcal{C}_o \cap S_1} \|\nabla u\|^2 = \lambda^2 \quad \text{where } \mathcal{C}_o = \{u \in C_o(E) \big| |\nabla u| \in L^2(E)\}.$$

9.2. Let $f \in L^2(E)$. Prove that the minimum

$$\min_{\{f_1, \dots, f_n\} \in \mathbb{C}^n\}} \|f - \sum_{i=1}^n f_i u_i\|$$

is achieved for $f_i = \langle f, u_i \rangle$. *Hint:* Compute the derivatives

$$\frac{\partial}{\partial f_i} \|f - \sum_{i=1}^n f_i u_i\|^2.$$

9.3. Prove Bessel's inequality $\sum_{i=1}^n f_i^2 \leq \|f\|^2$.

9.4. Prove Parseval's identity $\sum_{i=1}^\infty f_i^2 = \|f\|^2$.

10c Questions of Completeness

10.1. If $K(x; \cdot) \in L^2(E)$ uniformly in $x \in E$, then (10.2) gives another proof that to each eigenvalue λ_i there correspond only finitely many linearly independent eigenfunctions. *Hint:* If n_i is the number of linearly independent eigenfunctions corresponding to the eigenvalue λ_i, then $n_i \leq C|\lambda_i|E|$.

10.2. Prove that $\left\{\sqrt{\frac{2}{L}}\sin\frac{n\pi}{L}x\right\}$ is a complete orthonormal system in $L^2(0,L)$. *Hint:* Compute the eigenvalues and eigenfunctions of the Laplacian in one dimension, over $(0,L)$ with homogeneous Dirichlet data on $x=0$ and $x=L$.

10.3. Let m be an even positive integer, and let $C^m_{\text{odd}}(0,L)$ denote the space of functions in $C^m(0,L)$ whose even order derivatives vanish at $x=0$ and $x=L$, i.e.,

$$\frac{\partial^j}{\partial x^j}\varphi(0^+)=\frac{\partial^j}{\partial x^j}\varphi(L^-)=0 \quad \text{for all even integers} \quad 0\le j\le m.$$

Denoting by $\{v_n\}$ the complete orthonormal system in $L^2(0,L)$, of the previous problem prove that if $\varphi\in C^m_{\text{odd}}(0,L)$

$$|\langle\varphi,v_n\rangle|\le\frac{2L^j}{(n\pi)^j}\Big\|\frac{\partial^j}{\partial x^j}\varphi\Big\|_{\infty,[0,L]} \quad \text{for all } 0\le j\le m.$$

As a consequence

$$\Big\|\varphi-\sum_{n=1}^{j-1}\langle\varphi,v_n\rangle v_n\Big\|_{\infty,[0,L]}\le\frac{\text{const}}{j^m}\Big\|\frac{\partial^j}{\partial x^j}\varphi\Big\|_{\infty,[0,L]}.$$

10.1c Periodic Functions in $\mathbb{R}^N$

A function $f:\mathbb{R}^N\to\mathbb{R}$ is periodic of period 1 if $f(x+n)=f(x)$ for all $x\in\mathbb{R}^N$ and every N-tuple of integers $n\in\mathbb{Z}^N$.

Let $\mathbf{Q}=(0,1)^N$ denote the unit cube in $\mathbb{R}^N$. Every $f\in L^2(\mathbf{Q})$ can be regarded as the restriction to $\mathbf{Q}$ of a periodic function in $\mathbb{R}^N$ of period 1. If f is periodic of period 1, there exists a constant γ such that $f+\gamma$ is periodic of period 1 and has zero average over $\mathbf{Q}$. Consider the space

$$L^2_p(\mathbf{Q})=\left\{f\in L^2(\mathbf{Q})\Big|\int_{\mathbf{Q}}f\,dx=0\right\}$$

where the subscript p denotes "periodic function". An orthogonal basis for $L^2_p(\mathbf{Q})$ is found by solving the eigenvalue problem

$$-\Delta u=\lambda u \text{ in } \mathbf{Q}, \qquad \text{and} \qquad \frac{\partial u}{\partial\mathbf{n}}=0 \text{ on } \partial\mathbf{Q}.$$

Verify that

$$\prod_{j=1}^N\cos n_j\pi x_j, \qquad \prod_{j=1}^N\sin n_j\pi x_j; \qquad n=(n_1,\dots,n_N)\in\mathbb{Z}^N,\ |n|^2=\sum_{j=1}^N n_j^2$$

are eigenfunctions for the eigenvalues $\lambda=(\pi|n|)^2$. Any complex linear combination of these is still an eigenfunction. Prove that the system $\{e^{i\pi\langle n,x\rangle}\}$ for $n\in\mathbb{Z}^N$ is a complete orthogonal basis for $L^2_p(\mathbf{Q})$.

10.2c The Poisson Equation with Periodic Boundary Conditions

Consider the Neumann problem

$$u \in C^2(\mathbf{Q}) \cap C^1(\bar{\mathbf{Q}}), \quad \Delta u = f \in C^1(\bar{\mathbf{Q}}), \quad \frac{\partial u}{\partial \mathbf{n}} = 0 \text{ on } \partial \mathbf{Q}.$$

The necessary and sufficient condition for solvability is that f has zero average over $\mathbf{Q}$, which we assume. Write

$$f = \sum_{n \in \mathbb{Z}^N} \hat{f}_n e^{i\pi \langle n,x \rangle}, \qquad \text{where} \quad \hat{f}_n = \langle f, e^{i\pi \langle n,x \rangle} \rangle$$

and seek a solution of the type

$$u = \sum_{n \in \mathbb{Z}^N} \hat{u}_n e^{i\pi \langle n,x \rangle}, \qquad \hat{u}_n = \langle u, e^{i\pi \langle n,x \rangle} \rangle.$$

Prove that

$$\hat{u}_n = \frac{\hat{f}_n}{-\pi |n|^2} \quad \text{for all} \quad n \in \mathbb{Z}^N - \{0\}.$$

11c The Eigenvalue Problem for the Laplacian

11.1. A linear operator $A : L^2(E) \to L^2(E)$ is *positive* if

$$\langle Af - Ag, f - g \rangle \geq 0 \quad \text{for all } f, g \in L^2(E).$$

The operator A generated by the Green's function for the Laplacian with homogeneous Dirichlet data on ∂E, is positive in the sense that $\langle Af, f \rangle > 0$ for all $f \in L^2(E)$, $f \neq 0$. Assume first that $f \in C_o^\eta(E)$ for some $\eta \in (0,1)$. Then the function Af is the unique solution of the problem

$$u \in C^2(E) \cap C(\bar{E}), \quad -\Delta u = f, \quad \text{in } E, \quad u\big|_{\partial E} = 0.$$

Therefore

$$\langle Af, f \rangle = \langle u, -\Delta u \rangle = \|\nabla u\|^2.$$

Prove the positivity of A for general $f \in L^2(E)$.

11.2. Prove that if A is a symmetric, positive, compact operator in $L^2(E)$, then its eigenvalues are positive.

5

The Heat Equation

1 Preliminaries

Consider a material homogeneous body occupying a region $E \subset \mathbb{R}^N$ with boundary ∂E of class C^1 and outward unit normal $\mathbf{n}$. Identify the body with E and denote by $k > 0$ its dimensionless conductivity. The temperature distribution $(x,t) \to u(x,t)$ satisfies the second-order parabolic equation

$$u_t = k\Delta u \quad \text{for } x \in E \text{ and } t \in (t_1, t_2) \tag{1.1}$$

where $(t_1, t_2) \subset \mathbb{R}$ is some time interval of observation. By changing the time scale, we may assume that $k = 1$. Set formally

$$H(\cdot) = \frac{\partial}{\partial t} - \Delta, \qquad H^*(\cdot) = \frac{\partial}{\partial t} + \Delta.$$

The formal operators $H(\cdot)$ and $H^*(\cdot)$ are called the heat operator and the adjoint heat operator respectively. If $0 < T < \infty$ denote by E_T the cylindrical domain $E \times (0, T]$, and if $E = \mathbb{R}^N$, let S_T denote the *strip* $\mathbb{R}^N \times (0, T]$. The heat operator and its adjoint are well defined for functions in the class

$$\mathcal{H}(E_T) = \{u : E_T \to \mathbb{R} \mid u_t, u_{x_i x_j} \in C(E_T),\ i,j = 1, \dots, N\}.$$

Information on the thermal status of the body is gathered at the boundary of E over an interval of time $(0, T]$. That is, one might be given the temperature or the heat flux at $\partial E \times (0, T)$. Physically relevant problems consist in finding the temperature distribution in E for $t \geq 0$, from information on $\partial E \times (0, T)$ and the knowledge of the temperature $x \to u_o(x)$ at time $t = 0$. This leads to the following boundary value problems:

E. DiBenedetto, *Partial Differential Equations: Second Edition*,
Cornerstones, DOI 10.1007/978-0-8176-4552-6_6,

1.1 The Dirichlet Problem

Find $u \in \mathcal{H}(E_T) \cap C(\bar{E}_T)$ satisfying

$$\begin{aligned} H(u) &= 0 \quad \text{in } E_T \\ u\big|_{\partial E \times [0,T]} &= g \in C(\partial E \times (0,T]) \\ u(\cdot, 0) &= u_o \in C(\bar{E}). \end{aligned} \tag{1.2}$$

1.2 The Neumann Problem

Find $u \in \mathcal{H}(E_T) \cap C^1(\bar{E}_T)$ satisfying

$$\begin{aligned} &H(u) = 0 \text{ in } E_T \\ &Du \cdot \mathbf{n} = g \in C(\partial E \times (0,T)) \\ &u(\cdot, 0) = u_o \in C(\bar{E}). \end{aligned} \tag{1.3}$$

where D denotes the gradient with respect to the space variables only.

1.3 The Characteristic Cauchy Problem

Find $u \in \mathcal{H}(S_T) \cap C(\bar{S}_T)$ satisfying

$$\begin{aligned} H(u) &= 0 \quad \text{in } S_T \\ u(\cdot, 0) &= u_o \in C(\mathbb{R}^N) \cap L^\infty(\mathbb{R}^N). \end{aligned} \tag{1.4}$$

The initial datum in (1.4) is taken in the topology of the uniform convergence over compact sets $K \subset \mathbb{R}^N$, that is, $\|u(\cdot,t) - u_o\|_{\infty,K} \to 0$, as $t \to 0$, for all such K. In (1.4) the data are assigned on the characteristic surface $t = 0$. The Cauchy–Kowalewski theorem fails to hold in such a circumstance. Even if u_o is analytic, a solution of (1.4) near $t = 0$, that is for small positive and negative times, in general cannot be found. Indeed, changing t into $-t$ does not preserve (1.1) and the PDE distinguishes between solutions forward and backward in time. This corresponds to the physical fact that heat conduction is, in general, irreversible, i.e., given $x \to u_o(x)$, we may predict future temperatures, but we cannot in general determine the thermal status that generated that particular temperature distribution.

2 The Cauchy Problem by Similarity Solutions

The PDE $H(u) = 0$ is invariant by linear transformations $\bar{x} = hx$, $\bar{t} = h^2 t$ for $h \neq 0$. These are transformations that leave invariant the ratio $\xi = |x|^2/t$. This suggests looking for solutions u that are "separable" in the variables t

and ξ, that is, solutions of the form $u(x,t) = h(t)f(\xi)$. Substituting this in the PDE $H(u) = 0$ gives

$$th'f - 2Nhf' = h\xi[4f'' + f'].$$

Setting each side equal to zero yields

$$f(\xi) = \exp(-\xi/4), \qquad h(t) = t^{-N/2}$$

up to multiplicative constants. These remarks imply that a solution of $H(u) = 0$ in $\mathbb{R}^N \times (0,\infty)$ is given by

$$\Gamma(x,t) = \frac{1}{(4\pi t)^{N/2}} e^{-|x|^2/4t} \tag{2.1}$$

where the multiplicative constant $(4\pi)^{-N/2}$ has been chosen to satisfy the normalization (Section 2.1c of the Complements)

$$\frac{1}{[4\pi(t-s)]^{N/2}} \int_{\mathbb{R}^N} e^{-\frac{|x-y|^2}{4(t-s)}} dy = 1 \tag{2.2}$$

for all $x \in \mathbb{R}^N$ and all $s < t$.

Remark 2.1 The function Γ is called the *heat kernel* or the *fundamental solution* of the heat equation. It satisfies

$$\begin{aligned} (x,t) &\to \Gamma(x,t) \in C^\infty(\mathbb{R}^N \times \mathbb{R}^+) \\ x &\to \Gamma(x,t) \ \text{ is analytic for } \ t > 0. \end{aligned}$$

Let $H_{(\eta,\tau)}$ and $H^*_{(\eta,\tau)}$ denote respectively the heat operator and its adjoint with respect to the variables $\eta \in \mathbb{R}^N$ and $\tau \in \mathbb{R}$. By direct calculation

$$\left.\begin{aligned} H_{(x,t)}\Gamma(x-y;t-s) &= 0 \\ H^*_{(y,s)}\Gamma(x-y;t-s) &= 0 \end{aligned}\right. \quad \text{for } \ s < t < \infty. \tag{2.3}$$

Assume that $u \in \mathcal{H}(S_T)$ is a solution of the Cauchy problem (1.4) satisfying

$$\int_{\mathbb{R}^N} |u(x,t)|dx < \infty \quad \text{ for all } \ 0 \le t \le T \tag{2.4}$$

and the asymptotic decay

$$\left.\begin{aligned} \limsup_{r\to\infty} \left| \int_{|y|=r} \Gamma(x-y;t) Du \cdot \frac{y}{r} d\sigma \right| &= 0 \\ \limsup_{r\to\infty} \left| \int_{|y|=r} uD\Gamma(x-y;t) \cdot \frac{y}{r} d\sigma \right| &= 0 \end{aligned}\right. \quad \text{for all } \ 0 \le t < T \tag{2.5}$$

where $d\sigma$ denotes the surface measure on the sphere $|y| = r$. Multiply the first of (1.4) viewed in the variables (y, s) by $\Gamma(x - y; t - s)$, and integrate by parts in $dyds$ over the cylindrical domain $B_r \times (0, t - \varepsilon)$ for $\varepsilon \in (0, t)$. Letting $r \to \infty$ with the aid of (2.4) and (2.5), we arrive at

$$\int_{\mathbb{R}^N} u(y, t - \varepsilon)\varepsilon^{-N/2} e^{-\frac{|x-y|^2}{4\varepsilon}} dy = \int_{\mathbb{R}^N} u_o(y) t^{-N/2} e^{-\frac{|x-y|^2}{4t}} dy. \tag{2.6}$$

We let $\varepsilon \to 0$ as follows. Fix $\sigma > 0$ and write

$$\begin{aligned}\int_{\mathbb{R}^N} u(y, t - \varepsilon)\varepsilon^{-N/2} e^{-\frac{|x-y|^2}{4\varepsilon}} dy &= \int_{|y-x|>\sigma} u(y, t - \varepsilon)\varepsilon^{-N/2} e^{-\frac{|x-y|^2}{4\varepsilon}} dy \\ &\quad + \int_{|y-x|\le\sigma} u(y, t - \varepsilon)\varepsilon^{-N/2} e^{-\frac{|x-y|^2}{4\varepsilon}} dy \\ &= I_\varepsilon^{(1)} + I_\varepsilon^{(2)}.\end{aligned}$$

As $\varepsilon \to 0$, the first integral on the right-hand side tends to zero. We rewrite the second integral as

$$\begin{aligned}I_\varepsilon^{(2)} &= \int_{|y-x|<\sigma} [u(y, t - \varepsilon) - u(x, t)]\varepsilon^{-N/2} e^{-\frac{|x-y|^2}{4\varepsilon}} dy \\ &\quad + u(x, t) \int_{|y-x|<\sigma} \varepsilon^{-N/2} e^{-\frac{|x-y|^2}{4\varepsilon}} dy \\ &= [u(x, t) + O(\sigma + \varepsilon)] \int_{|y|<\sigma} \varepsilon^{-N/2} e^{-\frac{|y|^2}{4\varepsilon}} dy\end{aligned}$$

where $O(\sigma + \varepsilon)$ denotes a quantity that tends to zero as $(\sigma + \varepsilon) \to 0$. Write

$$\varepsilon^{-N/2} \int_{|y|<\sigma} e^{-\frac{|y|^2}{4\varepsilon}} dy = \varepsilon^{-N/2} \int_{\mathbb{R}^N} e^{-\frac{|y|^2}{4\varepsilon}} dy - \varepsilon^{-N/2} \int_{|y|>\sigma} e^{-\frac{|y|^2}{4\varepsilon}} dy.$$

The first integral can be computed from (2.2) with $x = 0$ and $(t - s) = \varepsilon$, i.e.,

$$\varepsilon^{-N/2} \int_{\mathbb{R}^N} e^{-\frac{|y|^2}{4\varepsilon}} dy = (4\pi)^{N/2}.$$

To estimate the second integral, introduce the change of variables $y = 2\sqrt{\varepsilon}\eta$, whose Jacobian is $(4\varepsilon)^{N/2}$. This gives

$$\varepsilon^{-N/2} \int_{|y|>\sigma} e^{-\frac{|y|^2}{4\varepsilon}} dy = \int_{|\eta|>\frac{\sigma}{2\sqrt{\varepsilon}}} e^{-|\eta|^2} d\eta.$$

This integrals tends to zero as $\varepsilon \to 0$, for $\sigma > 0$ fixed. Combine these calculations in (2.6), and let $\varepsilon \to 0$, while $\sigma > 0$ remains fixed, to obtain

$$(4\pi)^{N/2} u(x, t) = t^{-N/2} \int_{\mathbb{R}^N} u_o(y) e^{-\frac{|x-y|^2}{4t}} dy + O(\sigma).$$

Letting $\sigma \to 0$ gives the representation formula

$$u(x,t) = \frac{1}{(4\pi t)^{N/2}} \int_{\mathbb{R}^N} e^{-\frac{|x-y|^2}{4t}} u_o(y)dy = \int_{\mathbb{R}^N} \Gamma(x-y;t)u_o(y)dy. \tag{2.7}$$

Therefore every solution of the Cauchy problem satisfying the decay conditions (2.4)–(2.5) must be represented as in (2.7). Now consider (2.7), regardless of its derivation process. If $u_o \in C(\mathbb{R}^N) \cap L^\infty(\mathbb{R}^N)$, the integral on the right-hand side is convergent and defines a function u that satisfies the decay conditions (2.4)–(2.5). Moreover, by Remark 2.1, $u(x,t) \in C^\infty(S_T)$ and

$$x \to u(x,t) \text{ is locally analytic in } \mathbb{R}^N \text{ for all } 0 < t \le T. \tag{2.8}$$

Theorem 2.1 *Let $u_o \in C(\mathbb{R}^N) \cap L^\infty(\mathbb{R}^N)$. Then u defined by (2.7) is a solution to the Cauchy problem (1.4). Moreover, u is bounded in $\mathbb{R}^N \times \mathbb{R}^+$, and it is the only bounded solution to the Cauchy problem (1.4).*

Proof (existence) By construction $H(u) = 0$ in S_T. Moreover

$$\|u(\cdot,t)\|_{\infty,\mathbb{R}^N} \le \|u_o\|_{\infty,\mathbb{R}^N} \int_{\mathbb{R}^N} \Gamma(x-y;t)dy = \|u_o\|_{\infty,\mathbb{R}^N}. \tag{2.9}$$

Therefore u defined by (2.7) is bounded in $\mathbb{R}^N \times \mathbb{R}^+$. It remains to show that the initial datum is taken in the topology of uniform convergence over compact subsets of $\mathbb{R}^N$. Fix a compact set $K \subset \mathbb{R}^N$, recall the normalization (2.2), and write for $x \in K$

$$u(x,t) - u_o(x) = \frac{1}{(4\pi t)^{N/2}} \int_{\mathbb{R}^N} [u_o(y) - u_o(x)] e^{-\frac{|x-y|^2}{4t}} dy.$$

Divide the domain of integration on the right-hand side into $|x-y| < \sigma$ and $|x-y| \ge \sigma$ where $\sigma > 0$ is arbitrary but fixed. As $t \to 0$, the integral extended over $|x-y| > \sigma$ tends to zero and the one extended over $|x-y| < \sigma$ is majorized by

$$\sup_{\substack{x\in K \\ |x-y|<\sigma}} |u_o(y) - u_o(x)| \int_{\mathbb{R}^N} \Gamma(x-y;t)dy.$$

Therefore, for arbitrary $\sigma > 0$

$$\lim_{t\to 0} \|u(x,t) - u_o(x)\|_{\infty,K} \le \sup_{\substack{x\in K \\ |x-y|<\sigma}} |u_o(y) - u_o(x)|. \qquad \blacksquare$$

The proof of uniqueness will make use of the maximum principle discussed in the next sections. A first form of such a principle can be read from (2.9), that is, the supremum of $|u(\cdot,t)|$ at all instants $t > 0$ is no larger than the supremum of $|u_o|$.

Remark 2.2 Suppose that in (2.7), u_o is non-negative, not identically zero, and supported in the ball B_ε of radius $\varepsilon > 0$ centered at some point in $\mathbb{R}^N$. Then $u(x,t)$ is strictly positive for all $(x,t) \in S_T$. In particular, the *initial disturbance*, confined in B_ε, for however small ε, is felt by the solution at any $|x|$ however large, and any positive t, however small. Thus the initial disturbance propagates with infinite speed.

2.1 The Backward Cauchy Problem

Let $S^T = \mathbb{R}^N \times (-T, 0)$, and consider the problem of finding $u \in \mathcal{H}(S^T) \cap C(\bar{S}^T)$ satisfying

$$\begin{aligned} H(u) &= 0 \quad \text{in } S^T \\ u(\cdot, 0) &= u_o \in C(\mathbb{R}^N) \cap L^\infty(\mathbb{R}^N). \end{aligned} \tag{2.10}$$

The backward problem (2.10) is ill-posed in the sense that unlike the forward problem (1.4), it is not solvable in general within the class of bounded solutions. Indeed, if a bounded, continuous solution did exist for every choice of data $u_o \in C(\mathbb{R}^N) \cap L^\infty(\mathbb{R}^N)$, we would have by (2.7) and Theorem 2.1

$$u_o(x) = \int_{\mathbb{R}^N} \Gamma(x - y; T) u(y, -T)\, dy \tag{2.11}$$

and this would contradict (2.8), if for example, u_o is merely continuous.

3 The Maximum Principle and Uniqueness (Bounded Domains)

Let E be a bounded open subset of $\mathbb{R}^N$ and let $\partial_* E_T = \partial E_T - E \times \{T\}$ denote the parabolic boundary of E_T.

Theorem 3.1 *Let $u \in \mathcal{H}(E_T) \cap C(\bar{E}_T)$ satisfy $H(u) \le 0 (\ge 0)$ in E_T. Then*

$$\sup_{E_T} u = \sup_{\partial_* E_T} u \qquad \Big(\inf_{E_T} u = \inf_{\partial_* E_T} u\Big).$$

Proof We prove the statement only for $H(u) \le 0$. Let $\varepsilon \in (0, T)$ be arbitrary but fixed, and consider the function

$$\bar{\Omega}_{T-\varepsilon} \ni (x,t) \to v(x,t) = u(x,t) - \varepsilon t$$

which satisfies $H(v) < -\varepsilon < 0$ in $\bar{E}_{T-\varepsilon}$. Since v is continuous in $\bar{E}_{T-\varepsilon}$, it achieves its maximum at some $(x_o, t_o) \in \bar{E}_{T-\varepsilon}$. If $(x_o, t_o) \notin \partial_* E_{T-\varepsilon}$, then $H(v)(x_o, t_o) \ge 0$, contradicting $H(v) < 0$. Thus $(x_o, t_o) \in \partial_* E_{T-\varepsilon}$ and

$$u(x,t) \le 2\varepsilon T + \sup_{\partial_* E_T} u \qquad \text{for all } (x,t) \in \bar{E}_{T-\varepsilon} \text{ for all } \varepsilon > 0. \qquad \blacksquare$$

Corollary 3.1 *Let $u \in \mathcal{H}(E_T) \cap C(\bar{E}_T)$ satisfy $H(u) = 0$ in E_T. Then*

$$\|u\|_{\infty,E_T} = \|u\|_{\infty,\partial_* E_T}.$$

Remark 3.1 Theorem 3.1 is a weak maximum principle since it does not exclude that u might obtain its extremal values also at some other points in $\bar{E}_T$. For example, u could be identically constant in E_T. A strong maximum principle would assert that this is the only other possibility.

3.1 A Priori Estimates

Denote by λ the diameter of E. After a rotation and translation, we may, if necessary, arrange the coordinate axes so that

$$\text{for all } x = (x_1, \dots, x_N) \in \bar{E}, \quad x_1^o - \lambda \le x_1 \le x_1^o$$

for some $x^o = (x_1^o, \dots, x_N^o) \in \partial E$. This is possible since the heat operator is invariant under rotations and translations of the space variables. Let $u \in \mathcal{H}(E_T) \cap C(\bar{E}_T)$ be such that $\|H(u)\|_{\infty,E_T} < \infty$ and construct the two functions

$$w_\pm(x,t) = \|u\|_{\infty,\partial_* E_T} + e^\lambda[1 - e^{(x_1 - x_1^o)}]\|H(u)\|_{\infty,E_T} \pm u.$$

One verifies that $H(w_\pm) \ge 0$ in E_T and that

$$w_\pm \big|_{\partial_* E_T} \ge \|u\|_{\infty,\partial_* E_T} \pm u \big|_{\partial_* E_T}.$$

Therefore $w_\pm \ge 0$, by Theorem 3.1. This gives the following a priori estimate.

Corollary 3.2 *Let $u \in \mathcal{H}(E_T) \cap C(\bar{E}_T)$. Then*

$$\|u\|_{\infty,E_T} \le \|u\|_{\infty,\partial_* E_T} + (e^{\operatorname{diam}(E)} - 1)\|H(u)\|_{\infty,E_T}.$$

3.2 Ill-Posed Problems

A boundary value problem for $H(u) = 0$ with data prescribed on the whole boundary of E_T in general is not well-posed. For example, consider the rectangle $R = [0 < x < 1] \times [0 < t < 1]$, and let $\varphi \in C(\partial R)$ be non-constant and such that it takes an absolute maximum on the open line segment $[0 < x < 1] \times [t = 1]$. Then the problem

$$u \in \mathcal{H}(R), \quad u_t - u_{xx} = 0 \text{ in } R, \quad u\big|_{\partial R} = \varphi$$

cannot have a solution, for it would violate Theorem 3.1.

3.3 Uniqueness (Bounded Domains)

Corollary 3.3 *There exists at most one solution $u \in \mathcal{H}(E_T) \cap C(\bar{E}_T)$ of the boundary value problem*

$$u_t - \Delta u = f \in C(\bar{E}_T), \quad u\big|_{\partial_* E_T} = g \in C(\partial_* E_T).$$

Proof If u and v are solutions, $w = u - v$ solves

$$w_t - \Delta w = 0 \text{ in } E_T, \quad w\big|_{\partial_* E_T} = 0$$

and hence $w \equiv 0$ by Theorem 3.1. ■

4 The Maximum Principle in $\mathbb{R}^N$

Results analogous to Theorem 3.1 are possible in $\mathbb{R}^N$ if one imposes some conditions on the behavior of $x \to u(x,t)$ as $|x| \to \infty$. Such conditions are dictated by the solution formula (2.7). For such a formula to have a meaning, u_o does not have to be regular or bounded. It would suffice to require the convergence of the integral on the right-hand side for $0 < t \le T$. The next proposition gives some sufficient conditions for this to occur.

Proposition 4.1 *Assume that $u_o \in L^1_{\text{loc}}(\mathbb{R}^N)$ and satisfies the growth condition*

$$\begin{cases} \text{there exist positive constants } C_o, \alpha_o, r_o \text{ such that} \\ |u_o(x)| \le C_o e^{\alpha_o |x|^2} \text{ for almost all } |x| \ge r_o. \end{cases} \tag{4.1}$$

Then (2.7) defines a function $u \in C^\infty(S_T)$ for every $T \in \big(0, \frac{1}{4\alpha_o}\big)$. Moreover, $H(u) = 0$ in S_T, and for every $\varepsilon \in \big(0, \frac{1}{4\alpha_o}\big)$, there exists positive constants α, C, and r depending upon α_o, C_o, r_o, N, and ε such that

$$\begin{aligned} &|u(x,t)| \le \Gamma\big(\varepsilon; \frac{\varepsilon^2}{2N}\big) \|u_o\|_{1,B_{r_o}} + C e^{\alpha|x|^2} \\ &\text{for all } |x| > r \text{ and for all } 0 < t < \frac{1}{4\alpha_o} - \varepsilon. \end{aligned} \tag{4.2}$$

Proof Fix $\varepsilon \in \big(0, \frac{1}{4\alpha_o}\big)$ and $|x| > r_o + \varepsilon$, and write the integral in (2.7) as

$$\int_{|y| \le r_o} \Gamma(x-y;t) u_o(y) dy + \int_{|y| > r_o} \Gamma(x-y;t) u_o(y) dy = J_1 + J_2.$$

For $|x - y| > \varepsilon$

$$|J_1| \le \sup_{t \ge 0} \Gamma(\varepsilon; t) \|u_o\|_{1,B_{r_o}} = \Gamma\big(\varepsilon; \frac{\varepsilon^2}{2N}\big) \|u_o\|_{1,B_{r_o}}.$$

In estimating J_2 we perform the change of variables $y - x = 2\sqrt{t}\eta$, of Jacobian $(4t)^{N/2}$, and use (4.1) to estimate $|u_o(y)|$ from above. This gives

$$|J_2| \le C_o \pi^{-N/2} \int_{|y|>r_o} e^{-|\eta|^2} e^{\alpha_o |x+2\sqrt{t}\eta|^2} d\eta.$$

By the Schwarz inequality, for all $\delta > 0$

$$|x + 2\sqrt{t}\eta|^2 \le \big(1 + \frac{1}{\delta}\big)|x|^2 + 4(1+\delta)t|\eta|^2.$$

Therefore, for all $|x| > r = r_o + \varepsilon$

$$|J_2| \le C_o \pi^{-N/2} e^{\alpha_o(1+1/\delta)|x|^2} \int_{\mathbb{R}^N} e^{-(1-4\alpha_o(1+\delta)t)|\eta|^2} d\eta.$$

The integral on the right-hand side is convergent if

$$t < \frac{1}{4\alpha_o(1+\delta)} = \frac{1}{4\alpha_o} - \varepsilon.$$

This defines the choice of δ. Therefore $|J_2| \le C e^{\alpha|x|^2}$, where

$$C = C_o \pi^{-N/2} \int_{\mathbb{R}^N} e^{-(1-4\alpha_o(1+\delta)t)|\eta|^2} d\eta \quad \text{and} \quad \alpha = \alpha_o\big(1 + \frac{1}{\delta}\big). \qquad \blacksquare$$

In deriving a maximum principle for solutions of the heat equation in S_T, we require that such solutions satisfy a behavior of the type (4.2) as $|x| \to \infty$, but we make no further reference to the representation formula (2.7).

Theorem 4.1 *Let $u \in \mathcal{H}(S_T) \cap C(\bar{S}_T)$ satisfy $H(u) \ge 0$ in S_T and $u(\cdot, 0) \ge 0$. Assume moreover that*

$$\begin{cases} \textit{there exist positive constants } C, \alpha, r \textit{ such that} \\ u(x,t) \ge -Ce^{\alpha|x|^2} \textit{ for all } |x| \ge r \textit{ and all } 0 \le t \le T. \end{cases} \tag{4.3}$$

Then $u \ge 0$ in S_T.

Proof Choose $\beta > \alpha$ so large that $T > \frac{1}{8\beta} \overset{\text{def}}{=} T_1$. We first prove that $u \ge 0$ in the strip S_{T_1}. The function

$$v(x,t) = \frac{1}{(1-4\beta t)^{N/2}} e^{\beta|x|^2/(1-4\beta t)}$$

satisfies $H(v) = 0$, and $v(x,t) \ge e^{\beta|x|^2}$, in S_{T_1}. Let $\varepsilon > 0$ be arbitrary but fixed and set $w = u + \varepsilon v$. In view of the arbitrariness of ε, it will suffice to show that $w \ge 0$ in S_{T_1}. The function w satisfies $H(w) \ge 0$ in S_{T_1}, $w(\cdot, 0) \ge 0$, and

$$\liminf_{|x|\to\infty} w(x,t) \ge 0, \qquad \text{uniformly in } t \in [0, T_1].$$

Therefore, having fixed $(x_o, t_o) \in S_{T_1}$ and $\sigma > 0$, there exists $\rho > |x_o|$ such that $w(x,t) \ge -\sigma$ for $|x| \ge \rho$ for all $t \in [0, T_1]$. On the (bounded) cylinder

$Q = [|x| < \rho] \times (0, T_1)$ the function $\bar{w} = w + \sigma$ satisfies $H(\bar{w}) \ge 0$ in Q and $\bar{w} \ge 0$ on the parabolic boundary $\partial_* Q$ of Q. Therefore $\bar{w} \ge 0$ in Q, by Theorem 3.1. In particular, $w(x_o, t_o) \ge -\sigma$ for all $\sigma > 0$. Therefore $w \ge 0$ in S_{T_1}i, since $(x_o, t_o) \in S_{T_1}$ is arbitrary. To conclude the proof we repeat the argument in adjacent non-overlapping strips of width not exceeding $\frac{1}{8\beta}$, up to cover the whole of S_T. ■

Theorem 4.2 *Let $u \in \mathcal{H}(S_T) \cap C(\bar{S}_T)$ satisfy $H(u) \le 0$ in S_T and*

$$\begin{cases} \text{there exist positive constants } C, \alpha, r \text{ such that} \\ u(x,t) \le Ce^{\alpha|x|^2} \text{ for all } |x| \ge r \text{ and all } 0 \le t \le T. \end{cases} \tag{4.4}$$

Then

$$u(x,t) \le \sup_{\mathbb{R}^N} u(\cdot, 0) \quad \text{for all } (x,t) \in S_T.$$

Proof We may assume that $u(\cdot, 0) \in L^\infty(\mathbb{R}^N)$; otherwise, the statement is trivial. Assume first that T is so small that $4\alpha T < 1$ and consider the (bounded) cylinder $Q = [|x| < \rho] \times (0, T)$. The function

$$w = u - \frac{\varepsilon}{[4\pi(T-t)]^{N/2}} e^{|x|^2/4(T-t)}, \qquad \varepsilon > 0$$

satisfies $H(w) \le 0$ in S_T, and $w(x, 0) \le \sup_{\mathbb{R}^N} u(\cdot, 0)$ for $|x| < \rho$. Moreover, for $|x| = \rho$,

$$w \Big|_{|x|=\rho} \le Ce^{\alpha\rho^2} - \varepsilon(4\pi T)^{-N/2} e^{\rho^2/4T}.$$

Therefore, since $4T < 1/\alpha$, having fixed $\varepsilon > 0$, the parameter ρ can be chosen so large that $w|_{|x|=\rho} \le 0$. The conclusion now follows from Theorem 3.1 and the arbitrariness of ε. If $4\alpha T \ge 1$, subdivide S_T into finitely many strips of width less than $\frac{1}{4\alpha}$. ■

4.1 A Priori Estimates

Proposition 4.2 *Let $u \in \mathcal{H}(S_T) \cap C(\bar{S}_T)$ satisfy (4.3) and (4.4). Then*

$$\|u\|_{\infty, S_T} \le \|u(\cdot, 0)\|_{\infty, \mathbb{R}^N} + T\|H(u)\|_{\infty, S_T}.$$

Proof Assume that $\|u_o\|_{\infty, \mathbb{R}^N}$ and $\|H(u)\|_{\infty, S_T}$ are finite; otherwise, the statement is trivial. The functions

$$w_\pm = \|u(\cdot, 0)\|_{\infty, \mathbb{R}^N} + t\|H(u)\|_{\infty, S_T} \pm u$$

satisfy $H(w_\pm) \ge 0$ in S_T and $w_\pm(\cdot, 0) \ge 0$. Moreover, both $\bar{w}_\pm$ satisfy the asymptotic behavior (4.3). Therefore $w_\pm \ge 0$ in S_T, by Theorem 4.1. ■

Remark 4.1 The functional dependence of this estimate is optimal. Indeed, the estimate holds with equality for the function $u = 1 + t$.

4.2 About the Growth Conditions (4.3) and (4.4)

The conclusion of Theorem 4.1 fails if (4.3) is replaced by

$$u(x,t) \ge -ke^{\beta|x|^{2+\varepsilon}} \qquad \text{for any } \varepsilon > 0.$$

However Theorem 4.2 continues to hold for a growth slightly faster than (4.4). Precisely (S. Tacklind [146])

$$u(x,t) \le Ce^{\alpha|x|h(|x|)} \qquad \text{as } |x| \to \infty$$

where $h(\cdot)$ is positive non-decreasing and satisfies the optimal condition

$$\int^{\infty} \frac{ds}{h(s)} = +\infty.$$

5 Uniqueness of Solutions to the Cauchy Problem

Consider the class of functions $w \in \mathcal{H}(S_T)$ satisfying the growth condition

$$\begin{cases} \text{there exist positive constants } C, \alpha, r \text{ such that} \\ |u(x,t)| \le Ce^{\alpha|x|^2} \text{ for all } |x| \ge r \text{ and all } 0 \le t \le T. \end{cases} \tag{5.1}$$

Let $u, v \in \mathcal{H}(S_T) \cap C(\bar{S}_T)$ be solutions of the Cauchy problem (1.4) with initial data $u_o, v_o \in C(\mathbb{R}^N) \cap L^\infty(\mathbb{R}^N)$. If both u and v satisfy (5.1), then by Proposition 4.2,

$$\|u - v\|_{\infty, S_T} \le \|u_o - v_o\|_{\infty, \mathbb{R}^N}.$$

This inequality represents both a uniqueness and a stability result. Namely:

(i). *Uniqueness*: solutions of the Cauchy problem (1.4) are unique within the class (5.1).

(ii). *Stability*: within such a class, small variations on the data, measured in the norm of $L^\infty(\mathbb{R}^N)$, yield small variations on the solution measured in the same norm.

Proof (of Theorem 2.1 (Uniqueness)) If $u_o \in L^\infty(\mathbb{R}^N) \cap C(\mathbb{R}^N)$, the function u defined by the representation formula (2.7) is bounded by virtue of (2.9). It solves the heat equation in S_T, and it satisfies (5.1) by virtue of Proposition 4.1. Therefore, it is the only *bounded* solution of the Cauchy problem (1.4). ■

5.1 A Counterexample of Tychonov ([155])

The growth condition (5.1) is essential for uniqueness, as shown by the following counterexample due to Tychonov.

Proposition 5.1 *There exists a non-identically zero solution to the Cauchy problem*

$$u_t = u_{xx} \quad in \ \mathbb{R} \times (0, \infty), \quad u(x, 0) = 0.$$

Proof For $z \in \mathbb{C}$, let

$$\varphi(z) = \begin{cases} e^{-1/z^2} & \text{for } z \neq 0 \\ 0 & \text{for } z = 0 \end{cases}$$

and define

$$u(x,t) = \begin{cases} \sum\limits_{n=0}^{\infty} \dfrac{d^n}{dt^n}\varphi(t)\dfrac{x^{2n}}{(2n)!} & \text{for } t > 0 \\ 0 & \text{for } t = 0. \end{cases} \tag{5.2}$$

Proceeding formally

$$\lim_{t \to 0} u(x,t) = \sum_{n=0}^{\infty} \frac{d^n}{dt^n}\varphi(t) \big|_{t=0} \frac{x^{2n}}{(2n)!} = 0 \tag{i}$$

$$\begin{aligned} \frac{\partial^2 u}{\partial x^2} &= \sum_{n=0}^{\infty} \frac{d^n}{dt^n}\varphi(t) 2n(2n-1)\frac{x^{2n-2}}{(2n)!} \\ &= \sum_{n=1}^{\infty} \frac{d^n}{dt^n}\varphi(t)\frac{x^{2(n-1)}}{(2(n-1))!} \qquad \text{(ii)} \\ &= \sum_{n=0}^{\infty} \frac{d^{n+1}}{dt^{n+1}}\varphi(t)\frac{x^{2n}}{(2n)!} = \frac{\partial u}{\partial t}. \end{aligned}$$

■

These calculations become rigorous after we prove the following

Lemma 5.1 *The series in (5.2) and (i)–(ii) are uniformly convergent in a neighborhood of every point of* $\mathbb{R} \times \mathbb{R}^+$.

Proof The function $z \to \varphi(z)$ is holomorphic in $\mathbb{C} - \{0\}$. We identify the t-axis as the real axis of the complex plane. If $t > 0$ is fixed, the circle

$$\gamma = \{z \in \mathbb{C} \mid z = t + \frac{t}{2}e^{i\theta}\}, \quad 0 < \theta \leq 2\pi$$

does not meet the origin, and by the Cauchy formula ([18] page 72)

$$\frac{d^n}{dt^n}\varphi(t) = \frac{n!}{2\pi i}\int_\gamma \frac{\varphi(z)}{(z-t)^{n+1}}dz \quad \text{for all } n \in \mathbb{N}.$$

From this

$$\left|\frac{d^n}{dt^n}\varphi(t)\right| \leq \frac{n!}{2\pi}\int_\gamma \frac{e^{-Re(z^{-2})}}{|z-t|^{n+1}}|dz| = \frac{n!}{2\pi}\left(\frac{2}{t}\right)^n \int_0^{2\pi} e^{-Re(z^{-2})}d\theta.$$

For $z \in \gamma$

$$z^2 = t^2\left(1 + \frac{1}{2}e^{i\theta}\right)^2 \quad \text{and} \quad \frac{1}{z^2} = \frac{1}{t^2}\frac{\left(1 + \frac{1}{4}e^{-2i\theta} + e^{-i\theta}\right)}{\left|\left(1 + \frac{1}{2}e^{i\theta}\right)^2\right|^2}.$$

From this $\mathrm{Re}(z^{-2}) \ge (2t)^{-2}$ and

$$\Big|\frac{d^n}{dt^n}\varphi(t)\Big| \le n!\left(\frac{2}{t}\right)^n e^{-1/4t^2}, \qquad n \in \mathbb{N}.$$

Fix $a > 0$. For all $|x| < a$, the series in (5.2) is majorized, term by term, by the uniformly convergent series

$$e^{-1/4t^2}\sum_{n=0}^{\infty}\left(\frac{1}{t}\right)^n\frac{(a^2)^n}{n!} = e^{-1/4t^2}e^{a^2/t}.$$

Here we have used the Stirling inequality

$$\frac{2^n n!}{(2n)!} \le \frac{1}{n!}.$$

■

Remark 5.1 The function in (5.2) can also be defined for $t < 0$. Therefore the *backward* Cauchy problem

$$u_t - \Delta u = 0 \;\text{ in }\; \mathbb{R}^N \times (-\infty, 0), \quad u(x,0) = 0$$

fails in general to have a unique solution.

6 Initial Data in $L^1_{\rm loc}(\mathbb{R}^N)$

The Cauchy problem for the heat equation can be solved uniquely for rather coarse initial data, for example $u_o \in L^1_{\rm loc}(\mathbb{R}^N)$, provided they satisfy the growth condition (4.1). The solution will exist only within the strip S_T for $0 < T < \frac{1}{4\alpha_o}$, and the initial datum is taken in the sense of $L^1_{\rm loc}(\mathbb{R}^N)$, i.e.,

$$\|u(\cdot,t) - u_o\|_{1,K} \to 0 \;\text{ as }\; t \to 0, \;\text{ for all compact }\; K \subset \mathbb{R}^N. \tag{6.1}$$

Theorem 6.1 *Let $u_o \in L^1_{\rm loc}(\mathbb{R}^N)$ satisfy (4.1). Then (2.7) defines a solution of the Cauchy problem*

$$\begin{aligned} H(u) &= 0 \quad \text{in } S_T \text{ for } 0 < T < \frac{1}{4\alpha_o} \\ u(\cdot,0) &= u_o \quad \text{in the sense of } L^1_{\rm loc}(\mathbb{R}^N). \end{aligned} \tag{6.2}$$

Such a solution is unique within the class (5.1).

Proof Fix $\rho \ge r_o$, where r_o is the constant in the growth condition (4.1). For almost all $x \in B_\rho$, write

$$|u(x,t) - u_o(x)| \le \int_{\mathbb{R}^N} \Gamma(x-y;t)|u_o(y) - u_o(x)|dy.$$

Integrating in dx over B_ρ

$$\begin{aligned}\int_{B_\rho} |u(x,t) - u_o(x)|dx &\le \int_{B_\rho}\int_{\mathbb{R}^N} \Gamma(x-y;t)|u_o(y) - u_o(x)|dydx \\ &= \int_{B_\rho}\int_{|x-y|\le\sigma} \Gamma(x-y;t)|u_o(y) - u_o(x)|dydx \\ &\quad + \int_{B_\rho}\int_{|x-y|>\sigma} \Gamma(x-y;t)|u_o(y) - u_o(x)|dydx \\ &= I_1 + I_2.\end{aligned}$$

Let h be a vector in $\mathbb{R}^N$ of size $|h| \le \sigma$. Then the first integral is estimated by

$$I_1 \le \sup_{\substack{h\in\mathbb{R}^N \\ |h|\le\sigma}} \int_{B_\rho} |u_o(x+h) - u_o(x)|dx.$$

The second integral is estimated by

$$I_2 \le \pi^{-N/2} \int_{|\eta|>\sigma/2\sqrt{t}} e^{-|\eta|^2} \int_{B_\rho} |u_o(x+2\sqrt{t}\eta) - u_o(x)|dxd\eta.$$

For all $t > 0$ and $\eta \in \mathbb{R}^N$ such that $2\sqrt{t}|\eta| < 2\rho$, estimate

$$\int_{B_\rho} |u_o(x+2\sqrt{t}\eta) - u_o(x)|dx \le 2\|u_o\|_{1,B_{2\rho}}.$$

If $2\sqrt{t}|\eta| \ge 2\rho$, by the growth condition (4.1)

$$\int_{B_\rho} |u_o(x+2\sqrt{t}\eta) - u_o(x)|dx \le \|u_o\|_{1,B_{2\rho}} + C|B_\rho| \sup_{x\in B_\rho} e^{2\alpha_o|x|^2+8\alpha_o t|\eta|^2}.$$

Therefore for $\rho > r_o$ fixed

$$\begin{aligned}\|u(\cdot,t) - u_o\|_{1,B_\rho} &\le \text{const}(\rho) \int_{|\eta|>\sigma/2\sqrt{t}} e^{-(1-8\alpha_o t)|\eta|^2} d\eta \\ &\quad + \sup_{\substack{h\in\mathbb{R}^N \\ |h|\le\sigma}} \|u_o(x+h) - u_o(x)\|_{1,B_\rho}.\end{aligned}$$

The proof is concluded by recalling that the translation $T_h u_o = u_o(\cdot + h)$ is continuous in $L^1_{\text{loc}}(\mathbb{R}^N)$ ([31], Chapter IV, Section 20). ∎

6.1 Initial Data in the Sense of $L^1_{\rm loc}(\mathbb{R}^N)$

Part of the definition of a solution to the Cauchy problem (1.4) or (6.2) is to make precise in what sense the initial data are taken. The notion (6.1) is the weakest unambiguous requirement for data $u_o \in L^1_{\rm loc}(\mathbb{R}^N)$. If (6.2) holds, then there exists a sequence of times $\{t_n\} \to 0$ such that

$$u(x,t_n) \to u_o(x) \quad \text{for almost all } \; x \in \mathbb{R}^N.$$

and one might be tempted to take this as the sense in which $u(\cdot,t)$ takes its datum at $t=0$. Such a definition might, however, generate ambiguity. Indeed, the uniqueness may be lost, as shown by the following examples. The function Γ satisfies the heat equation in S_∞ and the growth condition (5.1). Moreover, $\Gamma(\cdot,t) \to 0$ a.e. in $\mathbb{R}^N$, as $t \to 0$, and yet $\Gamma \not\equiv 0$. For such a "solution" the identically zero initial datum is not taken in the sense of $L^1_{\rm loc}(\mathbb{R}^N)$. Indeed, for all $\rho > 0$

$$\int_{B_\rho} \Gamma(x,t)dx = \pi^{-N/2} \int_{|\eta|<\rho/2\sqrt{t}} e^{-|\eta|^2} d\eta \to 1 \qquad \text{as } \; t \to 0.$$

Even more striking is the following example in one space dimension. The function

$$v(x,t) = \frac{x}{4\sqrt{\pi}t^{3/2}} e^{-\frac{x^2}{4t}}$$

is a solution of the heat equation in $\mathbb{R}\times\mathbb{R}^+$ satisfies all the previous properties, and in addition, $v(x,t) \to 0$ as $t \to 0$ for *all* $x \in \mathbb{R}$. And yet $v \not\equiv 0$. One checks that for all $\rho > 0$

$$\sqrt{t} \int_{-\rho}^{\rho} |v(x,t)| dx \to \frac{1}{\sqrt{\pi}} \qquad \text{as } t \to 0$$

that is, the initial datum $u_o = 0$ is not taken in the sense of $L^1_{\rm loc}(\mathbb{R}^N)$.

7 Remarks on the Cauchy Problem

7.1 About Regularity

Let u_o be locally analytic in $\mathbb{R}^N$ and assume that it satisfies the growth condition (4.1). Then formula (2.7) defines the unique solution, within the class (5.1), to the Cauchy problem (1.4) in S_T for $0 < T < \frac{1}{4\alpha_o}$. Such a solution is locally analytic in the space variables. It is also analytic in the time variable within $\mathbb{R}^N \times (\varepsilon, T)$ for all $\varepsilon \in (0,T)$. Having in mind the Cauchy–Kowalewski theorem, it is natural to ask whether u is analytic in the x and t variables up to $t=0$. This is in general false, as shown by the following argument.

If u were analytic in t up to $t=0$, then u, u_t, Δu would have, in a right neighborhood of the origin, the absolutely convergent series representations

$$u(x,t) = \sum_{n=0}^{\infty} \varphi_n(x) t^n \tag{7.1}$$

$$u_t(x,t) = \sum_{n=1}^{\infty} \varphi_n(x) n t^{n-1}$$

$$\Delta u(x,t) = \sum_{n=0}^{\infty} \Delta\varphi_n(x) t^n$$

with analytic coefficients φ_n. This in the equation $H(u) = 0$ gives

$$\varphi_{n+1} = \frac{1}{(n+1)} \Delta\varphi_n, \quad n = 0, 1, \ldots$$

From this, by iteration, starting from $\varphi_o = u_o$

$$\varphi_n = \frac{\Delta^n u_o}{n!}, \quad n = 0, 1, 2, \ldots$$

where $\Delta^0 = \mathbb{I}$ and $\Delta^n = \Delta^{n-1}\Delta$, for $n \in \mathbb{N}$. Putting this in (7.1) gives a representation of u in the form

$$u(x,t) = \sum_{n=0}^{\infty} \frac{\Delta^n u_o}{n!} t^n.$$

From the uniform convergence, it follows that

$$\frac{\Delta^n u_o(x)}{n!} t^n \to 0 \quad \text{as } n \to \infty \tag{7.2}$$

for (x,t) fixed in the domain of uniform convergence. Let $D^n u_o$ denote the generic derivative of u_o, of order n . Expanding u_o about x within a ball of radius t we must have

$$\frac{|D^n u_o(x)|}{n!} t^n \to 0 \quad \text{as } n \to \infty. \tag{7.3}$$

Now there exist locally analytic initial data u_o satisfying (7.3) but not (7.2).[1]

7.2 Instability of the Backward Problem

We have already remarked that the backward Cauchy problem (2.10) in general does not have a solution. If it does, the datum u_o must by analytic by Remark 2.1 and the representation formula (2.11). Stability however might be lost, as shown by the following example, due to Hadamard ([62]):

$$u(x,t) = \varepsilon e^{-t/\varepsilon^2} \sin\left(\frac{x}{\varepsilon}\right), \quad \varepsilon > 0$$

solves (2.10) with $N = 1$ and $u_o(x) = \varepsilon \sin(x/\varepsilon)$. As $\varepsilon \to 0$, $u_o \to 0$ in the $L^\infty(\mathbb{R})$-norm. Yet for all $t < 0$, for all intervals $(-\rho, \rho)$, and for all $0 < p \le \infty$

$$\|u(\cdot,t)\|_{p,(-\rho,\rho)} \to \infty \quad \text{as } \varepsilon \to 0.$$

---

[1]Give examples of such functions in $\mathbb{R}$. *Hint:* Attempt e^{x^2} or $\ln(1+x^2)$, or a variant of these.

8 Estimates Near $t = 0$

Let $u_o \in L^\infty_{\rm loc}(\mathbb{R}^N)$ satisfy the growth condition (4.1), and let u be defined by (2.7) in the strip S_T for $0 < T < \frac{1}{4\alpha_o}$. We will study the behavior of $u(\cdot, t)$ and $|Du(\cdot, t)|$ as $t \to 0$. Since u_o is locally bounded, in (4.1) we may take $r_o = 0$, by possibly modifying the constant C_o. We will also investigate the behavior of $u_t(\cdot, t)$ and $u_{x_i x_j}(\cdot, t)$ as $t \to 0$, under the more stringent assumption that $u_o \in C^\delta_{\rm loc}(\mathbb{R}^N)$ for some $\delta \in (0, 1)$.

Proposition 8.1 *Let $u_o \in L^\infty_{\rm loc}(\mathbb{R}^N)$ satisfy (4.1) with $r_o = 0$. For all $\rho > 0$, there exist constants A_ℓ, for $\ell = 0, 1$, depending only on ρ, N, α_o, and C_o, such that*

$$|u(x,t)| \le A_o, \qquad |Du(x,t)| \le A_1 t^{-1/2} \tag{8.1}$$

for $(x,t) \in B_\rho \times (0, T]$.

Proposition 8.2 *Let $u_o \in C^\delta_{\rm loc}(\mathbb{R}^N)$, for some $\delta > 0$, satisfy (4.1) with $r_o = 0$. For all $\rho > 0$ there exists a constant A depending only on ρ, N, α_o, C_o, δ, and the Hölder constant of u_o over B_ρ such that*

$$|u_t(x,t)| + |u_{x_i x_j}(x,t)| \le A t^{\delta/2 - 1} \tag{8.2}$$

for $(x,t) \in B_\rho \times (0, T]$ and for all $i, j = 1, \dots, N$.

Proof (Proposition 8.1) Both estimates will follow from estimating

$$J_\ell = \int_{\mathbb{R}^N} \left(\frac{|x-y|}{2t} \right)^\ell \Gamma(x-y;t) |u_o(y)| dy \qquad \text{for } \ell = 0, 1.$$

The change of variable $y - x = 2\sqrt{t}\eta$ yields

$$\begin{aligned} J_\ell &= \frac{1}{\pi^{N/2}(\sqrt{t})^\ell} \int_{\mathbb{R}^N} |\eta|^\ell e^{-|\eta|^2} |u_o(x + 2\sqrt{t}\eta)| d\eta \\ &\le \frac{C_o e^{2\alpha_o |x|^2}}{\pi^{N/2}(\sqrt{t})^\ell} \int_{\mathbb{R}^N} |\eta|^\ell e^{-(1 - 8\alpha_o t)|\eta|^2} d\eta. \end{aligned}$$

Thus if $|x| < \rho$ and t is so small that $(1 - 8\alpha_o t) \ge \frac{1}{2}$

$$J_\ell \le \frac{\text{const}(C_o, N, \alpha_o)}{t^{\ell/2}} \int_0^\infty r^{N-1+\ell} e^{-\frac{1}{2}r^2} dr. \qquad \blacksquare$$

Proof (Proposition 8.2) First one computes

$$\begin{aligned} 0 = \frac{\partial^2}{\partial x_i \partial x_j} \int_{\mathbb{R}^N} \Gamma(x-y;t) dy &= \int_{\mathbb{R}^N} \frac{\partial^2}{\partial x_i \partial x_j} \Gamma(x-y;t) dy \\ &= \int_{\mathbb{R}^N} \frac{\partial^2}{\partial y_i \partial y_j} \Gamma(x-y;t) dy. \end{aligned}$$

Then for $|x| < \rho$ and $0 < t \le T$

$$|u_t(x,t)| + |u_{x_i x_j}(x,t)| \le 2 \sum_{h,k=1}^{N} \left| \int_{\mathbb{R}^N} \frac{\partial^2}{\partial y_h \partial y_k} \Gamma(x-y;t)(u_o(y) - u_o(x)) dy \right|$$

$$\le 2N \int_{\mathbb{R}^N} \left(\frac{|x-y|^2}{4t^2} + \frac{1}{2t} \right) \Gamma(x-y;t) |u_o(y) - u_o(x)| dy$$

$$= N \int_{|y|<2\rho} \left(\frac{|x-y|^2}{2t^2} + \frac{1}{t} \right) \Gamma(x-y;t) |u_o(y) - u_o(x)| dy$$

$$+ N \int_{|y|>2\rho} \left(\frac{|x-y|^2}{2t^2} + \frac{1}{t} \right) \Gamma(x-y;t) |u_o(y) - u_o(x)| dy$$

$$= H_1 + H_2.$$

Since $u_o \in C^\delta_{\rm loc}(\mathbb{R}^N)$

$$H_1 \le 2N h_o (4\pi t)^{-N/2} \int_{|y|<2\rho} \left(\frac{|x-y|^{2+\delta}}{4t^2} + \frac{|x-y|^\delta}{2t} \right) e^{-\frac{|x-y|^2}{4t}} dy$$

where h_o is the Hölder constant of u_o over $B_{2\rho}$. Perform the change of variables $y - x = 2\sqrt{t}\eta$, and majorize the resulting integral by extending it to the whole of $\mathbb{R}^N$ to get

$$H_1 \le \frac{\tilde{A}}{t^{1-\delta/2}} \int_0^\infty |\eta|^{N-1} (|\eta|^{2+\delta} + |\eta|^\delta) e^{-|\eta|^2} d|\eta|$$

where $\tilde{A} = 4N h_o \pi^{-N/2} \omega_N 2^\delta$. To estimate H_2, perform the same change of variables to get

$$H_2 \le 4C_o N \pi^{-N/2} e^{2\alpha_o |x|^2} \int_{|\eta|>\rho/2\sqrt{t}} \frac{(|\eta|^2 + 1)}{t} e^{-(1-8\alpha_o t)|\eta|^2} d\eta.$$

If t is so small that $(1 - 8\alpha_o t) \ge \frac{1}{2}$, this gives $H_2 \le \hat{A} t^{\delta/2 - 1}$, where

$$\hat{A} = \sup_{t \in (0, 1/4\alpha_o)} 4C_o N \pi^{-N/2} e^{2\alpha_o \rho^2} \int_{|\eta|>\rho/2\sqrt{t}} \frac{(|\eta|^2 + 1)}{t^{\delta/2}} e^{-\frac{1}{2}|\eta|^2} d\eta. \qquad \blacksquare$$

9 The Inhomogeneous Cauchy Problem

Consider the problem of finding $u \in \mathcal{H}(S_T)$ satisfying

$$H(u) = f \text{ in } S_T, \quad u(\cdot, 0) = u_o. \tag{9.1}$$

Assume that u_o is in $L^1_{\rm loc}(\mathbb{R}^N)$ and satisfies the growth condition (4.1). The initial datum in (9.1) is taken in the sense of $L^1_{\rm loc}(\mathbb{R}^N)$. On the forcing term f, we assume

$$f(\cdot,t) \in C^{\delta}_{\text{loc}}(\mathbb{R}^N) \quad \text{for some } \delta > 0 \text{ uniformly in } t > 0. \tag{9.2}$$

Moreover $f(\cdot,t)$ is required to satisfy the same growth condition as (4.1), uniformly in t, i.e.,

$$|f(x,t)| \le C_o e^{\alpha_o |x|^2} \quad \text{for } |x| > r_o \text{ uniformly in } t \ge 0. \tag{9.3}$$

Theorem 9.1 *Let (9.2) and (9.3) hold. Then there exists a solution to the inhomogeneous Cauchy problem (9.1) in the strip S_T for $0 < T < \frac{1}{4\alpha_o}$. Moreover, the solution is unique within the class (5.1) and is represented by*

$$u(x,t) = \int_{\mathbb{R}^N} \Gamma(x-y;t)u_o(y)dy + \int_0^t \int_{\mathbb{R}^N} \Gamma(x-y;t-s)f(y,s)dyds. \tag{9.4}$$

Proof Since the heat operator is linear, u can be constructed as the sum of the solution of the homogeneous Cauchy problem ($f = 0$) and the solution of (9.1) with $u_o = 0$. Thus it suffices to take $u_o = 0$ in (9.1). The family of homogeneous Cauchy problems

$$\begin{aligned} &(x,t;s) \to v(x,t;s) \in \mathcal{H}(\mathbb{R}^N \times [0 < s < t \le T]) \\ &v_t - \Delta v = 0 \text{ in } \mathbb{R}^N \times (s,T) \\ &v(\cdot,s;s) = f(\cdot,s) \end{aligned}$$

has, for all $0 < s < t \le T$, the unique bounded solution

$$v(x,t;s) = \int_{\mathbb{R}^N} \Gamma(x-y;t-s)f(y,s)dy$$

valid for $0 < t - s < T$. We claim that the function

$$u(x,t) = \int_0^t v(x,t;s)ds$$

solves (9.1), with $u_o = 0$. To show this, first observe that by virtue of the estimates of the previous section and assumption (9.2) and (9.2), the integrals

$$\int_0^t v(x,t;s)dx, \qquad \int_0^t v_t(x,t;s)ds, \qquad \int_0^t v_{x_i x_j}(x,t;s)ds \tag{9.5}$$

are uniformly convergent over compact subsets of $\mathbb{R}^N$. The convergence of the first integral implies that $u(\cdot,t) \to 0$ as $t \to 0$, in the sense of $L^1_{\text{loc}}(\mathbb{R}^N)$. Moreover, by direct calculation

$$\begin{aligned} u_t &= v(x,t;t) + \int_0^t v_t(x,t;s)ds \\ &= f(x,t) + \int_0^t \Delta v(x,t;s)ds = f(x,t) + \Delta u(x,t) \end{aligned}$$

where the calculation of the derivatives under the integral is justified by the uniform convergence of the integrals in (9.5). Thus u is a solution of (9.1) with $u_o = 0$. Such a solution is unique in view of (9.3). ■

This method is a particular case of the Duhamel principle. See Section 3.1c of the Problems and Complements of Chapter 6.

10 Problems in Bounded Domains

Let E be a bounded region of $\mathbb{R}^N$ with boundary ∂E of class C^1 and consider the Dirichlet problem (1.2). If $g = 0$, the problem is referred to as the homogeneous Dirichlet problem. We may solve such a homogeneous problem by separation of variables, i.e., by seeking solutions of the form $u(x,t) = X(x)T(t)$. Using the PDE, we find

$$T'(t) = -\lambda T(t), \quad t > 0; \qquad \Delta X = -\lambda X, \quad X \big|_{\partial E} = 0. \tag{10.1}$$

The second of these is solved by an infinite sequence of pairs (λ_n, v_n), where $\{\lambda_n\}$ is an increasing sequence of positive numbers and $\{v_n\}$ is a sequence of functions that form a complete orthonormal set in $L^2(E)$ (Section 11 of Chapter 4). In particular, the initial datum u_o, regarded as an element of $L^2(E)$, can be expanded as

$$u_o(x) = \sum_{i=1}^{\infty} \langle u_o, v_i\rangle v_i(x) \quad \text{with} \quad \|u_o\|_{2,E}^2 = \sum |\langle u_o, v_i\rangle|^2.$$

Then with λ_n determined by (10.1), one has $T_n(t) = T_{o,n}e^{-\lambda_n t}$, where $T_{o,n}$ are selected to satisfy the initial condition u_o. This gives approximate solutions of the form

$$u_n(x,t) = \sum_{i=1}^{n} T_{o,i}e^{-\lambda_i t}v_i(x), \quad T_{o,i} = \langle u_o, v_i\rangle.$$

Lemma 10.1 *The sequence $\{u_n(\cdot,t)\}$, is Cauchy in $L^2(E)$, uniformly in t.*

Proof Fix $\varepsilon > 0$ and let $n_o = n_o(\varepsilon)$ be such that

$$\sum_{i>n_o}^{\infty} |\langle u_o, v_i\rangle|^2 < \varepsilon. \tag{10.2}$$

Next for all $m > n > n_o$ and all $0 \le t \le T$

$$\|u_m(\cdot,t) - u_n(\cdot,t)\|_{2,E}^2 \le \Big\| \sum_{i=n}^{m} \langle u_o, v_i\rangle e^{-\lambda_i t}v_i(x)\Big\|_{2,E}^2 \le \sum_{i>n_o}^{\infty} |\langle u_o, v_i\rangle|^2 < \varepsilon. \qquad \blacksquare$$

Thus, formally, a solution to the homogeneous Dirichlet problem (1.2) is

$$u(x,t) = \sum_{i=1}^{\infty} \langle u_o, v_i\rangle e^{-\lambda_i t}v_i(x) \tag{10.3}$$

where the convergence of the series is meant in the sense of $L^2(E)$, uniformly in $t \in [0,T]$. It remains to interpret in what sense the PDE is satisfied and in what sense u takes the boundary data.

Lemma 10.2 *Let u be defined by (10.3). Then $t \to u(\cdot,t)$ is continuous in $L^2(E)$, Moreover, $u(\cdot,t)$ takes the initial datum u_o in the sense of $L^2(E)$*

$$\|u(\cdot,t) - u_o\|_{2,E} \to 0 \quad \textit{as}\, t \to 0. \tag{10.4}$$

Finally, $u(\cdot,t)$ *satisfies the decay estimate*

$$\|u(\cdot,t)\|_{2,E} \le e^{-\lambda_1 t}\|u_o\|_{2,E} \tag{10.5}$$

where λ_1 *is the first eigenvalue of the Laplacian in* E.

Proof From the definitions

$$u(x,t) - u_o(x) = \sum_{i=1}^{\infty} T_{o,i}(e^{-\lambda_i t} - 1)v_i(x).$$

Fix $\varepsilon > 0$ and choose n_o as in (10.2). Then

$$\begin{aligned}\|u(\cdot,t) - u_o\|_{2,E}^2 &= \sum_{i=1}^{\infty} |\langle u_o, v_i\rangle|^2(e^{-\lambda_i t} - 1)^2\\ &\le \sum_{i=1}^{n_o} |\langle u_o, v_i\rangle|^2(e^{-\lambda_i t} - 1)^2 + \varepsilon\\ &\le (1 - e^{-\lambda_{n_o} t})^2\|u_o\|_{2,E}^2 + \varepsilon.\end{aligned}$$

Therefore, letting $t \to 0$ gives

$$\limsup_{t\to 0} \|u(\cdot,t) - u_o\|_{2,E} \le \sqrt{\varepsilon}.$$

This proves (10.4) and also that $t \to u(\cdot,t)$ is continuous at $t = 0$, in the topology of $L^2(E)$. The continuity at every $t \in [0,T]$ is proved in a similar fashion. The decay estimate (10.5) follows from the representation (10.3), Parseval's identity, and the fact that $\{\lambda_n\}$ is an increasing sequence. ∎

Remark 10.1 This construction procedure as well as Lemmas 10.1 and 10.2 require only that the initial datum u_o be in $L^2(E)$.

10.1 The Strong Solution

Assume that $N \le 3$ and that the initial datum u_o is in $C^2(\bar{E})$ and satisfies $u_o = 0$ and $Du_o = 0$ on ∂E. Then, by Corollary 10.2 of Chapter 4, the series in (10.3) is absolutely and uniformly convergent. This implies that u satisfies the homogeneous boundary data on ∂E, in the sense of continuous functions. Also, the series

$$\sum_{i=0}^{\infty} \langle u_o, v_i\rangle \frac{d}{dt} e^{-\lambda_i t} v_i(x) \quad \text{and} \quad \sum_{i=0}^{\infty} \langle u_o, v_i\rangle e^{-\lambda_i t} \Delta v_i(x)$$

are absolutely and uniformly convergent. Therefore, the heat operator $H(\cdot)$ can be applied term by term in (10.3) to give

$$H(u) = \sum_{i=0}^{\infty} \langle u_o, v_i\rangle H[e^{-\lambda_i t} v_i(x)] = 0.$$

We conclude that if $1 \le N \le 3$, and if u_o satisfies the indicated regularity properties, then u as defined by (10.3) is a solution of the homogeneous Dirichlet problem (1.2).

10.2 The Weak Solution and Energy Inequalities

If $N > 3$, or if $u_o \in L^2(E)$, we will interpret the PDE in a weak sense. By construction, each u_n satisfies

$$\begin{aligned} u_{n,t} - \Delta u_n &= 0 \ \text{ in } E_T \\ u_n(\cdot, t)\big|_{\partial E} &= 0 \\ u_n(\cdot, 0) &= \sum_{i=0}^{n} \langle u_o, v_i \rangle v_i. \end{aligned} \tag{10.6}$$

Let $\varphi \in C^2(\bar{E}_T)$ vanish on ∂E for all t. Multiply the PDE satisfied by u_n by φ and integrate by parts over E_t to obtain

$$\int_E (u_n\varphi)(t)dx - \int_0^t \int_E [u_n\varphi_t + u_n \Delta\varphi]\, dx\, dt = \int_E (u_{o,n}\varphi)(x,0)dx.$$

Letting $n \to \infty$ gives

$$\int_E (u\varphi)(t)dx - \int_0^t \int_E uH^*(\varphi)\, dx\, dt = \int_E u_o\varphi(x,0)dx. \tag{10.7}$$

In this limiting process we use Lemma 10.1, which is valid for all $N \ge 1$. We regard (10.7), as a weak notion of a solution of the homogeneous Dirichlet problem (1.2), and we call u a *weak solution.*

Lemma 10.3 *Weak solutions in the sense of (10.5), (10.7) are unique.*

Proof The difference $w = u_1 - u_2$ of any two solutions satisfies (10.3), and in particular

$$\int_E (w\varphi)(t)dx - \int_0^t \int_E w\Delta\varphi\, dx\, d\tau = 0 \tag{10.8}$$

for all $\varphi \in C_o^1(E)$ independent of t. Since $w \in L^2(E_T)$, it must have for a.e. $t \in (0,T)$ a representation in terms of the eigenfunctions $\{v_n\}$, i.e.,

$$w(x,t) = \lim_{n\to\infty} \sum_{i=0}^{n} a_i(t)v_i(x) \qquad \text{for a.e. } t \in (0,T).$$

In (10.8) choose $\varphi = v_i$ to get

$$a_i(t) + \lambda_i \int_0^t a_i(s)ds = 0 \qquad \text{for all } i \in \mathbb{N}.$$

Thus $a_i(\cdot) = 0$ for all $i \in \mathbb{N}$, and $w = 0$. ∎

Remark 10.2 The choice $\varphi = v_i$ is admissible if $v_i \in C_o^1(E)$. By Corollary 11.2 of Chapter 3, the eigenfunctions v_i are Hölder continuous in $\bar{E}$. By the Schauder estimates of Section 9 of Chapter 2, $v_i \in C^{2+\eta}(E)$, and by a bootstrap argument, $v_i \in C^\infty(E)$. Actually v_i are of class $C_o^{1+\eta}$ up to ∂E. Such an estimate up to the boundary, has been indicated in Section 9 of the Problems and Complements of Chapter 2.

Remark 10.3 If $u_o \in C_o^1(E)$, and u is smooth enough, we may take $u = \varphi$ in (10.7) to obtain the energy identity

$$\frac{1}{2}\|u(t)\|_{2,E}^2 - \frac{1}{2}\|u_o\|_{2,E}^2 + \int_0^t \int_E |Du|^2 dx\, dt = 0.$$

This identity also contains a statement of uniqueness since the PDE is linear. Indeed, $u_o = 0$ implies $u(\cdot, t) = 0$.

11 Energy and Logarithmic Convexity

Let E be a bounded open set in $\mathbb{R}^N$ with boundary ∂E of class C^1 and let $u \in \mathcal{H}(E_T) \cap C(\bar{E}_T)$ be a solution of the homogeneous Dirichlet problem (1.2). The quantity

$$\mathcal{E}(t) = \|u(\cdot, t)\|_{2,E}^2$$

is the thermal energy of the body E at time t.

Proposition 11.1 *For every* $0 \le t_1 < t < t_2 \le T$

$$\mathcal{E}(t) \le [\mathcal{E}(t_1)]^{\frac{t_2 - t}{t_2 - t_1}} [\mathcal{E}(t_2)]^{\frac{t - t_1}{t_2 - t_1}}. \tag{11.1}$$

Proof Assume first that u is sufficiently regular as to justify the formal calculations below. Multiply the first of (1.2) by $u(\cdot, t)$ and integrate by parts over E, taking into account that $u(\cdot, t)$ vanishes on ∂E. This gives

$$\mathcal{E}' = 2\int_E u\Delta u\, dx = -2\int_E Du \cdot Du\, dx.$$

From this

$$\mathcal{E}'' = -4\int_E Du_t \cdot Du\, dx = 4\int_E u_t \Delta u\, dx = 4\int_E u_t^2 dx.$$

From this and Hölder's inequality

$$\mathcal{E}'^2 = \left(2\int_E u u_t\, dx\right)^2 \le \mathcal{E}\mathcal{E}''.$$

First assume that $\mathcal{E}(t) > 0$ for all $t \in [t_1, t_2]$. Then the function $t \to \ln \mathcal{E}(t)$ is well defined and convex in such an interval, since

$$\frac{d^2}{dt^2} \ln \mathcal{E} = \frac{\mathcal{E}''\mathcal{E} - \mathcal{E}'^2}{\mathcal{E}^2} \ge 0.$$

Therefore, for all $t_1 < t < t_2$

$$\ln \mathcal{E}(t) \le \frac{t_2 - t}{t_2 - t_1} \ln \mathcal{E}(t_1) + \frac{t - t_1}{t_2 - t_1} \ln \mathcal{E}(t_2).$$

If $\mathcal{E}(t) \ge 0$, replacing it with $\mathcal{E}_\varepsilon = \mathcal{E} + \varepsilon$ for $\varepsilon > 0$ proves (11.1) for $\mathcal{E}_\varepsilon$. Then let $\varepsilon \to 0$. These calculations can be made rigorous by working with the approximate solutions $\{u_n\}$ of (10.5) and then by letting $n \to \infty$. ∎

Remark 11.1 The energy $\mathcal{E}(\cdot)$ can also be defined for solutions of the homogeneous Neumann problem (1.2), and (11.1) holds for it ([116]).

11.1 Uniqueness for Some Ill-Posed Problems

Corollary 11.1 *There exists at most one solution to the homogeneous backward Dirichlet problem*

$$\begin{aligned} u \in \mathcal{H}(E_T) \cap C(\bar{E}_T), &\quad H(u) = 0 \;\; in \;\; E_T \\ u(\cdot, T) = u_T \in C(\bar{E}), &\quad u(\cdot, t) \big|_{\partial E} = 0. \end{aligned}$$

Proof It suffices to show that $u_T = 0$ implies $u(\cdot, t) = 0$. This follows from (11.1) with $t_2 = T$. ∎

12 Local Solutions

We have observed that solutions of the Cauchy problem representable by (2.7) are analytic in the space variables and C^∞ in time for $t > 0$. It turns out that this is also the case for every local solution of the heat equation in a space-time cylindrical domain E_T. Let $Q_\rho = B_\rho \times (-\rho^2, 0)$ denote the cylinder with "vertex" at the origin, height ρ^2, and transversal cross section the ball B_ρ. For $(x_o, t_o) \in \mathbb{R}^{N+1}$, we let $(x_o, t_o) + Q_\rho$ denote the box congruent to Q_ρ and with "vertex" at (x_o, t_o), i.e.,

$$(x_o, t_o) + Q_\rho = [|x - x_o| < \rho] \times [t_o - \rho^2, t_o].$$

If $(x_o, t_o) \in E_T$, we let $\rho > 0$ be so small that $(x_o, t_o) + Q_{4\rho} \subset E_T$. We also denote the integral average of $|u|$ over $(x_o, t_o) + Q_{4\rho}$ by

$$\fint_{(x_o,t_o)+Q_{4\rho}} |u| \, dy \, ds = \frac{1}{|Q_{4\rho}|} \int_{(x_o,t_o)+Q_{4\rho}} |u| \, dy \, ds.$$

Proposition 12.1 (Gevrey [52]) *Let $u \in \mathcal{H}(E_T)$ be a solution of the heat equation in E_T. There exist constants γ and C depending only on N such that for every box $(x_o, t_o) + Q_{4\rho} \subset E_T$*

$$\sup_{(x_o,t_o)+Q_\rho} |D^\alpha u| \le \gamma \frac{C^{|\alpha|} |\alpha|!}{\rho^{|\alpha|}} \fint_{(x_o,t_o)+Q_{4\rho}} |u| \, dy \, ds \tag{12.1}$$

for all multi-indices α. Moreover

$$\sup_{(x_o,t_o)+Q_\rho} \Big| \frac{\partial^k u}{\partial t^k} \Big| \le \gamma \frac{C^{2k} (2k)!}{\rho^{2k}} \fint_{(x_o,t_o)+Q_{4\rho}} |u| \, dy \, ds \tag{12.2}$$

for all positive integers k.

Proof It suffices to prove only (12.1), since

$$\frac{\partial^k}{\partial t^k} u = \Delta^k u.$$

After a translation, we may assume that (x_o, t_o) coincides with the origin and u is a solution of the heat equation in $Q_{4\rho}$. Construct a non-negative smooth cutoff function ζ in $Q_{4\rho}$ satisfying

$$\begin{aligned} &\zeta = 1 \text{ in } Q_{2\rho}, \quad |D\zeta| \le \frac{1}{4\rho}, \quad |\zeta_{y_i y_j}| \le \frac{1}{\rho^2}\, i,j = 1,\dots,N \\ &0 \le \zeta_t \le \frac{1}{\rho^2}, \quad \zeta(y,s) = 0 \text{ for } |y| \ge 4\rho \text{ and } \ s \le -(4\rho)^2. \end{aligned}$$

The function

$$w = \begin{cases} u\zeta & \text{in } [|y| \le 4\rho] \times (-(4\rho)^2, 0) \\ 0 & \text{otherwise} \end{cases}$$

coincides with u within $Q_{2\rho}$ and satisfies

$$H(w) = uH(\zeta) - 2Du \cdot D\zeta \stackrel{\text{def}}{=} f \quad \text{in } \mathbb{R}^N \times (-(4\rho)^2, 0].$$

Therefore, it can be viewed as the unique solution of the inhomogeneous Cauchy problem

$$H(w) = f \text{ in } \mathbb{R}^N \times (-(4\rho)^2, 0], \quad w(\cdot, -(4\rho)^2) = 0.$$

By Theorem 9.1

$$w(x,t) = \int_{-(4\rho)^2}^{t} \int_{\mathbb{R}^N} \Gamma(x-y; t-s) f(y,s)\, dy\, ds.$$

From this, after an integration by parts

$$\begin{aligned} w(x,t) &= \int_{-(4\rho)^2}^{t} \int_{\mathbb{R}^N} u(y,s) \big[\Gamma(x-y;t-s) H^*(\zeta) \\ &\qquad\qquad + 2D\Gamma(x-y;t-s) \cdot D\zeta \big]\, dy\, ds \\ &= \int_{-(4\rho)^2}^{-(2\rho)^2} \int_{|y|<4\rho} \Gamma(x-y;t-s) \zeta_t\, u\, dy\, ds \\ &\quad + \int_{-(4\rho)^2}^{t} \int_{2\rho<|y|<4\rho} \Gamma(x-y;t-s) \Delta\zeta\, u\, dy\, ds \\ &\quad + 2 \int_{-(4\rho)^2}^{t} \int_{2\rho<|y|<4\rho} D\Gamma(x-y;t-s) \cdot D\zeta\, u\, dy\, ds. \end{aligned}$$

Observe that in these integrals, if $|x| < \rho$ and $-\rho^2 \le t \le 0$, the kernel is not singular. Take the space derivatives of any order of both sides, for x and t in such a range, and use the properties of the cutoff function ζ to obtain

$$
\begin{aligned}
\sup_{Q_\rho}|D^\alpha u| &\le \frac{1}{\rho^2}\int_{-(4\rho)^2}^{-(2\rho)^2}\int_{|y|<4\rho}|D^\alpha\Gamma(x-y;t-s)||u|\,dy\,ds\\
&+\frac{1}{\rho^2}\int_{-(4\rho)^2}^{t}\int_{2\rho<|y|<4\rho}|D^\alpha\Gamma(x-y;t-s)||u|\,dy\,ds\\
&+\frac{1}{\rho}\int_{-(4\rho)^2}^{t}\int_{2\rho<|y|<4\rho}\sum_{i=1}^{n}|D^\alpha\Gamma_{x_i}(x-y;t-s)||u|\,dy\,ds\\
&= J_1+J_2+J_3.
\end{aligned}
$$

In the estimates to follow we denote by C and γ generic positive constants that can be different in different contexts. These may be quantitatively determined a priori only in terms of N and are independent of the multi-index α.

Lemma 12.1 *There exists a positive constant C such that*

$$
|D^\alpha\Gamma(x-y;t-s)|\le C^{|\alpha|}\left[\left(\frac{\rho}{t-s}\right)^{|\alpha|}+\frac{|\alpha|!}{\rho^{|\alpha|}}\right]\Gamma(x-y;t-s)
$$

for all $(x,t)\in Q_\rho$ and $(y,s)\in Q_{4\rho}$, and for every multi-index α.

Assuming the lemma for the moment, we proceed to estimate J_i. In estimating J_1 observe that within the domain of integration $t-s>\rho^2$. Therefore

$$
|D^\alpha\Gamma(x-y;t-s)|\le\gamma\frac{C^{|\alpha|}|\alpha|!}{\rho^{|\alpha|+N}}
$$

and

$$
J_1\le\gamma\frac{C^{|\alpha|}|\alpha|!}{\rho^{|\alpha|}}\frac{1}{|Q_{4\rho}|}\int_{Q_{4\rho}}|u|\,dy\,ds.
$$

The estimation of J_2 and J_3 hinges on the supremum of the function

$$
g(\tau)=\frac{1}{\tau^m}e^{-A/\tau},\qquad \tau>0
$$

where A and m are given positive constants. The supremum of g is achieved for $\tau=A/m$, and

$$
|g(\tau)|\le\left(\frac{m}{A}\right)^m e^{-m}\qquad\text{for all }\ \tau\ge 0.
$$

Within the domain of integration of J_2 and J_3 one has $|x-y|>\rho$, provided $|x|<\rho$. Therefore

$$
\begin{aligned}
|D^\alpha\Gamma(x-y;t-s)| &\le \gamma C^{|\alpha|}\frac{\rho^{|\alpha|}}{[4(t-s)]^{|\alpha|+N/2}}e^{-\rho^2/4(t-s)}\\
&\quad+\gamma C^{|\alpha|}\frac{|\alpha|!}{\rho^{|\alpha|}}\frac{1}{[4(t-s)]^{N/2}}e^{-\rho^2/4(t-s)}\\
&\le\gamma\frac{C^{|\alpha|}}{\rho^{|\alpha|+N}}|\alpha|^{|\alpha|}e^{-|\alpha|}+\gamma\frac{C^{|\alpha|}|\alpha|!}{\rho^{|\alpha|+N}}.
\end{aligned}
$$

By Stirling's inequality $m^m e^{-m} \le \gamma m!$. Therefore by modifying the constants C and γ

$$|D^\alpha \Gamma(x-y;t-s)| \le \gamma \frac{C^{|\alpha|}|\alpha|!}{\rho^{|\alpha|}} \frac{1}{\rho^N}$$

for $(x,t) \in Q_\rho$ and $(y,s) \in Q_{4\rho} - Q_{2\rho}$. With this estimate in hand, we deduce that for $(x,t) \in Q_\rho$ and all multi-indices α

$$J_2 \le \gamma \frac{C^{|\alpha|}|\alpha|!}{\rho^{|\alpha|}} \frac{1}{|Q_{4\rho}|} \int_{Q_{4\rho}} |u|\, dy\, ds.$$

As for J_3, the previous calculations give

$$J_3 \le \gamma \frac{C^{|\alpha|+1}|\alpha|!(|\alpha|+1)}{\rho^{|\alpha|}} \frac{1}{|Q_{4\rho}|} \int_{Q_{4\rho}} |u|\, dy\, ds.$$

Now the constant C can be further modified so that

$$C^{|\alpha|+1}(|\alpha|+1) \le \bar{C}^{|\alpha|} \qquad \text{for all multi-indices } \alpha$$

and the theorem follows. ■

Proof (Lemma 12.1) Fix a multi-index α of size $|\alpha| = n$ and let β be a multi-index of size $|\beta| = n+1$. Then

$$D^\beta \Gamma = D^\alpha \Gamma_{x_i} = D^\alpha \frac{(x-y)_i}{2(t-s)} \Gamma$$

for some $i = 1, \dots, N$. From this

$$2|D^\beta \Gamma| \le \left(\frac{\rho}{t-s}\right) |D^\alpha \Gamma| + \frac{n-1}{(t-s)} |D^{\bar{\alpha}} \Gamma|$$

where $\bar{\alpha}$ is a multi-index of size $|\bar{\alpha}| = n-1$. The lemma holds for $n = 1$. By induction, assuming that it does hold for multi-indices α of size $|\alpha| \le n$, we show that it continues to hold for multi-indices β of size $|\beta| = n+1$. Using the induction hypothesis

$$\frac{2}{\Gamma}|D^\beta \Gamma| \le C^n \left[\left(\frac{\rho}{t-s}\right)^{n+1} + \left(\frac{\rho}{t-s}\right) \frac{2n!}{\rho^n} + \left(\frac{\rho}{t-s}\right)^n \frac{n-1}{C\rho} \right].$$

By Young's inequality

$$\left(\frac{\rho}{t-s}\right) \frac{2n!}{\rho^n} \le \frac{1}{n+1} \left(\frac{\rho}{t-s}\right)^{n+1} + \frac{2^{\frac{n+1}{n}} n}{n+1} \frac{(n!)^{\frac{n+1}{n}}}{\rho^{n+1}}$$

$$\left(\frac{\rho}{t-s}\right)^n \frac{n-1}{C\rho} \le \frac{n}{n+1} \left(\frac{\rho}{t-s}\right)^{n+1} + \frac{(n-1)^{n+1}}{n+1} \frac{1}{C^{n+1}\rho^{n+1}}.$$

Using Stirling's inequality and choosing C sufficiently large

$$\frac{(n-1)^{n+1}}{n+1}\frac{1}{C^{n+1}} \le (n+1)!.$$

This in turn implies

$$\frac{|D^\beta \Gamma|}{\Gamma} \le \frac{C^n}{2}\left[2\left(\frac{\rho}{t-s}\right)^{n+1} + \left(1+\frac{4n}{(n+1)^2}(n!)^{1/n}\right)\frac{(n+1)!}{\rho^{n+1}}\right].$$

The number $(n!)^{1/n}$ is the geometric mean of the first n integers, which is majorized by its arithmetic mean. Therefore

$$\frac{4n}{(n+1)^2}(n!)^{1/n} \le \frac{4n}{(n+1)^2}\frac{\sum_{i=1}^n i}{n} = \frac{2n}{n+1}.$$

These remarks in the previous inequality yield

$$\frac{|D^\beta \Gamma|}{\Gamma} \le \frac{3}{2}C^n\left[\left(\frac{\rho}{t-s}\right)^{n+1} + \frac{(n+1)!}{\rho^{n+1}}\right].$$

It remains to choose C so that $\frac{3}{2}C^n \le C^{n+1}$. ∎

12.1 Variable Cylinders

To simplify the symbolism, let us assume that (x_o, t_o) coincides with the origin. The estimates of Theorem 12.1 give information on $D^\alpha u$ on the cylinder Q_ρ in terms of the L^1-norm of u over the larger box $Q_{2\rho}$. The proof could be repeated with minor variations to derive a similar statement for any pair of boxes Q_ρ and $Q_{\sigma\rho}$ for $\sigma \in (0,1)$. Tracing the constant dependence on σ gives:

Proposition 12.2 *Let u be a solution of the heat equation in Q_ρ. There exist constants C and γ, depending only on N, such that for every multi-index α, for every non-negative integer k, and for all $\sigma \in (0,1)$*

$$\|D^\alpha u\|_{\infty, Q_{\sigma\rho}} \le \gamma \frac{C^{|\alpha|}|\alpha|!}{(1-\sigma)^{N+2+|\alpha|}\rho^{|\alpha|}} \int_{Q_\rho} |u|\,dy\,ds \tag{12.3}$$

$$\|\frac{\partial^k}{\partial t^k}u\|_{\infty, Q_{\sigma\rho}} \le \gamma \frac{C^k(2k)!}{(1-\sigma)^{N+2+2k}\rho^{2k}} \int_{Q_\rho} |u|\,dy\,ds. \tag{12.4}$$

Remark 12.1 Estimates (12.3)–(12.4) hold for any pair of boxes $(x_o, t_o)+Q_\rho$ and $(x_o, t_o)+Q_{\sigma\rho}$ contained in Ω_T.

12.2 The Case $|\alpha| = 0$

We state explicitly the estimate of Proposition 12.2 for the case $|\alpha| = 0$.

Corollary 12.1 *Let $u \in \mathcal{H}(E_T)$ be a local solution of the heat equation in E_T. There exists a constant C depending only on N such that for every box $(x_o, t_o) + Q_\rho$ contained in E_T and all $\sigma \in (0,1)$*

$$\sup_{(x_o,t_o)+Q_{\sigma\rho}} |u| \le \frac{C}{(1-\sigma)^{N+2}} \fint_{(x_o,t_o)+Q_\rho} |u|\, dy\, ds. \tag{12.5}$$

These estimates have a number of consequences for local or global non-negative solutions. In the next two sections we present some of them.

13 The Harnack Inequality

Non-negative local solutions of the heat equation in E_T satisfy an inequality similar to the Harnack estimate valid for non-negative harmonic functions(Section 5.1 of Chapter 2). This inequality can be stated as follows. For $\rho > 0$ consider the box $\mathcal{Q}_\rho = B_\rho \times (-\rho^2, \rho^2)$, with its "center" at the origin. If $(x_o, t_o) \in E_T$, let

$$(x_o, t_o) + \mathcal{Q}_\rho = [|x - x_o| < \rho] \times (t_o - \rho^2, t_o + \rho^2)$$

be the box congruent to $\mathcal{Q}_\rho$ and centered at (x_o, t_o).

Theorem 13.1 *Let $u \in \mathcal{H}(E_T)$ be a non-negative solution of the heat equation in E_T. There exists a constant c depending only upon N such that for every box $(x_o, t_o) + \mathcal{Q}_{4\rho} \subset E_T$*

$$\inf_{|x-x_o|<\rho} u(x, t_o + \rho^2) \ge c\, u(x_o, t_o). \tag{13.1}$$

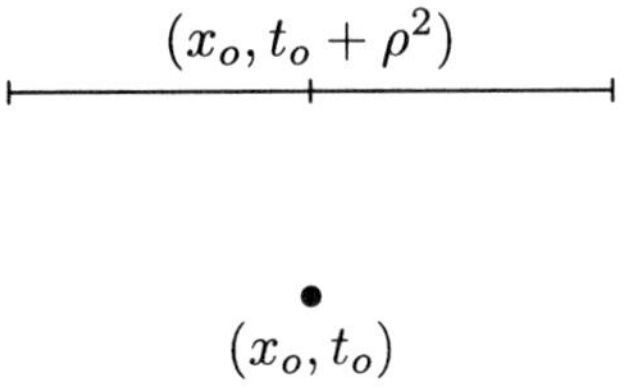

$(x_o, t_o + \rho^2)$

(x_o, t_o)

Fig. 13.1.

Such an estimate can be given different equivalent forms. We illustrate one of them, assuming for simplicity of notation that $(x_o, t_o) = (0,0)$. To distinguish between the upper part and the lower part of $\mathcal{Q}_\rho$, let us set

$$\mathcal{Q}_\rho^- = B_\rho \times (-\rho^2, 0), \qquad \mathcal{Q}_\rho^+ = B_\rho \times (0, \rho^2).$$

Fix $\sigma \in (0,1)$, and inside $\mathcal{Q}_\rho^+$ and $\mathcal{Q}_\rho^-$ construct the two sub-boxes

$$\mathcal{Q}_{\sigma\rho}^- = B_{\sigma\rho} \times (-\sigma\rho^2, 0), \qquad \mathcal{Q}_{\sigma\rho}^* = B_{\sigma\rho} \times \big((1-\sigma)\rho^2, \rho^2\big).$$

Theorem 13.2 *Let $u \in \mathcal{H}(Q_{4\rho})$ be a non-negative solution of the heat equation in $Q_{4\rho}$. For every $\sigma \in (0,1)$ there exists a constant c depending only upon N and σ such that*

$$\inf_{Q^*_{\sigma\rho}} u \geq c \sup_{Q^-_{\sigma\rho}} u. \tag{13.2}$$

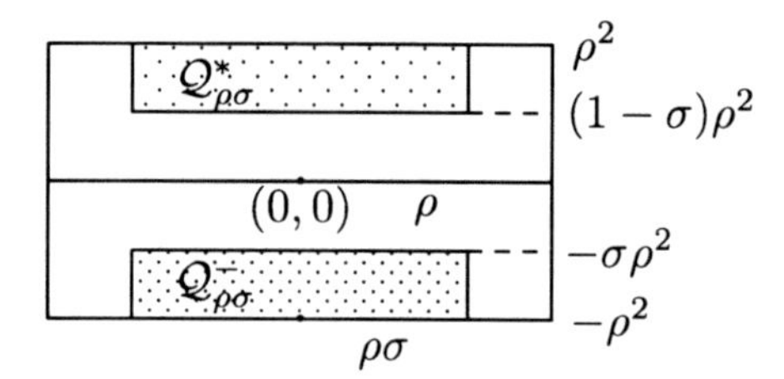

Fig. 13.2.

In the case of harmonic functions, the main tool in the proof of the Harnack estimate was the explicit Poisson representation formula of the solution of the Dirichlet problem for the Laplacian over a ball (formula (3.9) of Chapter 2). The corresponding Dirichlet problem for the heat equation over cylinders whose cross section is a sphere does not have an explicit solution formula. However, local representations will play a major role via the regularity results of Proposition 12.1.

The form (13.1) of the Harnack estimate is due independently to Pini ([118]) and Hadamard ([63]). The form (13.2) was introduced by Moser in a more general context ([110]). The proof we present here, based on an idea of Landis ([93]), is "non-linear" in nature, and its main ideas can be applied to a large class of parabolic equations, including degenerate ones ([30], Chapters 6–7, and [34]). Alternative forms of the parabolic Harnack inequality that resemble the mean value property of harmonic functions are in [35].

13.1 Compactly Supported Sub-Solutions

For given positive numbers M, r, b, and $(x,t) \in S_\infty$, consider the function

$$\psi(x,t) = \frac{M\, r^{2b}}{(t+r^2)^b}(4-|z|^2)^2_+, \qquad \text{where } |z|^2 = \frac{|x|^2}{t+r^2}.$$

One verifies that $\psi \in \mathcal{H}(S_\infty) \cap C(\bar{S}_\infty)$, and it vanishes identically outside the paraboloid $|z| < 2$.

Lemma 13.1 *The number $b > 0$ can be chosen so that $H(\psi) \leq 0$ in S_∞, for all $M > 0$.*

Proof By direct calculation

$$H(\psi) = \frac{Mr^{2b}}{(t+r^2)^{b+1}}(4-|z|^2)_+ \left(-b(4-|z|^2)_+ + 4N - 2\frac{|z|^4}{4-|z|^2}\right).$$

For $\frac{4N}{N+1} \le |z|^2 < 4$, we have $H(\psi) \le 0$. For $|z|^2 \le \frac{4N}{N+1}$

$$H(\psi) \le \frac{Mr^{2b}}{(t+r^2)^{b+1}}(1-|z|^2)_+ \left(-\frac{4b}{N+1} + 4N\right).$$

To prove the lemma, choose $b = N(N+1)$. ∎

We will consider a version of ψ "centered" at points $(x_o, t_o) \in \mathbb{R}^{N+1}$. Precisely

$$\Psi_{(x_o,t_o)}(x,t) = \psi(x - x_o, t - t_o) = \frac{Mr^{2b}}{[t - t_o + r^2]^b}\left(4 - \frac{|x - x_o|^2}{t - t_o + r^2}\right)_+^2.$$

Corollary 13.1 *Let* $b = N(N+1)$. *Then*

$$H\big(\Psi_{(x_o,t_o)}\big) \le 0 \quad \textit{in } \mathbb{R}^N \times (t_o, \infty).$$

13.2 Proof of Theorem 13.1

We may assume that $(x_o, t_o) = (0,0)$, that $\rho = 1$, and that $u(x_o, t_o) = 1$. This is achieved by the change of variables

$$x \to \frac{x - x_o}{\rho}, \qquad t \to \frac{t - t_o}{\rho^2}, \qquad u \to \frac{u}{u(0,0)}.$$

Thus we have to show that if u is a solution of the heat equation in the box $\mathcal{Q}_4 = B_4 \times (-4,4)$ such that $u(0,0) = 1$, then $u(x,1) \ge c$, for all $x \in B_1$, for a positive constant c depending only on N. To prove this we proceed in three steps.

13.2.1 Locating the Supremum of u in Q_1

For $s \in [0,1)$, consider the family of nested and expanding boxes

$$Q_s = [|x| < s] \times (-s^2, 0]$$

and the non-decreasing family of numbers

$$M_s = \sup_{Q_s} u, \qquad N_s = (1-s)^{-\xi}$$

where ξ is a positive constant to be chosen later. One checks that $M_o = N_o = 1$, and as $s \to 1$

$$\lim_{s\to 1} M_s < \infty \quad \text{and} \quad \lim_{s \to 1} N_s = \infty.$$

Therefore the equation $M_s = N_s$ has roots. Denote by s_o the largest of such roots, so that

$$\sup_{Q_{s_o}} u = M_{s_o} = (1 - s_o)^{-\xi} \quad \text{and } M_s < (1-s)^{-\xi} \quad \text{for } s > s_o.$$

Since $u \in C(\bar{\mathcal{Q}}_4)$, the supremum M_{s_o} is achieved at some $(x_o, t_o) \in \bar{Q}_{s_o}$, i.e., $u(x_o, t_o) = (1 - s_o)^{-\xi}$.

13.2.2 Positivity of u over a Ball

We show next that the "largeness" of u at (x_o, t_o) spreads over a small ball centered at x_o at the level t_o.

Lemma 13.2 *There exists $\varepsilon > 0$ depending only upon N and independent of s_o such that*

$$u(x, t_o) \ge \frac{1}{2}(1 - s_o)^{-\xi} \quad \textit{for all } |x - x_o| < \varepsilon \frac{1 - s_o}{2}.$$

Proof Costruct the box with "vertex" at (x_o, t_o) and radius $\frac{1}{2}(1 - s_o)$

$$(x_o, t_o) + Q_{\frac{1}{2}(1-s_o)} = \left[|x - x_o| < \frac{1 - s_o}{2}\right] \times \left[t_o - \left(\frac{1 - s_o}{2}\right)^2, t_o\right].$$

By construction, $(x_o, t_o) + Q_{\frac{1}{2}(1-s_o)} \subset Q_{\frac{1}{2}(1+s_o)}$, and by the definition of M_s and N_s

$$\sup_{(x_o,t_o)+Q_{\frac{1}{2}(1-s_o)}} u \le \sup_{Q_{\frac{1}{2}(1+s_o)}} u \le N_{\frac{1+s_o}{2}} = 2^\xi (1 - s_o)^{-\xi}.$$

Apply Proposition 12.2 with $|\alpha| = 1$ over the pair of boxes $(x_o, t_o) + Q_{\frac{1}{8}(1-s_o)}$ and $(x_o, t_o) + Q_{\frac{1}{2}(1-s_o)}$, to obtain

$$\sup_{(x_o,t_o)+Q_{\frac{1}{8}(1-s_o)}} |Du| \le \frac{C}{1 - s_o} \sup_{(x_o,t_o)+Q_{\frac{1}{2}(1-s_o)}} u$$

for a constant C dependent only upon N. Let $\varepsilon \in (0, \frac{1}{4})$ to be chosen later. Then for all $|x - x_o| < \varepsilon \frac{1}{2}(1 - s_o)$

$$\begin{aligned} u(x, t_o) &\ge u(x_o, t_o) - \varepsilon \frac{1 - s_o}{2} \sup_{(x_o,t_o)+Q_{\frac{1}{8}(1-s_o)}} |Du| \\ &\ge (1 - s_o)^{-\xi}(1 - 2^\xi C\varepsilon). \end{aligned}$$

To prove the lemma we choose ε small enough that $1 - 2^\xi C\varepsilon = \frac{1}{2}$. ∎

13.2.3 Expansion of the Positivity Set

The point (x_o, t_o) being fixed, consider the comparison function $\Psi_{(x_o,t_o)}$ for the choice of parameters

$$M = \frac{1}{2}(1 - s_o)^{-\xi}, \qquad r = \varepsilon \frac{1 - s_o}{2}.$$

By Lemma 13.2

$$u(x,t_o) \ge \frac{1}{2}(1-s_o)^{-\xi} \ge \Psi_{(x_o,t_o)}(x,t_o), \qquad \text{for } |x-x_o|<r.$$

Therefore, by the maximum principle, $u \ge \Psi_{(x_o,t_o)}$ in the box $B_4 \times [t_o,4)$. In particular, for $t=1$ and $|x|<1$

$$u(x,1) \ge 2^{1-4b}(1-s_o)^{-\xi+b}.$$

The knowledge of s_o is only qualitative. We render the estimate independent of s_o by choosing $\xi = b$. This gives $u(x,1) \ge 2^{1-4b} = c$. ■

14 Positive Solutions in S_T

We have shown that uniqueness for the Cauchy problem (1.4) holds within the class of functions satisfying the growth condition (5.1). However, the representation formula (2.7) is well defined for initial data $u_o \in L^1_{\text{loc}}(\mathbb{R}^N)$ for which the integral is convergent. This suggests that we consider the problem of uniqueness within the class of functions $u \in \mathcal{H}(S_T)$ such that

$$\int_{\mathbb{R}^N} |u(y,s)|\Gamma(x-y;t-s)dy < \infty \qquad \text{for } s \in (0,t). \tag{14.1}$$

It turns out that uniqueness for the Cauchy problem holds for functions in such a class. More important, every non-negative solution of the heat equation in S_T satisfies (14.1). Therefore, uniqueness for the Cauchy problem (1.4) holds within the class of non-negative solutions. This was observed by Widder in one space dimension ([164]). Here we give a different proof, valid in any space dimension.

Theorem 14.1 *Let $u \in \mathcal{H}(S_T)$ satisfy*

$$H(u) = 0 \ \text{ in } S_T, \qquad \text{and} \qquad u(\cdot,t) \to 0 \ \text{ in } L^1_{\text{loc}}(\mathbb{R}^N) \ \text{ as } t \to 0.$$

Then, if u satisfies (14.1), it vanishes identically in S_T.

Proof Let $(x,t) \in S_T$ be arbitrary but fixed. For $\rho > 2|x|$ consider the balls $B_{2\rho}$ and let $y \to \zeta(y) \in C^2_o(B_{2\rho})$ be a non-negative cutoff function in $B_{2\rho}$ satisfying

$$\zeta = 1 \ \text{ in } B_\rho, \quad |D\zeta| \le \frac{1}{\rho}, \quad |\zeta_{y_iy_j}| \le \frac{2}{\rho}^2.$$

For $\delta > 0$ let

$$h_\delta(u) = \begin{cases} 1 & \text{if } u > \delta \\ \dfrac{u}{\delta} & \text{if } |u| \le \delta \\ -1 & \text{if } u < -\delta. \end{cases} \tag{14.2}$$

Multiply the PDE by $h_\delta(u)\zeta(y)\Gamma(x-y;t-s)$, and integrate by parts in $dy\,ds$ over the cylindrical domain $B_{2\rho}\times(\tau,t-\varepsilon)$ for $0<\tau<t-\varepsilon$ and $0<\varepsilon<t$. This gives

$$\begin{aligned}\int_{B_{2\rho}\times\{t-\varepsilon\}}&\left(\int_0^u h_\delta(\xi)d\xi\right)\Gamma(x-y;\varepsilon)\zeta(y)dy\\ &+\frac{1}{\delta}\int_\tau^{t-\varepsilon}\int_{B_{2\rho}}|Du|^2\Gamma(x-y;t-s)\chi_{[}|u|<\delta]\zeta(y)\,dy\,ds\\ &=\int_{B_{2\rho}\times\{\tau\}}\left(\int_0^u h_\delta(\xi)d\xi\right)\Gamma(x-y;t-\tau)\zeta(y)dy\\ &+\int_\tau^{t-\varepsilon}\left(\int_0^u h_\delta(\xi)d\xi\right)(\Gamma\Delta\zeta+2D\Gamma\cdot D\zeta)\,dy\,ds.\end{aligned}$$

As $\tau\to0$, the first integral on the right-hand side tends to zero, since it is majorized by

$$\text{const}\int_{B_{2\rho}}|u(y,\tau)|dy\to0\ \text{ as }\ \tau\to0.$$

Discard the second term on the left-hand side since it is non-negative and let first $\delta\to0$ and then $\tau\to0$ to obtain

$$\begin{aligned}\int_{B_{2\rho}}&|u(y,t-\varepsilon)|\Gamma(x-y;\varepsilon)\zeta(y)dy\\ &\le\frac{2}{\rho^2}\int_0^{t-\varepsilon}\int_{B_{2\rho}}|u(y,s)|\Gamma(x-y;t-s)\,dy\,ds\\ &+\frac{2}{\rho}\int_0^{t-\varepsilon}\int_{B_{2\rho}-B_\rho}|u(y,s)||D\Gamma(x-y;t-s)|\,dy\,ds.\end{aligned}$$

The right-hand side of this inequality tends to zero as $\rho\to\infty$. This is obvious for the first term in view of (14.1). The second term is majorized by

$$\begin{aligned}\frac{1}{\rho}\int_0^{t-\varepsilon}&\int_{\rho<|y|<2\rho}|u(y,s)|\Gamma(x-y;t-s)\frac{|x-y|}{t-s}\,dy\,ds\\ &\le\frac{4}{\varepsilon}\int_0^{t-\varepsilon}\int_{\rho<|y|<2\rho}|u(y,s)|\Gamma(x-y;t-s)\,dy\,ds.\end{aligned}$$

Letting $\rho\to\infty$ gives

$$\int_{B_r}|u(y,t-\varepsilon)|\Gamma(x-y;\varepsilon)dy=0$$

for all $r>2|x|$ and all $\varepsilon\in(0,t)$. Finally, we let $\varepsilon\to0$. Arguing as in Section 2, in the derivation of the representation formula (2.7), gives $u(x,t)=0$. ■

14.1 Non-Negative Solutions

Theorem 14.2 *Let $u \in \mathcal{H}(S_T)$ be a non-negative solution of*

$$H(u) = 0 \text{ in } S_T, \quad \text{and} \quad u(\cdot, t) \to 0 \text{ in } L^1_{\text{loc}}(\mathbb{R}^N) \text{ as } t \to 0.$$

Then u vanishes identically in S_T.

It will suffice to prove:

Proposition 14.1 *Let $u \in \mathcal{H}(S_T)$ be a non-negative solution of the heat equation in S_T. Then $\forall (x_o, t_o) \in S_T$*

$$\int_{\mathbb{R}^N} u(y, s)\Gamma(x_o - y; t_o - s)dy \le u(x_o, t_o) \quad \text{for all } 0 < s < t_o. \tag{14.3}$$

Proof Fix $(x_o, t_o) \in S_T$ and $s \in (0, t_o)$ and introduce the change of variables

$$\tau = \frac{t - s}{t_o - s}, \qquad \eta = \frac{y - x_o}{\sqrt{t_o - s}}.$$

The function

$$U(\eta, \tau) = u\left(x_o + \sqrt{t_o - s}\eta, s + (t_o - s)\tau\right)$$

satisfies the heat equation in $\mathbb{R}^N \times [0, 1]$. For such a function, (14.3) becomes

$$\int_{\mathbb{R}^N} U(\eta, 0)\Gamma(\eta; 1)d\eta \le U(0, 1).$$

Thus it will be enough to prove that if $u \in \mathcal{H}(\bar{S}_1)$ is a non-negative solution of the heat equation in S_1 such that $u(\cdot, 0) \in C^2(\mathbb{R}^N)$, then

$$\int_{\mathbb{R}^N} u(y, 0)\Gamma(y; 1)dy \le u(0, 1). \tag{14.4}$$

To prove (14.4) fix $\rho > 0$ and consider the Cauchy problem

$$H(v) = 0 \text{ in } S_1, \quad v(x, 0) = \begin{cases} \zeta(x)u(x, 0) & \text{if } |x| < 2\rho \\ 0 & \text{otherwise} \end{cases} \tag{14.5}$$

where $x \to \zeta(x) \in C_o^\infty(B_{2\rho})$ is non-negative and equals one on the ball B_ρ. Since the initial datum is compactly supported in $B_{2\rho}$, the unique bounded solution of (14.5) is given by

$$v(x, t) = \int_{|y|<2\rho} \zeta(y)u(y, 0)\Gamma(x - y; t)dy.$$

Lemma 14.1 *$u \ge v$ in $\bar{S}_1$.*

Assuming this fact for the moment, it follows from the representation of v and the structure of the cutoff function ζ that

$$u(0,1) \geq \int_{B_\rho} u(y,0)\Gamma(y;1)dy.$$

This proves (14.4), since $\rho > 0$ is arbitrary. ∎

Proof (Lemma 14.1) The statement would follow from the maximum principle if u satisfied the growth condition (5.1). The positivity of u will replace such information. Let n_o be a positive integer larger than 2ρ, and for $n \geq n_o$, consider the sequence of homogeneous Dirichlet problems

$$\begin{aligned} H(v_n) &= 0 \text{ in } Q_n = B_n \times (0,1) \\ v_n \Big|_{|y|=n} &= 0 \\ v_n(x,0) &= \begin{cases} \zeta(x)u(x,0) & \text{if } |x| < 2\rho \\ 0 & \text{otherwise.} \end{cases} \end{aligned} \tag{14.6}$$

We regard the functions v_n as defined in the whole of S_1 by defining them to be zero outside Q_n. By the maximum principle applied over the bounded domains Q_n

$$0 \leq v_n \leq v_{n+1} \leq \|u(\cdot,0)\|_{\infty,B_{2\rho}} \quad \text{and} \quad v_n \leq u \tag{14.7}$$

for all $n \geq n_o$. By the second of these, the proof of the lemma reduces to showing that the increasing sequence $\{v_n\}$ converges to the unique solution of (14.5) uniformly over compact subsets of S_1. Consider compact subsets of the type $K = \bar{B}_R \times [\varepsilon, 1-\varepsilon]$ for $\varepsilon \in (0, \frac{1}{2})$ and $R \geq 2\rho$. By the estimates of Proposition 12.1 and the uniform upper bound of the first of (14.7), for every multi-index α and every positive integer k, there exists a constant C depending only on N, ε, R, $|\alpha|$, k and independent of n such that

$$\|D^\alpha v_n\|_{\infty,K} + \left\| \frac{\partial^k}{\partial t^k} v_n \right\|_{\infty,K} \leq C \quad \text{for all } n \geq 2R.$$

It follows, by a diagonalization process, that $\{v_n\} \to w$, uniformly over compact subsets of S_1, where $w \in C^\infty(S_1)$ and satisfies the heat equation. It remains to prove that

$$w(\cdot,t) \to \zeta u(\cdot,0) \text{ in } L^1_{\text{loc}}(\mathbb{R}^N) \text{ as } t \to 0.$$

For this, rewrite (14.6) as

$$\begin{gathered} f_n = v_n - \zeta u(x,0), \quad f_{n,t} - \Delta f_n = \Delta\zeta u(x,0) \text{ in } S_1 \\ f_n\Big|_{|x|=n} = 0, \quad f_n(x,0) = 0. \end{gathered}$$

Let $h_\delta(\cdot)$ be the approximation to the Heaviside function introduced in (14.2). Multiply the PDE by $h_\delta(f_n)$ and integrate over $B_n \times (0,t)$ for $t \in (0,1)$ to obtain

$$\int_{B_n\times\{t\}}\left(\int_0^{f_n} h_\delta(\xi)d\xi\right)dy + \frac{1}{\delta}\int_0^t\int_{B_n}|Df_n|^2\chi[|f_n|<\delta]\,dy\,ds$$
$$= \int_0^t\int_{B_{2\rho}} \Delta\zeta(y)u(y,0)h_\delta(f_n)\,dy\,ds.$$

Discard the second term on the left-hand side, which is non-negative, and let $\delta\to 0$ to get

$$\int_{B_R}|v_n(y,t)-\zeta(y)u(y,0)|dy \le t|B_{2\rho}|\|\Delta\zeta u(x,0)\|_{\infty,B_{2\rho}}.$$

Letting $n\to\infty$

$$\|w(y,t)-\zeta(y)u(y,0)\|_{1,B_R} \le t|B_{2\rho}|\|\Delta\zeta u(x,0)\|_{\infty,B_{2\rho}}.$$

By the first of (14.7), w is bounded; therefore by uniqueness of bounded solutions of the Cauchy problem, $w=v$. ∎

Remark 14.1 The Tychonov function defined in (5.2) is of variable sign.

Problems and Complements

2c Similarity Methods

2.1c The Heat Kernel Has Unit Mass

To verify (2.2), disregard momentarily the factor $\pi^{-N/2}$, and introduce the change of variables $y-x=2\sqrt{(t-s)}\eta$, whose Jacobian is $t[4(t-s)]^{N/2}$. This transforms the integral into

$$\begin{aligned}
\int_{\mathbb{R}^N} e^{-|\eta|^2}d\eta &= \int_{\mathbb{R}^N} e^{\eta_1^2+\dots+\eta_N^2}\,d\eta_1\cdots d\eta_N \\
&= \prod_{j=1}^N \int_{\mathbb{R}} e^{-\eta_j^2}d\eta_j = \left(\int_{\mathbb{R}} e^{-s^2}ds\right)^N \\
&= \left(\int_{\mathbb{R}} e^{-\eta_1^2}d\eta_1\int_{\mathbb{R}} e^{-\eta_2^2}d\eta_2\right)^{N/2} = \left(\int_{\mathbb{R}^2} e^{-|\eta|^2}d\eta\right)^{N/2} \\
&= \left(2\pi\int_0^\infty re^{-r^2}dr\right)^{N/2} = \pi^{N/2}.
\end{aligned}$$

2.2c The Porous Media Equation

Find similarity solutions for the non-linear evolution equation

$$u_t - \Delta u^m = 0 \quad u \geq 0, \quad m \geq 1.$$

This equation arises in the filtration of a fluid in a porous medium ([137]). Similarity solutions were derived independently by Barenblatt ([8]), and Pattle ([115]). Attempt solutions of the form $u(x,t) = h(t)f(\xi)$, where $\xi = \frac{|x|^2}{t^\sigma}$ and σ is a positive number to be found. Derive and solve ODE's for $h(\cdot)$ and $f(\cdot)$, to arrive at

$$\Gamma_m(x,t) = \frac{1}{t^{N/\kappa}}\left[1 - c\gamma_m\left(\frac{|x|^2}{t^{2/\kappa}}\right)\right]_+^{\frac{1}{m-1}}, \quad t > 0$$

$$\gamma_m = \frac{m-1}{2\kappa}, \quad \kappa = N(m-1) + 2$$

where $c > 0$ is an arbitrary constant. Show that as $m \to 1$, $\Gamma_m(x,t)$ tends to the fundamental solution of the heat equation. Find the constant c such that the total mass of Γ_m is 1, i.e.,

$$\int_{\mathbb{R}^N} \Gamma_m(x-y;t-\tau)dy = 1 \qquad (c = 4\pi).$$

Show that if $m > 1$, possible solutions to the Cauchy problem

$$\begin{aligned} u_t - \Delta u^m &= 0 \text{ in } S_T, \quad u \geq 0 \\ u(\cdot,0) &= u_o \in C(\mathbb{R}^N) \cap L^\infty(\mathbb{R}^N) \end{aligned}$$

cannot be represented as the convolution of Γ_m with the initial datum u_o.
Attempt to find similarity solutions when $0 < m < 1$.

2.3c The p-Laplacean Equation

Carry on the same analysis for the non-linear evolution equation

$$u_t - \operatorname{div}|Du|^{p-2}Du = 0 \quad p > 2.$$

A version of this equation arises in modeling certain non-Newtonian fluids ([90]). Then for $p = 2$ this reduces to the heat equation. The similarity solutions are

$$\Gamma_p(x,t) = \frac{1}{t^{N/\lambda}}\left[1 - c\gamma_p\left(\frac{|x|}{t^{1/\lambda}}\right)^{\frac{p}{p-1}}\right]_+^{\frac{p-1}{p-2}}, \quad t > 0$$

$$\gamma_p = \left(\frac{1}{\lambda}\right)^{\frac{1}{p-1}}\frac{p-2}{p}, \quad \lambda = N(p-2) + p.$$

Prove that $\Gamma_p \to \Gamma$ as $p \to 2$. Find the constant c so that Γ_p has mass 1. Attempt to find similarity solutions when $1 < p < 2$.

2.4c The Error Function

Prove that the unique solution of the Cauchy problem

$$u_t - u_{xx} = 0 \text{ in } \mathbb{R} \times \mathbb{R}^+, \quad u(x,0) = \begin{cases} 1 & \text{if } x \geq 0 \\ 0 & \text{if } x < 0 \end{cases}$$

is given by

$$u(x,t) = \frac{1}{2}\left[1 + E\left(\frac{x}{\sqrt{4t}}\right)\right], \qquad \text{where} \qquad E(s) = \frac{2}{\sqrt{\pi}} \int_0^s e^{-r^2} dr.$$

The function $s \to E(s)$ is the *error function.*

2.5c The Appell Transformation ([7])

Let u be a solution of the heat equation in $\mathbb{R} \times \mathbb{R}^+$. Then

$$w(x,t) = \Gamma(x,t) u\left(\frac{x}{t}, -\frac{1}{t}\right)$$

is also a solution of the heat equation in $\mathbb{R} \times \mathbb{R}^+$.

2.6c The Heat Kernel by Fourier Transform

For $f \in L^1(\mathbb{R}^N)$, let $\hat{f}$ denote its Fourier transform

$$\hat{f}(x) \stackrel{\text{def}}{=} \frac{1}{(2\pi)^{N/2}} \int_{\mathbb{R}^N} f(y) e^{-i\langle x,y\rangle} dy.$$

Here i is the imaginary unit and $\langle x,y\rangle = x_j y_j$. In general, assuming that $f \in L^1(\mathbb{R}^N)$ or even that f is compactly supported in $\mathbb{R}^N$, does not guarantee that $\hat{f} \in L^1(\mathbb{R}^N)$, as shown by the following examples.

2.1. Compute the Fourier transform of the characteristic function of the unit interval in $\mathbb{R}^1$. Show that $x \to (\chi_{[0,1]})^\wedge(x) \notin L^1(\mathbb{R})$.

2.2. Let $N = 1$, and let m be a positive integer larger than 2. Compute the Fourier transform of

$$f(x) = \begin{cases} 0 & \text{for } x < 1 \\ x^{-m} & \text{for } x \geq 1 \end{cases}$$

and show that $\hat{f} \notin L^1(\mathbb{R})$.

2.7c Rapidly Decreasing Functions

These examples show that $L^1(\mathbb{R}^N)$ is not closed under the operation of Fourier transform, and raise the question of finding a class of functions that is closed under such an operation. The class of smooth and rapidly decreasing functions in $\mathbb{R}^N$, or the Schwartz class, is defined by ([139])

$$\mathcal{S}_N \stackrel{\text{def}}{=} \left\{ \begin{array}{c} f \in C^\infty(\mathbb{R}^N) \mid \sup_{x\in\mathbb{R}^N} |x|^m |D^\alpha f(x)| < \infty \\ \text{for all } m \in \mathbb{N} \text{ and all multi-indices } \alpha \text{ of size } |\alpha| \geq 0 \end{array} \right\}.$$

Proposition 2.1c $f \in \mathcal{S}_N \Longrightarrow \hat{f} \in \mathcal{S}_N$.

Proof For $f \in \mathcal{S}_N$ and multi-indices α and β, compute

$$\begin{aligned} x^\beta D^\alpha \hat{f}(x) &= \frac{x^\beta}{(2\pi)^{N/2}} \int_{\mathbb{R}^N} f(y) D_x^\alpha e^{-i\langle x,y\rangle} dy \\ &= \frac{(-i)^{|\alpha|}}{(2\pi)^{N/2}} \int_{\mathbb{R}^N} x^\beta y^\alpha f(y) e^{-i\langle x,y\rangle} dy \\ &= \frac{(-i)^{|\alpha|-|\beta|}}{(2\pi)^{N/2}} \int_{\mathbb{R}^N} y^\alpha f(y) D_y^\beta e^{-i\langle x,y\rangle} dy \\ &= \frac{(-i)^{|\alpha+\beta|}}{(2\pi)^{N/2}} \int_{\mathbb{R}^N} D^\beta [y^\beta D^\alpha f(y)] e^{-i\langle x,y\rangle} dy. \end{aligned}$$

■

2.8c The Fourier Transform of the Heat Kernel

Proposition 2.2c *Let* $\varphi(x) = e^{-\frac{1}{2}|x|^2}$. *Then* $\hat{\varphi} = \varphi$.

Proof Assume first that $N = 1$. One verifies that φ and $\hat{\varphi}$ satisfy the same ODE

$$\varphi' + x\varphi = 0, \qquad \hat{\varphi}' + x\hat{\varphi} = 0, \qquad x \in \mathbb{R}.$$

Therefore $\hat{\varphi} = C\varphi$ for a constant C. From (2.2) with $t - s = \frac{1}{2}$ and $N = 1$

$$\frac{1}{\sqrt{2\pi}} \int_{\mathbb{R}} e^{-\frac{1}{2}y^2} dy = \hat{\varphi}(0) = 1.$$

Since also $\varphi(0) = 1$, we conclude that $C = 1$, and the proposition follows in the case of one dimension. If $N \geq 2$, by Fubini's theorem

$$\begin{aligned} \hat{\varphi}(x) &= \frac{1}{(2\pi)^{N/2}} \int_{\mathbb{R}^N} e^{-\frac{1}{2}|y|^2} e^{-i\langle x,y\rangle} dy \\ &= \prod_{j=1}^N \frac{1}{\sqrt{2\pi}} \int_{\mathbb{R}} e^{-\frac{1}{2}y_j^2} e^{-ix_j y_j} dy_j \\ &= \prod_{j=1}^N \hat{\varphi}(x_j) = \prod_{j=1}^N \varphi(x_j) = \varphi(x). \end{aligned}$$

■

2.3. Prove the rescaling formula $\hat{\psi}(\varepsilon x) = \varepsilon^{-N}\hat{\psi}(x/\varepsilon)$, valid for all $\psi \in \mathcal{S}_N$ and all $\varepsilon > 0$.

2.4. Verify the formula

$$\left(e^{-|x-y|^2(t-\tau)}\right)^{\wedge} = \frac{1}{[2(t-\tau)]^{N/2}} e^{-|x-y|^2/4(t-\tau)}$$

for all $t - \tau > 0$ fixed.

2.9c The Inversion Formula

Theorem 2.1c *Let $f \in \mathcal{S}_N$. Then*

$$f(x) = \frac{1}{(2\pi)^{N/2}} \int_{\mathbb{R}^N} \hat{f}(y) e^{i\langle x,y\rangle} dy.$$

Proof The formula follows by computing the limit

$$\frac{1}{(2\pi)^{N/2}} \int_{\mathbb{R}^N} \hat{f}(y) e^{i\langle x,y\rangle} dy = \lim_{\tau \to t} \frac{1}{(2\pi)^{N/2}} \int_{\mathbb{R}^N} \hat{f}(y) e^{-|y|^2(t-\tau)} e^{i\langle x,y\rangle} dy.$$

The integral on the right-hand side is computed by repeated application of Fubini's theorem:

$$\begin{aligned}
&\frac{1}{(2\pi)^{N/2}} \int_{\mathbb{R}^N} \hat{f}(y) e^{-|y|^2(t-\tau)} e^{i\langle x,y\rangle} dy \\
&\quad = \frac{1}{(2\pi)^N} \int_{\mathbb{R}^N} f(\eta) e^{-i\langle y,\eta\rangle} e^{-|y|^2(t-\tau)} e^{i\langle x,y\rangle} dy d\eta \\
&\quad = \frac{1}{(2\pi)^{N/2}} \int_{\mathbb{R}^N} f(\eta) \left(\frac{1}{(2\pi)^{N/2}} \int_{\mathbb{R}^N} e^{-|y|^2(t-\tau)} e^{-i\langle \eta - x, y\rangle} dy \right) d\eta \\
&\quad = \frac{1}{(2\pi)^{N/2}} \int_{\mathbb{R}^N} f(\eta) \left(e^{-|y|^2(t-\tau)} \right)^{\wedge} (\eta - x) d\eta \\
&\quad = \frac{1}{[4\pi(t-\tau)]^{N/2}} \int_{\mathbb{R}^N} f(\eta) e^{-|x-\eta|^2/4(t-\tau)} d\eta.
\end{aligned}$$

Therefore

$$\frac{1}{(2\pi)^{N/2}} \int_{\mathbb{R}^N} \hat{f}(y) e^{i\langle x,y\rangle} dy = \lim_{\tau \to t} \int_{\mathbb{R}^N} \Gamma(x-\eta; t-\tau) f(\eta) d\eta = f(x)$$

where the last limit is computed by the same technique leading to the representation formula (2.7). ∎

3c The Maximum Principle in Bounded Domains

Let E be a bounded domain in $\mathbb{R}^N$ with smooth boundary ∂E.

3.1. Let u be a solution of the Dirichlet problem (1.2) with $g = 0$. Prove that

$$\|u(\cdot,t)\|_{\infty,E} \le \frac{1}{(4\pi t)^{N/2}}\|u_o\|_{1,E}.$$

3.2. State and prove a maximum principle for $u \in \mathcal{H}(E_T)\cap C(\bar{E}_T)$ satisfying

$$H(u) = \mathbf{v}\cdot Du + c \ \text{ in } \ E_T.$$

where $\mathbf{v} \in \mathbb{R}^N$ and $c \in \mathbb{R}$ are given.

3.3. Discuss a possible maximum principle for $H(u) = \lambda u$ for $\lambda \in \mathbb{R}$.

3.4. Let $f \in C(\mathbb{R}^+)$ and consider the boundary value problem

$$\begin{aligned} u \in \mathcal{H}(E_\infty) &\cap C(\bar{E}_\infty) \\ u_t - \Delta u &= f(t)\left(u - \frac{|x|^2}{2N}\right) - 1 \ \text{ in } \ B_1 \times \mathbb{R}^+ \\ u(\cdot,t)\big|_{\partial_* B_1} &= \frac{1}{2N}. \end{aligned}$$

Prove that this problem has at most one solution, the solution is non-negative and satisfies

$$0 \le u(x,t) \le \frac{1}{2N}\exp\left(\int_0^t f(s)ds\right) + \frac{|x|^2}{2N}.$$

In particular, if $f \le 0$ then $u(x,t) \le 1/N$.

3.5. In the previous problem assume that

$$f(t) \le -\frac{C}{1+t} \quad \text{for all } \ t \ge t_*$$

for some $C > 0$ and some $t_* \ge 0$. Prove that

$$\lim_{t\to\infty} u(x,t) = \frac{|x|^2}{2N}.$$

Moreover, if $u(\cdot,0) = |x|^2/2N$, then $u(\cdot,t) = u(\cdot,0)$, for all f.

3.6. Let $f \in C(\bar{E}_T)$ and $\alpha \in (0,1)$. Prove that a non-negative solution of $H(u) = u^\alpha$ in E_T satisfies

$$\|u\|_{\infty,E_T} \le \frac{1}{1-\alpha}\|u\|_{\infty,\partial_* E_T} + \left[(e^{\text{diam}(E)} - 1)\|f\|_{\infty,E_T}\right]^{\frac{1}{1-\alpha}}.$$

3.1c The Blow-Up Phenomenon for Super-Linear Equations

Consider non-negative classical solutions of

$$u_t - \Delta u = u^\alpha \quad \text{in } E \times \mathbb{R}^+, \qquad \text{for some } \alpha \ge 1$$

that are bounded on the parabolic boundary of E_T, say

$$\sup_{\partial_* E_\infty} u \le M \qquad \text{for some } M > 0.$$

Prove that if $\alpha = 1$, then $u \le Me^t$. Therefore if $\alpha \in [0,1)$ the solution remains bounded for all $t \ge 0$, and if $\alpha = 1$, it remains bounded for all $t \ge 0$ with bound increasing with t. If $\alpha > 1$, an upper bound is possible only for finite times.

Lemma 3.1c *Let $\alpha > 1$. Then*

$$u(x,t) \le \frac{M}{[1-(\alpha-1)M^{\alpha-1}t]^{1/(\alpha-1)}}.$$

Proof (Hint) Divide the PDE by u^α and introduce the function

$$w = u^{1-\alpha} + (\alpha - 1)t.$$

Using that $\alpha > 1$, prove that $H(w) \ge 0$ in E_∞. Therefore, by the maximum principle

$$\frac{1}{u^{\alpha-1}} + (\alpha-1)t \ge \frac{1}{M^{\alpha-1}}. \qquad \blacksquare$$

Remark 3.1c This estimate is *stable* as $\alpha \to 1$ in the sense that as $\alpha \to 1$, the right-hand side converges to the corresponding exponential upper bound valid for $\alpha = 1$.

3.1.1c An Example for $\alpha = 2$

Even though the boundary data are uniformly bounded, the solution might indeed *blow-up* at interior points of E in finite time, as shown by the following example ([48]).

$$\begin{aligned} u_t - u_{xx} = u^2 \ \text{ in } (0,1)\times(0,\infty), \quad & u(\cdot,0) = u_o \\ u(0,t) = h_o(t), \quad u(1,t) = h_1(t), \quad & \text{for all } t \ge 0. \end{aligned} \tag{3.1c}$$

Assume that

$$u_o,\ h_o,\ h_1 \ge c = \frac{c_1}{c_2}$$

for positive constants c_1 and c_2 to be chosen. These constants can be chosen such that (3.1c) has no solution that remains bounded for finite times. Introduce the comparison function

$$v = \frac{c_1}{c_2 - x(1-x)t}.$$

By direct calculation

$$\begin{aligned} v_t - v_{xx} &= \frac{c_1 x(1-x)}{[c_2 - x(1-x)t]^2} + \frac{2c_1 t}{[c_2 - x(1-x)t]^2} - \frac{2c_1 t^2 (1-2x)^2}{[c_2 - x(1-x)t]^3} \\ &\le \frac{v^2}{c_1}\left(\frac{1}{4} + 2t\right). \end{aligned}$$

Taking $t \in (0, 4c_2)$ and choosing c_1 sufficiently large, this last term is majorized by v^2. Therefore

$$v_t - v_{xx} \le v^2 \quad \text{in } (0,1) \times (0, 4c_2).$$

Fix any time $T \in (0, 4c_2)$ and consider the domain $E_T = (0,1) \times (0,T)$. If u is a solution of (3.1c), the function $w = (v-u)e^{-\lambda t}$ for $\lambda > 0$ satisfies

$$w_t - w_{xx} \le -(\lambda - (v+u))w \;\text{ in }\; E_T, \quad w\big|_{\partial_* E_T} \le 0.$$

Therefore, by choosing λ sufficiently large, the maximum principle implies that $w \le 0$ in E_T.

3.2c The Maximum Principle for General Parabolic Equations

Let $\mathcal{L}_o(\cdot)$ be the differential operator introduced in (4.1c) of the Complements of Chapter 2. By using a technique similar to that of Theorem 4.1c, prove

Theorem 3.1c *Let $u \in \mathcal{H}(E_T) \cap C(\bar{E}_T)$ and let $c \le 0$. then*

$$u_t - \mathcal{L}_o(u) \le 0 \;\text{ in }\; E_T \quad \Longrightarrow \quad u(x,t) \le \sup_{\partial_* E} u \;\text{ in }\; E_T.$$

3.6. The maximum principle gives one-sided estimates for merely sub(super)-solutions of the heat equation. An important class of sub(super)-solutions is determined as follows. Let $u \in \mathcal{H}(E_T)$ be a solution of the heat equation in E_T. Prove that for every convex(concave) function $\varphi(\cdot) \in C^2(\mathbb{R})$, the composition $\varphi(u)$ is a sub(super)-solution of the heat equation in E_T.

4c The Maximum Principle in $\mathbb{R}^N$

4.1. Show that $u = 0$ is the only solution of the Cauchy problem

$$u \in \mathcal{H}(S_T) \cap C(\bar{S}_T) \cap L^2(S_T), \quad u_t - \Delta u = 0 \text{ in } S_T, \quad u(\cdot, 0) = 0. \tag{4.1c}$$

Hint: Let $x \to \zeta(x) \in C_o^2(B_{2\rho})$ be a non-negative cutoff function in $B_{2\rho}$ satisfying

$$\zeta = 1 \text{ in } B_\rho, \quad |D\zeta| \le \begin{cases} 0 & \text{if } |x| < \rho \\ \dfrac{2}{\rho} & \text{if } \rho < |x| < 2\rho, \end{cases}$$

$$|\zeta_{x_i x_j}| \le \begin{cases} 0 & \text{if } |x| < \rho \\ \dfrac{4}{\rho^2} & \text{if } \rho < |x| < 2\rho. \end{cases}$$

Multiply the PDE by $u\zeta^2$ and integrate over $B_{2\rho} \times (0,t)$ to derive

$$\int_{B_{2\rho}} u^2(t)\zeta^2 dx + 2\int_0^t \int_{B_{2\rho}} |Du|^2\zeta^2 dx\, ds = 4\int_0^t \int_{B_{2\rho}} \zeta u Du D\zeta dx\, ds.$$

By the Cauchy–Schwarz inequality, the last integral is majorized by

$$2\int_0^t \int_{B_{2\rho}} |Du|^2\zeta^2 dx\, ds + 2\int_o^t \int_{B_{2\rho}} u^2|D\zeta|^2 dx\, ds$$

and

$$2\int_0^t \int_{B_{2\rho}} u^2|D\zeta|^2 dx\, ds \le \frac{8}{\rho^2}\int_0^t \int_{\rho<|x|<2\rho} u^2|D\zeta|^2 dx\, ds.$$

Combine these estimates and let $\rho \to \infty$.

4.2. Prove that the same conclusion holds if in (4.1c) one replaces $L^2(S_T)$ with $L^1(S_T)$. *Hint:* Let $h_\delta(\cdot)$ be the approximation to the Heaviside function introduced in (14.2). Multiply the PDE by $h_\delta(u)\zeta$ and integrate over $B_\rho \times (0,t)$ to obtain

$$\begin{aligned} \int_{B_{2\rho}\times\{t\}} \left(\int_0^u h_\delta(\xi)d\xi\right) \zeta dx + \frac{1}{\delta}\int_0^t \int_{B_{2\rho}} |Du|^2\chi(|u| < \delta)\zeta\, dx\, ds \\ = -\int_0^t \int_{B_{2\rho}} D\left(\int_0^u h_\delta(\xi)d\xi\right) D\zeta\, dx\, ds. \end{aligned}$$

The last term is transformed and majorized by

$$\int_0^t \int_{B_{2\rho}} \left(\int_0^u h_\delta(\xi)d\xi\right) \Delta\zeta\, dx\, ds \le \frac{\text{const}}{\rho^2}\int_0^t \int_{\rho<|x|<2\rho} |u|\, dx\, ds.$$

Combining these estimates and letting $\delta \to 0$ we arrive at

$$\int_{B_\rho\times\{t\}} |u| dx ds \le \frac{\text{const}}{\rho^2}\int_0^t \int_{\rho<|x|<2\rho} |u|\, dx\, ds.$$

To conclude, let $\rho \to \infty$.

4.3. Prove that the same conclusion holds if u satisfies either one of the weaker conditions

$$\frac{u}{(1+|x|)} \in L^2(S_T), \qquad \frac{u}{(1+|x|^2)} \in L^1(S_T).$$

4.1c A Counterexample of the Tychonov Type

Prove that the function

$$u(x,t) = \int_0^\infty [e^{xy}\cos(xy+2ty^2) + e^{-xy}\cos(xy-2ty^2)]ye^{-y^{4/3}}\cos y^{4/3}dy$$

is another non-trivial solution of the Cauchy problem in $\mathbb{R}\times\mathbb{R}^+$ with vanishing initial data ([131]).

7c Remarks on the Cauchy Problem

7.1. Write down the explicit solution of

$$\begin{aligned} u_t - \Delta u &= u + \mathbf{b}\cdot\nabla u + e^t\sin(x_1 - b_1 t) \ \text{ in } \ S_T \\ u(x,0) &= |x| \end{aligned}$$

for a given $\mathbf{b}\in\mathbb{R}^N$. *Hint:* The function $v(x,t) = u(x-\mathbf{b}t, t)$ satisfies the PDE with $\mathbf{b}=0$.

7.2. Using the reflection technique, solve the homogeneous mixed boundary value problems

$$\begin{aligned} u_t - u_{xx} &= 0 \ \text{ in } \ \mathbb{R}^+\times\mathbb{R}^+ \\ u_x(0,t) &= 0 \ \text{ for } \ t>0 \\ u(x,0) &= u_o \in C^1(\bar{\mathbb{R}}^+) \\ u_{o,x}(0) &= 0 \end{aligned} \qquad \begin{aligned} u_t - u_{xx} &= 0 \ \text{ in } \ \mathbb{R}^+\times\mathbb{R}^+ \\ u(0,t) &= 0 \ \text{ for } \ t>0 \\ u(x,0) &= u_o \in C(\bar{\mathbb{R}}^+) \\ u_o(0) &= 0. \end{aligned}$$

where u_o is bounded in $\mathbb{R}^+$.

7.3. Solve the inhomogeneous mixed boundary value problems

$$\begin{aligned} u_t - u_{xx} &= \ \text{ in } \ \mathbb{R}^+\times\mathbb{R}^+ \\ u_x(0,t) &= h(t) \in C^1(\mathbb{R}^+) \\ u(x,0) &= u_o \in C^1(\bar{\mathbb{R}}^+) \\ u_{o,x}(0) &= h(0) \end{aligned} \qquad \begin{aligned} u_t - u_{xx} &= 0 \ \text{ in } \ \mathbb{R}^+\times\mathbb{R}^+ \\ u(0,t) &= h(t) \in C^1(\mathbb{R}^+) \\ u(x,0) &= u_o \in C(\bar{\mathbb{R}}^+) \\ u_o(0) &= h(0). \end{aligned}$$

12c On the Local Behavior of Solutions

Proposition 12.1c *Let $u \in \mathcal{H}(E_T)$ be a local solution of the heat equation in E_T. For every $p>0$ there exists a constant C, depending only on N and p such that, for all $(x_o,t_o)+Q_\rho \subset E_T$*

$$\sup_{(x_o,t_o)+Q_\rho}|u| \le C\left(\fint_{(x_o,t_o)+Q_{2\rho}}|u|^p dy\,ds\right)^{1/p}.$$

Proof The case $p=1$ is the content of Corollary 12.1. The case $p>1$ follows from this and Hölder's inequality. To prove the estimate for $0<p<1$, one may assume that $(x_o,t_o)=(0,0)$. Consider the increasing sequence of radii $\{\rho_n\}$, the family of nested expanding boxes $\{Q_n\}$, and the non-decreasing sequence of numbers $\{M_n\}$, defined by

$$\rho_n=\rho\sum_{i=0}^{n}2^{-i},\quad Q_n=B_{\rho_n}\times(-\rho_n^2,0),\quad M_n=\sup_{Q_n}|u|,\quad n=0,1,\dots.$$

Apply Corollary 12.1 to the pair of cylinders Q_n and Q_{n+1} to obtain

$$M_n\le C2^{(n+1)(N+2)}\fint_{Q_{n+1}}|u|\,dy\,ds.$$

Fix $p\in(0,1)$. Then by Young's inequality, for all $\delta>0$

$$\begin{aligned}M_n&\le C2^{(n+1)(N+2)}M_{n+1}^{1-p}\fint_{Q_{n+1}}|u|^p dy\,ds\\&\le\delta M_{n+1}+p\delta^{1-\frac{1}{p}}\left(C2^{(n+2)(N+2)}\right)^{1/p}\left(\fint_{Q_{2\rho}}|u|^p dy\,ds\right)^{1/p}.\end{aligned}$$

Setting

$$K=p\delta^{1-\frac{1}{p}}\left(C2^{(N+2)}\right)^{1/p},\qquad b=2^{\frac{N+2}{p}}$$

we arrive at the recursive inequalities

$$M_n\le\delta M_{n+1}+b^nK\left(\fint_{Q_{2\rho}}|u|^p dy\,ds\right)^{1/p}.$$

By iteration

$$M_o\le\delta^nM_{n+1}+bK\sum_{i=0}^{n}(\delta b)^i\left(\fint_{Q_{2\rho}}|u|^p dy\,ds\right)^{1/p}.$$

Choose δ small enough that $\delta b=\frac{1}{2}$, so that the series $\sum_{i=0}^{\infty}(\delta b)^i$ is convergent. Then let $n\to\infty$. ∎

6

The Wave Equation

1 The One-Dimensional Wave Equation

Consider the hyperbolic equation in two variables

$$u_{tt} - c^2 u_{xx} = 0. \tag{1.1}$$

The variable t stands for time, and *one-dimensional* refers to the number of space variables. A general solution of (1.1) in a *convex* domain $E \subset \mathbb{R}^2$, is given by

$$u(x,t) = F(x - ct) + G(x + ct) \tag{1.2}$$

where $s \to F(s), G(s)$ are of class C^2 within their domain of definition. Indeed, the change of variables

$$\xi = x - ct, \qquad \eta = x + ct \tag{1.3}$$

transforms E into a convex domain $\tilde{E}$ of the (ξ, η)-plane, and in terms of ξ and η, equation (1.1) becomes

$$U_{\xi\eta} = 0 \quad \text{where} \quad U(\xi, \eta) = u\Big(\frac{\xi + \eta}{2}, \frac{\eta - \xi}{2c}\Big).$$

Therefore $U_\xi = F'(\xi)$ and

$$U(\xi, \eta) = \int F'(\xi) d\xi + G(\eta).$$

Rotating the axes back of an angle $\theta = \arctan(c^{-1})$, maps $\tilde{E}$ into E back in the (x, t)-plane and

$$u(x,t) = F(x - ct) + G(x + ct).$$

The graphs of $\xi \to F(\xi)$ and $\eta \to G(\eta)$ are called *undistorted waves* propagating to the right and left respectively (right and left here refer to the positive

E. DiBenedetto, *Partial Differential Equations: Second Edition*,
Cornerstones, DOI 10.1007/978-0-8176-4552-6_7,

orientation of the x- and t-axes). The two lines obtained from (1.3) by making ξ and η constants are called characteristic lines. Write them in the parametric form

$$x_1(t) = ct + \xi, \quad x_2(t) = -ct + \eta, \qquad \text{for } t \in \mathbb{R}$$

and regard the abscissas $t \to x_i(t)$ for $i = 1, 2$ as points traveling on the x-axis, with velocities $\pm c$ respectively.

1.1 A Property of Solutions

Consider any parallelogram of vertices A, B, C, D with sides parallel to the characteristics $x = \pm ct + \xi$ and contained in some *convex* domain $E \subset \mathbb{R}^2$.

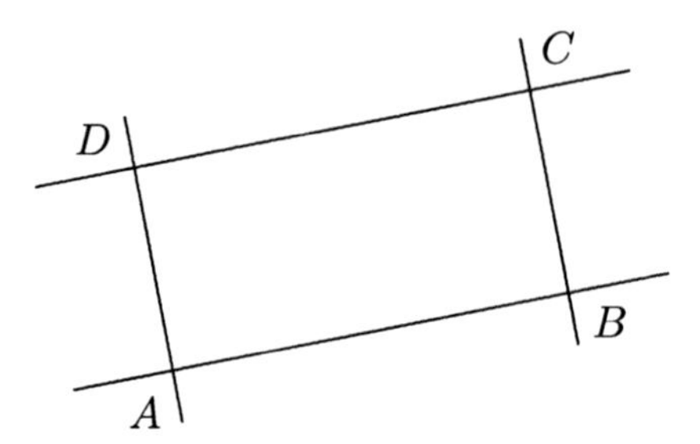

Fig. 1.1.

We call it a characteristic parallelogram. Let

$$A = (x, t), \quad B = (x + cs, t + s)$$
$$C = (x + cs - c\tau, t + s + \tau), \quad D = (x - c\tau, t + \tau)$$

be the coordinates of the vertices of a characteristic parallelogram, where s and τ are positive parameters. If a function $u \in C(E)$ is of the form (1.2), for two continuous functions $F(\cdot)$ and $G(\cdot)$, then

$$\begin{aligned}
u(A) &= F(x - ct) + G(x + ct) \\
u(C) &= F(x - 2c\tau - ct) + G(x + 2cs + ct) \\
u(B) &= F(x - ct) + G(x + 2cs + ct) \\
u(D) &= F(x - 2c\tau - ct) + G(x + ct).
\end{aligned}$$

Therefore

$$u(A) + u(C) = u(B) + u(D). \tag{1.4}$$

Therefore any solution of (1.1) satisfies (1.4). Vice versa if $u \in C^2(E)$ is of the form (1.2) for F and G of class C^2 and satisfies (1.4) for any characteristic parallelogram, rewrite (1.4) as

$$[u(x,t) - u(x+cs, t+s)] = [u(x-c\tau, t+\tau) - u(x+cs-c\tau, t+s+\tau)].$$

Using the Taylor formula one verifies that u satisfies the PDE (1.1). Since (1.4) only requires that u be continuous, it might be regarded as some sort of *weak* formulation of (1.1).

2 The Cauchy Problem

On the non-characteristic line $t = 0$, prescribe the shape and speed of the undistorted waves, and seek to determine the shape and speed of the solution of (1.1), for all the later and previous times. Formally, seek to solve the Cauchy problem

$$\begin{aligned} u_{tt} - c^2 u_{xx} &= 0 \qquad \text{in } \mathbb{R}^2 \\ u(\cdot, 0) &= \varphi \qquad \text{in } \mathbb{R} \\ u_t(\cdot, 0) &= \psi \qquad \text{in } \mathbb{R} \end{aligned} \tag{2.1}$$

for given $\varphi \in C^2(\mathbb{R})$ and $\psi \in C^1(\mathbb{R})$. According to (1.2) one has to determine the form of F and G from the initial data, i.e.,

$$F + G = \varphi, \quad F' + G' = \varphi', \quad -F' + G' = \frac{1}{c}\psi.$$

From this

$$F' = \frac{1}{2}\varphi' - \frac{1}{2c}\psi, \qquad G' = \frac{1}{2}\varphi' + \frac{1}{2c}\psi.$$

This, in turn, implies

$$\begin{aligned} F(\xi) &= \frac{1}{2}\varphi(\xi) - \frac{1}{2c}\int_0^\xi \psi(s)ds + c_1 \\ G(\eta) &= \frac{1}{2}\varphi(\eta) + \frac{1}{2c}\int_0^\eta \psi(s)ds + c_2 \end{aligned}$$

for two constants c_1 and c_2. Therefore

$$u(x,t) = \frac{1}{2}[\varphi(x-ct) + \varphi(x+ct)] + \frac{1}{2c}\int_{x-ct}^{x+ct} \psi(s)ds \tag{2.2}$$

since, in view of the second of (2.1), $c_1 + c_2 = 0$. Formula (2.2) is the explicit d'Alembert representation of the *unique* solution of the Cauchy problem (2.1). The right-hand side of (2.2) is well defined whenever $\varphi \in C_{\text{loc}}(\mathbb{R})$ and $\psi \in L^1_{\text{loc}}(\mathbb{R})$. However, in such a case, the corresponding function $(x,t) \to u(x,t)$ need not satisfy the PDE in the classical sense. For this reason, (2.2) might be regarded as some sort of *weak* solution of the Cauchy problem (2.1) whenever the data satisfy merely the indicated reduced regularity.

Remark 2.1 (Domain of Dependence) The value of u at (x,t) is determined by the restriction of the initial φ and ψ, data to the interval $[x-ct, x+ct]$. If the initial speed ψ vanishes on such an interval, then $u(x,t)$ depends only n the datum φ at the points $x \pm ct$ of the x-axis.

Remark 2.2 (Propagation of Disturbances) The value of the initial data $\varphi(\xi)$, $\psi(\xi)$ at a point ξ of the x-axis is felt by the solution only at points (x,t) within the sector

$$[x-ct \le \xi] \cap [x+ct \ge \xi].$$

If $\psi \equiv 0$, it is felt only at points of the characteristic curves $x = \pm ct + \xi$.

Remark 2.3 (Well-Posedness) The Cauchy problem (2.1) is well-posed in the sense of Hadamard, i.e., (a) there exists a solution; (b) the solution is unique; (c) the solution is stable. Statement (c) asserts that small perturbations of the data φ and ψ yield small changes in the solution u. This is also referred to as *continuous dependence* on the data. Such a statement becomes precise only when a topology is introduced to specify the meaning of "small" and "continuous".

Since the problem is linear, to prove (c) it will suffice to show that "small data" yield "small solutions". As a smallness condition on φ and ψ, take

$$\|\varphi\|_{\infty,\mathbb{R}}, \quad \|\psi\|_{\infty,\mathbb{R}} < \varepsilon \ \text{ for some } \varepsilon > 0.$$

Then formula (2.2) gives that the solution u corresponding to such data satisfies

$$\|u(\cdot,t)\|_{\infty,\mathbb{R}} \le (1+t)\varepsilon.$$

This proves the continuous dependence on the data in the topology of $L^\infty(\mathbb{R})$. If in addition, the initial velocity ψ is compactly supported in $\mathbb{R}$, say in the interval $(-L, L)$, then

$$\|u\|_{\infty,\mathbb{R}^2} < \Big(1 + \frac{L}{c}\Big)\varepsilon.$$

3 Inhomogeneous Problems

Let $f \in C^1(\mathbb{R}^2)$ and consider the inhomogeneous Cauchy problem

$$\begin{aligned} u_{tt} - c^2 u_{xx} &= f && \text{in } \mathbb{R}^2 \\ u(\cdot,0) &= \varphi && \text{in } \mathbb{R} \\ u_t(\cdot,0) &= \psi && \text{in } \mathbb{R}. \end{aligned} \tag{3.1}$$

The solution of (3.1) can be constructed by superposing the unique solution of (2.1) with a solution of

$$\begin{aligned} v_{tt} - c^2 v_{xx} &= f && \text{in } \mathbb{R}^2 \\ v(\cdot,0) = v_t(\cdot,0) &= 0 && \text{in } \mathbb{R}. \end{aligned} \tag{3.2}$$

To solve the latter, introduce the change of variables (1.3), which transforms (3.2) into

$$U_{\xi\eta}(\xi,\eta) = -\frac{1}{4c^2}F(\xi,\eta), \quad \text{where} \quad F(\xi,\eta) = f\Big(\frac{\xi+\eta}{2}, -\frac{\xi-\eta}{2c}\Big).$$

The initial conditions translate into

$$U(s,s) = U_\xi(s,s) = U_\eta(s,s) = 0 \qquad \forall s \in \mathbb{R}.$$

Integrate the transformed PDE in the first variable, over the interval (η, ξ). Taking into account the initial conditions

$$U_\eta(\xi,\eta) = -\frac{1}{4c^2}\int_\eta^\xi F(s,\eta)ds.$$

Next integrate in the second variable, over (ξ,η). This gives

$$U(\xi,\eta) = \frac{1}{4c^2}\int_\xi^\eta \int_\xi^z F(s,z)ds\, dz. \tag{3.3}$$

In (3.3) perform the change of variables

$$-\frac{s-z}{2c} = \tau, \qquad \frac{s+z}{2} = \sigma$$

whose Jacobian is $2c$. The domain of integration is transformed into

$$x - ct = \xi < \sigma - c\tau < \sigma + c\tau < \eta = x + ct.$$

Therefore, in terms of x and t, (3.3) gives the unique solution of (3.2) in the form

$$v(x,t) = \frac{1}{2c}\int_0^t \int_{x-c(t-\tau)}^{x+c(t-\tau)} f(\sigma,\tau)d\sigma\, d\tau. \tag{3.4}$$

Remark 3.1 (Duhamel's Principle ([38])) Consider the one-parameter family of initial value problems

$$\begin{aligned} v_{tt} - c^2 v_{xx} &= 0 && \text{in } \mathbb{R}\times(\tau,\infty) \\ v(\cdot,\tau) &= 0 && \text{in } \mathbb{R} \\ v_t(\cdot,\tau) &= f(\cdot,\tau) && \text{in } \mathbb{R}. \end{aligned}$$

By the d'Alembert formula (2.2)

$$v(x,t;\tau) = \frac{1}{2c}\int_{x-c(t-\tau)}^{x+c(t-\tau)} f(\sigma,\tau)d\sigma.$$

Therefore, it follows from (3.4), that the solution of (3.2) is given by "superposing" $\tau \to v(x,t;\tau)$ for $\tau \in (0,t)$. This is a particular case of Duhamel's principle (see Section 3.1c of the Complements).

Remark 3.2 It follows from the solution formula (3.4) that if $x \to f(x,t)$ is odd about some x_o, then $x \to v(x,t)$ is also odd about x_o for all $t \in \mathbb{R}$. In particular, $u(x_o,t) = 0$ for all $t \in \mathbb{R}$.

4 A Boundary Value Problem (Vibrating String)

A string of length L vibrates with its end-points kept fixed. Let $(x,t) \to u(x,t)$ denote the vertical displacement at time t of the point $x \in (0,L)$. Assume that at time $t = 0$ the shape of the string and its speed are known, say $\varphi, \psi \in C^2[0,L]$. At all times $t \in \mathbb{R}$ the phenomenon is described by the boundary value problem

$$\begin{aligned} u_{tt} &= c^2 u_{xx} && \text{in } (0,L)\times\mathbb{R} \\ u(0,\cdot) &= u(L,\cdot) = 0 && \text{in } \mathbb{R} \\ u(\cdot,0) &= \varphi,\ u_t(\cdot,0) = \psi && \text{in } (0,L). \end{aligned} \tag{4.1}$$

The data φ and ψ are required to satisfy the compatibility conditions

$$\varphi(0) = \varphi(L) = \psi(0) = \psi(L) = 0.$$

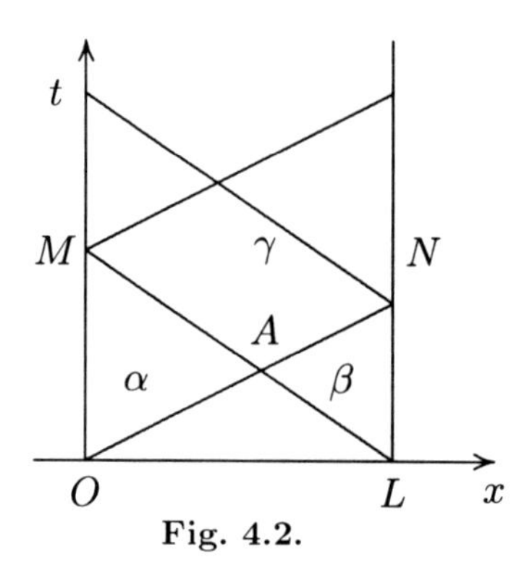

Fig. 4.2.

At each point of $[0,L] \times \mathbb{R}$, the solution $u(x,t)$ of (4.1), can be determined by making use of the solution formula (2.2) for the Cauchy problem, and formula (1.4). First draw the characteristic $x = ct$ originating at $(0,0)$, and the characteristic $x = -ct + L$ originating at $(L,0)$, and let A be their intersection. As they intersect the vertical axes $x = 0$ and $x = L$, reflect them by following the characteristic of opposite slope, as in Figure 4.2. The solution $u(x,t)$ is determined for all (x,t) in the closed triangle OAL by means of (2.2). Every point P of the triangle OAM is a vertex of a parallelogram with sides parallel to the characteristics, and such that of the three remaining vertices, two lie on the characteristic $x = ct$, where u is known, and the other is on the vertical line $x = 0$, where $u = 0$. Thus $u(P)$ can be calculated from (1.4). Analogously u can be computed at every point of the closure of LAN. We may now proceed in this fashion to determine u progressively at every point of the closure of the regions α, β, etc.

4.1 Separation of Variables

Seek a solution of (4.1) in the form $u(x,t) = X(x)T(t)$. The equation yields

$$\begin{array}{ll} T'' = c^2 \lambda T & \text{in } \mathbb{R} \\ X'' = \lambda X & \text{in } (0, L) \end{array} \qquad \lambda \in \mathbb{R}. \tag{4.2}$$

The first of these implies that only negative values of λ yield bounded solutions. Setting $\lambda = -\gamma^2$, the second gives the one-parameter family of solutions

$$X(x) = C_1 \sin \gamma x + C_2 \cos \gamma x.$$

These will satisfy the boundary conditions at $x = 0$ and $x = L$ if $C_2 = 0$ and $\gamma = n\pi/L$ for $n \in \mathbb{N}$. Therefore, the functions

$$X_n(x) = \sin \frac{n\pi}{L} x, \qquad n \in \mathbb{N}$$

represent a family of solutions for the second of (4.2). With the indicated choice of γ, the first of (4.1) gives

$$T_n(t) = A_n \sin\left(\frac{n\pi c}{L} t\right) + B_n \cos\left(\frac{n\pi c}{L} t\right).$$

The solutions $u_n = X_n T_n$ can be superposed to give the general solution in the form

$$u(x,t) = \sum_{n=1}^{\infty} \left[A_n \sin\left(\frac{n\pi c}{L} t\right) + B_n \cos\left(\frac{n\pi c}{L} t\right)\right] \sin\left(\frac{n\pi}{L} x\right). \tag{4.3}$$

The numbers A_n and B_n are called the *Fourier coefficients* of the series in (4.3), and are computed from the initial conditions, i.e.,

$$\sum_{n=1}^{\infty} B_n \sin \frac{\pi n}{L} x = \varphi(x), \qquad \sum_{n=1}^{\infty} A_n \frac{n\pi c}{L} \sin \frac{n\pi}{L} x = \psi(x).$$

Since the system $\left\{\sin \frac{n\pi x}{L}\right\}$ is orthogonal and complete in $L^2(0, L)$ (**10.2** of the Problem and Complements of Chapter 4), one computes

$$A_n = \frac{2}{n\pi c} \int_0^L \sin \frac{n\pi x}{L} \psi(x) dx, \quad B_n = \frac{2}{L} \int_0^L \sin \frac{n\pi x}{L} \varphi(x) dx. \tag{4.4}$$

Remark 4.1 We have assumed $\varphi, \psi \in C^2[0, L]$. Actually, the method leading to (4.3) requires only that φ and ψ be in $L^2(0, L)$. Therefore, one might define the solutions obtained by (4.3) as weak solutions of (4.1), whenever merely $\varphi, \psi \in L^2(0, L)$. The PDE, however, need not be satisfied in the classical sense.

Remark 4.2 The nth term in (4.3) is called the nth mode of vibration or the nth *harmonic*. We rewrite the nth harmonic as

$$G_n \sin \frac{n\pi}{L} x \cos \frac{n\pi c}{L}(t - \tau_n)$$

where G_n and τ_n are two new constants called *amplitude* and *phase angle* respectively. The solution u can be thought of as the superposition of independent harmonics, each vibrating with amplitude G_n, phase angle τ_n, and frequency $\nu_n = n\pi c/L$.

The method of separation of variables and the principle of superposition were introduced by D. Bernoulli ([9, 10]), even though not in the context of a formal PDE. In the context of the wave equation, the method was suggested, on a more formal basis by d'Alembert; it was employed by Poisson and developed by Fourier [42].

4.2 Odd Reflection

We describe another method to solve (4.1) by referring to the Cauchy problem (2.1). If the initial data φ and ψ are odd with respect to $x = 0$, then u is odd with respect to $x = 0$. Analogously, if φ and ψ are odd about $x = L$, the same holds for u. It follows that the solution of the Cauchy problem (2.1) with φ and ψ odd about both points $x = 0$ and $x = L$ must be zero at $x = 0$ and $x = L$, for all $t \in \mathbb{R}$, i.e., it satisfies the boundary conditions at $x = 0$ and $x = L$ prescribed by (4.1). This suggests constructing a solution of (4.1) by converting it into an initial value problem (a Cauchy problem) with initial data given by the odd extension of φ and ψ about both $x = 0$ and $x = L$. For φ, such an extension is given by

$$\tilde{\varphi}(x) = \begin{cases} \varphi(x - nL) & \text{for } x \in \big(nL, (n+1)L\big) \ \ n \in \mathbb{Z} \text{ even} \\ -\varphi((n+1)L - x) & \text{for } x \in \big(nL, (n+1)L\big) \ \ n \in \mathbb{Z} \text{ odd.} \end{cases}$$

An analogous formula holds for $\tilde{\psi}$. Then the solution of (4.1) is given by the restriction to $(0, L) \times \mathbb{R}$ of

$$\tilde{u}(x,t) = \frac{1}{2}[\tilde{\varphi}(x - ct) + \tilde{\varphi}(x + ct)] + \frac{1}{2c}\int_{x-ct}^{x+ct} \tilde{\psi}(s)ds$$

constructed by the d'Alembert formula.

Remark 4.3 Even if φ and ψ are in $C^2[0, L]$, their odd extensions might fail to be of class C^2 across $x = nL$. However, for $(x,t) \in (0, L) \times \mathbb{R}$, the points $x \pm ct$ are in the interior of some interval $\big(nL, (n+1)L\big)$ for some $n \in \mathbb{N}$, so that u is actually a classical solution of (4.1).

4.3 Energy and Uniqueness

Let $u \in C^2([0, L] \times \mathbb{R})$ be a solution of (4.1). The quantity

$$\mathcal{E}(t) = \int_0^L (u_t^2 + c^2 u_x^2)(x,t)dx \tag{4.5}$$

is called the energy of the system at the instant t. Multiplying the first of (4.1) by u_t, integrating by parts over $(0, L)$, and using the boundary conditions at $x = 0$ and $x = L$ gives

$$\frac{d}{dt}\int_0^L (u_t^2 + c^2 u_x^2)(x,t)dx = \mathcal{E}'(t) = 0.$$

Thus $\mathcal{E}(t) = \mathcal{E}(0)$ for all $t \in \mathbb{R}$, and the energy is conserved. Also, if $\varphi = \psi = 0$, then $u = 0$ in $(0, L) \times \mathbb{R}$. In view of the linearity of the PDE one concludes that C^2 solutions of (4.1) are unique.

4.4 Inhomogeneous Problems

Let $f \in C^1\big((0,L) \times \mathbb{R}\big)$, and consider the inhomogeneous boundary value problem

$$\begin{aligned} u_{tt} - c^2 u_{xx} &= f && \text{in } (0,L) \times \mathbb{R} \\ u(0,\cdot) = u(L,\cdot) &= 0 && \text{in } \mathbb{R} \\ u(\cdot,0) = \varphi,\ u_t(\cdot,0) &= \psi && \text{in } (0,L). \end{aligned} \tag{4.6}$$

The solution $u(x,t)$ represents the position, at point x and at time t, of a string vibrating under the action of a load f applied at time t at its points $x \in (0, L)$. The solution of (4.6) can be constructed by superposing the unique solution of (4.1) with the unique solution of

$$\begin{aligned} v_{tt} - c^2 v_{xx} &= f && \text{in } (0,L) \times \mathbb{R} \\ v(0,\cdot) = v(L,\cdot) &= 0 && \text{in } \mathbb{R} \\ v(\cdot,0) = v_t(\cdot,0) &= 0 && \text{in } (0,L). \end{aligned}$$

This, in turn, can be solved by reducing it to an initial value problem, through an odd reflection of $x \to f(x,t)$, for all $t \in \mathbb{R}$, about $x = 0$ and $x = L$, as suggested by Remark 3.1.

5 The Initial Value Problem in N Dimensions

Introduce formally the d'Alembertian

$$\Box \overset{\text{def}}{=} \frac{\partial^2}{\partial t^2} - c^2 \Delta.$$

and, given $\varphi \in C^3(\mathbb{R}^N)$ and $\psi \in C^2(\mathbb{R}^N)$, consider the Cauchy problem

$$\begin{aligned} \Box u &= 0 && \text{in } \mathbb{R}^N \times \mathbb{R} \\ u(\cdot,0) &= \varphi && \text{in } \mathbb{R}^N. \\ u_t(\cdot,0) &= \psi \end{aligned} \tag{5.1}$$

If $N \geq 3$, the problem (5.1) can be solved by the Poisson method of spherical means, and if $N = 2$ by the Hadamard method of descent.

5.1 Spherical Means

Let ω_N denote the measure of the unit sphere in $\mathbb{R}^N$ and let $d\omega$ denote the surface measure on the unit sphere of $\mathbb{R}^N$, that is, the infinitesimal solid angle in $\mathbb{R}^N$. If $v \in C(\mathbb{R}^N)$, the spherical mean of v at x of radius ρ is

$$\begin{aligned} M(v;x,\rho) &= \frac{1}{\text{meas}[\partial B_\rho(x)]} \int_{\partial B_\rho(x)} v d\sigma \\ &= \frac{1}{\omega_N \rho^{N-1}} \int_{|x-y|=\rho} v(y) d\sigma(y) = \frac{1}{\omega_N} \int_{|\nu|=1} v(x+\rho\nu) d\omega \end{aligned}$$

where ν ranges over the unit sphere of $\mathbb{R}^N$.

Remark 5.1 The function $\rho \to M(v;x,\rho)$ can be defined in all of $\mathbb{R}$ by an even reflection about the origin since

$$\int_{|\nu|=1} v(x+\rho\nu) d\omega = \int_{|\nu|=1} v(x-\rho(-\nu)) d\omega = \int_{|\nu|=1} v(x-\rho\nu) d\omega.$$

Remark 5.2 If $v \in C^s(\mathbb{R}^N)$ for some $s \in \mathbb{N}$, then $x \to M(v;x,\rho) \in C^s(\mathbb{R}^N)$.

Remark 5.3 Knowing $(x,\rho) \to M(v;x,\rho)$ permits one to recover $x \to v(x)$, since

$$\lim_{\rho \to 0} M(v;x,\rho) = v(x) \quad \text{for all } x \in \mathbb{R}^N.$$

5.2 The Darboux Formula

Assume that $v \in C^2(\mathbb{R}^N)$. By the divergence theorem

$$\begin{aligned} \int_{|x-y|<\rho} \Delta v(y) dy &= \int_{|x-y|=\rho} \nabla v(y) \cdot \nu d\sigma(y) \\ &= \rho^{N-1} \int_{|\nu|=1} \nabla v(x+\rho\nu) \cdot \nu d\omega \\ &= \rho^{N-1} \frac{d}{d\rho} \int_{|\nu|=1} v(x+\rho\nu) d\omega. \end{aligned}$$

Therefore

$$\begin{aligned} \frac{\partial}{\partial\rho} M(v;x,\rho) &= \frac{1}{\omega_N \rho^{N-1}} \int_{|x-y|<\rho} \Delta v(y) dy \\ &= \frac{1}{\omega_N \rho^{N-1}} \int_0^\rho r^{N-1} \Delta_x \int_{|\nu|=1} v(x+r\nu) d\omega dr. \end{aligned}$$

Multiplying by ρ^{N-1} and taking the derivative with respect to ρ yields

$$\frac{\partial}{\partial\rho}\left(\rho^{N-1}\frac{\partial}{\partial\rho}M(v;x,\rho)\right) = \Delta_x(\rho^{N-1}M(v;x,\rho)).$$

This, in turn, gives Darboux's formula

$$\left(\frac{\partial^2}{\partial\rho^2} + \frac{N-1}{\rho}\frac{\partial}{\partial\rho}\right)M(v;x,\rho) = \Delta_x M(v;x,\rho) \tag{5.2}$$

valid for all $v \in C^2(\mathbb{R}^N)$.

5.3 An Equivalent Formulation of the Cauchy Problem

Let $u \in C^2(\mathbb{R}^N \times \mathbb{R})$ be a solution of (5.1). Then for all $x \in \mathbb{R}^N$ and for all $\rho > 0$

$$\begin{aligned}\Delta_x M(u;x,\rho) &= \frac{1}{\omega_N}\int_{|\nu|=1}\Delta_x u(x+\rho\nu,t)d\omega\\ &= \frac{1}{c^2\omega_N}\int_{|\nu|=1}\frac{\partial^2}{\partial t^2}u(x+\rho\nu,t)\,d\omega = \frac{1}{c^2}\frac{\partial^2}{\partial t^2}M(u;x,\rho).\end{aligned}$$

Therefore, setting

$$M(\rho,t) = M(u(x,t);x,\rho)$$

and recalling Remarks 5.1 and 5.3, one concludes that $u \in C^2(\mathbb{R}^N \times \mathbb{R})$ is a solution of (5.1) if and only if

$$\begin{aligned}\frac{\partial^2}{\partial t^2}M(\rho,t) &= c^2\left(\frac{\partial^2}{\partial\rho^2} + \frac{N-1}{\rho}\frac{\partial}{\partial\rho}\right)M(\rho,t)\\ M(\rho,0) &= M(\varphi;x,\rho) = M_\varphi(x,\rho)\\ M_t(\rho,0) &= M(\psi;x,\rho) = M_\psi(x,\rho).\end{aligned} \tag{5.3}$$

6 The Cauchy Problem in $\mathbb{R}^3$

If $N = 3$, the initial value problem (5.3) becomes, on multiplication by ρ

$$\begin{aligned}\frac{\partial^2}{\partial t^2}(\rho M(\rho,t)) &= c^2\frac{\partial^2}{\partial\rho^2}(\rho M(\rho,t)) \text{ in } \mathbb{R}\times\mathbb{R}\\ \rho M(\rho,0) &= \rho M_\varphi(x,\rho)\\ \rho M_t(\rho,0) &= \rho M_\psi(x,\rho).\end{aligned} \tag{6.1}$$

By the d'Alembert formula (2.2)

$$\rho M(\rho,t) = \frac{1}{2}[(\rho - ct)M_\varphi(x,\rho - ct) + (\rho + ct)M_\varphi(x,\rho + ct)] + \frac{1}{2c}\int_{\rho - ct}^{\rho + ct} sM_\psi(x,s)ds.$$

Differentiating with respect to ρ

$$\begin{aligned} M(\rho,t) + \rho M_\rho(\rho,t) &= \frac{1}{2}[M_\varphi(x,\rho - ct) + M_\varphi(x,\rho + ct)] \\ &+ \frac{1}{2}\left[(\rho - ct)\frac{\partial}{\partial \rho}M_\varphi(x,\rho - ct) + (\rho + ct)\frac{\partial}{\partial \rho}M_\varphi(x,\rho + ct)\right] \\ &+ \frac{1}{2c}\left[(\rho + ct)M_\psi(x,\rho + ct) - (\rho - ct)M_\psi(x,\rho - ct)\right]. \end{aligned}$$

Letting $\rho \to 0$ gives the solution formula for (5.1)

$$\begin{aligned} u(x,t) &= \frac{1}{8\pi}\left(\int_{|\nu|=1}\varphi(x + c\nu t)d\omega + \int_{|\nu|=1}\varphi(x - c\nu t)d\omega\right) \\ &+ \frac{1}{8\pi}\left(ct\int_{|\nu|=1}\nabla\varphi(x + \nu ct)\cdot\nu\, d\omega - ct\int_{|\nu|=1}\nabla\varphi(x - c\nu t)\cdot\nu\, d\omega\right) \\ &+ \frac{1}{8\pi}\left(t\int_{|\nu|=1}\psi(x + \nu ct)d\omega + t\int_{|\nu|=1}\psi(x - c\nu t)d\omega\right). \end{aligned} \tag{6.2}$$

From this and Remark 5.1

$$u(x,t) = \frac{1}{4\pi}\frac{\partial}{\partial t}\left(t\int_{|\nu|=1}\varphi(x + \nu ct)d\omega\right) + \frac{1}{4\pi}t\int_{|\nu|=1}\psi(x + \nu ct)d\omega. \tag{6.3}$$

This can be written in the equivalent form

$$4\pi c^2 u(x,t) = \frac{\partial}{\partial t}\left(\frac{1}{t}\int_{|x-y|=ct}\varphi(y)d\sigma\right) + \frac{1}{t}\int_{|x-y|=ct}\psi(y)d\sigma. \tag{6.4}$$

By carrying out the differentiation under the integral in (6.3)

$$4\pi c^2 u(x,t) = \frac{1}{t^2}\int_{|x-y|=ct}[t\psi(y) + \varphi(y) + \nabla\varphi\cdot(x - y)]d\sigma. \tag{6.5}$$

Theorem 6.1 *Let $N = 3$ and assume that $\varphi \in C^3(\mathbb{R}^3)$ and $\psi \in C^2(\mathbb{R}^3)$. Then there exists a unique solution to the Cauchy problem (5.1), and it is given by (6.2)–(6.5).*

Proof We have only to prove the uniqueness. If $u, v \in C^2(\mathbb{R}^3 \times \mathbb{R})$ are two solutions, the spherical mean of their difference

$$\tilde{M} = \frac{3}{4\pi} \int_{|\nu|=1} (u - v)(x + \rho\nu) d\omega$$

satisfies (6.1) with homogeneous data. By the uniqueness of solutions to the one-dimensional Cauchy problem, $\tilde{M} = 0$ for all $\rho > 0$. Thus $u = v$. ■

Formulas (6.2)–(6.5) are the *Kirchoff formulas*; they permit one to read the relevant properties of the solution u.

Remark 6.1 (Domain of Dependence) The solution at a point $(x, t) \in \mathbb{R}^{N+1}$ for $N = 3$ depends on the data φ and ψ and the derivatives φ_{x_i} on the sphere $|x - y| = ct$. Unlike the 1-dimensional case, the data in the interior of $B_{ct}(x)$ are not relevant to the value of u at (x, t).

Remark 6.2 (Regularity) In the case $N = 1$ the solution is as regular as the data. If $N = 3$, because of the t-derivative intervening in the representation (6.4), solutions of (5.1) are less regular than the data φ and ψ. In general, if $\varphi \in C^{m+1}(\mathbb{R}^3)$ and $\psi \in C^m(\mathbb{R}^3)$ for some $m \in \mathbb{N}$, then $u \in C^m(\mathbb{R}^3 \times \mathbb{R})$. Thus if φ and ψ are merely of class C^2 in $\mathbb{R}^3$, then $u_{x_i x_i}$ might blow-up at some point $(x, t) \in \mathbb{R}^3 \times \mathbb{R}$ even though $\varphi_{x_i x_j}$, and $\psi_{x_i x_j}$ are bounded. This is known as the *focussing effect*. In view of Remark 6.1, the set of singularities might become *compressed* for $t > 0$ into a smaller set called the *caustic*.

Remark 6.3 (Compactly Supported Data) In the remainder of this section we assume that the initial data φ and ψ are compactly supported, say in the ball $B_r(0)$, and discuss the stability in $L^\infty(\mathbb{R}^3)$ for all $t \in \mathbb{R}$. From the solution formula (6.3), it follows that $x \to u(x, t)$ is supported in the spherical annulus $(ct - r)_+ \le |x| \le r + ct$. A disturbance concentrated in $B_r(0)$ affects the solution only within such a spherical annulus.

Remark 6.4 (Decay for Large Times) We continue to assume that the data φ and ψ are supported in the ball $B_r(0)$. By Remark 6.3, the solution $x \to u(x, t)$ is also compactly supported in $\mathbb{R}^3$. The solution is also compactly supported in the t variable, in the following sense:

$$t \to u(x, t) = 0 \quad \text{if for fixed } |x|, \ |t| \text{ is sufficiently large.}$$

A stronger statement holds, i.e., $\|u\|_{\infty,\mathbb{R}}(t) \to 0$ as $t \to \infty$. Indeed, from (6.5), for large times

$$\|u(t)\|_{\infty,\mathbb{R}^3} \le \frac{(1 + c)r^2}{c^2 t} (\|\varphi\|_{\infty,\mathbb{R}^3} + \|\nabla\varphi\|_{\infty,\mathbb{R}^3} + \|\psi\|_{\infty,\mathbb{R}^3}) \tag{6.6}$$

since the sphere $|x - y| = ct$ intersects the support of the data, at most in a disc of radius r.

Remark 6.5 (Energy) Let $\mathcal{E}(t)$ denote the energy of the system at time t

$$\mathcal{E}(t) = \int_{\mathbb{R}^3} \left(u_t^2 + c^2|Du|^2\right) dx$$

where D denotes the gradient with respect to the space variables only. Multiplying the PDE $\Box u = 0$ by u_t and integrating by parts in $\mathbb{R}^3$ yields

$$\frac{d}{dt}\mathcal{E}(t) = 0.$$

The compactly supported nature of $x \to u(x,t)$ is employed here in justifying the integration by parts. The same result would hold for a solution $u \in C^2(\mathbb{R}^3 \times \mathbb{R})$ satisfying

$$|Du|(\cdot,t) \in L^2(\mathbb{R}^3) \quad \text{for all } t \in \mathbb{R}. \tag{6.7}$$

A consequence is

Lemma 6.1 *There exists at most one solution to the Cauchy problem (5.1) within the class (6.7).*

Also, taking into account (6.6) and Theorem 6.1,

Theorem 6.2 *Let $N = 3$ and assume that φ and ψ are supported in the ball B_r for some $r > 0$. Assume further that $\varphi \in C^3(\mathbb{R}^3)$ and $\psi \in C^2(\mathbb{R}^3)$. Then there exists a unique solution to the Cauchy problem (5.1), and it is given by (6.2)–(6.4). Moreover, such a solution is stable in $L^\infty(\mathbb{R}^3)$.*

Therefore, for smooth and compactly supported initial data, (5.1) is well-posed in the sense of Hadamard, in the topology of $L^\infty(\mathbb{R}^3)$.

7 The Cauchy Problem in $\mathbb{R}^2$

Consider the Cauchy problem for the wave equation in two space dimensions

$$\begin{aligned} u_{tt} - c^2(u_{x_1x_1} + u_{x_2x_2}) &= 0 && \text{in } \mathbb{R}^2 \times \mathbb{R} \\ u(\cdot,0) &= \varphi && \text{in } \mathbb{R}^2 \\ u_t(\cdot,0) &= \psi && \text{in } \mathbb{R}^2. \end{aligned} \tag{7.1}$$

Theorem 7.1 *Assume that $\varphi \in C^3(\mathbb{R}^2)$ and $\psi \in C^2(\mathbb{R}^2)$. Then the Cauchy problem (7.1) has the unique solution*

$$\begin{aligned} u(x_1,x_2,t) = {} & \frac{\partial}{\partial t}\left(\frac{1}{2\pi c}\int_{D_{ct}(x_1,x_2)} \frac{\varphi(y_1,y_2)dy_1dy_2}{\sqrt{c^2t^2 - (y_1-x_1)^2 - (y_2-x_2)^2}}\right) \\ & + \frac{1}{2\pi c}\int_{D_{ct}(x_1,x_2)} \frac{\psi(y_1,y_2)dy_1y_2}{\sqrt{c^2t^2 - (y_1-x_1)^2 - (y_2-x_2)^2}} \end{aligned} \tag{7.2}$$

where $D_{ct}(x_1,x_2)$ is the disc of center (x_1,x_2) and radius ct.

The Hadamard method of descent ([62]), consists in viewing the solution of (7.1) as an x_3-independent solution of (5.1) for $N = 3$, for which one has the explicit representations (6.2)–(6.5). Let S be the sphere in $\mathbb{R}^3$, of center $(x_1, x_2, 0)$ and radius ct

$$S = \left\{(y_1, y_2, y_3) \in \mathbb{R}^3 \mid (x_1 - y_1)^2 + (x_2 - y_2)^2 + y_3^2 = c^2t^2\right\}.$$

From (6.5)

$$\begin{aligned} u(x_1, x_2, t) &= u(x_1, x_2, 0, t) \\ &= \frac{\partial}{\partial t}\left(\frac{1}{4\pi c^2 t}\int_S \varphi(y_1, y_2) d\sigma\right) + \frac{1}{4\pi c^2 t}\int_S \psi(y_1, y_2)\, d\sigma. \end{aligned}$$

If $P = (y_1, y_2, y_3) \in S$ and if $\nu(P)$ is the outward unit normal to S at P, then for $|y_3| > 0$

$$\nu \cdot \frac{y_3}{|y_3|} = \frac{y_3}{ct} \qquad \text{and} \qquad d\sigma = \frac{ct}{|y_3|} dy_1 dy_2$$

where $dy = dy_1 dy_2$ is the Lebesgue measure in $\mathbb{R}^2$ and (y_1, y_2) ranges over the disc $(y_1 - x_1)^2 + (y_2 - x_2)^2 < (ct)^2$. Also

$$|y_3| = \sqrt{c^2t^2 - [(y_1 - x_1)^2 + (y_2 - x_2)^2]}.$$

Carry these remarks in the previous formula and denote by $x = (x_1, x_2)$ and $y = (y_1, y_2)$ points in $\mathbb{R}^2$ to obtain

$$\begin{aligned} u(x, t) = &\frac{\partial}{\partial t}\left(\frac{1}{2\pi c}\int_{|y-x|<ct} \frac{\varphi(y)}{\sqrt{c^2t^2 - |y - x|^2}} dy\right) \\ &+ \frac{1}{2\pi c}\int_{|y-x|<ct} \frac{\psi(y)}{\sqrt{c^2t^2 - |y - x|^2}} dy \end{aligned} \tag{7.3}$$

where we have used that as $P = (y_1, y_2, y_3) = (y, y_3)$ runs over S, y runs twice over the disc $|y - x| < ct$. Formula (7.3) is the Poisson formula for the solution of (7.1).

Remark 7.1 (Domain of Dependence) The solution u at a point $(x, t) \in \mathbb{R}^2 \times \mathbb{R}$ depends on the values of the initial data φ, $\nabla\varphi$, and ψ on the whole disc $|y - x| < ct$. This is in contrast to the three-dimensional case in which only the values on the sphere of center x and radius ct were relevant.

Remark 7.2 (Disturbances and The Huygens Principle) The values of the data φ, $\nabla\varphi$, and ψ at some $x_o \in \mathbb{R}^2$ (initial disturbances at x_o) will not affect a point x until time $ct(x) = |x - x_o|$, and will affect $u(x, t)$ at all further times $t > t(x)$. Therefore a signal starting at x_o at time $t = 0$ is received by x at $t = t(x)$ and keeps being "received" thereafter. This explains the propagation of circular waves in still water originating from a "nearly-a-point" disturbance. In the three-dimensional case, an initial disturbance

$\varphi(x_o)$, $\nabla\varphi(x_o)$, and $\psi(x_o)$ at $x_o \in \mathbb{R}^3$ reaches x at time $ct = |x - x_o|$ and will not affect $u(x,t)$ for all later times. This is a special case of the Huygens principle, which states that if $N \geq 3$ and N is odd, signals originating at some $x_o \in \mathbb{R}^N$ are received by an observer at $x \in \mathbb{R}^N$ only at a single instant.

8 The Inhomogeneous Cauchy Problem

Consider the inhomogeneous initial value problem

$$\begin{aligned} \Box u &= f \in C^2(\mathbb{R}^N \times \mathbb{R}) && \text{in } \mathbb{R}^N \times \mathbb{R},\ N = 2,3 \\ u(\cdot,0) &= \varphi \in C^3(\mathbb{R}^N) && \text{in } \mathbb{R}^N \\ u_t(\cdot,0) &= \psi \in C^2(\mathbb{R}^N) && \text{in } \mathbb{R}^N. \end{aligned} \tag{8.1}$$

The solution is the sum of the unique solution of (5.1) ($f = 0$), and

$$\begin{aligned} \Box v &= f && \text{in } \mathbb{R}^N \times \mathbb{R} \\ v(x,0) &= v_t(x,0) = 0 && \text{in } \mathbb{R}^N. \end{aligned} \tag{8.2}$$

The Duhamel principle permits one to reduce the solution of (8.2) to the solution of the family of homogeneous problems ($f = 0$)

$$\begin{aligned} \Box w(x,t;\tau) &= 0 && \text{in } \mathbb{R}^N \times (t > \tau) \\ w(\cdot,\tau;\tau) &= 0 && \text{in } \mathbb{R}^N \\ w_t(\cdot,\tau;\tau) &= f(\cdot,\tau) \end{aligned}$$

By Duhamel's principle, the solution of (8.2) is given by

$$v(x,t) = \int_0^t w(x,t;\tau)d\tau.$$

Indeed, by direct calculation

$$v_t(x,t) = \int_0^t w_t(x,t;\tau)d\tau$$

since $w(x,t;t) = 0$. Therefore $v(x,0) = v_t(x,0) = 0$. Next

$$\begin{aligned} v_{tt} &= w_t(x,t;t) + \int_0^t w_{tt}(x,t;\tau)d\tau \\ &= f(x,t) + c^2 \int_0^t \Delta w(x,t;\tau)d\tau = f + c^2 \Delta v \end{aligned}$$

so that (8.2) holds. If $N = 3$ and $t \geq 0$

$$v(x,t) = \frac{1}{4\pi c^2} \int_0^t \frac{1}{(t-\tau)} \int_{|x-y|=c(t-\tau)} f(y,\tau)d\sigma\, d\tau. \tag{8.3}$$

If $N = 2$ and $t \geq 0$

$$v(x,t) = \frac{1}{2\pi c} \int_0^t \int_{|x-y| \leq c(t-\tau)} \frac{f(y,\tau)}{\sqrt{c^2(t-\tau)^2 - |x-y|^2}} dy\, d\tau. \tag{8.4}$$

Remark 8.1 (Domain of Dependence) If $N = 3$, the value of v at a point (x,t), for $t > 0$, depends only on the values of the forcing term f on the surface of the truncated backward characteristic cone

$$[|x-y| = c(t-\tau)] \cap [0 \leq \tau \leq t].$$

If $N = 2$, the domain of dependence is the full truncated backward characteristic cone

$$[|x-y| < c(t-\tau)] \cap [0 \leq \tau \leq t].$$

Remark 8.2 (Disturbances) The effect of a source disturbance at a point (x_o, t_o) is not felt at x until the time

$$t(x) = t_o + \frac{1}{c}|x - x_o|.$$

Notice that $\frac{1}{c}|x - x_o|$ is the time it takes for an *initial* disturbance at x_o to affect x. Thus $f(x_o, t_o)$ can be viewed as an *initial datum* delayed to a time t_o. For this reason, the solution formulas (8.3), (8.4) are referred to as *retarded potentials*.

9 The Cauchy Problem for Inhomogeneous Surfaces

The methods introduced for the inhomogeneous initial value problem permit one to solve the following non-characteristic Cauchy problem

$$\begin{aligned} \Box u &= f && \text{in } \mathbb{R}^3 \times (t > \Phi) \\ u(\cdot, \Phi) &= \varphi && \text{in } \mathbb{R}^3 \\ u_t(\cdot, \Phi) &= \psi && \text{in } \mathbb{R}^3. \end{aligned} \tag{9.1}$$

The data φ, and ψ are now given on the surface $\Sigma = [t = \Phi]$. Such a surface must be non-characteristic in the sense that $c|\nabla\Phi| \neq 1$ in $\mathbb{R}^3$. We require that Σ is nearly flat, in the sense

$$c\|\nabla\Phi\|_{\infty,\mathbb{R}^3} < 1. \tag{9.2}$$

To convey the main ideas of the technique, we will assume that φ, ψ, and Φ are as smooth as needed to carry out the calculations below. Finally, without loss of generality, we may assume that $\Phi \geq 0$.

9.1 Reduction to Homogeneous Data on $t = \Phi$

First consider the problem of finding $v \in C^3(\mathbb{R}^3 \times \mathbb{R})$, a solution of

$$\begin{gathered}(\Box v - f)\big|_{t=\Phi} = (\Box v - f)_t\big|_{t=\Phi} = (\Box v - f)_{tt}\big|_{t=\Phi} = 0 \\ v(\cdot, \Phi) = \varphi, \quad v_t(\cdot, \Phi) = \psi.\end{gathered} \tag{9.3}$$

Lemma 9.1 *Let (9.2) hold. Then there exists a solution to problem (9.3).*

Proof Seek v of the form

$$v(x,t) = \sum_{i=0}^{4} a_i(x)(t - \Phi(x))^i$$

where $x \to a_i(x)$, for $i = 1, \dots, 4$, are smooth functions to be calculated. The last two of (9.3) give $a_o = \varphi$ and $a_1 = \psi$. Next, by direct calculation

$$\begin{aligned}\Box v = &\sum_{i=2}^{4} i(i-1)a_i(x)(t - \Phi(x))^{i-2} - c^2 \sum_{i=0}^{4} \Delta a_i(x)(t - \Phi(x))^i \\ &- 2c^2 \sum_{i=1}^{4} i\nabla a_i(x)\nabla(t - \Phi(x))(t - \Phi(x))^{i-1} \\ &- c^2 \sum_{i=1}^{4} a_i(t - \Phi(x))^{i-1}\Delta(t - \Phi(x)) \\ &- c^2 \sum_{i=2}^{4} i(i-1)a_i(x)(t - \Phi(x))^{i-2}|\nabla(t - \Phi(x))|^2.\end{aligned}$$

From this and (9.2)–(9.3)

$$\begin{aligned}2(1 - c^2|\nabla\Phi|^2)a_2 &= c^2[\Delta(\varphi - a_1\Phi) + \Phi\Delta a_1] + f \\ 6(1 - c^2|\nabla\Phi|^2)a_3 &= c^2[\Delta(\psi - 2a_2\Phi) + 2\Phi\Delta a_2] + f_t \\ 24(1 - c^2|\nabla\Phi|^2)a_4 &= 2c^2[\Delta(a_2 - 3a_3\Phi) + 3\Phi\Delta a_3] + f_{tt}.\end{aligned}$$

■

9.2 The Problem with Homogeneous Data

Look for a solution of (9.1) of the form $w = u - v$, and set $F = f - \Box v$. Then w satisfies

$$\begin{aligned}\Box w &= F &&\text{in } \mathbb{R}^3 \times (t > \Phi) \\ w(\cdot, \Phi) &= w_t(\cdot, \Phi) = 0 &&\text{in } \mathbb{R}^3.\end{aligned} \tag{9.4}$$

By the construction process of the solution of (9.3) $F = F_t = F_{tt} = 0$ on $t = \Phi$, so that the function

$$F_o(x,t) = \begin{cases} F(x,t) & \text{for } t \ge \Phi(x) \\ 0 & \text{for } t \le \Phi(x) \end{cases}$$

is of class C^2 in $\mathbb{R}^3 \times \mathbb{R}$. Then solve

$$\begin{aligned} \Box \bar{w} &= F_o && \text{in } \mathbb{R}^3 \times (t > 0) \\ \bar{w}(x,0) &= \bar{w}_t(x,0) = 0 && \text{in } \mathbb{R}^3 \end{aligned}$$

whose solution is given by the representation formula (8.3). The restriction of $\bar{w}$ to $[t > \Phi]$ is the solution of (9.4). This will follow from (8.3) and the next lemma.

Lemma 9.2 *Let (9.2) hold. Then* $\bar{w}(x,t) = 0$ *for* $t \le \Phi(x)$.

Proof In (8.3), written for $\bar{w}$ and F_o, fix x and $t \le \Phi(x)$. For all y on the lateral surface of the backward truncated characteristic cone

$$[|x-y| = c(t-\tau)] \cap [0 \le \tau < t \le \Phi(x)]$$

we must have $\tau < \Phi(y)$. Indeed, if not

$$|x-y| \le c(\Phi(x) - \Phi(y)) \le c|\nabla\Phi(\xi)||x-y|$$

for some ξ on the line segment $\tau x + (1-\tau)y$ for $\tau \in (0,1)$. In view of (9.2) this yields a contradiction. Since F_o vanishes for (y,τ) such that $\tau \le \Phi(y)$, the lemma follows. ∎

The solution obtained this way is unique. This is shown as in Theorem 6.1. Unlike the Cauchy–Kowalewski theorem, the data are not required to be analytic and the solution is global. Analytic data would yield analytic solutions only near Σ.

10 Solutions in Half-Space. The Reflection Technique

Consider the initial boundary value problem

$$\begin{aligned} \Box u &= f && \text{in } (\mathbb{R}^2 \times \mathbb{R}^+) \times \mathbb{R} \\ u(\cdot,0) &= \varphi && \text{for } x_3 \ge 0 \\ u_t(\cdot,0) &= \psi && \text{for } x_3 \ge 0 \\ u(x_1,x_2,0,t) &= h(x_1,x_2,t) && \text{for } x_3 = 0,\ t \ge 0. \end{aligned} \tag{10.1}$$

If the data are sufficiently smooth, and there is a solution of class C^3 in the closed half-space $\mathbb{R}^2 \times [x_3 \ge 0] \times [t \ge 0]$, the following compatibility conditions must be satisfied

$$\begin{aligned} h(x_1,x_2,0) &= \varphi(x_1,x_2) \\ h_t(x_1,x_2,0) &= \psi(x_1,x_2) \\ c^2\Delta\varphi + f(x_1,x_2,0,t) &= h_{tt}(x_1,x_2,0) \\ c^2\Delta\psi + f_t(x_1,x_2,0,t) &= h_{ttt}(x_1,x_2,0). \end{aligned} \tag{10.2}$$

Assume henceforth that (10.2) are satisfied and reduce the problem to one with homogeneous data on the hyperplane $x_3 = 0$.

10.1 An Auxiliary Problem

First find a solution $v \in C^3(\mathbb{R}^3 \times \mathbb{R})$ of the problem

$$\begin{gathered} v\big|_{x_3=0} = h,\ v_{x_3}\big|x_3 = 0 = v_{x_3x_3}\big|_{x_3=0} = 0 \\ (\Box v - f)\big|_{x_3=0} = (\Box v - f)x_3\big|_{x_3=0} = (\Box v - f)_{x_3x_3}\big|_{x_3=0} = 0. \end{gathered} \tag{10.3}$$

Lemma 10.1 *There exists a smooth solution to (10.3).*

Proof Look for solutions of the form

$$v(x,t) = h(x_1,x_2,t) + \sum_{i=2}^{4} a_{i-1}(x_1,x_2,t)x_3^i$$

and calculate

$$\begin{aligned} \Box v - f = (h_{tt} - c^2\Delta h)(x_1,x_2,t) + \sum_{i=2}^{3}[\Box a_{i-1}(x_1,x_2,t)]x_3^i \\ - c^2\sum_{i=2}^{4} i(i-1)a_{i-1}x_3^{i-2} - f(x,t). \end{aligned}$$

Therefore the conditions (10.3) yield

$$2c^2a_1 = \Box h - f\big|_{x_3=0},\ 6c^2a_2 = -f_{x_3}\big|_{x_3=0},\ 24c^2a_3 = -f_{x_3x_3}\big|_{x_3=0}. \qquad \blacksquare$$

10.2 Homogeneous Data on the Hyperplane $x_3 = 0$

Set $w = u - v$ and $F = f - \Box v$. Then

$$\begin{aligned} \Box w &= F && \text{in } (\mathbb{R}^2 \times [x_3 > 0]) \times [t > 0] \\ w(\cdot,0) &= \varphi_o \stackrel{\text{def}}{=} \varphi - v(\cdot,0) && \text{in } \mathbb{R}^2 \times [x_3 \ge 0] \\ w_t(\cdot,0) &= \psi_o \stackrel{\text{def}}{=} \psi - v_t(\cdot,0) && \text{in } \mathbb{R}^2 \times [x_3 \ge 0] \\ w\big|_{x_3=0} &= 0 && \text{for } x_3 = 0,\ t \ge 0. \end{aligned}$$

Let $\tilde{F}$, $\tilde{\varphi}_o$, and $\tilde{\psi}_o$ be the odd extensions of F, φ_o, and ψ_o about $x_3 = 0$, and consider the problem

$$\begin{aligned} \Box\tilde{w} &= \tilde{F} && \text{in } \mathbb{R}^3 \times \mathbb{R} \\ \tilde{w}(\cdot,0) &= \tilde{\varphi}_o(x) && \text{in } \mathbb{R}^3 \\ \tilde{w}_t(\cdot,0) &= \tilde{\psi}_o(x) && \text{in } \mathbb{R}^3. \end{aligned}$$

If this problem has a smooth solution $\tilde{w}$, it must be odd about $x_3 = 0$, that is, $\tilde{w}(x_1,x_2,0,t) = 0$, so that the restriction of $\tilde{w}$ to $x_3 \ge 0$ is the unique solution of the indicated problem with homogeneous data on $x_3 = 0$. To establish the existence of $\tilde{w}$ we have only to check that $\tilde{\varphi}_o \in C^3(\mathbb{R}^3)$, $\tilde{\psi}_o \in C^2(\mathbb{R}^3)$ and $\tilde{F} \in C^2(\mathbb{R}^3 \times \mathbb{R})$. For this it will suffice to check that

$$\begin{aligned} F = F_{x_3} = F_{x_3 x_3} &= 0 \\ \tilde{\varphi}_o = \tilde{\varphi}_{0,x_3} = \tilde{\varphi}_{0,x_3 x_3} &= 0 \qquad \text{for } x_3 = 0. \\ \tilde{\psi}_o = \tilde{\psi}_{0,x_3} = \tilde{\psi}_{0,x_3 x_3} &= 0 \end{aligned}$$

These conditions follow from the definition of odd reflection about $x_3 = 0$, the compatibility conditions (10.2), and the construction (10.3) of the auxiliary function v.

11 A Boundary Value Problem

Let E be a bounded open set in $\mathbb{R}^N$ with smooth boundary ∂E and consider the initial boundary value problem

$$\begin{aligned} \Box u &= 0 && \text{in } E \times \mathbb{R}^+ \\ u(\cdot, t)\big|_{\partial E} &= 0 && \text{in } \mathbb{R}^+ \\ u(\cdot, 0) &= \varphi && \text{in } E \\ u_t(\cdot, 0) &= \psi && \text{in } E. \end{aligned} \tag{11.1}$$

Here $u(x,t)$ represents the displacement, at the point x at time t, of a vibrating ideal body, kept at rest at the boundary at ∂E. By the energy method, (11.1) has at most one solution. To find such a solution we use an N-dimensional version of the method of separation of variables of Section 4.1. Solutions of the type $T(t)X(x)$ yield

$$\begin{aligned} -\Delta X_n &= \lambda_n X && \text{in } E \\ X_n &= 0 && \text{on } \partial E \end{aligned} \qquad n \in \mathbb{N} \tag{11.2}$$

and

$$T_n''(t) = -c^2 \lambda_n T_n(t) \quad \text{for } t > 0, \quad n \in \mathbb{N}. \tag{11.3}$$

The next proposition is a consequence of Theorem 11.1 of Section 11 of Chapter 4.

Proposition 11.1 *There exists an increasing sequence $\{\lambda_n\}$ of positive numbers and a sequence of corresponding functions $\{v_n\} \subset C^2(E)$ satisfying (11.2). Moreover $\{v_n\}$ form a complete orthonormal system in $L^2(E)$.*

Using this fact, write the solution u as

$$u(x,t) = \sum T_n(t) v_n(x) \tag{11.4}$$

and deduce that the initial conditions to be associated to (11.3) are derived from (11.4) and the initial data in (11.1), i.e.,

$$T_n(0) = \int_E v_n \varphi \, dx, \qquad T_n'(0) = \int_E v_n \psi \, dx.$$

Thus

$$T_n(t) = \int_E \Big[\psi \frac{\sin(c\sqrt{\lambda_n}t)}{c\sqrt{\lambda_n}} + \varphi \cos(c\sqrt{\lambda_n}t)\Big] v_n \, dx.$$

Even though the method is elegant and simple, the eigenvalues and eigenfunctions for the Laplace operator in E can be calculated explicitly only for domains with a simple geometry (see Section 8 of the Problems and Complements of Chapter 3). The approximate solutions

$$u_n(x,t) = \sum_{i=1}^{n} T_i(t) v_i(x)$$

satisfy, for all $i \in \mathbb{N}$, the approximating problems

$$\begin{aligned} \Box u_n &= 0 && \text{in } E \times \mathbb{R} \\ u_n(\cdot,t)\big|_{\partial E} &= 0 && \text{in } \mathbb{R} \\ u_n(x,0) = \varphi_n(x) &\stackrel{\text{def}}{=} \sum_{i=1}^{n} \langle \varphi, v_i \rangle v_i(x) && \text{in } E \\ u_{n,t}(x,0) = \psi_n(x) &\stackrel{\text{def}}{=} \sum_{i=1}^{n} \langle \psi, v_i \rangle v_i(x) && \text{in } E \end{aligned} \tag{11.5}$$

The function $u(\cdot,t)$ defined by (11.4) is meant as the limit of $u_n(\cdot,t)$ in $L^2(E)$, uniformly in $t \in \mathbb{R}$. The PDE in (11.1) and the initial data are verified in the following weak sense. Let f be any function in $C^2(\bar{E} \times \mathbb{R})$, and vanishing on ∂E. Multiply the PDE in (11.5) by any such f and integrate by parts over $E \times (0,t)$, where $t \in \mathbb{R}$ is arbitrary but fixed. This gives

$$\begin{aligned} \int_E u_n(x,t) f(x,t) \, dx &+ \int_0^t \int_E u_n(x,t)(x,\tau) \Box f \, dx \, d\tau \\ &= \int_E \psi_n f(x,0) \, dx - \int_E \varphi_n f_t(x,0) \, dx. \end{aligned}$$

Letting $n \to \infty$ gives the weak form of (11.1)

$$\begin{aligned} \int_E u(x,t) f(x,t) \, dx &+ \int_0^t \int_E u(x,t)(x,\tau) \Box f \, dx \, dt \\ &= \int_E \psi f(x,0) \, dx - \int_E \varphi f_t(x,0) \, dx \end{aligned}$$

for all $f \in C^2(\bar{E} \times \mathbb{R})$ vanishing on ∂E.

12 Hyperbolic Equations in Two Variables

The most general linear hyperbolic equation in two variables $x = (x_1, x_2)$ takes the form

$$\mathcal{L}(u) = \frac{\partial^2 u}{\partial x_1 \partial x_2} + \mathbf{b} \cdot \nabla u + cu = f \tag{12.1}$$

where $\mathbf{b} = (b_1, b_2)$ and c, f are given continuous functions in $\mathbb{R}^2$. For this, the characteristics are the lines $x_i = (\text{const})_i$ for $i = 1, 2$. If $\mathbf{b} = c = f = 0$, then, up to a change of variables, (12.1) can be rewritten in the form of the wave equation

$$v_{tt} - v_{xx} = 0 \quad \text{in } \mathbb{R}^2 \tag{12.2}$$

where

$$\begin{aligned} x_1 &= x - t \\ x_2 &= x + t \end{aligned} \qquad \text{and} \qquad v(x,t) = u(x-t, x+t).$$

Therefore if v is prescribed on the characteristics $x \pm t = \text{const}$, the method of the characteristic parallelograms of Section 1.1 permits one to solve (12.2) in the whole of $\mathbb{R}^2$.

13 The Characteristic Goursat Problem

The characteristic Goursat problem consists in finding $u \in C^2(\mathbb{R}^2)$ satisfying[1]

$$\mathcal{L}(u) = f \ \text{ in } \mathbb{R}^2, \qquad u\Big|_{x_i=0} = \varphi_i \in C^2(\mathbb{R}), \quad i = 1, 2. \tag{13.1}$$

Theorem 13.1 *There exists a unique solution to the characteristic Goursat problem (13.1).*

13.1 Proof of Theorem 13.1: Existence

Setting $\nabla u = (w_1, w_2) = \mathbf{w}$, by virtue of (12.1)

$$\frac{\partial}{\partial x_2} w_1 = \frac{\partial}{\partial x_1} w_2 = f - \mathbf{b} \cdot \mathbf{w} - cu.$$

Integrate the first of these equations over $(0, x_2)$ and the second over $(0, x_1)$. Taking into account the data φ_i on the characteristics $x_i = 0$, $i = 1, 2$, recast (13.1) into the equivalent form

$$\begin{aligned} w_1(x) &= \varphi_2'(x_1) + \int_0^{x_2} (f - \mathbf{b} \cdot \mathbf{w} - cu)(x_1, s)ds \\ w_2(x) &= \varphi_1'(x_2) + \int_0^{x_1} (f - \mathbf{b} \cdot \mathbf{w} - cu)(s, x_2)ds \\ u(x) &= \varphi_2(x_1) + \int_0^{x_2} w_2(x_1, s)ds. \end{aligned} \tag{13.2}$$

[1] The problem is also referred to as the Darboux–Goursat problem. For $\mathcal{L}(\cdot)$ linear, the problem was posed and solved by Darboux, [25](Tome II, pages 91-94). The non-linear case of $u_{x_1 x_2} = F(x_1, x_2, u, x_{x_1}, u_{x_2})$ was solved by E. Goursat, [60, Vol. 3 part I]. See also J. Hadamard, [61](pages 107–108).

The last equation could be equivalently replaced by

$$u(x) = \varphi_1(x_2) + \int_0^{x_1} w_1(s, x_2)ds.$$

To solve (13.2), define

$$u_o = \varphi_2, \quad w_{1,o} = \varphi_2', \quad w_{2,o} = \varphi_1'$$

and recursively, for $n = 0, 1, \dots$

$$\begin{aligned} w_{1,n+1}(x) &= \varphi_2'(x_1) + \int_0^{x_2} [f - \mathbf{b} \cdot (w_{1,n}, w_{2,n}) - cu_n](x_1, s)ds \\ w_{2,n+1}(x) &= \varphi_1'(x_2) + \int_0^{x_1} [f - \mathbf{b} \cdot (w_{1,n}, w_{2,n}) - cu_n](s, x_2)ds \\ u_{n+1} &= \varphi_2(x_1) + \int_0^{x_2} w_{2,n}(x_1, s)ds. \end{aligned} \tag{13.2$_n$}$$

A solution of (13.2) can be found by letting $n \to \infty$ in $(13.2)_n$, provided the sequences $\{u_n\}$ and $\{w_{i,n}\}$ for $i = 1, 2$ are uniformly convergent over compact subsets of $\mathbb{R}^2$. For this it suffices to prove that the telescopic series

$$u_o + \sum(u_n - u_{n-1}) \quad \text{and} \quad w_{i,o} + \sum(w_{i,n} - w_{i,n-1}) \tag{13.3}$$

are absolutely and uniformly convergent on compact subsets $K \subset \mathbb{R}^2$. Having fixed one such K, one may assume that it is a square about the origin with sides parallel to the coordinate axes, and such that $\text{meas}(K) \le 1$. Set

$$\begin{gathered} V_n = (u_n, w_{1,n}, w_{2,n}), \quad |x| = |x_1| + |x_2| \\ \|V_n - V_{n-1}\| = |u_n - u_{n-1}| + |w_{1,n} - w_{1,n-1}| + |w_{2,n} - w_{2,n-1}| \\ C_K = 1 + \|\mathbf{b}\|_{\infty,K} + \|c\|_{\infty,K} + \|f\|_{\infty,K}, \quad A_K = 1 + \|V_o\|_{\infty,K}. \end{gathered}$$

Lemma 13.1 *For all $x \in K$ and all $n \in \mathbb{N}$*

$$\|V_n - V_{n-1}\|(x) \le A_K(2C_K)^n \frac{|x|^n}{n!}. \tag{13.4}$$

Proof From $(13.2)_{n=0}$

$$\begin{aligned} w_{1,1} - w_{1,o} &= \int_0^{x_2} [f - \mathbf{b} \cdot (w_{1,o}, w_{2,o}) - cu_o](x_1, s)ds \\ w_{2,1} - w_{2,o} &= \int_0^{x_1} [f - \mathbf{b} \cdot (w_{1,o}, w_{2,o}) - cu_o](s, x_2)ds \\ u_1 - u_o &= \int_0^{x_2} w_{2,o}(x_1, s)ds. \end{aligned}$$

From this

$$\begin{aligned}\|V_1 - V_o\|(x) \le \|f\|_{\infty,K} + (\|\mathbf{b}\|_{\infty,K} + \|c\|_{\infty,K})\|V_o\|_{\infty,K} \int_0^{x_2} ds \\ + \|f\|_{\infty,K} + (\|\mathbf{b}\|_{\infty,K} + \|c\|_{\infty,K})\|V_o\|_{\infty,K} \int_0^{x_1} ds \\ + \|V_o\|_{\infty,K} \int_0^{x_2} ds \\ \le A_K C_K |x|.\end{aligned}$$

Therefore (13.4) holds for $n = 1$. We show by induction that if it does hold for n it continues to hold for $n+1$. From (13.2), for all $x \in K$

$$\begin{aligned}&\|V_{n+1} - V_n\|(x) \\ &\quad \le C_K \left(\int_0^{x_2} \|V_n - V_{n-1}\|(x_1, s) ds + \int_0^{x_1} \|V_n - V_{n-1}\|(s, x_2) ds \right) \\ &\quad \le A_K \frac{2^n C_K^{n+1}}{(n-1)!} \left(\int_0^{x_2} |(x_1, s)|^n ds + \int_0^{x_1} |(s, x_2)|^n ds \right) \\ &\quad \le A_K (2C_K)^{n+1} \frac{|x|^{n+1}}{(n+1)!}.\end{aligned}$$

■

Returning to the absolute convergence of the series in (13.3), it follows from the lemma that for all $x \in K$

$$\|V_o\|(x) + \sum \|V_n - V_{n-1}\|(x) \le A_K \left(1 + \sum (2C_K)^n \frac{|x|^n}{n!} \right) = A_K e^{2C_K|x|}.$$ ■

13.2 Proof of Theorem 13.1: Uniqueness

Let us assume that there exist two locally bounded solutions of the system (13.2), say $(u^{(i)}, w_1^{(i)}, w_2^{(i)}) = V^{(i)}$ for $i = 1, 2$, and set

$$\|V^{(1)} - V^{(2)}\| = |u^{(1)} - u^{(2)}| + |w_1^{(1)} - w_1^{(2)}| + |w_2^{(1)} - w_2^{(2)}|.$$

Write the system (13.2) for $V^{(1)}$ and $V^{(2)}$, and subtract the resulting equations, to obtain for all $x \in K$

$$\|V^{(1)} - V^{(2)}\|(x) \le \|V^{(1)} - V^{(2)}\|_{\infty,K} B_K |x|$$

where $B_K = \|\mathbf{b}\|_{\infty,K} + \|c\|_{\infty,K}$. Since K is an arbitrary compact subset of $\mathbb{R}^2$, this implies $V^{(1)} = V^{(2)}$ identically. ■

13.3 Goursat Problems in Rectangles

Let $\alpha_1 < \beta_1$ and $\alpha_2 < \beta_2$, and let R be the rectangle $[\alpha_1, \beta_1] \times [\alpha_2, \beta_2]$. Prescribe data $\varphi_1 \in C^2[\alpha_1, \beta_1]$ and $\varphi_2 \in C^2[\alpha_2, \beta_2]$ on the segments $[\alpha_1, \beta_1]$ and $[\alpha_2, \beta_2]$, and consider the problem of finding $u \in C^2(R)$ satisfying

$$\begin{aligned} \mathcal{L}(u) &= f && \text{in } R \\ u(x_1, \alpha_2) &= \varphi_2(x_1) && \text{for } x_1 \in [\alpha_1, \beta_1] \\ u(\alpha_1, x_2) &= \varphi_1(x_2) && \text{for } x_2 \in [\alpha_2, \beta_2]. \end{aligned} \tag{13.5}$$

The same proof applies, and one may conclude that (13.5) has a unique solution. Analogously, there exists a unique solution to the characteristic problem

$$\begin{aligned} \mathcal{L}(u) &= f && \text{in } R \\ u(x_1, \beta_2) &= \varphi_2(x_1) && \text{for } x_1 \in [\alpha_1, \beta_1] \\ u(\beta_1, x_2) &= \varphi_1(x_2) && \text{for } x_2 \in [\alpha_2, \beta_2]. \end{aligned} \tag{13.6}$$

14 The Non-Characteristic Cauchy Problem and the Riemann Function

Let Γ be a regular curve in $\mathbb{R}^2$ whose tangent is nowhere parallel to either of the coordinate axes. For example

$$\Gamma = \begin{cases} x_1 = s & s \in \mathbb{R} \\ x_2 = h(s) \in C^1(\mathbb{R}) & \\ h'(s) < 0 & \text{for all } s \in \mathbb{R}. \end{cases}$$

Consider the problem of finding $u \in C^2(\mathbb{R}^2)$ satisfying

$$\mathcal{L}(u) = f \text{ in } \mathbb{R}^2, \quad u\big|_\Gamma = u_{x_2}\big|_\Gamma = 0 \tag{14.1}$$

where $\mathcal{L}(\cdot)$ is defined in (12.1). As an example, take the case $\mathbf{b} = c = 0$, and Γ is the line $x_2 = -x_1$. Then (14.1) reduces to the Cauchy problem for the wave equation

$$\begin{aligned} v_{tt} - v_{xx} &= \tilde{f} && \text{in } \mathbb{R}^2 \\ v(\cdot, 0) &= 0 && t = x_1 + x_2 \\ v_t(\cdot, 0) &= 0 && x = -x_1 + x_2, \end{aligned}$$

where

$$v(x,t) = u\Big(\frac{t-x}{2}, \frac{t+x}{2}\Big), \qquad \tilde{f}(x,t) = f\Big(\frac{t-x}{2}, \frac{t+x}{2}\Big).$$

This problem has a unique solution is given by the representation formula (3.4). We will prove that (14.1) has a unique solution and will exhibit a representation formula for it.

Through a point $x \in \mathbb{R}^2 - \Gamma$, draw two lines parallel to the coordinate axes and let E_x be the region enclosed by these lines and Γ, as in Figure 14.3

$$E_x = \{(\sigma, s) \mid h(\sigma) < s < x_2;\ \alpha < \sigma < x_1\}.$$

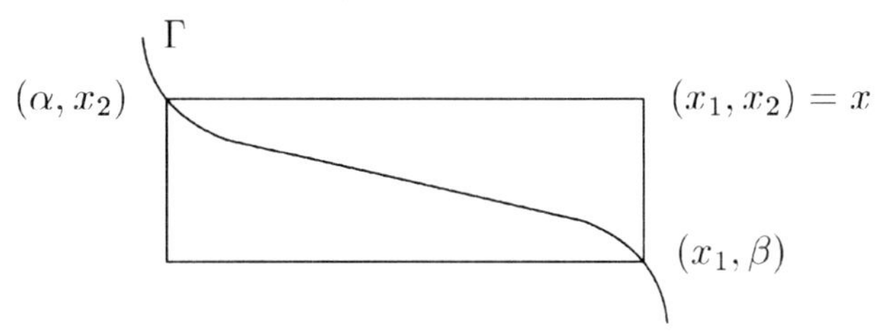

Fig. 14.3.

Let $\mathcal{L}^*(\cdot)$ denote the adjoint operator to $\mathcal{L}(\cdot)$

$$\mathcal{L}^*(v) = \frac{\partial^2 v}{\partial x_1 \partial x_2} - \operatorname{div}(\mathbf{b}v) + cv.$$

This is well defined if $b_i \in C^1(\mathbb{R}^2)$, which we assume henceforth. Let u, v be a pair of functions in $C^2(\mathbb{R}^2)$, and compute the quantity

$$\iint_{E_x} [v\mathcal{L}(u) - u\mathcal{L}^*(v)]dy.$$

The outward unit normal to E_x on Γ is $\mathbf{n} = (-h', 1)/\sqrt{1+h'^2}$. Therefore by Green's theorem

$$\begin{aligned}
&\iint_{E_x} [v\mathcal{L}(u) - u\mathcal{L}^*(v)]dy = (uv)(x) - (uv)(\alpha, x_2) \\
&- \int_\beta^{x_2} u[v_{x_2} - vb_1](x_1, s)ds - \int_\alpha^{x_1} u[v_{x_1} - vb_2](s, x_2)ds \qquad (14.2)\\
&- \int_\alpha^{x_1} vh'(s)[u_{x_2} + ub_1](s, h(s))ds - \int_\alpha^{x_1} u[vb_2 - v_{x_1}](s, h(s))ds.
\end{aligned}$$

If u is a solution to (14.1), then (14.2) reduces to

$$\begin{aligned}
\iint_{E_x} [vf - u\mathcal{L}^*(v)]dy = (uv)(x) &- \int_\beta^{x_2} u[v_{x_2} - b_1 v](x_1, s)ds \\
&- \int_\alpha^{x_1} u[v_{x_1} - b_2 v](s, x_2)ds.
\end{aligned} \qquad (14.3)$$

Next, in (14.3), we make a particular choice of the function v. For each fixed $x \in \mathbb{R}^2$, let $y \to \mathcal{R}(y; x) \in C^2(\mathbb{R}^2)$ satisfy

$$\begin{aligned}
\mathcal{L}_y^*[\mathcal{R}(y; x)] &= 0 \quad \text{in } \mathbb{R}^2 \\
\mathcal{R}(x_1, y_2; x) &= \exp\left[\int_{x_2}^{y_2} b_1(x_1, s)ds\right] \\
\mathcal{R}(y_1, x_2; x) &= \exp\left[\int_{x_1}^{y_1} b_2(s, x_2)ds\right].
\end{aligned} \qquad (14.4)$$

Such a function exists, and it can be constructed by the method of successive approximations of the previous section. The last two of (14.4) imply that $\mathcal{R}(x;x) = 1$. Therefore, writing (14.3) for $y \to v(y) = \mathcal{R}(y;x)$ yields the representation formula

$$u(x) = \iint_{E_x} \mathcal{R}(y;x) f(y) dy. \tag{14.5}$$

This formula, derived under the assumption that a solution of (14.1) exists, indeed does give the unique solution of such a non-characteristic problem, as can be verified by direct calculation. The function $y \to \mathcal{R}(y;x)$ is called the Riemann function ([129]), with pole at x, for the operator $\mathcal{L}(\cdot)$ in $\mathbb{R}^2$.[2]

Remark 14.1 The integral formula (14.3) and the Riemann function $\mathcal{R}(\cdot;\cdot)$ permit us to give a representation formula for the unique solution of the non-characteristic problem (12.1) with inhomogeneous data on Γ

$$\mathcal{L}(u) = f \text{ in } \mathbb{R}^2, \quad u\big|_\Gamma = \varphi, \; u_{x_2}\big|_\Gamma = \psi \tag{14.1$'$}$$

for given smooth functions in $\mathbb{R}$.

15 Symmetry of the Riemann Function

The Riemann function $y \to \mathcal{R}^*(y;x)$, with pole at x, for $\mathcal{L}^*(\cdot)$ satisfies

$$\begin{aligned} &\mathcal{L}[\mathcal{R}^*(y;x)] = 0 \text{ in } \mathbb{R}^2 \\ &\mathcal{R}^*(y_1, x_2; x) = \exp\left[-\int_{x_1}^{y_1} b_2(s, x_2) ds\right] \\ &\mathcal{R}^*(x_1, y_2; x) = \exp\left[-\int_{x_2}^{y_2} b_1(x_1, s) ds\right]. \end{aligned} \tag{15.1}$$

It follows from this that $\mathcal{R}^*(x;x) = 1$.

Lemma 15.1 $\mathcal{R}(y;x) = \mathcal{R}^*(x;y)$.

Proof Let $x = (x_1, x_2)$ and $y = (y_1, y_2)$ be fixed in $\mathbb{R}^2$ and be such that the line through them is not parallel to either coordinate axis. Without loss of generality may assume that $y_1 < x_1$ and $y_2 < x_2$, and construct the rectangle

$$Q_{x,y} = [y_1 < s < x_1] \times [y_2 < \tau < x_2].$$

[2] For an N-dimensional version of the Riemann function, see Hadamard [61].

By Green's theorem, for every pair of functions $u, v \in C^2(\mathbb{R}^2)$

$$\begin{aligned}\iint_{Q_{x,y}} [v\mathcal{L}(u) - u\mathcal{L}^*(v)]dsd\tau &= (uv)(x) - (uv)(y) \\ &- \int_{y_2}^{x_2} v[u_{x_2} + b_1 u](y_1, \tau)d\tau - \int_{y_1}^{x_1} v[u_{x_1} + b_2 u](s, y_2)ds \\ &- \int_{y_2}^{x_2} u[v_{x_2} - b_1 v](x_1, \tau)d\tau - \int_{y_1}^{x_1} u[v_{x_1} - b_2 v](s, x_2)d\tau.\end{aligned} \tag{15.2}$$

Write this identity for $v = \mathcal{R}(\cdot; x)$ and $u = \mathcal{R}^*(\cdot; y)$. ∎

Remark 15.1 (The Characteristic Goursat Problem) The integral formula (15.2) and the Riemann function permit one to give a representation formula in terms of $\mathcal{R}(\cdot;\cdot)$ of the characteristic Goursat problems (13.5) and (13.6).

Problems and Complements

2c The d'Alembert Formula

2.1. Solve the Cauchy problems

$$\begin{aligned} u_{tt} - u_{xx} = f \ \text{ in } \mathbb{R}\times\mathbb{R} \\ u(\cdot, 0) = u_t(\cdot, 0) = 0 \end{aligned} \qquad \text{for} \qquad \begin{aligned} f(x,t) &= e^{x-t} \\ f(x,t) &= x^2. \end{aligned}$$

3c Inhomogeneous Problems

3.1c The Duhamel Principle ([38])

A linear differential operator with constant coefficients and of order $n \in \mathbb{N}$ in the space variables $x = (x_1, \dots, x_N)$ is defined by

$$\mathcal{L}(w) = \sum_{|\alpha|\le n} A_\alpha D^\alpha w, \quad A_\alpha \in \mathbb{R}, \quad w \in C^n(\mathbb{R}^N).$$

Let $f \in C(\mathbb{R}^{N+1})$, and for a positive integer $m \ge 2$ let

$$(x,t;\tau) \to v(x,t;\tau), \quad x \in \mathbb{R}^N,\ t \in (\tau, \infty),\ \tau \in \mathbb{R}$$

be a family of solutions of the homogeneous Cauchy problems

$$\begin{aligned}
&\frac{\partial^m}{\partial t^m} v = \mathcal{L}(u) && \text{in } \mathbb{R}^N \times (\tau, \infty),\ m \ge 2 \\
&\frac{\partial^j}{\partial t^j} v(\cdot, \tau; \tau) = 0 && \text{for } j = 0, 1, \dots, m-2 \\
&\frac{\partial^{m-1}}{\partial t^{m-1}} v(\cdot, \tau; \tau) = f(\cdot, \tau)
\end{aligned} \tag{3.1c}$$

parametrized with $\tau \in \mathbb{R}$. Then, the inhomogeneous Cauchy problem

$$\begin{aligned}
&\frac{\partial^m}{\partial t^m} u = \mathcal{L}(u) + f(x,t) && \text{in } \mathbb{R}^N \times \mathbb{R} \\
&\frac{\partial^j}{\partial t^j} u(\cdot, 0) = 0 && \text{for } j = 0, 1, \dots, m-1
\end{aligned} \tag{3.2c}$$

has a solution given by

$$u(x,t) = \int_0^t v(x,t;\tau) d\tau \qquad (x,t) \in \mathbb{R}^N \times \mathbb{R}. \tag{3.3c}$$

Formulate a general Duhamel's principle if $m = 1$.

4c Solutions for the Vibrating String

4.1. Solve the boundary value problems

$$\begin{aligned}
u_{tt} - u_{xx} = f \ \text{ in } \ (0,L) \times \mathbb{R} \qquad & \qquad f(x,t) = e^x \\
u(0,\cdot) = u(L,\cdot) = 0 \qquad & \text{for} \quad f(x,t) = \sin \pi x \\
u(\cdot, 0) = u_t(\cdot, 0) = 0 \qquad & \qquad f(x,t) = x^2.
\end{aligned}$$

4.2. Solve the boundary value problem

$$\begin{aligned}
u_{tt} - u_{xx} = x \ \text{ in } \ (0,1) \times \mathbb{R} \\
u(\cdot, 0) = x^2(1-x),\ u_t(\cdot, 0) = 0 \\
u_x(0, \cdot) = 0,\ u(1, \cdot) = 0.
\end{aligned}$$

4.3. Let $\beta \in \mathbb{R}$ be a given constant. Solve

$$\begin{aligned}
u_{tt} - u_{xx} = \beta(2u_t - \beta u) \ \text{ in } \ (0,1) \times \mathbb{R} \\
u(\cdot, 0) = \varphi \in C^2(0,1),\ u_t(\cdot, 0) = 0 \\
u(0, \cdot) = u(1, \cdot) = 0.
\end{aligned}$$

4.4. Solve the previous problem for $\varphi \in C(0,1)$ but not necessarily of class $C^2(0,1)$. Take, for example

$$\varphi(x) = \begin{cases} 2hx & \text{for } x \in (0, \frac{1}{2}) \\ 2h(1-x) & \text{for } x \in (\frac{1}{2}, 1) \end{cases}$$

where h is a given positive constant.

4.5. Let $a \in \mathbb{R}$, and consider the boundary value problem

$$\begin{aligned} u_{tt} + au_t - u_{xx} &= 0 & &\text{in } (0,1) \times (t > 0) \\ u(0,t) = u(1,t) &= 0 & &\text{for } t > 0 \\ u(\cdot, 0) = \varphi,\ u_t(\cdot, 0) &= \psi & &\text{in } (0,1). \end{aligned}$$

Find an expression for the energy $\mathcal{E}(t)$, introduced in (4.5) in terms of u_t. *Hint:* Setting

$$f(t) = \int_0^t \int_0^1 u_t^2(x,s)\,dx\,ds$$

derive a differential inequality $f' \le A - Bf$, for suitable constants A, B.

4.6. In the previous problem take

$$a = 1, \quad \varphi(x) = \sin \pi x + 2 \sin 5\pi x, \quad \psi = 0.$$

Write down the explicit solution. Find constants c_1 and c_2 such that $|u| + |u_t| \le c_1 e^{c_2 t}$.

4.7. Solve by the separation of variables

$$\begin{aligned} u_{tt} - u_{xx} &= \cos 2t \\ u(0,t) = u(1,t) &= 0 & &\text{in } (0,1) \times \mathbb{R}^+ \\ u(x,0) &= 0 & &\text{for } t > 0 \\ u_t(x,0) &= \sum_{n=1}^{\infty} \sin 2n\pi x. & &\text{in } (0,1) \end{aligned}$$

4.8. Let u be the solution of (4.1) defined in (4.3)–(4.4). Discuss questions of convergence of the formal approximating solutions

$$u_n = \sum_{j=1}^{n} (A_j \sin j\pi t + B_j \cos j\pi t) \sin j\pi x.$$

Take $L = c = 1$ and verify that for all $p, q, j \in \mathbb{N}$

$$\left\| \frac{\partial^p}{\partial x^p} \frac{\partial^q}{\partial t^q} u_n \right\|_{\infty, (0,1) \times \mathbb{R}} \le \sum_{j=1}^{n} (j\pi)^{p+q} (|A_j| + |B_j|).$$

4.9. Let m be a positive even integer, and let $C^m_{\text{odd}}(0,1)$ be defined as in **10.3** of the Problems and Complements of Chapter 4. Assume that φ and ψ are in $C^m_{\text{odd}}(0,1)$, and prove that $u_n \to u$ in $C^m[(0,1) \times \mathbb{R}]$.

6c Cauchy Problems in $\mathbb{R}^3$

6.1. Solve the Cauchy problem

$$\Box u = 0 \text{ in } \mathbb{R}^3 \times \mathbb{R}, \quad u(x,0) = |x|^2, \quad u_t(x,0) = x_3 \text{ in } \mathbb{R}^3.$$

6.2. Find a space-independent solution of $\Box u = e^{-t}$ in $\mathbb{R}^3 \times \mathbb{R}$, and use it to find the solution of

$$\Box u = e^{-t} \text{ in } \mathbb{R}^3 \times \mathbb{R}, \quad u(\cdot,0) = x_1, \quad u_t(\cdot,0) = x_2 x_3.$$

6.1c Asymptotic Behavior

6.3. Let u be the solution of

$$\Box u = 0 \text{ in } \mathbb{R}^3 \times \mathbb{R}, \quad u(\cdot,0) = 0, \quad u_t(\cdot,0) = |x|^k$$

for some $k > 0$. Compute the limit of $u(0,t)$ as $t \to \infty$. Prove that as $|x| \to \infty$ the solution has the form

$$u(x,t) = a(x,t)(1 + |x|^k).$$

Find $a(x,t)$ and prove that

$$\lim_{|x|\to\infty} |u(x,t) - a(x,t)(1 + |x|)| = 0$$

uniformly on compact intervals of the t-axis.

6.4. Let u_ε be the unique solution of $\Box u_\varepsilon = 0$ in $\mathbb{R}^3 \times \mathbb{R}$, with initial data

$$u_\varepsilon(\cdot,0) = 0, \qquad \frac{\partial}{\partial t} u_\varepsilon(x,0) = \begin{cases} e^{\frac{-\varepsilon^2}{\varepsilon^2 - |x|^2}} & \text{for } |x| < \varepsilon \\ 0 & \text{for } |x| \geq \varepsilon. \end{cases}$$

Study the limit of u_ε, as $\varepsilon \to 0$, in some appropriate topology.

6.2c Radial Solutions

6.5. Let B be the unit ball about the origin in $\mathbb{R}^3$ and consider the problem (internal vibrations of a contracted sphere)

$$\begin{aligned} u_{tt} - \Delta u &= 0 && \text{in } B \times \mathbb{R} \\ u(\cdot,t)\big|_{\partial B} &= 0 && \text{for } t \in \mathbb{R} \\ u(x,0) &= 0 && \text{in } B \\ u_t(x,0) &= \cos\frac{\pi}{2}|x| && \text{in } B. \end{aligned} \tag{6.1c}$$

Find a radial solution of (6.1c) by the following steps:
(i) Set $|x| = \rho$ and recast the problem as

$$\begin{aligned} &u_{tt} - u_{\rho\rho} - \frac{2}{\rho} u_\rho = 0 && \text{in } (0,1) \times \mathbb{R} \\ &u(1,t) = u_\rho(0,t) = 0 && \text{for } t \in \mathbb{R} \\ &u(\rho,0) = 0,\ u_t(\rho,0) = \cos\frac{\pi}{2}\rho && \text{for } \rho \in (0,1). \end{aligned} \tag{6.2c}$$

(ii) Let v be the symmetric extension of $u(\cdot,t)$ about the origin

$$v(\rho,t) = \begin{cases} u(\rho,t) & \text{for } \quad 0 < \rho < 1 \\ u(-\rho,t) & \text{for } \ -1 < \rho < 0. \end{cases}$$

Verify that $v_\rho(0,t) = 0$, and that v solves

$$\begin{aligned} &v_{tt} - v_{\rho\rho} - \frac{2}{\rho} v_\rho = 0 && \text{in } (-1,1) \times \mathbb{R} \\ &v(-1,t) = v(1,t) = 0 && \text{for } t \in \mathbb{R} \\ &v(\rho,0) = 0,\ v_t(\rho,0) = \cos\frac{\pi}{2}\rho && \text{for } \rho \in (-1,1). \end{aligned} \tag{6.3c}$$

(iii) Solve (6.3c) and verify that

$$v(0,t) = t\cos\frac{\pi}{2}t \quad \text{for } |t| < 1.$$

6.6. Now consider (6.2c) in the whole of $\mathbb{R}^3$, i.e.,

$$\begin{aligned} u_{tt} - \Delta u &= 0 && \text{in } \mathbb{R}^3 \times \mathbb{R} \\ u(x,0) &= 0 && \text{in } \mathbb{R}^3 \\ u_t(x,0) &= \cos\frac{\pi}{2}|x| && \text{in } \mathbb{R}^3. \end{aligned}$$

Write down the explicit solution and check that $u = v$.

6.7. Prove that all radial solutions of the wave equation in $\mathbb{R}^3 \times \mathbb{R}$ are of the form

$$u(x,t) = \frac{F(|x| - ct) + G(|x| + ct)}{|x|}$$

for functions $F(\cdot)$ and $G(\cdot)$ of class $C^2(\mathbb{R})$.

6.8. Write down the explicit solution of

$$\Box u = 0 \ \text{ in } \mathbb{R}^3 \times \mathbb{R}, \quad u(\cdot.0) = 0,\ u_t(\cdot,0) = \psi \tag{6.4c}$$

where ψ is radial.

6.9. In the case $c = 1$ and

$$\psi(|x|) = \begin{cases} 1 & \text{for } |x| < 1 \\ 0 & \text{for } |x| \ge 1 \end{cases}$$

prove that the unique solution of (6.4c) is

$$u(x,t) = \begin{cases} t & \text{if } 0 < |x| < 1-t, \qquad t \in (0,1] \\ \dfrac{1-(|x|-t)^2}{4|x|} & \text{if } |1-t| < |x| < t+1, \quad t \geq 0 \\ 0 & \text{if } 0 \leq |x| \leq t+1. \end{cases}$$

In particular, the solution is discontinuous at $(0,1)$.

6.3c Solving the Cauchy Problem by Fourier Transform

We will use here notions and techniques introduced in Sections 2.6c–2.9c of the Problems and Complements of Chapter 5. Consider the Cauchy problem

$$v_{tt} - \Delta v = 0 \text{ in } \mathbb{R}^N \times \mathbb{R}^+, \quad v(\cdot, 0) = 0, \quad v_t(\cdot, 0) = \psi. \tag{6.5c}$$

We assume that ψ is in the class of the rapidly decreasing functions or the Schwartz class $\mathcal{S}_N$, and seek a solution $v(\cdot, t)$ in the same class with respect to the space variables. Taking the Fourier transform of the PDE in (6.5c) with respect to the space variables gives

$$\hat{v}_{tt} + |y|^2 \hat{v} = 0, \quad \hat{v}(y, 0) = 0, \quad \hat{v}_t(y, 0) = \hat{\psi}.$$

This can be solved explicitly to give

$$\hat{v}(y,t) = \frac{\sin|y|t}{|y|} \hat{\psi}(y).$$

Prove that

$$\sup_{y \in \mathbb{R}^N} \left| D^\alpha \frac{\sin|y|t}{|y|} \right| < \infty, \qquad \text{for every multi-index } \alpha.$$

Deduce that $\hat{v}(\cdot, t) \in \mathcal{S}_N$. By the inversion formula obtain the solution of (6.5c) in the form

$$v(x,t) = \frac{1}{(2\pi)^N} \int_{\mathbb{R}^N} \frac{\sin|y|t}{|y|} \hat{\psi}(y) e^{i\langle x, y \rangle} dy. \tag{6.6c}$$

Now consider the general Cauchy problem

$$u_{tt} - \Delta u = 0 \text{ in } \mathbb{R}^N \times \mathbb{R}^+, \quad u(\cdot, 0) = \varphi, \quad u_t(\cdot, 0) = \psi \tag{6.7c}$$

with both φ and ψ in $\mathcal{S}_N$. Verify that if w solves (6.5c) with the initial condition $w_t(\cdot, 0) = \varphi$, then the solution of (6.7c) is given by

$$u(x,t) = v(x,t) + w_t(x,t).$$

It follows from (6.6c) that the solution of (6.7c) can be represented by the formula

$$u(x,t) = \frac{1}{(2\pi)^N} \int_{\mathbb{R}^N} \left(\frac{\sin|y|t}{|y|} \hat{\psi}(y) + \cos|y|t \, \hat{\varphi}(y) \right) e^{i\langle x, y \rangle} dy.$$

6.3.1c The 1-Dimensional Case

If $N = 1$, by the inversion formula

$$
\begin{aligned}
u(x,t) &= \frac{1}{\sqrt{2\pi}} \int_{\mathbb{R}} \frac{e^{iy(x+t)} + e^{iy(x-t)}}{2} \hat{\varphi}(y)dy \\
&\quad + \frac{1}{\sqrt{2\pi}} \int_{\mathbb{R}} \frac{e^{iy(x+t)} - e^{iy(x-t)}}{2iy} \hat{\psi}(y)dy \\
&= \frac{1}{2}[\varphi(x+t) + \varphi(x-t)] + \frac{1}{2} \int_{x-t}^{x+t} \left(\frac{1}{\sqrt{2\pi}} \int_{\mathbb{R}} e^{i\eta y} \hat{\psi}(y)dy \right) d\eta \\
&= \frac{1}{2}[\varphi(x+t) + \varphi(x-t)] + \frac{1}{2} \int_{x-t}^{x+t} \psi(y)dy.
\end{aligned}
$$

6.3.2c The Case $N = 3$

We refer to the representation formula (6.6c). Let $N = 3$ and prove the formula

$$
\frac{\sin |y|t}{|y|} = \frac{1}{4\pi t} \int_{|\eta|=t} e^{i\langle \eta, y\rangle} d\sigma = \frac{t}{4\pi} \int_{|\eta|=1} e^{it\langle \eta, y\rangle} d\sigma.
$$

Hint: If T is a rotation matrix in $\mathbb{R}^3$, then

$$
\int_{|\eta|=t} e^{i\langle \eta, y\rangle} d\sigma(\eta) = \int_{|\eta|=t} e^{i\langle \eta, Ty\rangle} d\sigma(y).
$$

Next choose T such that $Ty = |y|(0,0,1)$ and compute the integral by introducing polar coordinates. By the Fubini theorem and the inversion formula, one computes from (6.6c)

$$
\begin{aligned}
v(x,t) &= \frac{1}{4\pi t} \int_{\mathbb{R}^3} \left[\frac{1}{(2\pi)^{3/2}} \int_{|\eta|=t} e^{i\langle y, \eta\rangle} d\sigma \right] \hat{\psi}(y) e^{i\langle x, y\rangle} dy \\
&= \frac{1}{4\pi t} \int_{|\eta|=t} \psi(x+\eta) d\sigma = \frac{1}{4\pi t} \int\limits_{|x-y|=t} \psi(y) d\sigma(y).
\end{aligned}
$$

The solution of (6.7c) is given by

$$
u(x,t) = \frac{\partial}{\partial t} \left(\frac{1}{4\pi t} \int_{|x-y|=t} \varphi(y) d\sigma(y) \right) + \frac{1}{4\pi t} \int_{|x-y|=t} \psi(y) d\sigma(y).
$$

7c Cauchy Problems in $\mathbb{R}^2$ and the Method of Descent

7.1. Solve the Cauchy problem

$$
u_{tt} - \Delta u = 0 \text{ in } \mathbb{R}^2 \times \mathbb{R}, \quad u(x,0) = |x|^2, \quad u_t(x,0) = 1 \text{ in } \mathbb{R}^2.
$$

7.2. Write down the explicit solution of

$$u_{tt} - \Delta u = 0 \text{ in } \mathbb{R}^2 \times \mathbb{R}, \quad u(\cdot, 0) = p(\cdot), \quad u_t(\cdot, 0) = 0$$

where p is a homogeneous polynomial of degree 10 in x_1 and x_2. Write down $u(0,0,t)$ in the case $p(x_1, x_2) = (x_1^2 + x_2^2)^5$.

7.3. Solve the problem

$$u_{tt} - \Delta u = 0 \text{ in } \mathbb{R}^2 \times \mathbb{R}, \quad u(x, 0) = 0, \quad u_t(x, 0) = \sin(x_1 + x_2).$$

Find the solution in the form $u(x,t) = a(t) \sin(x_1 + x_2)$.

7.4. Recover the d'Alembert formula (2.2) from the Poisson formula (7.3) and the method of descent.

7.1c The Cauchy Problem for $N = 4, 5$

In the Darboux formula (5.2) take $N = 5$ and let

$$w(x, \rho, t) = \rho^2 \frac{\partial}{\partial \rho} M(u; x, \rho, t) + 3\rho M(u; x, \rho, t).$$

Verify that $w_{tt} = c^2 w_{\rho\rho}$ and solve (5.1) for $N = 5$. Prove that the solution of $(5.1)'$ is given by

$$\begin{aligned} u(x,t) &= \left(\frac{1}{3}t^2 \frac{\partial}{\partial t} + t\right)\left(\frac{1}{\omega_5 (ct)^4} \int_{|x-y|=ct} \psi(y) d\sigma\right) \\ &+ \frac{\partial}{\partial t}\left(\frac{1}{3}t^2 \frac{\partial}{\partial t} + t\right)\left(\frac{1}{\omega_5 (ct)^4} \int_{|x-y|=ct} \varphi(y) d\sigma\right). \end{aligned}$$

Use the previous result and the method of descent to solve (5.1) for $N = 4$.

8c Inhomogeneous Cauchy Problems

8.1c The Wave Equation for the N and $(N + 1)$-Laplacian

Denote by Δ_N the Laplacian with respect to the N variables $x = (x_1, \dots, x_N)$ and by Δ_{N+1} the Laplacian with respect to the $(N + 1)$ variables (x, x_{N+1}). Let $k \in \mathbb{R}$ be a given constant and let $u \in C^2(\mathbb{R}^N \times \mathbb{R})$ be a solution of

$$u_{tt} = c^2 \Delta_N u - k^2 u \quad \text{in } \mathbb{R}^N \times \mathbb{R}.$$

Then for any two given constants A and B, the function

$$v(x, x_{N+1}, t) = \left[A \cos\left(\frac{k}{c} x_{N+1}\right) + B \sin\left(\frac{k}{c} x_{N+1}\right)\right] u(x,t)$$

solves

$$v_{tt} = c^2 \Delta_{N+1} v \quad \text{in } \mathbb{R}^{N+1} \times \mathbb{R}. \tag{8.1c}$$

Similarly, if u solves

$$u_{tt} = c^2 \Delta_N u + k^2 u \quad \text{in } \mathbb{R}^N \times \mathbb{R}$$

then

$$v(x, x_{N+1}, t) = \left[A \cosh\left(\frac{k}{c} x_{N+1}\right) + B \sinh\left(\frac{k}{c} x_{N+1}\right)\right] u(x,t)$$

solves (8.1c). Use these remarks and the method of descent to solve the Cauchy problems

$$\begin{aligned} u_{tt} &= c^2 \Delta_2 u \pm \lambda^2 u && \text{in } \mathbb{R}^2 \times \mathbb{R} \\ u(\cdot, 0) &= \varphi \in C^3(\mathbb{R}^2) && \text{in } \mathbb{R}^2 \\ u_t(\cdot, 0) &= \psi \in C^2(\mathbb{R}^2) && \text{in } \mathbb{R}^2 \end{aligned}$$

where $\lambda \in \mathbb{R}$ is a given constant.

8.1.1c The Telegraph Equation

Solve the Cauchy problems

$$\begin{aligned} u_{tt} &= u_{xx} \pm \lambda^2 u && \text{in } \mathbb{R} \times \mathbb{R} \\ u(\cdot, 0) &= \varphi \in C^3(\mathbb{R}) && \text{in } \mathbb{R} \\ u_t(\cdot, 0) &= \psi \in C^2(\mathbb{R}) && \text{in } \mathbb{R}. \end{aligned} \tag{T$_\pm$}$$

The equation $(\mathrm{T})_-$ is called the telegraph equation. Set

$$B^+(s) = \frac{2}{\pi} \int_0^{\pi/2} \cos(s \sin\theta) d\theta, \quad B^-(s) = \frac{2}{\pi} \int_0^{\pi/2} \cosh(s \sin\theta) d\theta$$

where s is a real parameter. Prove that the solutions $u_\pm$ of $(\mathrm{T})_\pm$ are

$$\begin{aligned} u_\pm(x,t) &= \frac{1}{2} \int_{x-t}^{x+t} \psi(s) B^\pm\left(\lambda \sqrt{t^2 - (x-s)^2}\right) ds \\ &\quad + \frac{\partial}{\partial t}\left[\frac{1}{2} \int_{x-t}^{x+t} \varphi(s) B^\pm\left(\lambda\sqrt{t^2 - (x-s)^2}\right) ds\right]. \end{aligned}$$

The functions $B^\pm(\cdot)$ are the Bessel functions of order zero ([12]).

8.2c Miscellaneous Problems

8.1. Let a, b, c be given constants. Solve

$$\begin{aligned} u_{tt} &= u_{xx} + a u_x + b u_t + c u \quad \text{in } \mathbb{R} \times \mathbb{R} \\ u(\cdot, 0) &= \varphi \in C^3(\mathbb{R}), \quad u_t(\cdot, 0) = \psi \in C^2(\mathbb{R}). \end{aligned}$$

Hint: Reduce the problem to the previous one by exponential shifts in x and t. The solution is

$$2u(x,t) = e^{\frac{bt-ax}{2}} \int_{x-t}^{x+t} e^{\frac{as}{2}} \left(\psi(s) - \frac{b}{2}\varphi(s) \right) A\left(\lambda\sqrt{t^2-(x-s)^2} \right) ds$$
$$+ e^{\frac{bt-ax}{2}} \frac{\partial}{\partial t} \left(\int_{x-t}^{x+t} e^{\frac{as}{2}} \psi(s) A\left(\lambda\sqrt{t^2-(x-s)^2} \right) ds \right)$$

where

$$\text{if } a^2 \geq b^2 + 4c, \quad \lambda = \sqrt{\frac{a^2-b^2}{4} - c}, \quad A(s) = B^-(s)$$

$$\text{if } a^2 \leq b^2 + 4c, \quad \lambda = \sqrt{c - \frac{a^2-b^2}{4}}, \quad A(s) = B^+(s).$$

8.2. Solve

$$u_{tt} = u_{xx} + au_x + bu_t + cu + f \text{ in } \mathbb{R} \times \mathbb{R}$$

$$u(\cdot,0) = \varphi \in C^3(\mathbb{R}), \quad u_t(\cdot,0) = \psi \in C^2(\mathbb{R}), \quad f \in C^2(\mathbb{R}^2).$$

8.3. Let β be a given constant. Solve the boundary value problem

$$u_{tt} - u_{xx} = -\beta u_t \text{ in } (0,1) \times \mathbb{R}^+, \quad u(0,\cdot) = u(1,\cdot) = 0$$

$$u(\cdot,0) = \varphi \in C^2(0,1) \cap C[0,1], \quad u_t(\cdot,0) = 0.$$

8.4. Solve the problem

$$\begin{aligned} u_{tt} - u_{xx} - \frac{2}{x}u_x &= 0 && \text{in } (-1,1) \times \mathbb{R} \\ u(-1,\cdot) = u(1,\cdot) &= 0 && \text{in } \mathbb{R} \\ u(x,0) &= 0 && \text{for } x \in (-1,1) \\ u_t(x,0) &= \cos\frac{\pi}{2}x && \text{for } x \in (-1,1). \end{aligned}$$

Verify that the solution is symmetric about the origin, that $u_x(0,t) = 0$ for all t, and moreover

$$u(0,t) = t\cos\frac{\pi}{2}x \quad \text{for } |t| < 4.$$

Hint: $\Box(xu) = 0$.

8.5. Solve explicitly

$$\begin{aligned} u_{tt} - (u_{xx} + u_{yy}) &= 0 && \text{in } \left(-\frac{\pi}{2}, \frac{\pi}{2}\right) \times \mathbb{R} \times \mathbb{R}^+ \\ u\left(\pm\frac{\pi}{2}, y, t\right) &= 0 && \text{for } t > 0 \text{ and } y \in \mathbb{R} \\ u(x,y,0) &= \cos x \cos y && \text{in } \mathbb{R}^2 \\ u_t(x,y,0) &= 0 && \text{in } \mathbb{R}^2 \end{aligned}$$

8.6. Solve the problem

$$u_{xy} = xy \text{ in } y > x, \quad u(s,s) = 0,\ u(s,-s) = s^3$$

$$\nabla u(s,s)\cdot(-1,1) = \begin{cases} \sqrt{2}s(1-s) & \text{if } s \in (0,1) \\ 0 & \text{otherwise.} \end{cases}$$

10c The Reflection Technique

10.1. Find the solution of the boundary value problem

$$\begin{aligned} u_{tt} - u_{xx} &= 0 && \text{in } \mathbb{R}^+ \times \mathbb{R}^+ \\ u(0,\cdot) &= h && \text{for } t > 0 \\ u(\cdot,0) &= \varphi && \text{for } y > 0 \\ u_t(\cdot,0) &= \psi && \text{for } y > 0 \end{aligned}$$

where the data h, φ, ψ are smooth and satisfy the compatibility conditions

$$h(0) = \varphi(0), \quad h'(0) = \psi(0), \quad h''(0) = c^2\varphi''(0).$$

10.2. Transform the problem

$$u_{tt} - u_{xx} = 0, \quad \text{in } [x > 0] \times [t > 0]$$

$$u(x,0) = \begin{cases} 0 & \text{in } [0 \le x \le 1] \cup [x \ge 2] \\ (2-x)(x-1) & \text{in } [1 \le x \le 2] \end{cases}$$

$$u_t(x,0) = 0$$

into another one in the whole of $\mathbb{R}$. Find the times t such that $u(3,t) \neq 0$. Find the extrema of $x \to u(x,10)$.

11c Problems in Bounded Domains

11.1c Uniqueness

Let $E \subset \mathbb{R}^N$ be bounded, open, and with boundary ∂E of class C^1. Prove that there exists at most one solution to the boundary value problem

$$\begin{aligned} u_{tt} - \Delta u + k(x,t)u &= 0, && \text{in } E \times \mathbb{R} \\ \frac{\partial}{\partial \mathbf{n}} u + q(x,t)u &= 0, && \text{on } \partial E \times \mathbb{R} \\ u(\cdot,0) &= \varphi, && \text{in } E \\ u_t(\cdot,0) &= \psi, && \text{in } E \end{aligned}$$

where φ and ψ are smooth and $k(\cdot,\cdot)$ and $q(\cdot,\cdot)$ are bounded and non-negative in their domain of definition.

11.2c Separation of Variables

11.1. Let $R = [0,1] \times [0,\pi]$, and consider the problem

$$\begin{aligned} u_{tt} - \Delta u &= 0, \quad \text{in } R \times \mathbb{R}^+ \\ u(\cdot,t)\big|_{\partial R} &= 0 \\ u(x,0) &= 0 \\ u_t(x,0) &= f(x_1)g(x_2) \end{aligned}$$

where $f(0) = f(1) = g(0) = g(\pi) = 0$. Solve by the separation of variables. In particular, write down the explicit solution for the data

$$f(x) = \begin{cases} x & \text{if } 0 \le x \le \frac{3}{4} \\ -3(x-1) & \text{if } \frac{3}{4} < x \le 1, \end{cases} \qquad g(y) = \sum_{n=1}^{\infty} \sin ny.$$

11.3. Solve (11.1) for $E = [0,1]^3$ in terms of the eigenvalues and eigenfunctions (λ_i^2, v_i) of the Laplacian in E

$$\Delta v_i = \lambda_i^2 v_i. \tag{11.1c}$$

11.3-(i). Solve (11.1c) by the separation of variables. Denote by x, y, z the coordinates in $\mathbb{R}^3$ and seek a solution of the form $v_i = X_i(x)W_i(y,z)$. Then

$$X_i'' = -\xi_i^2 X_i \quad \text{and} \quad \Delta_{(y,z)} W_i = -\nu_i^2 W_i$$

where ξ_i and ν_i are positive numbers linked by $\xi_i^2 + \nu_i^2 = \lambda_i^2$.

11.3-(ii). Find W_i of the form $W_i(y,z) = Y_i(y)Z_i(z)$. Then

$$Y_i'' = -\eta_i^2 Y_i \quad \text{and} \quad Z_i'' = -\zeta_i^2 Z_i$$

where η_i and ζ_i are positive numbers linked by $\eta_i^2 + \zeta_i^2 = \nu_i^2$.

11.3-(iii). Verify that for all triples (m,n,ℓ) of positive integers, the pairs

$$\lambda_i^2 = \pi^2\left(m^2 + n^2 + \ell^2\right), \quad v_i = \sin \pi m \, \sin \pi n \, \sin \pi \ell$$

are eigenvalues and eigenfunctions of (11.1c). Prove that these are all the eigenvalues and eigenfunctions of (11.1c).

12c Hyperbolic Equations in Two Variables

12.1c The General Telegraph Equation

Let $u(s,t)$ be the intensity of electric current in a conductor, considered as a function of t and the distance s from a fixed point of the conductor. Let α denote the capacity and β the induction coefficients. Then

$$u_{tt} - c^2 u_{ss} + (\alpha + \beta) u_t + \alpha\beta u = 0 \text{ in } \mathbb{R} \times \mathbb{R}.$$

Setting

$$e^{1/2(\alpha+\beta)t} u(s,t) = v(x,y), \qquad x = s + ct,\ y = s - ct$$

transforms the equation into

$$v_{xy} + \lambda v = 0, \qquad \lambda = \left(\frac{\alpha - \beta}{4c}\right)^2.$$

14c Goursat Problems

14.1. Prove that (14.5) is the unique solution of (14.1).

14.2. Prove that (14.1)$'$ has a unique solution and give a representation formula in terms of $\mathcal{R}(\cdot;\cdot)$.

14.3. Give a representation formula for the characteristic Goursat problems (13.5) and (13.6) in terms of $\mathcal{R}(\cdot;\cdot)$.

14.4. Use the method of successive approximations of Section 13, to find the Riemann function for the operator

$$\mathcal{L}(v) = \frac{\partial^2}{\partial x_1 \partial x_2} v + \mathbf{b} \cdot \nabla v + cv$$

where b_1, b_2, and c are constants.

14.1c The Riemann Function and the Fundamental Solution of the Heat Equation

The fundamental solution of the heat equation $u_y = u_{xx}$ can be recovered as the limit, as $\varepsilon \to 0$, of the Riemann function for the hyperbolic equation ([61], 145–147).

$$u_{xx} + \varepsilon u_{xy} - u_y = 0.$$

The change of variables $\xi = y$ and $\eta = x - \frac{1}{\varepsilon} y$ transforms this equation into

$$\varepsilon u_{\xi\eta} + \frac{1}{\varepsilon} u_\eta - u_\xi = 0.$$

Using the previous problem, show that the Riemann function, with pole at the origin, for such an equation is given by

$$\mathcal{R}\big[(\xi,\eta);(0,0)\big] = e^{\frac{\xi}{\varepsilon^2} - \frac{\eta}{\varepsilon}} J_o\left(2\sqrt{\frac{\xi\eta}{\varepsilon^3}}\right)$$

where $J_o(\cdot)$ is the Bessel function of order zero. Returning to the original coordinates

$$\mathcal{R}\big[(x,y);(0,0)\big] = e^{\frac{2y}{\varepsilon^2}-\frac{x}{\varepsilon}} J_o\left(\frac{2}{\varepsilon^2}\sqrt{y(\varepsilon x - y)}\right).$$

Let $\varepsilon \to 0$ to recover the fundamental solution of the heat equation in one space dimension with pole at the origin. For the asymptotic behavior of $J_o(s)$ as $s \to \infty$, see [12].

7

Quasi-Linear Equations of First-Order

1 Quasi-Linear Equations

A first-order quasi-linear PDE is an expression of the form

$$a_i(x, u(x))u_{x_i} = a_o(x, u(x)) \tag{1.1}$$

where x ranges over a region $E \subset \mathbb{R}^N$, the function u is in $C^1(E)$, and $(x, z) \to a_i(x, z)$ are given smooth functions of their arguments. Introduce the vector $\mathbf{a} = (a_1, \dots, a_N)$, and rewrite (1.1) as

$$(\mathbf{a}, a_o) \cdot (\nabla u, -1) = 0. \tag{1.1'}$$

Thus if u is a solution of (1.1), the vector $(\mathbf{a}, a_o)$ is tangent to the graph of u at each of its points. For this reason, the graph of u is called an *integral surface* for (1.1). More generally, an N-dimensional surface Σ of class C^1 is an integral surface for (1.1) if for every point $P = (x, z) \in \Sigma$, the vector $\big(\mathbf{a}(P), a_o(P)\big)$ is tangent to Σ at P. The curves

$$\begin{gathered}(-\delta, \delta) \ni t \to \begin{cases} \dot{x}_i(t) = a_i(x(t), z(t)), & i = 1, \dots, N \\ \dot{z}(t) = a_o(x(t), z(t)) \end{cases} \\ (x(0), z(0)) = (x_o, z_o) \in E \times \mathbb{R}\end{gathered} \tag{1.2}$$

defined for some $\delta > 0$, are the characteristics associated to (1.1), originating at (x_o, z_o). The solution of (1.2) is local in t, and the number δ that defines the interval of existence might depend upon (x_o, z_o). For simplicity we assume that there exists some $\delta > 0$ such that the range of the parameter t is $(-\delta, \delta)$, for all $(x_o, z_o) \in E \times \mathbb{R}$.

Proposition 1.1 *An N-dimensional hypersurface Σ is an integral surface for (1.1) if and only if it is the union of characteristics.*

Proof Up to possibly relabeling the coordinate variables and the components a_i, represent Σ, locally as $z = u(x)$ for some u of class C^1. For $(x_o, z_o) \in \Sigma$, let $t \to (x(t), z(t))$ be the characteristic trough (x_o, z_o), set

E. DiBenedetto, *Partial Differential Equations: Second Edition*,
Cornerstones, DOI 10.1007/978-0-8176-4552-6_8,

$$w(t) = z(t) - u(x(t))$$

and compute

$$\begin{aligned}\dot{w} = \dot{z} - u_{x_i}\dot{x}_i &= a_o(x,z) - a_i(x,z)u_{x_i}(x) \\ &= a_o(x, u(x) + w) - a_i(x, u(x) + w)u_{x_i}(x).\end{aligned}$$

Since $(x_o, z_o) \in \Sigma$, $w(0) = 0$. Therefore w satisfies the initial value problem

$$\begin{aligned}\dot{w} &= a_o(x, u(x) + w) - a_i(x, u(x) + w)u_{x_i}(x) \\ w(0) &= 0.\end{aligned} \tag{1.3}$$

This problem has a unique solution. If Σ, represented as $z = u(x)$, is an integral surface, then $w = 0$ is a solution of (1.3) and therefore is its only solution. Thus $z(t) = u(x(t), t)$, and Σ is the union of characteristics. Conversely, if Σ is the union of characteristics, then $w = 0$, and (1.3) implies that Σ is an integral surface. ■

2 The Cauchy Problem

Let $s = (s_1, \ldots, s_{N-1})$ be an $(N-1)$-dimensional parameter ranging over the cube $Q_\delta = (-\delta, \delta)^{N-1}$. The Cauchy problem associated with (1.1) consists in assigning an $(N-1)$-dimensional hypersurface $\Gamma \subset \mathbb{R}^{N+1}$ of parametric equations

$$Q_\delta \ni s \to \begin{cases} x = \xi(s) = (\xi_1(s), \ldots, \xi_N(s)) \\ z = \zeta(s), \quad (\xi(s), \zeta(s)) \in E \times \mathbb{R} \end{cases} \tag{2.1}$$

and seeking a function $u \in C^1(E)$ such that $\zeta(s) = u(\xi(s))$ for $s \in Q_\delta$ and the graph $z = u(x)$ is an integral surface of (1.1).

2.1 The Case of Two Independent Variables

If $N = 2$, then s is a scalar parameter and Γ is a curve in $\mathbb{R}^3$, say for example

$$(-\delta, \delta) \ni s \to \mathbf{r}(s) = (\xi_1, \xi_2, \zeta)(s).$$

Any such a curve is non-characteristic if the two vectors $(a_1, a_2, \zeta)(\mathbf{r}(s))$ and $(\xi_1', \xi_2', \zeta')(s)$ are not parallel for all $s \in (-\delta, \delta)$. The projection of $s \to \mathbf{r}(s)$ into the plane $[z = 0]$ is the planar curve

$$(-\delta, \delta) \ni s \to \mathbf{r}_o(s) = (\xi_1, \xi_2)(s)$$

of tangent vector $(\xi_1', \xi_2')(s)$. The projections of the characteristics through $\mathbf{r}(s)$ into the plane $[z = 0]$ are called *characteristic projections*, and have tangent vector $(a_1, a_2)(\mathbf{r}(s))$. We impose on $s \to \mathbf{r}(s)$ that its projection into the plane $[z = 0]$ be nowhere parallel to the characteristic projections, that is, the two vectors $(a_1, a_2)(\mathbf{r}(s))$ and $(\xi_1', \xi_2')(s)$ are required to be linearly independent for all $s \in (-\delta, \delta)$.

2.2 The Case of N Independent Variables

Returning to Γ as given in (2.1), one may freeze all the components of s but the ith, and consider the map

$$s_i \to (\xi_1, \dots, \xi_N, \zeta)(s_1, \dots, s_i, \dots, s_{N-1}).$$

This is a curve traced on Γ with the tangent vector

$$\left(\frac{\partial \xi}{\partial s_i}, \frac{\partial \zeta}{\partial s_i}\right) = \left(\frac{\partial \xi_1}{\partial s_i}, \dots, \frac{\partial \xi_N}{\partial s_i}, \frac{\partial \zeta}{\partial s_i}\right).$$

Introduce the $(N-1) \times (N+1)$ matrix

$$\left(\frac{\partial \xi(s)}{\partial s} \; \frac{\partial \zeta(s)}{\partial s}\right) = \left(\frac{\partial \xi_j(s)}{\partial s_i} \; \frac{\partial \zeta(s)}{\partial s_i}\right).$$

The $(N-1)$-dimensional surface Γ is non-characteristic if the vectors

$$(\mathbf{a}, a_o)(\xi(s), \zeta(s)), \qquad \left(\frac{\partial \xi}{\partial s_i}, \frac{\partial \zeta}{\partial s_i}\right)(s)$$

are linearly independent for all $s \in Q_\delta$, equivalently, if the $N \times (N+1)$ matrix

$$\begin{pmatrix} \mathbf{a}(\xi(s), \zeta(s)) & a_o(\xi(s), \zeta(s)) \\ \dfrac{\partial \xi(s)}{\partial s} & \dfrac{\partial \zeta(s)}{\partial s} \end{pmatrix} \tag{2.2}$$

has rank N. We impose that the characteristic projections be nowhere parallel to $s \to \xi(s)$, that is

$$\det \begin{pmatrix} \mathbf{a}(\xi(s), \zeta(s)) \\ \dfrac{\partial \xi(s)}{\partial s} \end{pmatrix} \neq 0 \quad \text{for all } \; s \in Q_\delta. \tag{2.3}$$

Thus we require that the first $N \times N$ minor of the matrix (2.2) be non-trivial.

3 Solving the Cauchy Problem

In view of Proposition 1.1, the integral surface Σ is constructed as the union of the characteristics drawn from points $(\xi, \zeta)(s) \in \Gamma$, that is, Σ is the surface

$$(-\delta, \delta) \times Q_\delta \ni (t, s) \to \big(x(t, s), z(t, s)\big)$$

given by

$$\begin{aligned} \frac{d}{dt} x(t, s) &= \mathbf{a}(x(t, s), z(t, s)), \qquad x(0, s) = \xi(s) \\ \frac{d}{dt} z(t, s) &= a_o(x(t, s), z(t, s)), \qquad z(0, s) = \zeta(s). \end{aligned} \tag{3.1}$$

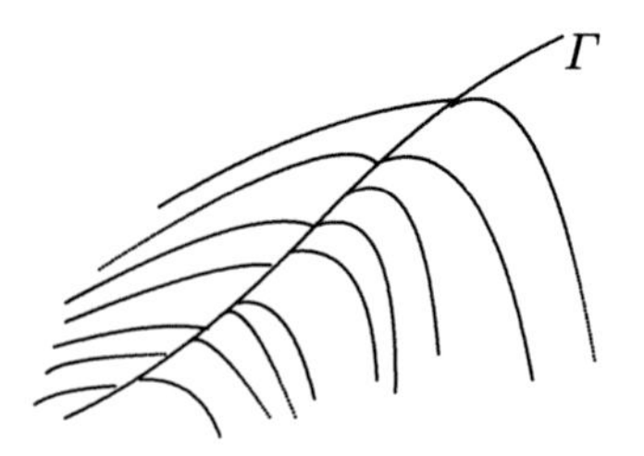

Fig. 3.1.

The solutions of (3.1) are local in t. That is, for each $s \in Q_\delta$, (3.1) is solvable for t ranging in some interval $(-t(s), t(s))$. By taking δ smaller if necessary, we may assume that $t(s) = \delta$ for all $s \in Q_\delta$. If the map

$$\mathcal{M} : (-\delta, \delta) \times Q_\delta \ni (t, s) \to x(t, s)$$

is invertible, then there exist functions $\mathbf{S} : E \to Q_\delta$ and $T : E \to (-\delta, \delta)$ such that $s = \mathbf{S}(x)$ and $t = T(x)$ and the unique solution of the Cauchy problem (1.1), (2.1) is given by

$$u(x) = z(t, s) \stackrel{\text{def}}{=} z(T(x), \mathbf{S}(x)).$$

The invertibility of $\mathcal{M}$ must be realized in particular at Γ, so that the determinant of the Jacobian matrix

$$J = \begin{pmatrix} \dfrac{dx(0, s)}{dt} & \dfrac{\partial \xi(s)}{\partial s} \end{pmatrix}^t$$

must not vanish, that is, Γ cannot contain characteristics. In view of (3.1) for $t = 0$, this is precisely condition (2.3).

The actual computation of the solution involves solving (3.1), calculating the expressions of s and t in terms of x, and substituting them into the expression of $z(t, s)$. The method is best illustrated by some specific examples.

3.1 Constant Coefficients

In (1.1), assume that the coefficients a_i for $i = 0, \dots, N$ are constant. The characteristics are lines of parametric equations

$$x(t) = x_o + \mathbf{a}t, \qquad z(t) = z_o + a_o, t \qquad t \in \mathbb{R}.$$

The first N of these are the characteristic projections. It follows from $(1.1)'$ that the function $f(x, z) = u(x) - z$ is constant along such lines. If Γ is given as in (2.1), the integral surface is

$$\begin{aligned} x(t, s) &= \xi(s) + \mathbf{a}t \\ z(t, s) &= \zeta(s) + a_o t \end{aligned} \qquad s = (s_1, \dots, s_{N-1}). \tag{3.2}$$

The solution $z = u(x)$ is obtained from the last of these upon substitution of s and t calculated from the first N. As an example, let $N = 2$ and let Γ be a curve in the plane $x_2 = 0$, say

$$\Gamma = \{\xi_1(s) = s;\ \xi_2(s) = 0;\ \zeta(s) \in C^1(\mathbb{R})\}.$$

The characteristics are the lines of symmetric equations

$$\frac{x_1 - x_{1,o}}{a_1} = \frac{x_2 - x_{2,o}}{a_2} = \frac{z - z_o}{a_o}, \qquad (x_{1,o}, x_{2,o}, z_o) \in \mathbb{R}^3$$

with the obvious modifications if some of the a_i are zero. The characteristic projections are the lines

$$a_2 x_1 = a_1 x_2 + \text{const}.$$

These are not parallel to the projection of Γ on the plane $z = 0$, provided $a_2 \neq 0$, which we assume. Then (3.2) implies

$$x_2 = a_2 t, \quad \xi_1(s) = s = x_1 - \frac{a_1}{a_2} x_2$$

and the solution is given by

$$u(x_1, x_2) = \zeta\Big(x_1 - \frac{a_1}{a_2} x_2\Big) + \frac{a_o}{a_2} x_2.$$

3.2 Solutions in Implicit Form

Consider the quasi-linear equation

$$\mathbf{a}(u) \cdot \nabla u = 0, \qquad \mathbf{a} = (a_1, a_2, \dots, a_N) \tag{3.3}$$

where $a_i \in C(\mathbb{R})$, and $a_N \neq 0$. The characteristics through points $(x_o, z_o) \in \mathbb{R}^{N+1}$ are the lines

$$x(t) = x_o + \mathbf{a}(z_o) t \quad \text{lying on the hyperplane } z = z_o.$$

A solution u of (3.3) is constant along these lines. Consider the Cauchy problem with data on the hyperplane $x_N = 0$, i.e.,

$$u(x_1, \dots, x_{N-1}, 0) = \zeta(x_1, \dots, x_{N-1}) \in C^1(\mathbb{R}^{N-1}).$$

In such a case the hypersurface Γ is given by

$$\mathbb{R}^{N-1} \ni s \to \begin{cases} x_i = s_i, \quad i = 1, \dots, N-1 \\ x_N = 0 \\ z(s) = \zeta(s). \end{cases} \tag{3.4}$$

Setting $\bar{x} = (x_1, \dots, x_{N-1})$, and $\bar{\mathbf{a}} = (a_1, \dots, a_{N-1})$, the integral surface associated with (3.3) and Γ as in (3.4), is

$$
\begin{aligned}
\bar{x}(t,s) &= s + \bar{\mathbf{a}}(\zeta(s))t \\
x_N(t,s) &= a_N(\zeta(s))t \\
z(t,s) &= \zeta(s).
\end{aligned} \tag{3.5}
$$

From the first two compute

$$
s = \bar{x} - \frac{\bar{\mathbf{a}}(\zeta(s))}{a_N(\zeta(s))} x_N.
$$

Since a solution u of (3.3) must be constant along $z(s,t) = \zeta(s)$, we have $\zeta(s) = u(x)$. Substitute this in the expression of s, and substitute the resulting s into the third of (3.5). This gives the solution of the Cauchy problem (3.3)–(3.4) in the implicit form

$$
u(x) = \zeta\Big(\bar{x} - \frac{\bar{\mathbf{a}}(u(x))}{a_N(u(x))} x_N\Big) \tag{3.6}
$$

as long as this defines a function u of class C^1. By the implicit function theorem this is the case in a neighborhood of $x_N = 0$. In general, however, (3.6) fails to give a solution global in x_N.

4 Equations in Divergence Form and Weak Solutions

Let $(x,u) \to \mathbf{F}(x,u)$ be a measurable vector-valued function in $\mathbb{R}^N \times \mathbb{R}$ and consider formally, equations of the type

$$
\operatorname{div} \mathbf{F}(x,u) = 0 \qquad \text{in } \mathbb{R}^N. \tag{4.1}
$$

The equation (3.3) can be written in this form for $\mathbf{F}(u) = \int^u \mathbf{a}(\sigma)d\sigma$. A measurable function u is a *weak solution* of (4.1) if $\mathbf{F}(\cdot,u) \in [L^1_{\text{loc}}(\mathbb{R}^N)]^N$, and

$$
\int_{\mathbb{R}^N} \mathbf{F}(x,u) \cdot \nabla\varphi \, dx = 0 \quad \text{for all } \varphi \in C_o^\infty(\mathbb{R}^N). \tag{4.2}
$$

This is formally obtained from (4.1) by multiplying by φ and integrating by parts. Every classical solution is a weak solution. Every weak solution such that $\mathbf{F}(\cdot,u)$ is of class C^1 in some open set $E \subset \mathbb{R}^N$ is a classical solution of (4.1) in E. Indeed, writing (4.2) for all $\varphi \in C_o^\infty(E)$ implies that (4.1) holds in the classical sense within E. Weak solutions could be classical in sub-domains of $\mathbb{R}^N$. In general, however, weak solutions fail to be classical in the whole of $\mathbb{R}^N$ as shown by the following example. Denote by (x,y) the coordinates in $\mathbb{R}^2$ and consider the Burgers equation ([14, 15]

$$
\frac{\partial}{\partial y} u + \frac{1}{2} \frac{\partial}{\partial x} u^2 = 0. \tag{4.3}
$$

One verifies that the function

$$u(x,y) = \begin{cases} -\frac{2}{3}\big(y + \sqrt{3x + y^2}\big) & \text{for } 4x + y^2 > 0 \\ 0 & \text{for } 4x + y^2 < 0 \end{cases}$$

solves the PDE in the weak form

$$\int_{\mathbb{R}^2} \{u\varphi_y + \tfrac{1}{2}u^2\varphi_x\}dxdy = 0 \quad \text{for all } \varphi \in C_o^\infty(\mathbb{R}^2).$$

The solution is discontinuous across the parabola $3x + y^2 = 0$.

4.1 Surfaces of Discontinuity

Let $\mathbb{R}^N$ be divided into two parts, E_1 and E_2, by a smooth surface Γ of unit normal $\boldsymbol{\nu}$ oriented, say, toward E_2. Let $u \in C^1(\bar{E}_i)$ for $i = 1,2$ be a weak solution of (4.1), discontinuous across Γ. Assume also that $\mathbf{F}(\cdot, u) \in C^1(\bar{E}_i)$, so that

$$\operatorname{div} \mathbf{F}(x,u) = 0 \quad \text{in } E_i \text{ for } i = 1,2$$

in the classical sense. Let $[\mathbf{F}(\cdot,u)]$ denote the jump of $\mathbf{F}(\cdot,u)$ across Γ, i.e.,

$$[\mathbf{F}(x,u)] = \lim_{E_1 \ni x \to \Gamma} \mathbf{F}(x,u) - \lim_{E_2 \ni x \to \Gamma} \mathbf{F}(x,u).$$

Rewrite (4.2) as

$$\int_{E_1} \mathbf{F}(x,u) \cdot \nabla\varphi dx + \int_{E_2} \mathbf{F}(x,u) \cdot \nabla\varphi dx = 0 \quad \text{for all } \varphi \in C_o^\infty(\mathbb{R}^N).$$

Integrating by parts with the aid of Green's theorem gives

$$\int_\Gamma \varphi[\mathbf{F}(x,u)] \cdot \boldsymbol{\nu} d\sigma = 0 \quad \text{for all } \varphi \in C_o^\infty(\mathbb{R}^N).$$

Thus if a weak solution suffers a discontinuity across a smooth surface Γ, then

$$[\mathbf{F}(x,u)] \cdot \boldsymbol{\nu} = 0 \quad \text{on } \Gamma. \tag{4.4}$$

Even though this equation has been derived globally, it has a local thrust, and it can be used to find possible local discontinuities of weak solutions.

4.2 The Shock Line

Consider the PDE in two independent variables

$$u_y + a(u)u_x = 0 \quad \text{for some } a \in C(\mathbb{R}) \tag{4.5}$$

and rewrite it as

$$\frac{\partial}{\partial y}R(u) + \frac{\partial}{\partial x}S(u) = 0 \quad \text{in } \mathbb{R}^2$$

where

$$R(u) = u \quad \text{and} \quad S(u) = \int^u a(s)ds.$$

More generally, $R(\cdot)$ and $S(\cdot)$ could be any two functions satisfying

$$S'(u) = a(u)R'(u).$$

Let u be a weak solution of (4.5) in $\mathbb{R}^2$, discontinuous across a smooth curve of parametric equations $\Gamma = \{x = x(t), y = y(t)\}$. Then, according to (4.4), Γ must satisfy the shock condition[1]

$$[R(u)]x' - [S(u)]y' = 0. \tag{4.6}$$

In particular, if Γ is the graph of a function $y = y(x)$, then $y(\cdot)$ satisfies the differential equation

$$y' = \frac{[R(u)]}{[S(u)]}.$$

As an example, consider the case of the Burgers equation (4.3). Let u be a weak solution of (4.3), discontinuous across a smooth curve Γ parametrized locally as $\Gamma = \{x = x(t), y = t\}$. Then (4.6) gives the differential equation of the shock line[2]

$$x'(t) = \frac{[u^+(x(t),t) + u^-(x(t),t)]}{2}, \qquad u^\pm = \lim_{x \to x(t)^\pm} u(x,t). \tag{4.7}$$

5 The Initial Value Problem

Denote by (x,t) points in $\mathbb{R}^N \times \mathbb{R}^+$, and consider the quasi-linear equation in $N+1$ variables

$$u_t + a_i(x,t,u)u_{x_i} = a_o(x,t,u) \tag{5.1}$$

with data prescribed on the N-dimensional surface $[t=0]$, say

$$u(x,0) = u_o(x) \in C^1(\mathbb{R}^N). \tag{5.2}$$

Using the $(N+1)$st variable t as a parameter, the characteristic projections are

$$\begin{aligned} &x_i'(t) = a_i(x(t),t,z(t)) && \text{for } i = 1,\dots,N \\ &x_{N+1} = t && \text{for } t \in (-\delta,\delta) \text{ for some } \delta > 0 \\ &x(0) = x_o \in \mathbb{R}^N. \end{aligned}$$

Therefore $t = 0$ is non-characteristic and the Cauchy problem (5.1)–(5.2) is solvable. If the coefficients a_i are constant, the integral surface is given by

---

[1]This is a special case of the Rankine–Hugoniot shock condition ([124, 76]).

[2]The notion of *shock* will be made more precise in Section 13.3.

$$x(t,s) = s + \mathbf{a}t, \quad z(t,s) = u_o(s) + a_o t$$

and the solution is

$$u(x,t) = u_o(x - \mathbf{a}t) + a_o t.$$

In the case $N = 1$ and $a_o = 0$, this is a traveling wave in the sense that the graph of u_o travels with velocity a_1 in the positive direction of the x-axis, keeping the same shape.

5.1 Conservation Laws

Let $(x,t,u) \to \mathbf{F}(x,t,u)$ be a measurable vector-valued function in $\mathbb{R}^N \times \mathbb{R}^+ \times \mathbb{R}$ and consider formally homogeneous, initial value problems of the type

$$\begin{aligned} u_t + \operatorname{div} \mathbf{F}(x,t,u) &= 0 \;\text{ in }\; \mathbb{R}^N \times \mathbb{R}^+ \\ u(\cdot, 0) &= u_o \in L^1(\mathbb{R}^N). \end{aligned} \tag{5.3}$$

These are called conservation laws. The variable t represents the time, and u is prescribed at some initial time $t = 0$.

Remark 5.1 The method of integral surfaces outlined in Section 2, gives solutions near the non-characteristic surface $t = 0$. Because of the physics underlying these problems we are interested in solutions defined only for positive times, that is defined only on one side of the surface carrying the data.

A function u is a weak solution of the initial value problem (5.3) if

(a) $u(\cdot,t) \in L^1_{\text{loc}}(\mathbb{R}^N)$ for all $t \ge 0$, and $F_i(\cdot,\cdot,u) \in L^1_{\text{loc}}(\mathbb{R}^N \times \mathbb{R}^+)$, for all $i = 1, \dots, N$

(b) the PDE is satisfied in the sense

$$\int_0^\infty \int_{\mathbb{R}^N} \big[u\varphi_t + \mathbf{F}(x,t,u) \cdot D\varphi\big] dxdt = 0 \tag{5.4}$$

for all $\varphi \in C_o^\infty(\mathbb{R}^N \times \mathbb{R}^+)$, and where D denotes the gradient with respect to the space variables only.

(c) the initial datum is taken in the sense of $L^1_{\text{loc}}(\mathbb{R}^N)$, that is

$$\lim_{\mathbb{R}^+ \ni t \to 0} \|u(\cdot,t) - u_o\|_{1,K} = 0 \tag{5.5}$$

for all compact sets $K \subset \mathbb{R}^N$.

6 Conservation Laws in One Space Dimension

Let $a(\cdot)$ be a continuous function in $\mathbb{R}$ and consider the initial value problem

$$u_t + a(u)u_x = 0 \text{ in } \mathbb{R}\times\mathbb{R}^+, \quad u(\cdot,0) = u_o \in C(\mathbb{R}). \tag{6.1}$$

The characteristic through $(x_o, 0, z_o)$, using t as a parameter, is

$$z(t) = z_o, \quad x(t) = x_o + a(z_o)t$$

and the integral surface is

$$x(s,t) = s + a(u_o(s))t, \quad z(s,t) = u_o(s).$$

Therefore the solution, whenever it is well defined, can be written implicitly as

$$u(x,t) = u_o(x - a(u)t). \tag{6.2}$$

The characteristic projections through points $(s,0)$ of the x-axis are the lines

$$x = s + a(u_o(s))t$$

and u remains constant along such lines. Two of these characteristic projections, say

$$x = s_i + a(u_o(s_i))t \quad i = 1,2, \quad \text{such that } a(u_o(s_1)) \neq a(u_o(s_2)) \tag{γ_i}$$

intersect at (ξ,η) given by

$$\begin{aligned} \xi &= \frac{a(u_o(s_1))s_2 - a(u_o(s_2))s_1}{a(u_o(s_1)) - a(u_o(s_2))} \\ \eta &= -\frac{s_1 - s_2}{a(u_o(s_1)) - a(u_o(s_2))}. \end{aligned} \tag{6.3}$$

Since u is constant along each of the γ_i, it must be discontinuous at (ξ,η), unless $u_o(s) = \text{const}$. Therefore the solution exists only in a neighborhood of the x-axis. It follows from (6.3) and Remark 5.1 that the solution exists for all $t > 0$ if the function $s \to a(u_o(s))$ is increasing. Indeed, in such a case, the intersection point of the characteristic lines γ_1 and γ_2 occurs in the half-plane $t < 0$. If $a(\cdot)$ and $u_o(\cdot)$ are differentiable, compute from (6.2)

$$\begin{aligned} u_t &= -\frac{u_o'(x - a(u)t)a(u)}{1 + u_o'(x - a(u)t)a'(u)t} \\ u_x &= \frac{u_o'(x - a(u)t)}{1 + u_o'(x - a(u)t)a'(u)t}. \end{aligned}$$

These are implicitly well defined if $a(\cdot)$ and $u_o(\cdot)$ are increasing functions, and when substituted into (6.1) satisfy the PDE for all $t > 0$. Rewrite the initial value problem (6.1) as

$$\begin{aligned} u_t + F(u)_x &= 0 \text{ in } \mathbb{R}\times\mathbb{R}^+ \\ u(\cdot,0) &= u_o \end{aligned} \qquad \text{where } F(u) = \int_0^u a(s)ds. \tag{6.4}$$

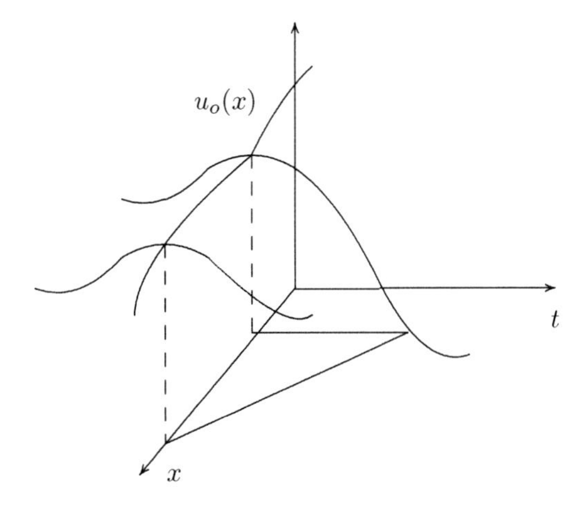

Fig. 6.1.

Proposition 6.1 *Let $F(\cdot)$ be convex and of class C^2, and assume that the initial datum $u_o(\cdot)$ is non-decreasing and of class C^1. Then the initial value problem (6.4) has a unique classical solution in $\mathbb{R} \times \mathbb{R}^+$.*

6.1 Weak Solutions and Shocks

If the initial datum u_o is decreasing, then a solution global in time is necessarily a weak solution. The shock condition (4.6) might be used to construct weak solutions, as shown by the following example. The initial value problem

$$u_t + \tfrac{1}{2}(u^2)_x = 0 \quad \text{in} \quad \mathbb{R} \times \mathbb{R}^+$$

$$u(x,0) = \begin{cases} 1 & \text{for } x < 0 \\ 1 - x & \text{for } 0 \le x \le 1 \\ 0 & \text{for } x \ge 1 \end{cases} \tag{6.5}$$

has a unique weak solution for $0 < t < 1$, given by

$$u = \begin{cases} 1 & \text{for } x < t \\ \dfrac{x-1}{t-1} & \text{for } t < x < 1 \\ 0 & \text{for } x \ge 1. \end{cases} \tag{6.6}$$

For $t > 1$ the geometric construction of (6.2) fails for the sector $1 < x < t$.

The jump discontinuity across the lines $x = 1$ and $x = t$ is 1. Therefore, starting at $(1, 1)$ we draw a curve satisfying (4.7). This gives the shock line $2x = t + 1$, and we define the weak solution u for $t > 1$ as

$$u = \begin{cases} 1 & \text{for } 2x < t + 1 \\ 0 & \text{for } 2x > t + 1. \end{cases} \tag{6.7}$$

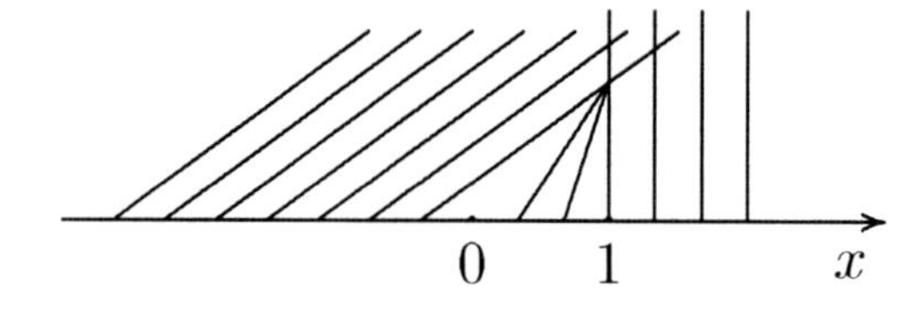

Fig. 6.2.

Remark 6.1 For $t > 1$ fixed, the solution $x \to u(x,t)$ drops from 1 to 0 as the increasing variable x crosses the shock line.

6.2 Lack of Uniqueness

If u_o is non-decreasing and somewhere discontinuous, then (6.4) has, in general, more than one weak solution. This is shown by the following Riemann problem:

$$u_t + \frac{1}{2}(u^2)_x = 0 \ \text{ in } \ \mathbb{R} \times \mathbb{R}^+$$
$$u(x,0) = \begin{cases} 0 & \text{for } x \le 0 \\ 1 & \text{for } x > 0. \end{cases} \tag{6.8}$$

No points of the sector $0 < x < t$ can be reached by characteristics originating from the x-axis and carrying the data (Figure 6.3). The solution is zero for $x < 0$, and it is 1 for $x > t$. Enforcing the shock condition (4.7) gives

$$u(x,t) = \begin{cases} 0 & \text{for } 2x < t \\ 1 & \text{for } 2x > t. \end{cases} \tag{6.9}$$

However, the continuous function

$$u(x,t) = \begin{cases} 0 & \text{for } x < 0 \\ \dfrac{x}{t} & \text{for } 0 \le x \le t \\ 1 & \text{for } x > t \end{cases} \tag{6.10}$$

is also a weak solution of (6.8).

7 Hopf Solution of The Burgers Equation

Insight into the solvability of the initial value problem (6.4) is gained by considering first the special case of the Burgers equation, for which $F(u) = \frac{1}{2}u^2$. Hopf's method [71] consists in solving first the regularized parabolic problems

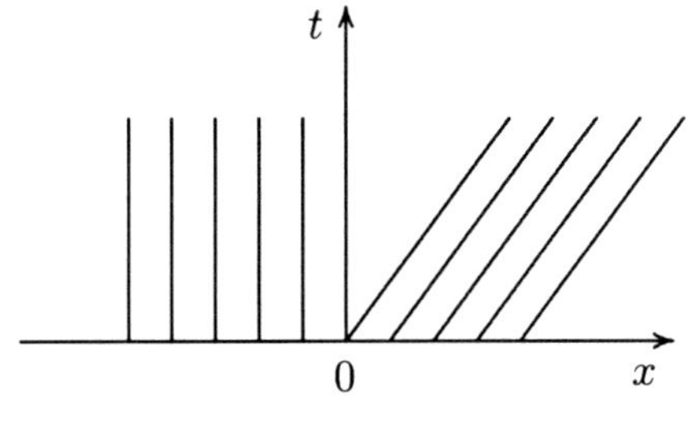

Fig. 6.3.

$$\begin{aligned} u_{n,t} - \frac{1}{n} u_{n,xx} &= -u_n u_{n,x} \quad \text{in } \mathbb{R} \times \mathbb{R}^+ \\ u_n(\cdot, 0) &= u_o \end{aligned} \tag{7.1}$$

and then letting $n \to \infty$ in a suitable topology. Setting

$$U(x,t) = \int_{x_o}^{x} u_n(y,t) dy$$

for some arbitrary but fixed $x_o \in \mathbb{R}$ transforms the Cauchy problem (7.1) into

$$\begin{aligned} U_t - \frac{1}{n} U_{xx} &= -\frac{1}{2} (U_x)^2 \quad \text{in } \mathbb{R} \times \mathbb{R}^+ \\ U(x,0) &= \int_{x_o}^{x} u_o(s) ds. \end{aligned}$$

Next, one introduces the new unknown function $w = e^{-\frac{n}{2} U}$ and verifies that w is a positive solution of the Cauchy problem

$$\begin{aligned} w_t - \frac{1}{n} w_{xx} &= 0 \quad \text{in } \mathbb{R} \times \mathbb{R}^+ \\ w(x,0) &= e^{-\frac{n}{2} \int_{x_o}^{x} u_o(s) ds}. \end{aligned} \tag{7.2}$$

Such a positive solution is uniquely determined by the representation formula

$$w(x,t) = \frac{1}{\sqrt{4\pi t}} \int_{\mathbb{R}} e^{-\frac{n}{2} \int_{x_o}^{y} u_o(s) ds} e^{-n \frac{|x-y|^2}{4t}} dy$$

provided u_o satisfies the growth condition[3]

$$|u_o(s)| \le C_o |s|^{1-\varepsilon_o} \qquad \text{for all } |s| \ge r_o$$

for some given positive constants C_o, r_o, and ε_o. The unique solutions u_n of (7.1) are then given explicitly by

$$u_n(x,t) = \int_{\mathbb{R}} \frac{(x-y)}{t} d\lambda_n(y)$$

[3]See (2.7), Section 14, and Theorem 2.1 of Chapter 5.

where $d\lambda_n(y)$ are the probability measures

$$d\lambda_n(y) = \frac{e^{-\frac{n}{2}\left(\int_{x_o}^{y} u_o(s)ds + \frac{|x-y|^2}{2t}\right)}}{\int_{\mathbb{R}} e^{-\frac{n}{2}\left(\int_{x_o}^{y} u_o(s)ds + \frac{|x-y|^2}{2t}\right)} dy} dy.$$

The a priori estimates needed to pass to the limit can be derived either from the parabolic problems (7.1)–(7.2) or from the explicit representation of u_n and the corresponding probability measures $\lambda_n(y)$. In either case they depend on the fact that $F(\cdot)$ is convex and $F' = a(\cdot)$ is strictly increasing.[4]

Because of the parabolic regularization (7.1), it is reasonable to expect that those solutions of (6.4) constructed in this way satisfy some form of the maximum principle.[5] It turns out that Hopf's approach, and in particular the explicit representation of the approximating solutions u_n and the corresponding probability measures $\lambda_n(y)$, continues to hold for the more general initial value problem (6.4). It has been observed that these problems fail, in general, to have a unique solution. It turns out that those solutions of (6.4) that satisfy the maximum principle, form a special subclass of solutions within which uniqueness holds. These are called *entropy* solutions.

8 Weak Solutions to (6.4) When $a(\cdot)$ is Strictly Increasing

We let $a(\cdot)$ be continuous and strictly increasing in $\mathbb{R}$, that is, there exists a positive constant L such that

$$a'(s) \ge \frac{1}{L} \quad \text{a.e. } s \in \mathbb{R}. \tag{8.1}$$

Assume that the initial datum u_o satisfies

$$\begin{aligned} &u_o \in L^\infty(\mathbb{R}) \cap L^1(-\infty, x) \text{ for all } x \in \mathbb{R} \\ &\limsup_{x \to -\infty} |u_o(x)| = 0 \\ &\inf_{x \in \mathbb{R}} \int_{-\infty}^{x} u_o(s)ds \ge -C \text{ for some } C > 0. \end{aligned} \tag{8.2}$$

For example, the datum of the Riemann problem (6.8) satisfies such a condition. The initial datum is not required to be increasing, nor in $L^1(\mathbb{R})$. Since $F(\cdot)$ is convex ([31], Chapter IV, Section 13)

[4] Some cases of non-convex F are in [79].

[5] By **3.2.** of the Problems and Complements of Chapter 5, the presence of the term $u_n u_{n,x}$ in (7.1) is immaterial for a maximum principle to hold.

$$F(u) - F(v) \ge a(v)(u - v) \quad u, v \in \mathbb{R}, \quad a(v) = F'(v) \tag{8.3}$$

and since F' is strictly increasing, equality holds only if $u = v$. This inequality permits one to solve (6.4) in a weak sense and to identify a class of solutions, called *entropy* solutions, within which uniqueness holds ([97, 99]).

8.1 Lax Variational Solution

To illustrate the method assume first that $F(\cdot)$ is of class C^2 and that u_o is regular, increasing and satisfies

$$u_o(x) = 0 \text{ for all } x < b \text{ for some } b < 0.$$

The geometric construction of (6.2) guarantees that a solution must vanish for $x < b$ for all $t > 0$. Therefore the function

$$U(x, t) = \int_{-\infty}^{x} u(s, t) ds$$

is well defined in $\mathbb{R} \times \mathbb{R}^+$. Integrating (6.4) in dx over $(-\infty, x)$ shows that U satisfies the initial value problem

$$U_t + F(U_x) = 0 \text{ in } \mathbb{R} \times \mathbb{R}^+, \quad U(x, 0) = \int_{-\infty}^{x} u_o(s) ds.$$

It follows from (8.3), with $u = U_x$ and all $v \in \mathbb{R}$, that

$$U_t + a(v) U_x \le a(v) v - F(v) \tag{8.4}$$

and equality holds only if $v = u(x, t)$. For $(x, t) \in \mathbb{R} \times \mathbb{R}^+$ fixed, consider the line of slope $1/a(v)$ through (x, t). Denoting by (ξ, τ) the variables, such a line has equation $x - \xi = a(v)(t - \tau)$, and it intersects the axis $\tau = 0$ at the abscissa

$$\eta = x - a(v) t. \tag{8.5}$$

The left-hand side of (8.4) is the derivative of U along such a line. Therefore

$$\frac{d}{d\tau} U\big(x - a(v)(t - \tau), \tau\big) = U_t + a(v) U_x \le a(v) v - F(v).$$

Integrating this over $\tau \in (0, t)$ gives

$$U(x, t) \le \int_{-\infty}^{\eta} u_o(s) ds + t\big[a(v) v - F(v)\big]$$

valid for all $v \in \mathbb{R}$, and equality holds only for $v = u(x, t)$. From (8.5) compute

$$v = a^{-1}\left(\frac{x - \eta}{t}\right) \tag{8.6}$$

and rewrite the previous inequality for $U(x,t)$ in terms of η only, that is

$$U(x,t) \le \Psi(x,t;\eta) \qquad \text{for all } \eta \in \mathbb{R} \tag{8.7}$$

where

$$\begin{aligned} \Psi(x,t;\eta) = &\int_{-\infty}^{\eta} u_o(s)ds \\ &+ t\Big\{\Big(\frac{x-\eta}{t}\Big)a^{-1}\Big(\frac{x-\eta}{t}\Big) - F\Big[a^{-1}\Big(\frac{x-\eta}{t}\Big)\Big]\Big\}. \end{aligned} \tag{8.8}$$

Therefore, having fixed (x,t), for that value of $\eta = \eta(x,t)$ for which v in (8.6) equals $u(x,t)$, equality must hold in (8.7). Returning now to $F(\cdot)$ convex and u_o satisfying (8.1)–(8.2), the arguments leading to (8.7) suggest the construction of the weak solution of (6.4) in the following two steps:

Step 1: For (x,t) fixed, minimize the function $\Psi(x,t;\eta)$, i.e., find $\eta = \eta(x,t)$ such that

$$\Psi(x,t;\eta(x,t)) \le \Psi(x,t;s) \qquad \text{for all } s \in \mathbb{R}. \tag{8.9}$$

Step 2: Compute $u(x,t)$ from (8.6), that is

$$u(x,t) = a^{-1}\Big(\frac{x-\eta(x,t)}{t}\Big). \tag{8.10}$$

9 Constructing Variational Solutions I

Proposition 9.1 *For fixed $t > 0$ and a.e. $x \in \mathbb{R}$ there exists a unique $\eta = \eta(x,t)$ that minimizes $\Psi(x,t;\cdot)$. The function $x \to u(x,t)$ defined by (8.10) is a.e. differentiable in $\mathbb{R}$ and satisfies*

$$\frac{u(x_2,t) - u(x_1,t)}{x_2 - x_1} \le \frac{L}{t} \qquad \textit{for a.e. } x_1 < x_2 \in \mathbb{R}. \tag{9.1}$$

Moreover, for a.e. $(x,t) \in \mathbb{R} \times \mathbb{R}^+$

$$|u(x,t)| \le \sqrt{\frac{2L}{t}}\Big(\int_{-\infty}^{x-a(o)t} u_o(s)ds - \inf_{y\in\mathbb{R}} \int_{-\infty}^{y} u_o(s)ds\Big)^{1/2}. \tag{9.2}$$

Proof The function $\eta \to \Psi(x,t;\eta)$ is bounded below. Indeed, by the expressions (8.6) and (8.8) and the assumptions (8.1)–(8.2)

$$\Psi(x,t;\eta) \ge \inf_{y\in\mathbb{R}} \int_{-\infty}^{y} u_o(s)ds + t[va(v) - F(v)] \ge -C + \frac{t}{2L}v^2 \tag{9.3}$$

for $\eta = x - a(v)t$. A minimizer can be found by a minimizing sequences $\{\eta_n\}$, that is one for which

$$\Psi(x,t;\eta_n) > \Psi(x,t;\eta_{n+1}) \quad \text{and} \quad \lim \Psi(x,t;\eta_n) = \inf_{\eta} \Psi(x,t;\eta).$$

By (9.3), the sequence $\{\eta_n\}$ is bounded. Therefore, a subsequence can be selected and relabeled with n such that $\{\eta_n\} \to \eta(x,t)$. Since $\Psi(x,t;\cdot)$ is continuous in $\mathbb{R}$

$$\lim_{n\to\infty} \Psi(x,t;\eta_n) = \Psi(x,t;\eta(x,t)) \le \Psi(x,t;\eta), \qquad \text{for all } \eta \in \mathbb{R}.$$

This process guarantees the existence of at least one minimizer for every fixed $x \in \mathbb{R}$. Next we prove that such a minimizer is unique, for a.e. $x \in \mathbb{R}$. ■

Let $H(x)$ denote the set of all the minimizers of $\Psi(x,t;\cdot)$, and define a function $x \to \eta(x,t)$ as an arbitrary selection out of $H(x)$.

Lemma 9.1 *If $x_1 < x_2$, then $\eta(x_1,t) < \eta(x_2,t)$.*

Proof (of Proposition 9.1 assuming Lemma 9.1) Since $x \to \eta(x,t)$ is increasing, it is continuous in $\mathbb{R}$ except possibly for countably many points. Therefore, $\eta(x,t)$ is uniquely defined for a.e. $x \in \mathbb{R}$. From (8.10) it follows that for a.e. $x_1 < x_2$ and some $\xi \in \mathbb{R}$

$$\begin{aligned} u(x_2,t) - u(x_1,t) &\le \frac{a^{-1'}(\xi)}{t}\{(x_2 - x_1) - [\eta(x_2,t) - \eta(x_1,t)]\} \\ &\le \frac{a^{-1'}(\xi)}{t}(x_2 - x_1) \le \frac{L}{t}(x_2 - x_1). \end{aligned}$$

This proves (9.1). To prove (9.2), write (9.3) for $\eta = \eta(x,t)$, the unique minimizer of $\Psi(x,t;\cdot)$. For such a choice, by (8.10), $v = u$. Therefore

$$\begin{aligned} \frac{t}{2L}u^2(x,t) &\le \Psi\big(x,t;\eta(x,t)\big) - \inf_{y\in\mathbb{R}} \int_{-\infty}^{y} u_o(s)ds \\ &\le \Psi(x,t;\eta) - \inf_{y\in\mathbb{R}} \int_{-\infty}^{y} u_o(s)ds \end{aligned}$$

for all $\eta \in \mathbb{R}$, since $\eta(x,t)$ is a minimizer. Taking $\eta = x - a(0)t$ and recalling the definitions (8.8) of $\Psi(x,t;\cdot)$ proves (9.2). ■

9.1 Proof of Lemma 9.1

Let $\eta_i = \eta(x_i,t)$ for $i = 1,2$. It will suffice to prove that

$$\Psi(x_2,t;\eta_1) < \Psi(x_2,t;\eta) \qquad \text{for all } \eta < \eta_1. \tag{9.4}$$

By minimality, $\Psi(x_1,t;\eta_1) \le \Psi(x_1,t;\eta)$ for all $\eta < \eta_1$. From this

$$\begin{aligned} \Psi(x_2,t;\eta_1) + &[\Psi(x_1,t;\eta_1) - \Psi(x_2,t;\eta_1)] \\ &\le \Psi(x_2,t;\eta) + [\Psi(x_1,t;\eta) - \Psi(x_2,t;\eta)]. \end{aligned}$$

Therefore inequality (9.4) will follow if the function

$$\eta \to L(\eta) = \Psi(x_1, t; \eta) - \Psi(x_2, t; \eta)$$

is increasing. Rewrite $L(\eta)$ in terms of $v_i = v_i(\eta)$ given by (8.6) with $x = x_i$ for $i = 1, 2$. This gives

$$L(\eta) = t[v_1 a(v_1) - F(v_1)] - t[v_2 a(v_2) - F(v_2)] = t \int_{v_2}^{v_1} s a'(s) ds.$$

From this one computes

$$\begin{aligned} L'(\eta) &= t \left[v_1 a'(v_1) \frac{\partial v_1}{\partial \eta} - v_2 a'(v_2) \frac{\partial v_2}{\partial \eta} \right] \\ &= a^{-1}\left(\frac{x_2 - \eta}{t}\right) - a^{-1}\left(\frac{x_1 - \eta}{t}\right) > 0. \end{aligned}$$

■

10 Constructing Variational Solutions II

For fixed $t > 0$, the minimizer $\eta(x, t)$ of $\Psi(x, t; \cdot)$ exists and is unique for a.e. $x \in \mathbb{R}$. We will establish that for all such (x, t)

$$a^{-1}\left(\frac{x - \eta(x,t)}{t}\right) = \lim_{n\to\infty} \int_{\mathbb{R}} a^{-1}\left(\frac{x - \eta}{t}\right) d\lambda_n(\eta) \tag{10.1}$$

where $d\lambda_n(\eta)$ are the probability measures on $\mathbb{R}$

$$d\lambda_n(\eta) = \frac{e^{-n\Psi(x,t;\eta)}}{\int_{\mathbb{R}} e^{-n\Psi(x,t;\eta)} d\eta} d\eta \qquad \text{for } n \in \mathbb{N}. \tag{10.2}$$

Therefore the expected solution

$$u(x, t) = a^{-1}\left(\frac{x - \eta(x, t)}{t}\right)$$

can be constructed by the limiting process (10.1). More generally, we will establish the following.

Lemma 10.1 *Let f be a continuous function in $\mathbb{R}$ satisfying the growth condition*

$$|f(v)| \le C_o |v| e^{c_o \int_o^v s a'(s) ds} \qquad \textit{for all } |v| \ge \gamma_o \tag{10.3}$$

for given positive constants C_o, c_o and γ_o. Then for fixed $t > 0$ and a.e. $x \in \mathbb{R}$

$$f[u(x, t)] = \lim_{n\to\infty} \int_{\mathbb{R}} f\left[a^{-1}\left(\frac{x - \eta}{t}\right)\right] d\lambda_n(\eta).$$

Proof Introduce the change of variables

$$v = a^{-1}\left(\frac{x-\eta}{t}\right), \qquad v_o = a^{-1}\left(\frac{x-\eta(x,t)}{t}\right) \tag{10.4}$$

and rewrite $\Psi(x,t;\eta)$ as

$$\Psi(v) = \int_{-\infty}^{x-a(v)t} u_o(s)ds + t[va(v) - F(v)]. \tag{10.5}$$

The probability measures $d\lambda_n(\eta)$ are transformed into the probability measures

$$d\mu_n(v) = \frac{e^{-n\Psi(v)}a'(v)}{\int_{\mathbb{R}} e^{-n\Psi(v)}a'(v)dv}dv = \frac{e^{-n[\Psi(v)-\Psi(v_o)]}a'(v)}{\int_{\mathbb{R}} e^{-n[\Psi(v)-\Psi(v_o)]}a'(v)dv}dv \tag{10.6}$$

and the statement of the lemma is equivalent to

$$f(v_o) = \lim_{n\to\infty}\int_{\mathbb{R}} f(v)d\mu_n(v)$$

where v_o is the unique minimizer of $v \to \Psi(v)$. For this it suffices to show that

$$I_n \overset{\text{def}}{=} \int_{\mathbb{R}} |f(v) - f(v_o)|d\mu_n(v) \to 0 \qquad \text{as } n \to \infty.$$

By (9.3) the function $\Psi(\cdot)$ grows to infinity as $|v| \to \infty$. Since v_o is the only minimizer, for each $\varepsilon > 0$ there exists $\delta = \delta(\varepsilon) > 0$ such that

$$\Psi(v) > \Psi(v_o) + \delta \qquad \text{for all } |v - v_o| > \varepsilon. \tag{10.7}$$

Moreover, the numbers ε and δ being fixed, there exists some positive number σ such that

$$\Psi(v) \le \Psi(v_o) + \tfrac{1}{2}\delta \qquad \text{for all } |v - v_o| < \sigma.$$

From this we estimate from below

$$\int_{\mathbb{R}} e^{-n[\Psi(v)-\Psi(v_o)]}a'(v)dv \ge \frac{1}{L}\int_{|v-v_o|<\sigma} e^{-n[\Psi(v)-\Psi(v_o)]}dv \ge \frac{2\sigma}{L}e^{-\frac{1}{2}n\delta}.$$

Therefore

$$d\mu_n(v) \le \frac{L}{2\sigma}e^{\frac{1}{2}n\delta}e^{-n[\Psi(v)-\Psi(v_o)]}a'(v)dv.$$

Next estimate I_n by using these remarks as

$$\begin{aligned} I_n &\le \int_{|v-v_o|<\varepsilon} |f(v) - f(v_o)|d\mu_n(v) \\ &\quad + \sup_{|v-v_o|<2\gamma}|f(v)|\frac{L}{2\sigma}e^{\frac{1}{2}n\delta}\int_{\varepsilon<|v-v_o|<2\gamma} e^{-n[\Psi(v)-\Psi(v_o)]}a'(v)dv \\ &\quad + \frac{L}{2\sigma}e^{\frac{1}{2}n\delta+n\Psi(v_o)}\int_{|v-v_o|>2\gamma} (|f(v)| + |f(v_o)|)e^{-n\Psi(v)}a'(v)dv \\ &= I_n^{(1)} + I_n^{(2)} + I_n^{(3)} \end{aligned}$$

where γ is a positive number to be chosen. Denoting by $\omega(\cdot)$ the modulus of continuity of f, estimate $I_n^{(1)} \le \omega(\varepsilon)$, since $d\mu_n(v)$ is a probability measure. The second term $I_n^{(2)}$ is estimated by means of (10.7) as

$$\begin{aligned} I_n^{(2)} &\le \sup_{|v-v_o|<2\gamma} |f(v)|a'(v)\frac{L}{2\sigma}e^{\frac{1}{2}n\delta}\int_{\varepsilon<|v-v_o|<2\gamma} e^{-n[\Psi(v)-\Psi(v_o)]}dv \\ &\le C(\gamma,\sigma,L)e^{-\frac{1}{2}n\delta} \end{aligned}$$

for a constant C depending only on the indicated quantities. Thus $I_n^{(2)} \to 0$ as $n\to\infty$. The last term $I_n^{(3)}$ is estimated using the lower bound

$$\Psi(v) \ge -C + t[va(v) - F(v)] \ge -C + t\int_0^v sa'(s)ds.$$

Also choose $\gamma \ge \gamma_o$, where γ_o is the constant in the growth condition (10.3). By choosing γ even larger if necessary, we may ensure that $[|v-v_o|>2\gamma] \subset [|v|>\gamma]$. For this choice

$$I_n^{(3)} \le \frac{C_oL}{\sigma}e^{n[\delta+C+\Psi(v_o)]}\int_{|v|>\gamma} e^{-(nt-c_o)\int_o^v sa'(s)ds}|v|a'(v)dv. \tag{10.8}$$

If n is so large that $nt-c_o>0$, the integral on the right-hand side of (10.8) can be computed explicitly, and estimated as follows

$$\begin{aligned} \int_{|v|>\gamma} e^{-(nt-c_o)\int_o^v sa'(s)ds}|v|a'(v)dv &= \int_\gamma^\infty \cdots dv + \int_{-\infty}^{-\gamma} \cdots dv \\ &= \frac{2}{nt-c_o}e^{-(nt-c_o)\int_o^\gamma sa'(s)ds} \le \frac{2}{nt-c_o}e^{-(nt-c_o)\frac{\gamma^2}{2L}}. \end{aligned}$$

This in (10.8) gives

$$I_n^{(3)} \le \frac{2C_oLe^{\frac{c_o\gamma^2}{2L}}}{\sigma(nt-c_o)}e^{-n\left(-\delta-C-\Psi(v_o)+\frac{t\gamma^2}{2L}\right)}.$$

The number $t>0$ being fixed, choose γ large enough that

$$-\delta - C - \Psi(v_o) + \frac{t\gamma^2}{2L} > 0.$$

Then let $n\to\infty$ to conclude that $\lim_{n\to\infty} I_n \le \omega(\varepsilon)$ for all $\varepsilon>0$. ∎

11 The Theorems of Existence and Stability

11.1 Existence of Variational Solutions

Theorem 11.1 (Existence) *Let the assumptions (8.1)–(8.2) hold, and let $u(\cdot,t)$ denote the function constructed in Sections 8–10. Then*

$$\|u(\cdot,t)\|_{\infty,\mathbb{R}} \le \|u_o\|_{\infty,\mathbb{R}} \quad \textit{for all } \ t>0. \tag{11.1}$$

The function u solves the initial value problem (6.4) in the weak sense

$$\int_0^t \int_{\mathbb{R}} [u\varphi_t + F(u)\varphi_x] dx\, d\tau = \int_{\mathbb{R}} u(x,t)\varphi(x,t)dx - \int_{\mathbb{R}} u_o(x)\varphi(x,0)dx \tag{11.2}$$

for all $\varphi \in C^1[\mathbb{R}^+; C_o^\infty(\mathbb{R})]$ and a.e. $t>0$. Moreover, u takes the initial datum u_o in the sense of $L^1_{\mathrm{loc}}(\mathbb{R})$, that is, for every compact subset $K \subset \mathbb{R}$

$$\lim_{t\to 0} \|u(\cdot,t) - u_o\|_{1,K} = 0. \tag{11.3}$$

Finally, if $u_o(\cdot)$ is continuous, then for all $t>0$

$$u(x,t) = u_o(x - a[u(x,t)]t) \qquad \textit{for a.e. } \ x \in \mathbb{R}. \tag{11.4}$$

11.2 Stability of Variational Solutions

Assuming the existence theorem for the moment, we establish that the solutions constructed by the method of Sections 8–10 are stable in $L^1_{\mathrm{loc}}(\mathbb{R})$. Let $\{u_{o,m}\}$ be a sequence of functions satisfying (8.2) and in addition

$$\begin{cases} \|u_{o,m}\|_{\infty,\mathbb{R}} \le \gamma \|u_o\|_{\infty,\mathbb{R}} & \text{for all } m, \text{ for some } \gamma>0 \\ u_{o,m} \to u_o \quad \text{weakly in } L^1(-\infty,x) & \text{for all } x \in \mathbb{R}. \end{cases} \tag{11.5}$$

Denote by u_m the functions constructed by the methods of Sections 8–10, corresponding to the initial datum $u_{o,m}$. Specifically, first consider the functions $\Psi_m(x,t;\cdot)$ defined as in (8.8), with u_o replaced by $u_{o,m}$. For fixed $t>0$, let $\eta_m(x,t)$ be a minimizer of $\Psi_m(x,t;\cdot)$. Such a minimizer is unique for almost all $x \in \mathbb{R}$. Then set

$$u_m(x,t) = a^{-1}\left(\frac{x - \eta_m(x,t)}{t}\right).$$

Theorem 11.2 (Stability in $L^1_{\mathrm{loc}}(\mathbb{R})$) *For fixed $t>0$ and all compact subsets $K \subset \mathbb{R}$*

$$\|u_m(\cdot,t) - u(\cdot,t)\|_{1,K} \to 0 \qquad \textit{as } \ m \to \infty.$$

Proof Denote by $\mathcal{E}_o$ and $\mathcal{E}_m$ the subsets of $\mathbb{R}$ where $u(\cdot,t)$ and $u_m(\cdot,t)$ are not uniquely defined. The set $\mathcal{E} = \bigcup \mathcal{E}_m$ has measure zero and $\{u_m(\cdot,t), u(\cdot,t)\}$ are all uniquely well defined in $\mathbb{R} - \mathcal{E}$. We claim that

$$\lim_{m\to\infty} u_m(x,t) = u(x,t) \qquad \text{and} \qquad \lim_{m\to\infty} \eta_m(x,t) = \eta(x,t)$$

for all $x \in \mathbb{R} - \mathcal{E}$, where $\eta(x,t)$ is the unique minimizer of $\Psi(x,t;\cdot)$. By (11.1) and the first of (11.5), $\{u_m(x,t)\}$ is bounded. Therefore also $\{\eta_m(x,t)\}$ is bounded, and a subsequence $\{\eta_{m'}(x,t)\}$ contains in turn a convergent subsequence, say for example $\{\eta_{m''}(x,t)\} \to \eta_o(x,t)$. By minimality

$$\Psi_{m''}(x,t;\eta_{m''}(x,t)) \le \Psi_{m''}(x,t;\eta(x,t)).$$

Letting $m'' \to \infty$

$$\Psi(x,t;\eta_o(x,t)) \le \Psi(x,t;\eta(x,t)).$$

Therefore $\eta_o(x,t) = \eta(x,t)$, since the minimizer of $\Psi(x,t;\cdot)$ is unique. Therefore any subsequence out of $\{\eta_m(x,t)\}$ contains in turn a subsequence convergent to the same limit $\eta(x,t)$. Thus the entire sequence converges to $\eta(x,t)$. Such a convergence holds for all $x \in \mathbb{R} - \mathcal{E}$, i.e., $\{u_m(\cdot,t)\} \to u(\cdot,t)$ a.e. in $\mathbb{R}$. Since $\{u_m(\cdot,t)\}$ is uniformly bounded in $\mathbb{R}$, the stability theorem in $L^1_{\text{loc}}(\mathbb{R})$ follows from the Lebesgue dominated convergence theorem. ∎

12 Proof of Theorem 11.1

12.1 The Representation Formula (11.4)

Let $d\lambda_n(\eta)$ and $d\mu_n(v)$ be the probability measures introduced in (10.2) and (10.6), and set

$$\begin{aligned} u_n(x,t) &= \int_{\mathbb{R}} a^{-1}\left(\frac{x-\eta}{t}\right) d\lambda_n(\eta) = \int_{\mathbb{R}} v d\mu_n(v) \\ F_n(x,t) &= \int_{\mathbb{R}} F\left[a^{-1}\left(\frac{x-\eta}{t}\right)\right] d\lambda_n(v) = \int_{\mathbb{R}} F(v) d\mu_n(v) \\ H_n(x,t) &= \ln \int_{\mathbb{R}} e^{-n\Psi(x,t;\eta)} d\eta = \ln \int_{\mathbb{R}} e^{-n\Psi(v)} a'(v) dv \end{aligned} \tag{12.1}$$

where the integrals on the right are computed from those on the left by the change of variables (10.4)–(10.5). From the definitions (8.8) and (10.5) of $\Psi(x,t;\eta)$ and $\Psi(v)$, and recalling that $F' = a(\cdot)$

$$\begin{aligned} \Psi_x(x,t;\eta) &= a^{-1}\left(\frac{x-\eta}{t}\right) = v \\ \Psi_t(x,t;\eta) &= -F\left[a^{-1}\left(\frac{x-\eta}{t}\right)\right] = -F(v). \end{aligned} \tag{12.2}$$

Then compute

$$\begin{aligned} \frac{\partial}{\partial x} H_n(x,t) &= -n \int_{\mathbb{R}} \Psi_x(x,t;\eta) d\lambda_n(\eta) = -n \int_{\mathbb{R}} v d\mu_n(v) \\ &= -n \int_{\mathbb{R}} u_o(x - a(v)t) d\mu_n(v) \\ \frac{\partial}{\partial t} H_n(x,t) &= -n \int_{\mathbb{R}} \Psi_t(x,t;\eta) d\lambda_n(\eta) = n \int_{\mathbb{R}} F(v) d\mu_n(v). \end{aligned}$$

Therefore

$$u_n(x,t) = -\frac{1}{n}\frac{\partial}{\partial x}H_n(x,t), \qquad F_n(x,t) = \frac{1}{n}\frac{\partial}{\partial t}H_n(x,t).$$

These imply

$$\frac{\partial}{\partial t}u_n + \frac{\partial}{\partial x}F_n = 0 \quad \text{in } \mathbb{R}\times\mathbb{R}^+ \tag{12.3}$$

and

$$u_n(x,t) = \int_{\mathbb{R}} u_o(x - a(v)t)d\mu_n(v). \tag{12.4}$$

Since $u_o \in L^\infty(\mathbb{R})$ and $d\mu_n(v)$ is a probability measure

$$\|u_n(\cdot,t)\|_{\infty,\mathbb{R}} \le \|u_o\|_{\infty,\mathbb{R}} \quad \text{for all } t > 0.$$

Therefore by Lemma 10.1 and Lebesgue's dominated convergence theorem, $\{u_n(\cdot,t)\}$ and $\{F_n(\cdot,t)\}$ converge to $u(\cdot,t)$ and $F[u(\cdot,t)]$ respectively in $L^1_{\mathrm{loc}}(\mathbb{R})$, for all $t > 0$. Moreover $\{u_n\}$ and $\{F_n\}$ converge to u and $F(u)$ respectively, in $L^1_{\mathrm{loc}}(\mathbb{R}\times\mathbb{R}^+)$. This also proves (11.1).

If u_o is continuous, the representation formula (11.4) follows from (12.4) and Lemma 10.1, upon letting $n \to \infty$. ■

12.2 Initial Datum in the Sense of $L^1_{\mathrm{loc}}(\mathbb{R})$

Assume first $u_o \in C(\mathbb{R})$. Then by the representation formula (11.4)

$$\lim_{t\to 0}\|u(\cdot,t) - u_o\|_{1,K} = \lim_{t\to 0}\int_K |u_o(x - a[u(x,t)]t) - u_o(x)|dx = 0$$

since u is uniformly bounded in K for all $t > 0$. If u_o merely satisfies (8.2), construct a sequence of smooth functions $u_{o,m}$ satisfying (11.5), and in addition $\{u_{o,m}\} \to u_o$ in $L^1_{\mathrm{loc}}(\mathbb{R})$. Such a construction may be realized through a mollification kernel $J_{1/m}$, by setting $u_{o,m} = J_{1/m} * u_o$. By the stability Theorem 11.2

$$\|u(\cdot,t) - u_m(\cdot,t)\|_{1,K} \to \qquad \text{for all } t > 0.$$

Moreover, since $u_{o,m}$ are continuous

$$\|u_m(\cdot,t) - u_{o,m}\|_{1,K} \to \qquad \text{as } t \to 0.$$

This last limit is actually uniform in m. Indeed

$$\begin{aligned}
\int_K |u_m(x,t) - u_{o,m}(x)|dx &= \int_K |u_{o,m}(x - a[u_m(x,t)]t) - u_{o,m}(x)|dx \\
&= \int_K |J_{1/m} * [u_o(x - a[u_m(x,t)]t) - u_o(x)]|dx \\
&\le \int_K |u_o(x - a[u_m(x,t)]t) - u_o(x)|dx.
\end{aligned}$$

Since $a[u_m(x,t)]$ is uniformly bounded in K for all $t>0$, the right-hand side tends to zero as $t\to 0$, uniformly in m.

Fix a compact subset $K\subset\mathbb{R}$ and $\varepsilon>0$. Then choose $t>0$ such that

$$\|u_m(\cdot,t)-u_{o,m}\|_{1,K}\le\varepsilon.$$

Such a time t can be chosen independent of m, in view of the indicated uniform convergence. Then we write

$$\|u(t)-u_o\|_{1,K}\le\|u(t)-u_m(t)\|_{1,K}+\|u_m(t)-u_{o,m}\|_{1,K}+\|u_{o,m}-u_o\|_{1,K}.$$

Letting $m\to\infty$ gives $\|u(\cdot,t)-u_o\|_{1,K}\le\varepsilon$. ∎

12.3 Weak Forms of the PDE

Multiply (12.3) by $\varphi\in C^1[\mathbb{R}^+;C_o^\infty(\mathbb{R})]$ and integrate over $(\varepsilon,t)\times\mathbb{R}$ for some fixed $\varepsilon>0$. Integrating by parts and letting $n\to\infty$ gives

$$\int_\varepsilon^t\int_{\mathbb{R}}[u\varphi_t+F(u)\varphi_x]dx\,d\tau=\int_{\mathbb{R}}u(x,t)\varphi(x,t)dx-\int_{\mathbb{R}}u(x,\varepsilon)\varphi(x,\varepsilon)dx.$$

Now (11.2) follows, since $u(\cdot,t)\to u_o$ in $L^1_{\rm loc}(\mathbb{R})$ as $t\to 0$. ∎

The following proposition provides another weak form of the PDE.

Proposition 12.1 *For all $t>0$ and a.e. $x\in\mathbb{R}$*

$$\begin{aligned}\int_{-\infty}^x u(s,t)ds&=\Psi[x,t;u(x,t)]\\&=\int_{-\infty}^{x-a[u(x,t)]t}u_o(s)ds+t[ua(u)-F(u)](x,t).\end{aligned}\tag{12.5}$$

Proof Integrate (12.3) in $d\tau$ over (ε,t) and then in ds over (k,x), where k is a negative integer, and in the resulting expression let $n\to\infty$. Taking into account the expression (12.1) of F_n and the second of (12.2), compute

$$\begin{aligned}\int_k^x u(s,t)ds-\int_k^x u(s,\varepsilon)ds&=\lim_{n\to\infty}\int_\varepsilon^t\int_{\mathbb{R}}[\Psi_\tau(x,\tau;\eta)-\Psi_\tau(k,\tau;\eta)]\,d\lambda_n(\eta)d\tau\\&=\Psi[x,t;u(x,t)]-\Psi[k,\varepsilon;u(k,\varepsilon)]\end{aligned}$$

by virtue of Lemma 10.1. To prove the proposition first let $\varepsilon\to 0$ and then $k\to-\infty$. ∎

13 The Entropy Condition

A consequence of (9.1) is that the variational solution claimed by Theorem 11.1 satisfies the *entropy condition*

$$\limsup_{0<h\to 0} [u(x+h,t) - u(x,t)] \le 0 \tag{13.1}$$

for all fixed $t > 0$ and a.e. $x \in \mathbb{R}$. The notion of a weak solution introduced in Section 5.1 does not require that (13.1) be satisfied. However, as shown by the examples in Section 6.2, weak solutions need not be unique. We will prove that weak solutions of the initial value problem (6.4) that in addition satisfy the entropy condition (13.1), are unique. The method, due to Kruzhkov [85], is N-dimensional and uses a notion of entropy condition more general than (13.1).

13.1 Entropy Solutions

Consider the initial value problem

$$\begin{aligned} u_t + \operatorname{div} \mathbf{F}(u) = 0 \ \text{ in } \ S_T = \mathbb{R}^N \times (0,T] \\ u(\cdot,0) = u_o \in L^1_{\text{loc}}(\mathbb{R}^N) \end{aligned} \tag{13.2}$$

where $\mathbf{F} \in [C^1(\mathbb{R})]^N$. A weak solution of (13.2), in the sense of (5.4)–(5.5), is an *entropy* solution if

$$\iint_{S_T} \text{sign}(u-k)\{(u-k)\varphi_t + [\mathbf{F}(u) - \mathbf{F}(k)] \cdot D\varphi\} dxdt \ge 0 \tag{13.3}$$

for all non-negative $\varphi \in C^1_o(S_T)$ and all $k \in \mathbb{R}$, where D denotes the gradient with respect to the space variables only.

The first notion of entropy solution is due to Lax [97, 99], and it amounts to (13.1). A more general notion, that would cover some cases of non-convex $F(\cdot)$, and would ensure stability, still in one space dimension, was introduced by Oleinik [113, 114]. A formal derivation and a motivation of Kruzhkov notion of entropy solution (13.3) is in Section 13c of the Problems and Complements. When $N = 1$ the Kruzhkov and Lax notions are equivalent, as we show next.

13.2 Variational Solutions of (6.4) are Entropy Solutions

Proposition 13.1 *Let u be the weak variational solution claimed by Theorem 11.1. Then for every convex function $\Phi \in C^2(\mathbb{R})$ and all non-negative $\varphi \in C^\infty_o(\mathbb{R} \times \mathbb{R}^+)$*

$$\iint_{\mathbb{R}\times\mathbb{R}^+} \left[\Phi(u)\varphi_t + \left(\int_k^u F'(s)\Phi'(s)ds \right) \varphi_x \right] dxdt \ge 0 \quad \textit{for all } k \in \mathbb{R}.$$

Corollary 13.1 *The variational solutions claimed by Theorem 11.1 are entropy solutions.*

Proof Apply the proposition with $\Phi(s) = |s-k|$, modulo an approximation procedure. Then $\Phi'(s) = \text{sign}(s-k)$ for $s \ne k$. ■

The proof of Proposition 13.1 uses the notion of Steklov averages of a function $f \in L^1_{loc}(\mathbb{R} \times \mathbb{R}^+)$. These are defined as

$$f_h(x,t) = \fint_x^{x+h} f(s,t)ds, \qquad f_\ell(x,t) = \fint_t^{t+\ell} f(x,\tau)d\tau$$

$$f_{h\ell}(x,t) = \fint_t^{t+\ell} \fint_x^{x+h} f(s,\tau)dsd\tau$$

for all $h \in \mathbb{R}$ and all $\ell \in \mathbb{R}$ such that $t + \ell > 0$. One verifies that as $h, \ell \to 0$

$$\begin{array}{llll} f_h(\cdot,t) \to f(\cdot,t) & \text{in } L^1_{loc}(\mathbb{R}) & & \text{a.e. } t \in \mathbb{R}^+ \\ f_\ell(x,\cdot) \to f(x,\cdot) & \text{in } L^1_{loc}(\mathbb{R}^+) & & \text{a.e. } x \in \mathbb{R} \\ f_{h\ell} \quad \to f & \text{in } L^1_{loc}(\mathbb{R} \times \mathbb{R}^+). & & \end{array}$$

Lemma 13.1 *The variational solutions of Theorem 11.1 satisfy the weak formulation*

$$\begin{aligned} \frac{\partial}{\partial t} u_{h\ell} + \frac{\partial}{\partial x} F_{h\ell}(u) = 0 \quad &\text{in } \mathbb{R} \times \mathbb{R}^+ \\ u_{h\ell}(\cdot,0) = \fint_0^\ell u_h(\cdot,\tau)d\tau.& \end{aligned} \tag{13.4}$$

Moreover, $u_{h\ell}(\cdot,0) \to u_o$ *in* $L^1_{loc}(\mathbb{R})$ *as* $h,\ell \to 0$.

Proof Fix $(x,t) \in \mathbb{R} \times \mathbb{R}^+$ and $h \in \mathbb{R}$ and $\ell > 0$. Integrate (12.3) in $d\tau$ over $(t, t+\ell)$ and in ds over $(x, x+h)$, and divide by $h\ell$. Letting $n \to \infty$ proves the lemma. ■

Proof (Proposition 13.1) Let $\Phi \in C^2(\mathbb{R})$ be convex and let $\varphi \in C_o^\infty(\mathbb{R} \times \mathbb{R}^+)$ be non-negative. Multiplying the first of (13.4) by $\Phi'(u_{h\ell})\varphi$ and integrating over $\mathbb{R} \times \mathbb{R}^+$, gives

$$\begin{aligned} -\iint_{\mathbb{R}\times\mathbb{R}^+} &[\Phi(u_{h\ell})\varphi_t - F'(u_{h\ell})\Phi'(u_{h\ell})u_{h\ell x}\varphi]\, dx\, dt \\ &= \iint_{\mathbb{R}\times\mathbb{R}^+} [F_{h,\ell}(u) - F(u_{h\ell})]\Phi''(u_{h\ell})u_{h\ell x}\varphi dx\, dt \\ &\quad + \iint_{\mathbb{R}\times\mathbb{R}^+} [F_{h,\ell}(u) - F(u_{h\ell})]\Phi'(u_{h\ell})\varphi_x dx\, dt. \end{aligned}$$

The second term on the left-hand side is transformed by an integration by parts and equals

$$\iint_{\mathbb{R}\times\mathbb{R}^+} \left(\int_k^{u_{h\ell}} F'(s)\Phi'(s)ds \right) \varphi_x dx\, dt$$

where k is an arbitrary constant. Then let $\ell \to 0$ and $h \to 0$ in the indicated order to obtain

$$\iint_{\mathbb{R}\times\mathbb{R}^+}\left[\Phi(u)\varphi_t+\left(\int_k^u F'(s)\Phi'(s)ds\right)\varphi_x\right]dx\,dt$$
$$=-\lim_{h\to 0}\iint_{\mathbb{R}\times\mathbb{R}^+}[F_h(u)-F(u_h)]\Phi''(u_h)(u_h)_x\varphi dx\,dt.$$

It remains to show that the right-hand side is non-negative. Since $F(\cdot)$ is convex, by Jensen's inequality $F_h(u)\ge F(u_h)$. By (9.1), for a.e. $(x,t)\in\mathbb{R}\times\mathbb{R}^+$

$$(u_h)_x=\frac{\partial}{\partial x}\fint_x^{x+h}u(s,t)ds=\frac{u(x+h,t)-u(x,t)}{h}\le\frac{L}{t}.$$

$$-\lim_{h\to 0}\iint_{\mathbb{R}\times\mathbb{R}^+}[F_h(u)-F(u_h)]\Phi''(u_h)(u_h)_x\varphi dx\,dt$$
$$\ge\lim_{h\to 0}\iint_{\mathbb{R}\times\mathbb{R}^+}\Phi''(u_h)[F_h(u)-F(u_h)]\varphi\frac{L}{\tau}dx\,dt=0. \qquad \blacksquare$$

13.3 Remarks on the Shock and the Entropy Conditions

Let u be an entropy solution of (13.2), discontinuous across a smooth hypersurface Γ. The notion (13.3) contains information on the nature of the discontinuities of u across Γ. In particular, it does include the shock condition (4.4) and a weak form of the entropy condition (13.1).

If $P\in\Gamma$, the ball $B_\rho(P)$ centered at P with radius ρ is divided by Γ, at least for small ρ, into B_ρ^+ and B_ρ^- as in Figure 13.1. Let $\boldsymbol{\nu}=(\nu_t;\nu_{x_1},\dots,\nu_{x_N})=(\nu_t;\boldsymbol{\nu}_x)$ denote the unit normal oriented toward B_ρ^+.

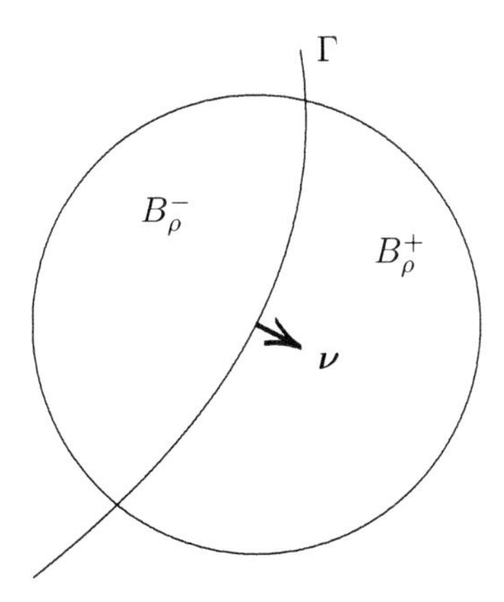

Fig. 13.1.

We assume that $u \in C^1(\bar{B}_\rho^\pm)$ and that it satisfies the equation in (13.2) in the classical sense in $B_\rho^\pm$. In (13.3) take a non-negative test function $\varphi \in C_o^\infty(B_\rho(P))$ and integrate by parts by means of Green's theorem. This gives, for all $k \in \mathbb{R}$

$$\begin{aligned}\int_\Gamma & \operatorname{sign}(u^+ - k)\{(u^+ - k)\nu_t + [\mathbf{F}(u^+) - \mathbf{F}(k)] \cdot \boldsymbol{\nu}_x\}\varphi d\sigma \\ & \le \int_\Gamma \operatorname{sign}(u^- - k)\{(u^- - k)\nu_t + [\mathbf{F}(u^-) - \mathbf{F}(k)] \cdot \boldsymbol{\nu}_x\}\varphi d\sigma\end{aligned}$$

where $d\sigma$ is the surface measure on Γ and $u^\pm$ are the limits of $u(x,t)$ as (x,t) tends to Γ from $B_\rho^\pm$. Since $\varphi \ge 0$ is arbitrary, this gives the pointwise inequality

$$\begin{aligned}\operatorname{sign}(u^+ - k)&\{(u^+ - k)\nu_t + [\mathbf{F}(u^+) - \mathbf{F}(k)] \cdot \boldsymbol{\nu}_x\} \\ &\le \operatorname{sign}(u^- - k)\{(u^- - k)\nu_t + [\mathbf{F}(u^-) - \mathbf{F}(k)] \cdot \boldsymbol{\nu}_x\}\end{aligned} \tag{13.5}$$

on Γ. If $k > \max\{u^+, u^-\}$, (13.5) implies

$$([u^+ - u^-], [\mathbf{F}(u^+) - \mathbf{F}(u^-)]) \cdot \boldsymbol{\nu} \ge 0$$

and if $k < \min\{u^+, u^-\}$

$$([u^+ - u^-], [\mathbf{F}(u^+) - \mathbf{F}(u^-)]) \cdot \boldsymbol{\nu} \le 0.$$

Therefore, the surface of discontinuity Γ must satisfy the shock condition (4.4). Next, in (13.5), take $k = \frac{1}{2}[u^+ + u^-]$, to obtain

$$\operatorname{sign}[u^+ - u^-]\big[\mathbf{F}(u^+) + \mathbf{F}(u^-) - 2\mathbf{F}(k)\big]\boldsymbol{\nu}_x \le 0. \tag{13.6}$$

This is an N-dimensional generalized version of the entropy condition (13.1).

Lemma 13.2 *If $N = 1$ and $F(\cdot)$ is convex, then (13.6) implies (13.1).*

Proof If $N = 1$, Γ is a curve in $\mathbb{R}^2$, and we may orient it, locally, so that $\boldsymbol{\nu} = (\nu_t, \nu_x)$, and $\nu_x \ge 0$. Since $F(\cdot)$ is convex, (8.3) implies that

$$F(u^\pm) - F(k) \ge F'(k)(u^\pm - k).$$

Adding these two inequalities gives

$$[F(u^+) + F(u^-) - 2F(k)]\nu_x \ge 0.$$

This in (13.6) implies $\operatorname{sign}[u^+ - u^-] \le 0$. ∎

14 The Kruzhkov Uniqueness Theorem

Theorem 14.1 *Let u and v be two entropy solutions of (13.2) satisfying in addition*

$$\left\| \frac{\mathbf{F}(u) - \mathbf{F}(v)}{u - v} \right\|_{\infty, S_T} \le M \quad \textit{for some } M > 0. \tag{14.1}$$

Then $u = v$.

Remark 14.1 The assumption (14.1) is satisfied if $\mathbf{F} \in C^1(\mathbb{R})$ and the solutions are bounded. In particular

Corollary 14.1 *There exists at most one bounded entropy solution to the initial value problem (6.4).*

14.1 Proof of the Uniqueness Theorem I

Lemma 14.1 *Let u and v be any two entropy solutions of (13.2). Then for every non-negative $\varphi \in C_o^\infty(S_T)$*

$$\iint_{S_T} \operatorname{sign}(u - v)\{(u - v)\varphi_t + [\mathbf{F}(u) - \mathbf{F}(v)] \cdot D\varphi\} dxdt \ge 0. \tag{14.2}$$

Proof For $\varepsilon > 0$, let J_ε be the Friedrichs mollifying kernels, and set

$$\delta_\varepsilon\left(\frac{x - y}{2}, \frac{t - \tau}{2}\right) = J_\varepsilon\left(\frac{t - \tau}{2}\right) J_\varepsilon\left(\frac{|x - y|}{2}\right).$$

Let $\varphi \in C_o^\infty(S_T)$ be non-negative and assume that its support is contained in the cylinder $B_R \times (s_1, s_2)$ for some $R > 0$ and $\varepsilon < s_1 < s_2 < T - \varepsilon$. Set

$$\lambda(x, t; y, \tau) = \varphi\left(\frac{x + y}{2}, \frac{t + \tau}{2}\right) \delta_\varepsilon\left(\frac{x - y}{2}, \frac{t - \tau}{2}\right). \tag{14.3}$$

The function λ is compactly supported in $S_T \times S_T$, with support contained in

$$\left[\frac{|x + y|}{2} < R\right] \cap \left[\frac{|x - y|}{2} < \varepsilon\right] \; ; \; \left[s_1 < \frac{|t + \tau|}{2} < s_2\right] \cap \left[\frac{|t - \tau|}{2} < \varepsilon\right].$$

The variables of integration in (13.3) are x and t. We take $k = v(y, \tau)$ for a.e. $(y, \tau) \in S_T$ and integrate in $dyd\tau$ over S_T. This gives

$$\begin{aligned} \iint_{S_T} \iint_{S_T} & \operatorname{sign}[u(x, t) - v(y, \tau)]\{[u(x, t) - v(y, \tau)]\lambda_t \\ & + [\mathbf{F}(u(x, t)) - \mathbf{F}(v(y, \tau))] \cdot \nabla_x \lambda\} dx\, dt\, dy\, d\tau \ge 0. \end{aligned}$$

Analogously, one may write (13.3) for v in the variables of integration y, τ, and take $k = u(x, t)$. Integrating in $dx\, dt$ over S_T gives an analogous inequality with λ_t and $D_x\lambda$ replaced by λ_τ and $D_y\lambda$. Adding these two inequalities gives

$$\begin{aligned}\iint_{S_T}\iint_{S_T} &\{|u(x,t)-v(y,\tau)|(\lambda_t+\lambda_\tau)+\operatorname{sign}[u(x,t)-v(y,\tau)] \\ &+[\mathbf{F}(u(x,t)-\mathbf{F}(v(y,\tau)]\cdot(D_x\lambda+D_y\lambda)\}dx\,dt\,dy\,d\tau\geq 0.\end{aligned}\tag{14.4}$$

To transform this integral, compute from (14.3)

$$\begin{aligned}\lambda_t+\lambda_\tau &= \varphi_t\Big(\frac{x+y}{2},\frac{t+\tau}{2}\Big)\delta_\varepsilon\Big(\frac{x-y}{2},\frac{t-\tau}{2}\Big) \\ D_x\lambda+D_y\lambda &= D\varphi\Big(\frac{x+y}{2},\frac{t+\tau}{2}\Big)\delta_\varepsilon\Big(\frac{x-y}{2},\frac{t-\tau}{2}\Big).\end{aligned}$$

Then, in the resulting integral, make the change of variables

$$\frac{x+y}{2}=\xi,\qquad \frac{t+\tau}{2}=s;\qquad\qquad \frac{x-y}{2}=\eta,\qquad \frac{t-\tau}{2}=\sigma.$$

The domain of integration is mapped into

$$\{[|\xi|<R]\times[s_1,s_2]\}\times\{\,[|\eta|<\varepsilon]\times[|\sigma|<\varepsilon]\}$$

and (14.4) is transformed into

$$\begin{aligned}&\iint_{S_T}\varphi_t(\xi,s)\Big\{\iint_{S_T}|u(\xi+\eta,s+\sigma)-v(\xi-\eta,s-\sigma)|\delta_\varepsilon(\eta,\sigma)d\eta\,d\sigma\Big\}d\xi\,ds \\ &\quad+\iint_{S_T}D\varphi(\xi,s)\cdot\Big\{\iint_{S_T}\operatorname{sign}[u(\xi+\eta,s+\sigma)-v(\xi-\eta,s-\sigma)] \\ &\quad\times[\mathbf{F}(u(\xi+\eta,s+\sigma)-\mathbf{F}(v(\xi-\eta,s-\sigma)]\delta_\varepsilon(\eta,\sigma)d\eta\,d\sigma\Big\}d\xi\,ds\geq 0.\end{aligned}$$

By the properties of mollifiers the integrals in $\{\cdots\}$ converge respectively to

$$|u(\xi,s)-v(\xi,s)|\quad\text{ and }\quad \operatorname{sign}[u(\xi,s)-v(\xi,s)][\mathbf{F}(u(\xi,s)-\mathbf{F}(v(\xi,s))]$$

for a.e. $(\xi,s)\in S_T$. Moreover, they are uniformly bounded in ε, for a.e. $(\xi,s)\in\operatorname{supp}(\varphi)$. Therefore (14.2) follows by letting $\varepsilon\to 0$ in the previous expression and passing to the limit under the integrals. ∎

14.2 Proof of the Uniqueness Theorem II

Fix $x_o\in\mathbb{R}^N$ and $R>0$ and construct the backward characteristic cone of "slope" M

$$[|x-x_o|<M(T-t)]\times[0<t<T].$$

The cross section of this cone with the hyperplane $t=$ const, for $0<t<T$, is the ball $|x-x_o|<M(T-t)$. The uniqueness theorem is a consequence of the following

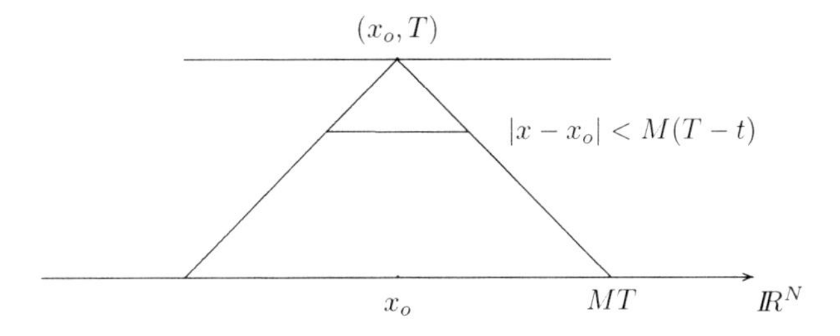

Fig. 14.1.

Proposition 14.1 *For all $x_o \in \mathbb{R}^N$ and for almost all $0 < \tau < t < T$*

$$\int_{|x-x_o|<M(T-t)} |u-v|(x,t)dx \le \int_{|x-x_o|<M(T-\tau)} |u-v|(x,\tau)dx. \tag{14.5}$$

Proof Assume $x_o = 0$ and in (14.2) take

$$\varphi(x,t) = \int_{t-t_1}^{t-\tau} J_\varepsilon(s)ds \int_{|x|-M(T-t)+\varepsilon}^{\infty} J_\varepsilon(s)ds \tag{14.6}$$

where $\varepsilon < \tau < t_1 < T - \varepsilon$ are arbitrary but fixed. Such a φ is admissible since it is non-negative, is in $C^\infty(S_T)$, and vanishes outside the truncated backward cone

$$\big[|x| < M(T-t)\big] \times \big[\tau - \varepsilon < t < t_1 + \varepsilon\big].$$

Compute

$$\begin{aligned} \varphi_t &= [J_\varepsilon(t-\tau) - J_\varepsilon(t-t_1)] \int_{|x|-M(T-t)+\varepsilon}^{\infty} J_\varepsilon(s)ds \\ &\quad - J_\varepsilon(|x| - M(T-t)+\varepsilon)M \int_{t-t_1}^{t-\tau} J_\varepsilon(s)ds \\ D\varphi &= -J_\varepsilon(|x| - M(T-t)+\varepsilon)\frac{x}{|x|} \int_{t-t_1}^{t-\tau} J_\varepsilon(s)ds. \end{aligned}$$

Put this in (14.2) and change the sign to obtain

$$\begin{aligned} &\iint_{S_T} [J_\varepsilon(t-t_1) - J_\varepsilon(t-\tau)]|u-v| \int_{|x|-M(T-t)+\varepsilon}^{\infty} J_\varepsilon(s)ds\,dx\,dt \\ &\le \iint_{S_T} \int_{t-t_1}^{t-\tau} J_\varepsilon(s)dsJ_\varepsilon(|x| - M(T-t)+\varepsilon)\{|\mathbf{F}(u) - \mathbf{F}(v)| - M|u-v|\}dx\,dt. \end{aligned}$$

By virtue of (14.1), the right-hand side is non-positive. Letting $\varepsilon \to 0$, by the properties of the mollifiers, the left-hand side converges to

$$\int_{|x|<M(T-t_1)} |u-v|(x,t_1)dx - \int_{|x|<M(T-\tau)} |u-v|(x,\tau)dx. \qquad \blacksquare$$

14.3 Stability in $L^1(\mathbb{R}^N)$

Let u and v be entropy solutions of (13.2) defined in the whole of S_∞. Fix $T > 0$ and rewrite (14.5) as

$$\int_{|y|<M(T-t)} |u-v|(x_o - y, t)dy \le \int_{|y|<M(T-\tau)} |u-v|(x_o - y, \tau)dy.$$

Integrating this in dx_o over $\mathbb{R}^N$ gives

$$\|u-v\|_{1,\mathbb{R}^N}(t) \le \left(\frac{T-\tau}{T-t}\right)^N \|u-v\|_{1,\mathbb{R}^N}(\tau).$$

Since the solutions u and v are global in time, let $T \to \infty$, and deduce that the function $t \to \|u-v\|_{1,\mathbb{R}^N}(t)$ is non-decreasing.

Theorem 14.2 *Let u and v be, global-in-time, entropy solutions of (13.2) originating from initial data u_o and v_o in $L^1(\mathbb{R}^N) \cap L^\infty(\mathbb{R}^N)$, and let (14.1) hold. Then*

$$\|u-v\|_{1,\mathbb{R}^N}(t) \le \|u_o - v_o\|_{1,\mathbb{R}^N} \qquad \textit{for a.e. } t > 0.$$

15 The Maximum Principle for Entropy Solutions

Proposition 15.1 *Let u and v be any two weak entropy solutions of (13.2). Then for all $x_o \in \mathbb{R}^N$*

$$\int_{|x-x_o|<M(T-t)} (u-v)_+(x,t)dx \le \int_{|x-x_o|<M(T-\tau)} (u-v)_+(x,\tau)dx$$

for a.e. $0 < \tau < t < T$.

Proof Since u and v are weak solutions of (13.2) in the sense of (5.4)–(5.5), starting from these, we may arrive at an analogue of (14.4) with equality and without the extra factor $\mathrm{sign}[u(x,t) - v(y,\tau)]$. Precisely, starting from (5.4) written for

$$u - v(y,\tau) \quad \text{and} \quad [\mathbf{F}(u(x,t)) - \mathbf{F}(v(y,\tau))]$$

choose φ and λ as in (14.3) and proceed as before to arrive at

$$\iint_{S_T}\iint_{S_T} \{[u(x,t) - v(y,\tau)](\lambda_t + \lambda_\tau) \\ + [\mathbf{F}(u(x,t)) - \mathbf{F}(v(y,\tau))]\cdot(D_x\lambda + D_y\lambda)\} dx\,dt\,dy\,d\tau = 0.$$

Add this to (14.4) and observe that

$$|u-v| + (u-v) = 2(u-v)_+$$

to obtain

$$\iint_{S_T}\iint_{S_T}\{2[u(x,t)-v(y,\tau)]_+(\lambda_t+\lambda_\tau)+\big(1+\text{sign}[u(x,t)-v(y,\tau)]\big)$$
$$\times[\mathbf{F}(u(x,t))-\mathbf{F}(v(y,\tau))]\cdot(D_x\lambda+D_y\lambda)\big\}dx\,dt\,dy\,d\tau\geq 0.$$

Proceeding as in the proof of Lemma 14.1, we arrive at the inequality

$$\iint_{S_T}\{2(u-v)_+\varphi_t+(1+\text{sign}\,[u-v])[\mathbf{F}(u)-\mathbf{F}(v)]\cdot D\varphi\}dx\,dt\geq 0$$

for all non-negative $\varphi\in C_o^\infty(S_T)$. Take φ as in (14.6) and let $\varepsilon\to 0$. ∎

Corollary 15.1 *Let u and v be two entropy solutions of (13.2), and let (14.1) hold. Then $u_o\geq v_o$ implies $u\geq v$ in S_T.*

Corollary 15.2 *Let u be an entropy solution of (13.2), and let (14.1) hold. Then*

$$\|u(\cdot,t)\|_{\infty,\mathbb{R}^N}\leq\|u_o\|_{\infty,\mathbb{R}^N}\qquad \textit{for a.e. } 0<t<T.$$

Corollary 15.3 *Assume that $u_o\in L^\infty(\mathbb{R}^N)$ and let $\mathbf{F}\in C^1(\mathbb{R})$. Then there exists at most one bounded, weak entropy solution to the initial value problem (13.2).*

Problems and Complements

3c Solving the Cauchy Problem

3.1. Solve $x\cdot\nabla u=\alpha$ with Cauchy data $h(\cdot)$ on the hyperplane $x_N=1$, that is

$$\begin{aligned}&\Gamma=\{\xi_i=s_i\},\quad i=1,\dots,N-1\\&\xi_N=1,\quad \zeta(s)=h(s)\in C^1(\mathbb{R}^{N-1}).\end{aligned}$$

Denote by $\bar{x}=(x_1,\dots,x_{N-1})$ points in $\mathbb{R}^{N-1}$ and by $(\bar{x},x_N)$, points in $\mathbb{R}^N$. An integral surface is given by

$$\bar{x}(s,t)=se^t,\quad x_N(s,t)=e^t,\quad z(s,t)=h(s)+\alpha t.$$

For $x_N>0$, we have $s=\bar{x}/x_N$, and the solution is given by

$$u(x)=h\left(\frac{\bar{x}}{x_N}\right)+\ln x_N^\alpha.$$

3.2. Let u solve the linear equation $a_i(x)u_{x_i} = \gamma u$ for some $\gamma \in \mathbb{R}$. Show that the general solution is given by $x \to u_o(x)u(x)$, where u_o is a solution of the associated homogeneous equation.

3.3. Show that the characteristic projections of

$$yu_x + xu_y = \gamma u, \quad \gamma > 0 \tag{3.1c}$$

are the curves

$$\begin{aligned} x(t) &= x_o \cosh t + y_o \sinh t \\ y(t) &= x_o \sinh t + y_o \cosh t \end{aligned} \qquad (x_o, y_o) \in \mathbb{R}^2$$

and observe in particular that the lines $x = \pm y$ are characteristic.

3.4. Show that the general solution of the homogeneous equation associated with (3.1c) is $f(x^2 - y^2)$, for any $f \in C^1(\mathbb{R})$.

3.5. Solve (3.1c) for Cauchy data $u(\cdot, 0) = h \in C^1(\mathbb{R})$. The integral surfaces are

$$x(s,t) = s\cosh t, \quad y(s,t) = s\sinh t, \quad z(s,t) = h(s)e^{\gamma t}.$$

The solution exists in the sector $|x| > |y|$, and it is given by

$$u(x,t) = h(\sqrt{x^2 - y^2}) \left(\frac{x+y}{x-y}\right)^{\gamma/2}.$$

This is discontinuous at $x = y$ and continuous but not of class C^1 at $x = -y$. Explain in terms of characteristics.

3.6. Consider (3.1c) with data on the characteristic $x = y$, that is

$$u(x,x) = h(x) \in C^1(\mathbb{R}).$$

In general, the problem is not solvable. Following the method of Section 3, we find the integral surfaces

$$\begin{aligned} x(s,t) &= s(\cosh t + \sinh t) \\ y(s,t) &= s(\cosh t + \sinh t) \\ z(s,t) &= h(s)e^{\gamma t}. \end{aligned}$$

From these compute

$$se^t = \frac{x+y}{2}, \qquad u(x,y) = \left(\frac{x+y}{2}\right)^{\gamma} \frac{h(s)}{s^\gamma}.$$

Therefore the problem is solvable only if $h(s) = Cs^\gamma$. It follows from **3.4** that $C = f(x^2 - y^2)$, for any $f \in C^1(\mathbb{R})$.

3.7. Show that the characteristic projections of $yu_x - xu_y = \gamma u$, are the curves

$$\begin{aligned} x(t) &= x_o \cos t + y_o \sin t \\ y(t) &= y_o \cos t - x_o \sin t \end{aligned} \qquad (x_o, y_o) \in \mathbb{R}^2.$$

Solve the Cauchy problem with data $u(x,0) = h(x)$. Show that if $\gamma = 0$ then the Cauchy problem is globally solvable only if $h(\cdot)$ is symmetric.

6c Explicit Solutions to the Burgers Equation

6.1. Verify that for $\lambda > 0$, the following are families of weak solutions to the Burgers equations in $\mathbb{R} \times \mathbb{R}^+$.

$$W^+(x,t) = \begin{cases} 0 & \text{for } x < 0 \\ \dfrac{x}{t} & \text{for } 0 \le x \le \sqrt{2\lambda t} \\ 0 & \text{for } x > \sqrt{2\lambda t}. \end{cases} \tag{6.1c}$$

$$W^-(x,t) = \begin{cases} 0 & \text{for } x < -\sqrt{2\lambda t} \\ \dfrac{x}{t} & \text{for } -\sqrt{2\lambda t} \le x \le 0 \\ 0 & \text{for } x > 0. \end{cases} \tag{6.2c}$$

$$U(x,t) = \begin{cases} 0 & \text{for } x < 0 \\ \dfrac{x}{t} & \text{for } 0 \le x \le \dfrac{\lambda}{2} \\ \dfrac{x-\lambda}{t} & \text{for } \dfrac{\lambda}{2} < x \le \lambda \\ 0 & \text{for } x > \lambda. \end{cases} \tag{6.3c}$$

6.2c Invariance of Burgers Equations by Some Transformation of Variables

Let φ be a solution of Burgers equation in $\mathbb{R}\times\mathbb{R}^+$. Verify that for all $a, b, c \in \mathbb{R}$ the following transformed functions are also solutions of Burgers equation:

$$u(x,t) = \varphi(x+a, t+b) \quad \text{for } t > -b \tag{i}$$

$$u(x,t) = a + \varphi(x - at, t) \tag{ii}$$

$$u(x,t) = a\varphi(bx, abt) \stackrel{\text{def}}{=} T_{a,b}\varphi \tag{iii}$$

$$u(x,t) = \frac{x}{t} + \frac{a}{t}\varphi\Big(\frac{bx}{t}, c - \frac{ab}{t}\Big) \quad \text{for } t > \frac{ab}{c}. \tag{iv}$$

6.2. Assume that a weak solution φ is known of the initial value problem

$$\varphi_t + \frac{1}{2}(\varphi^2)_x = 0 \text{ in } \mathbb{R} \times \mathbb{R}^+, \quad \varphi(\cdot,0) = \varphi_o.$$

where φ_o is subject to proper assumptions that would ensure existence of such a φ. Find a solutions of the initial value problems

$$u_t + \frac{1}{2}(u^2)_x = 0 \text{ in } \mathbb{R}\times\mathbb{R}^+; \qquad u_t + \frac{1}{2}(u^2)_x = 0 \text{ in } \mathbb{R}\times\mathbb{R}^+;$$

$$u(\cdot,0) = a + \varphi_o; \qquad u(\cdot,0) = \gamma x + \varphi_o.$$

A solution of the first is

$$u(x,t) = a + \varphi(x - at, t)$$

and a solution of the second is

$$u(x,t) = \frac{\gamma x}{1+\gamma t} + \frac{1}{1+\gamma t}\varphi\Big(\frac{x}{1+\gamma t}, \frac{t}{1+\gamma t}\Big).$$

Note that the initial values of these solutions do not satisfy the assumptions (8.2).

6.3. Prove that those solutions of Burgers equations for which $\varphi = T_{1,b}\varphi$ are of the form $f(x/t)$.

6.4. Prove that those solutions of Burgers equations for which $\varphi = T_{b,b}\varphi$ are of the form $f(\sqrt{x/t})/\sqrt{t}$.

6.3c The Generalized Riemann Problem

Consider the initial value problem

$$\begin{gathered} u_t + \frac{1}{2}(u^2)_x = 0 \;\text{ in }\; \mathbb{R}\times\mathbb{R}^+ \\ u(x,0) = \begin{cases} \alpha + px & \text{for } x < 0 \\ \beta + qx & \text{for } x > 0 \end{cases} \end{gathered} \tag{6.4c}$$

where α, β, p, q are given constants. Verify that if $\alpha \le \beta$, then the solution to (6.4c) is

$$u(x,t) = \begin{cases} \dfrac{\alpha + px}{1+pt} & \text{for } x \le \alpha t \\[2mm] \dfrac{x}{t} & \text{for } \alpha t \le x \le \beta t \\[2mm] \dfrac{\beta + qx}{1+qt} & \text{for } x \ge \beta t \end{cases} \qquad (\alpha \le \beta)$$

for all times $1 + (\alpha \wedge \beta)t > 0$. If $\alpha > \beta$, the characteristics from the left of $x = 0$ intersect the characteristics from the right. Let $x = x(t)$ be the line of discontinuity and verify that a weak solution is given by

$$u(x,t) = \begin{cases} \dfrac{\alpha + px}{1+pt} & \text{for } x < x(t) \\[2mm] \dfrac{\beta + qx}{1+qt} & \text{for } x > x(t) \end{cases} \qquad (\alpha > \beta)$$

where $x = x(t)$ satisfies the shock condition (4.7). Enforcing it gives

$$x'(t) = \frac{1}{2}\Big(\frac{\alpha + px(t)}{1+pt} + \frac{\beta + qx(t)}{1+qt}\Big).$$

Solve this ODE to find

$$x(t) = \frac{\alpha\sqrt{1+qt} + \beta\sqrt{1+qt}}{\sqrt{1+pt} + \sqrt{1+qt}}t.$$

13c The Entropy Condition

Solutions of (13.2) can be constructed by solving first the Cauchy problems

$$u_{\varepsilon,t} - \varepsilon\Delta u_\varepsilon + \operatorname{div}\mathbf{F}(u_\varepsilon) = 0 \text{ in } S_T$$
$$u_\varepsilon(\cdot,0) = u_o$$

and then letting $\varepsilon \to 0$. Roughly speaking, as $\varepsilon \to 0$, the term $\varepsilon\Delta u_\varepsilon$ "disappears" and the solution is found as the limit, in a suitable topology, of the net $\{u_\varepsilon\}$. The method can be made rigorous by estimating $\{u_\varepsilon\}$, uniformly in ε, in the class of functions of *bounded variation* ([157]).

In what follows we assume that a priori estimates have been derived that ensure that $\{u_\varepsilon\} \to u$ in $L^1_{\text{loc}}(S_T)$. Let $k \in \mathbb{R}$ and write the PDE as

$$\frac{\partial}{\partial t}(u_\varepsilon - k) - \varepsilon\Delta(u_\varepsilon - k) + \operatorname{div}[\mathbf{F}(u_\varepsilon) - \mathbf{F}(k)] = 0.$$

Let $h_\delta(\cdot)$ be the approximation to the Heaviside function introduced in (14.2) of Chapter 5. Multiply the PDE by $h_\delta(u_\varepsilon - k)\varphi$, where $\varphi \in C_o^\infty(S_T)$ is non-negative and integrate by parts over S_T to obtain

$$\begin{aligned}\iint_{S_T}\Big\{\frac{\partial}{\partial t}\Big(\int_0^{u_\varepsilon - k} h_\delta(s)ds\Big)\varphi dx\,dt + \varepsilon h'_\delta(u_\varepsilon - k)|Du_\varepsilon|^2\varphi \\ + \varepsilon h_\delta(u_\varepsilon - k)D(u_\varepsilon - k)\cdot D\varphi \\ + h_\delta(u_\varepsilon - k)[\mathbf{F}(u_\varepsilon) - \mathbf{F}(k)]\cdot D\varphi \\ + h'_\delta(u_\varepsilon - k)[\mathbf{F}(u_\varepsilon) - \mathbf{F}(k)]\cdot D(u_\varepsilon - k)\varphi\Big\}dx\,dt = 0.\end{aligned}$$

First let $\delta \to 0$ and then let $\varepsilon \to 0$. The various terms are transformed and estimated as follows.

$$\begin{aligned}\lim_{\varepsilon\to 0}\lim_{\delta\to 0}\iint_{S_T}\frac{\partial}{\partial t}\Big(\int_0^{u_\varepsilon - k} h_\delta(s)ds\Big)\varphi dx\,dt \\ = -\lim_{\varepsilon\to 0}\lim_{\delta\to 0}\iint_{S_T}\Big(\int_0^{u_\varepsilon - k} h_\delta(s)ds\Big)\varphi_t dx\,dt \\ = -\iint_{S_T}|u - k|\varphi_t dx\,dt.\end{aligned}$$

The second term on the left-hand side is non-negative and is discarded. Next

$$\begin{aligned}\lim_{\varepsilon\to 0}\lim_{\delta\to 0}\iint_{S_T}\varepsilon h_\delta(u_\varepsilon - k)D(u_\varepsilon - k)\cdot D\varphi dx\,dt \\ = \lim_{\varepsilon\to 0}\lim_{\delta\to 0}\iint_{S_T}\varepsilon D\Big(\int_0^{u_\varepsilon - k} h_\delta(s)ds\Big)\cdot D\varphi dx\,dt \\ = -\lim_{\varepsilon\to 0}\lim_{\delta\to 0}\iint_{S_T}\varepsilon\Big(\int_0^{u_\varepsilon - k} h_\delta(s)ds\Big)\Delta\varphi dx\,dt = 0.\end{aligned}$$

$$\lim_{\varepsilon\to 0}\lim_{\delta\to 0}\iint_{S_T} h_\delta(u_\varepsilon - k)[\mathbf{F}(u_\varepsilon) - \mathbf{F}(k)]\cdot D\varphi dx\,dt$$
$$= \iint_{S_T} \operatorname{sign}(u-k)[\mathbf{F}(u) - \mathbf{F}(k)]\cdot D\varphi dx\,dt.$$

The last term is transformed and estimated as

$$\iint_{S_T} \operatorname{div}\Big(\int_0^{u_\varepsilon} h'_\delta(s-k)[\mathbf{F}(s) - \mathbf{F}(k)]ds\Big)\varphi dx\,dt$$
$$= -\iint_{S_T}\Big(\int_0^{u_\varepsilon} h'_\delta(s-k)[\mathbf{F}(s) - \mathbf{F}(k)]ds\Big)\cdot D\varphi dx\,dt.$$

For $\varepsilon > 0$ fixed

$$\lim_{\delta\to 0} h'_\delta(s-k)[\mathbf{F}(s) - \mathbf{F}(k)] = 0 \qquad \text{a.e. } s\in(0,u_\varepsilon).$$

Moreover, by (14.1)

$$0 \le h'_\delta(s-k)[\mathbf{F}(s) - \mathbf{F}(k)] \le M.$$

Therefore by dominated convergence

$$\lim_{\delta\to 0}\iint_{S_T} h'_\delta(u_\varepsilon - k)[\mathbf{F}(u_\varepsilon) - \mathbf{F}(k)]\cdot D(u_\varepsilon - k)\varphi dx\,dt = 0.$$

Combining these remarks yields (13.3).

14c The Kruzhkov Uniqueness Theorem

The theorem of Kruzhkov holds for the following general initial value problem

$$\begin{aligned} u_t - \operatorname{div}\mathbf{F}(x,t,u) &= g(x,t,u) \;\text{ in }\; S_T \\ u(\cdot,0) &= u_o \in L^1_{\text{loc}}(\mathbb{R}^N). \end{aligned} \tag{14.1c}$$

A function $u\in L^\infty_{\text{loc}}(S_T)$ is an entropy solution of (14.1c) if for all $k\in\mathbb{R}$

$$\begin{aligned} \iint_{S_T} \operatorname{sign}(u-k)\big\{(u-k)\varphi_t + [\mathbf{F}(x,t,u) - \mathbf{F}(x,t,k)]\cdot D\varphi \\ + [F_{i,x_i}(x,t,u) + g(x,t,u)]\varphi\big\}dx\,dt \ge 0 \end{aligned} \tag{14.2c}$$

provided the various integrals are well defined. Assume

$$g, F_i \in C^1(S_T\times\mathbb{R}) \quad i = 1,\dots,N. \tag{14.3c}$$

Moreover

$$
\begin{aligned}
\left\| \frac{\mathbf{F}(x,t,u) - \mathbf{F}(x,t,v)}{u-v} \right\|_{\infty, S_T \times \mathbb{R}} &\le M_o \\
\sum_{i=1}^{N} \left\| \frac{F_{i,x_i}(x,t,u) - F_{i,x_i}(x,t,v)}{u-v} \right\|_{\infty, S_T \times \mathbb{R}} &\le M_1 \\
\left\| \frac{\mathbf{F}(x,t,u) - \mathbf{F}(x,t,v)}{u-v} \right\|_{\infty, S_T \times \mathbb{R}} &\le M_2 \\
\left\| \frac{g(x,t,u) - g(x,t,v)}{u-v} \right\|_{\infty, S_T \times \mathbb{R}} &\le M_3
\end{aligned}
\tag{14.4c}
$$

for given positive constants M_i, $i = 0, 1, 2, 3$. The initial datum is taken in the sense of $L^1_{\rm loc}(\mathbb{R}^N)$. Set $M = \max\{M_o, M_1, M_2, M_3\}$.

Theorem 14.1c *Let u and v be two entropy solutions of (14.1c) and let (14.3c)–(14.4c) hold. There exists a constant γ dependent only on N and the numbers M_i, $i = 0, 1, 2, 3$, such that for all $T > 0$ and all $x_o \in \mathbb{R}^N$*

$$
\int_{|x-x_o|<M(T-t)} |u-v|(x,t)dx \le e^{\gamma t} \int_{|x-x_o|<MT} |u_o - v_o| dx
$$

for a.e. $0 < t < T$.

8

Non-Linear Equations of First-Order

1 Integral Surfaces and Monge's Cones

A first-order non-linear PDE is an expression of the form

$$F(x, u, \nabla u) = 0 \tag{1.1}$$

where x ranges over a given region $E \subset \mathbb{R}^N$, the function u is in $C^1(E)$ and F is a given smooth real-valued function of its arguments. If u is a solution of (1.1), then its graph $\Sigma(u)$ is an *integral surface* for (1.1). Conversely, a surface Σ is an integral surface for (1.1) if it is the graph of a smooth function u solution of (1.1). For a fixed $(x, z) \in E \times \mathbb{R}$, consider the associated equation $F(x, z, p) = 0$ and introduce the set

$$\mathbf{P}(x, z) = \{\text{the set of all } p \in \mathbb{R}^N \text{ satisfying } \ F(x, z, p) = 0\}.$$

If $\Sigma(u)$ is an integral surface for (1.1), then for every $(x, z) \in \Sigma(u)$

$$z = u(x) \quad \text{and} \quad p = \nabla u(x). \tag{1.2}$$

Therefore solving (1.1) amounts to finding a function $u \in C^1(E)$ such that for all $x \in E$, among the pairs (x, z) there is one for which (1.2) holds. Let Σ be an integral surface for (1.1). For $(x_o, z_o) \in \Sigma$ consider the family of planes

$$z - z_o = p \cdot (x - x_o), \qquad p \in \mathbf{P}(x, z). \tag{1.3}$$

Since Σ is an integral surface, among these there must be one tangent Σ at (x_o, z_o). The envelope of such a family of planes is a cone $C(x_o, z_o)$, called Monge's cone with vertex at (x_o, z_o). Thus the integral surface Σ is tangent, at each of its points, to the Monge's cone with vertex at that point.[1]

[1] Gaspard Monge, Beaune, France 1746–1818 Paris, combined equally well his scientific vocation with his political aspirations. He took part in the French Revolution and became minister of the navy in the Robespierre government (1792). Mathematician and physicist of diverse interests, he contributed with Lavoisier to the chemical synthesis of water (1785), and with Bertholet and Vandermonde in identifying various metallurgical states of iron (1794). The indicated construction is in *Feuilles d'Analyse appliquée à la Géométrie*, lectures delivered at the École Polytechnique in 1801, and published by J. Liouville in 1850.

E. DiBenedetto, *Partial Differential Equations: Second Edition*,
Cornerstones, DOI 10.1007/978-0-8176-4552-6_9,

1.1 Constructing Monge's Cones

The envelope of the family of planes in (1.3) is that surface $\mathbb{S}$ tangent, at each of its points, to one of the planes of the family (1.3). Thus for each $(x, z) \in \mathbb{S}$, there exists $p = p(x)$ such that the corresponding plane in (1.3), for such a choice of p, has the same normal as $\mathbb{S}$. These remarks imply that the equation of $\mathbb{S}$ is

$$z - z_o = p(x) \cdot (x - x_o). \tag{1.4}$$

The tangency requirement can be written as

$$\underbrace{p_j(x) + p_{i,x_j}(x_i - x_{o,i})}_{\substack{j\text{th component of the}\\ \text{normal to } \mathbb{S} \text{ at } x}} \quad = \quad \underbrace{\qquad p_j(x) \qquad}_{\substack{j\text{th component of the normal}\\ \text{to the tangent plane at } x}} .$$

This gives the N equations

$$p_{x_j} \cdot (x - x_o) = 0, \qquad i = 1, \ldots, N.$$

Since $p(x) \in \mathbf{P}_o$, the vector-valued function $x \to p(x)$ must also satisfy

$$D_p F\big(x_o, z_o, p(x)\big) \cdot p_{x_j} = 0, \qquad j = 1, \ldots, N$$

where $D_p F = (F_{p_1}, \ldots, F_{p_N})$. It follows that for each x fixed in a neighborhood of x_o, the vectors $D_p F$ and $x - x_o$ are parallel, and there exists $\lambda(x)$ such that

$$\begin{aligned} D_p F\big(x_o, z_o, p(x)\big) &= \lambda(x)(x - x_o) \\ F\big(x_o, z_o, p(x)\big) &= 0. \end{aligned} \tag{1.5}$$

This is a non-linear system of $N + 1$ equations in the $N + 1$ unknowns $p_1(x), \ldots, p_N(x), \lambda(x)$. Solving it and putting the functions $x \to p(x)$ so obtained in (1.4) gives the equation of the envelope.

1.2 The Symmetric Equation of Monge's Cones

Eliminating λ from (1.4) and the first of (1.5) gives the *symmetric* equation of the cone

$$\frac{z - z_o}{p \cdot D_p F} = \frac{x_i - x_{o,i}}{F_{p_i}}, \qquad i = 1, \ldots, N. \tag{1.6}$$

This implies the Cartesian form of the Monge's cone $C(x_o, z_o)$

$$|z - z_o|^2 = \left(\frac{p \cdot D_p F}{|D_p F|} \right)^2 |x - x_o|^2. \tag{1.7}$$

Remark 1.1 If $F(x, z, p)$ is such that the "coefficient" of $|x - x_o|^2$ is constant, then the cone in (1.7) is circular and its axis is normal to the hyperplane $z = 0$. This is occurs for the first-order non-linear PDE $|\nabla u| = \text{const}$, which arises in geometric optics.

We stress however that the indicated "coefficient" depends on x via the functions $x \to p_j(x)$, and therefore $C(x_o, z_o)$ is not, in general, a circular cone, nor is its axis normal to the hyperplane $z = 0$.

2 Characteristic Curves and Characteristic Strips

Let ℓ denote the line of intersection between the cone in (1.6) and the hyperplane tangent to the integral surface Σ at (x_o, z_o). For $(x,z) \in \ell$, the vector $p(x)$ remains constant. Therefore, for infinitesimal increments dz and dx_i, along ℓ

$$\frac{dz}{p \cdot D_p F} = \frac{dx_1}{F_{p_1}} = \cdots = \frac{dx_N}{F_{p_N}}$$

where p and F_{p_i} are computed at $p(x)$, constant along ℓ. We conclude that ℓ has directions

$$(D_p F(x_o, z_o, p(x)), p \cdot D_p F(x_o, z_o, p(x)))$$

where $p(x)$ is computed on ℓ. Following these directions, starting from (x_o, z_o), trace a curve on the surface Σ. Such a curve, described in terms of a parameter $t \in (-\delta, \delta)$, for some $\delta > 0$, takes the form

$$\begin{array}{ll} \dot{x}(t) = D_p F\big(x(t), z(t), p(t)\big) & x(0) = x_o \\ \dot{z}(t) = p \cdot D_p F\big(x(t), z(t), p(t)\big) & z(0) = z_o. \end{array} \tag{2.1}$$

Here $p(t)$ is the solution of the system (1.5) with x_o and z_o replaced by $x(t)$ and $z(t)$, and computed at points x on the tangency line of the integral surface Σ with the Monge's cone with vertex at $\big(x(t), z(t)\big)$. The system (2.1) is not well defined, because the functions $t \to p_i(t)$ are in general not known. For quasi-linear equations, $F(x, z, p) = a_i(x, z)p_i$. In such a case $F_{p_i} = a_i(x, z)$ are independent of p and (2.1) are the characteristics originating at P_o (Section 1 of Chapter 7). Because of this analogy, we call the curves (2.1), *characteristics*. To render such a system well defined, observe that if Σ is an integral surface, then $p_i = u_{x_i}(x)$. From this and the first of (2.1)

$$\dot{p}_i = u_{x_i x_j} \dot{x}_j = u_{x_i x_j} F_{p_j}.$$

Also, from the PDE (1.1), by differentiation

$$F_{x_i} + F_u u_{x_i} + F_{p_j} u_{x_i x_j} = 0.$$

Therefore

$$\dot{p}_i = -F_{x_i} - F_u u_{x_i}, \qquad i = 1, \ldots, N.$$

Thus, the characteristics for the non-linear equation (1.1) are the curves

$$(-\delta, \delta) \ni t \to \Gamma(t) = \begin{cases} \dot{x}(t) = D_p F\big(x(t), z(t), p(t)\big) \\ \dot{z}(t) = p \cdot D_p F\big(x(t), z(t), p(t)\big) \\ \dot{p}(t) = -D_x F\big(x(t), z(t), p(t)\big) \\ \qquad\qquad -F_z\big(x(t), z(t), p(t)\big)p(t) \end{cases} \tag{2.2}$$

where $D_x F = (F_{x_1}, \dots, F_{x_N})$. For every choice of "initial" data

$$\big(x(0), z(0), p(0)\big) = (x_o, z_o, p_o) \in E \times \mathbb{R} \times \mathbb{R}^N$$

the system (2.2) has a unique solution, local it t, with the interval of existence depending, in general, on the initial datum. To simplify the presentation we assume that the interval of unique solvability is $(-\delta, \delta)$, for every choice of data (x_o, z_o, p_o).

2.1 Characteristic Strips

A solution of (2.2) can be thought of as a curve $t \to \big(x(t), z(t)\big) \in \mathbb{R}^{N+1}$ whose points are associated to an infinitesimal portion of the hyperplane trough them and normal $p(t)$. Putting together these portions along $t \to \big(x(t), z(t)\big)$, the function $t \to \Gamma(t)$ can be regarded as a strip of infinitesimal width, called a *characteristic strip*. These remarks suggest that integral surfaces are union of characteristic strips. Let Σ be a hypersurface in $\mathbb{R}^{N+1}$ given as the graph of $z = u(x) \in C^1(E)$, and for $(x_o, z_o) \in \Sigma$, let $t \to \Gamma_{(x_o,z_o)}(t)$ be the characteristic strip originating at (x_o, z_o), that is, the unique solution of (2.2) with data

$$x(0) = x_o, \quad z(0) = z_o = u(x_o), \quad p(0) = \nabla u(x_o). \tag{2.3}$$

The surface Σ is a union of characteristic strips if for every $(x_o, z_o) \in \Sigma$, the strip $t \to \Gamma_{(x_o,z_o)}(t)$ is contained in Σ, in the sense that

$$z(t) = u(x(t)) \quad \text{and} \quad p(t) = \nabla u(x(t)) \quad \text{for all } t \in (-\delta, \delta). \tag{2.4}$$

Proposition 2.1 *An integral surface for (1.1) is union of characteristic strips.*

Proof Let Σ be the graph of a solution $u \in C^1(E)$ of (1.1). Having fixed $x_o \in E$, let $t \to x(t)$ be the unique solution of

$$\dot{x}(t) = D_p F\big(x(t), u(x(t)), \nabla u(x(t))\big), \qquad x(0) = x_o.$$

One verifies that the $2N + 1$ functions

$$(-\delta, \delta) \ni t \to \quad x(t), \quad z(t) = u(x(t)), \quad p(x(t)) = \nabla u(x(t))$$

solve (2.2), with initial data (2.3). These are then characteristic strips. ∎

Remark 2.1 Unlike the case of quasi-linear equations, the converse does not hold, as (2.4) are not sufficient for one to conclude that Σ is an integral surface. Indeed, even though F is constant along $t \to \big(x(t), z(t)\big)$, the PDE (1.1) need not hold identically.

3 The Cauchy Problem

Let $s = (s_1, \dots, s_{N-1})$ be an $(N-1)$-dimensional parameter ranging over the cube $Q_\delta = (-\delta, \delta)^{N-1}$. The Cauchy problem associated with (1.1) consists in assigning an $(N-1)$-dimensional hypersurface $\Gamma \subset \mathbb{R}^{N+1}$ of parametric equations

$$Q_\delta \ni s \to \Gamma(s) = \begin{cases} x = \xi(s) = \big(\xi_1(s), \dots, \xi_N(s)\big) \\ z = \zeta(s), \quad \big(\xi(s), \zeta(s)\big) \in E \times \mathbb{R} \end{cases} \tag{3.1}$$

and seeking a function $u \in C^1(E)$ such that $\zeta(s) = u(\xi(s))$ for $s \in Q_\delta$ and such that the graph $z = u(x)$ is an integral surface of (1.1).

An integral surface for the Cauchy problem, must be a union of characteristic strips, and it must contain Γ. Therefore, one might attempt to construct it by drawing, from each point $(\xi(s), \zeta(s)) \in \Gamma$, a characteristic strip, a solution of (2.2), starting from the initial data

$$x(0) = \xi(s), \quad z(0) = \zeta(s), \quad p(0, s) = p(s) \in \mathbb{R}^N.$$

However, a surface that is a union of characteristic strips need not be an integral surface. Moreover, starting from a point on Γ, one may construct ∞^N characteristic strips, each corresponding to a choice of the initial vector $p(0, s) = p(s)$. A geometric construction of a solution to the Cauchy problem for (1.1), hinges on a criterion that would identify, for each $(\xi(s), \zeta(s)) \in \Gamma$, those initial data $p(0, s)$ for which the union of the corresponding characteristic strips, is indeed an integral surface.

3.1 Identifying the Initial Data $p(0, s)$

Set $(\xi(0), \zeta(0)) = (\xi_o, \zeta_o) \in \Gamma$, and assume that there exists a vector $\mathbf{p}_o$ such that

$$F(\xi_o, \zeta_o, p_o) = 0, \qquad D\zeta(0) = p_o \cdot \nabla\xi(0) \tag{3.2}$$

and in addition[2]

$$\det \begin{pmatrix} \nabla\xi(0) \\ D_p F(\xi_o, \zeta_o, p_o) \end{pmatrix} \neq 0.$$

Consider now the N-valued function

$$Q_\delta \times \mathbb{R}^N \ni (s, p) \to \Psi(s, p) = \begin{pmatrix} \nabla\zeta(s) - p \cdot \nabla\xi(s) \\ F\big(\xi(s), \zeta(s), p\big) \end{pmatrix}.$$

By (3.2) such a function vanishes for $(s, p) = (0, p_o)$. More generally, for $s \in Q_\delta$, we seek those vectors $p(s)$ for which Ψ vanishes, that is $\Psi(s, p(s)) = 0$.

[2] See Section 2.2 of Chapter 7 for symbolism and motivation.

By the implicit function theorem, this defines, locally, a smooth N-valued function $Q_\delta \ni s \to p(s)$ such that

$$\left.\begin{aligned} \nabla\zeta(s) &= p(\mathrm{s}) \cdot \nabla\xi(s) \\ F\big(\xi(s), \zeta(s), p(s)\big) &= 0 \end{aligned}\right. \qquad \text{for all } \; s \in Q_\delta. \tag{3.3}$$

Such a representation holds locally in a neighborhood of $s = 0$, which might be taken as Q_δ by possibly reducing δ. The vector $p(s)$, so identified, is the set of initial data $p(0, s) = p(s)$ to be taken in the construction of the characteristic strips.

3.2 Constructing the Characteristic Strips

The characteristic strips may now be constructed as the solutions of the system of ODEs

$$\begin{aligned} \frac{d}{dt}x(t,s) &= D_pF\big(x(t,s), z(t,s), p(t,s)\big) \\ \frac{d}{dt}z(t,s) &= p(t,s) \cdot D_pF\big(x(t,s), z(t,s), p(t,s)\big) \\ \frac{d}{dt}p(t,s) &= -D_xF\big(x(t,s), z(t,s), p(t,s)\big) \\ &\qquad -F_z\big(x(t,s), z(t,s), p(t,s)\big)p(t,s) \end{aligned} \tag{3.4}$$

with initial data given at each $s \in Q_\delta$

$$x(0,s) = \xi(s), \quad z(0,s) = \zeta(s), \quad p(0,s) = p(s). \tag{3.5}$$

The solution of (3.4)–(3.5) is local in t, that is, it exists in a time interval that depends on the initial data, or equivalently on the parameter $s \in Q_\delta$. By further reducing δ if needed, we may assume that (3.4)–(3.5) is uniquely solvable for $(t, s) \in (-\delta, \delta) \times Q_\delta$. Having solved such a system, consider the map

$$(-\delta, \delta) \times Q_\delta \ni (t, s) \to \big(x(t,s), z(t,s)\big).$$

This represents a surface $\Sigma \subset \mathbb{R}^{N+1}$, which by construction contains a local portion of Γ about (ξ_o, ζ_o).

Proposition 3.1 *The surface Σ is an integral surface for the Cauchy problem (1.1), (3.1).*

4 Solving the Cauchy Problem

To prove the proposition, we construct a function $x \to u(x)$ whose graph is Σ and that solves (1.1) in a neighborhood of (ξ_o, ζ_o). Observe first that by continuity, (3.2) continues to hold in a neighborhood of $s = 0$, i.e.,

$$\det \begin{pmatrix} \nabla\xi(s) \\ D_p F\big(\xi(s), \zeta(s), p(s)\big) \end{pmatrix} \neq 0 \qquad \text{for } s \in Q_\delta$$

where δ is further reduced if needed. Next consider the map

$$M : (-\delta, \delta) \times Q_\delta \ni (t, s) \to x(t, s).$$

From (3.4)–(3.5) and the previous remarks

$$\det \begin{pmatrix} D_s x(0, s) \\ x_t(0, s) \end{pmatrix} = \det \begin{pmatrix} \nabla\xi(s) \\ D_p F\big(\xi(s), \zeta(s), p(s)\big) \end{pmatrix} \neq 0$$

for all $s \in Q_\delta$. By continuity this continues to hold for $t \in (-\delta, \delta)$, where δ is further reduced if necessary. Therefore

$$\det \begin{pmatrix} x_t(t, s) \\ D_s x(t, s) \end{pmatrix} \neq 0 \quad \text{for all } (t, s) \in (-\delta, \delta) \times Q_\delta. \tag{4.1}$$

Therefore M is locally invertible in a neighborhood of ξ_o. In particular, there exist $\varepsilon = \varepsilon(\delta)$, a cube $Q_\varepsilon(\xi_o)$, and smooth functions T and S, defined in $Q_\varepsilon(\xi_o)$, such that $t = T(x)$ and $s = S(x)$ for $x \in Q_\varepsilon(\xi_o)$.

The function $x \to u(x)$ is constructed by setting

$$u(x) = z\big(T(x), S(x)\big) \quad \text{for } x \in Q_\varepsilon(\xi_o).$$

By construction, $u\big(\xi(s)\big) = \zeta(s)$ and $t \to F\big(x(t, s), z(t, s), p(t, s)\big)$ is constant for all $s \in Q_\delta$. Moreover, by (3.3), F is also constant along Γ. Therefore

$$F\big(x(t, s), z(t, s), p(t, s)\big) = 0 \quad \text{for all } (t, s) \in (-\delta, \delta) \times Q_\delta. \tag{4.2}$$

It remains to prove that

$$p\big(T(x), S(x)\big) = \nabla u(x) \quad \text{for all } x \in Q_\varepsilon(\xi_o). \tag{4.3}$$

From the definition of $u(\cdot)$

$$z_t(t, s) = \nabla u \cdot x_t(t, s), \qquad D_s z(t, s) = \nabla u \cdot D_s x(t, s). \tag{4.4}$$

These and the equations of the characteristic strips, yield

$$[\nabla u - p(t, s)] \cdot x_t(t, s) = 0 \quad \text{for all } (t, s) \in (-\delta, \delta) \times Q_\delta. \tag{4.5}$$

4.1 Verifying (4.3)

Lemma 4.1 *The relation (4.3) would follow from*

$$D_s z(t, s) = p(t, s) \cdot D_s x(t, s) \quad \textit{for all } (t, s) \in (-\delta, \delta) \times Q_\delta. \tag{4.6}$$

Proof Assuming (4.6) holds true, rewrite it as

$$\nabla u \cdot D_s x(t,s) = p(t,s) D_s x(t,s)$$

which follows by making use of the second of (4.4). Combining this with (4.5) gives the following linear homogeneous algebraic system in the unknowns $\nabla u - p(t,s)$

$$\begin{aligned}\big(\nabla u - p(t,s)\big) \cdot x_t(t,s) &= 0\\ \big(\nabla u - p(t,s)\big) \cdot D_s x(t,s) &= 0.\end{aligned}$$

By (4.1), this admits only the trivial solution for all $(t,s) \in (-\delta,\delta) \times Q_\delta$. ■

To establish (4.6), set

$$M(t,s) = D_s z(t,s) - p(t,s) \cdot D_s x(t,s)$$

and verify that by the first of (3.3), $M(0,s) = 0$ for all $s \in Q_\delta$. From (4.2)

$$D_p F \cdot D_s p + D_x F \cdot D_s x = -F_z D_s z.$$

Using this identity in s and the equations (3.4) of the characteristic strips, compute

$$\begin{aligned}M_t &= D_s z_t - p_t \cdot D_s x - p \cdot D_s x_t\\ &= D_s p D_p F + p \cdot D_p D_s F + D_x F \cdot D_s x + F_z p \cdot D_s x - p \cdot D_p D_s F\\ &= F_z p \cdot D_s x - F_z D_s z\\ &= -F_z (D_s z - p \cdot D_s x) = -F_z M.\end{aligned}$$

This has the explicit integral

$$M(t,s) = M(0,s) \exp\left(-\int_0^t F_z d\tau\right)$$

and gives $(t,s) \to M(t,s) = 0$, since $M(0,s) = 0$.

4.2 A Quasi-Linear Example in $\mathbb{R}^2$

Denote by (x,y) the coordinates in $\mathbb{R}^2$, and given two positive numbers A and ρ, consider the Cauchy problem

$$x u u_x - A u_y = 0, \qquad u(\rho, y) = y.$$

The surface Γ in (3.1) is the line $z = y$ in the plane $x = \rho$, which can be written in the parametric form

$$\xi_1(s) = \rho, \quad \xi_2(s) = s, \quad \zeta(s) = s, \quad s \in \mathbb{R}.$$

We solve the Cauchy problem in a neighborhood of $(\rho, 0, 0)$. The vector p_o satisfying (3.2) is $p_o = (0, 1)$. The system (3.4) takes the form

$$\begin{aligned} x_t(t,s) &= x(t,s)z(t,s) \\ y_t(t,s) &= -A \\ z_t(t,s) &= x(t,s)z(t,s)p_1(t,s) - Ap_2(t,s) \\ p_{1,t}(t,s) &= -z(t,s)p_1(t,s) - x(t,s)p_1^2(t,s) \\ p_{2,t}(t,s) &= x(t,s)p_1(t,s)p_2(t,s) \end{aligned}$$

with initial conditions

$$x(0,s) = \rho, \quad y(0,s) = s, \quad z(0,s) = s, \quad p_1(0,s) = 0, \quad p_2(0,s) = 1.$$

The solution is

$$z(t,s) = s, \quad y(t,s) = -At + s, \quad \ln \frac{x}{\rho} = st.$$

Eliminate the parameters s and t to obtain the solution in implicit form

$$u^2(x,t) - yu(x,y) = A \ln \frac{x}{\rho}.$$

The solution is analytic in the region $y^2 + 4A \ln(x/\rho) > 0$.

5 The Cauchy Problem for the Equation of Geometrical Optics

Let Φ_o be a surface in $\mathbb{R}^N$ with parametric equations $x = \xi(s)$ where s is a $(N-1)$-parameter ranging over some cube $Q_\delta \subset \mathbb{R}^{N-1}$. Consider the Cauchy problem for the *eikonal* equation ([32] Chapter 9, Section 8)

$$|\nabla u| = 1 \qquad u \Big|_{\Phi_o} = 0. \tag{5.1}$$

The function $x \to u(x)$ is the time it takes a light ray to reach x starting from a point source at the origin. The level sets $\Phi_t = [u = t]$ are the wave fronts of the light propagation, and the light rays are normal to these fronts. Thus Φ_o is an initial wave front, and the Cauchy problem seeks to determine the fronts Φ_t at later times t. The Monge's cones are circular, with vertical axis and their equation is (Section 1.2)

$$|z - t| = |x - y| \qquad \text{for every } y \in \Phi_t.$$

The characteristic strips are constructed from (3.4)–(3.5) as

$$\begin{aligned} &x_t(t,s) = p(t,s) && x(0,s) = \xi(s) && s \in Q_\delta \\ &z_t(t,s) = 1 && z(0,t) = 0 && t \in (-\delta, \delta) \\ &p_t(t,s) = 0 && p(0,s) = p(s). && \end{aligned} \tag{5.2}$$

Computing the initial vectors $p(s)$ from (3.3) gives

$$p(s) \cdot \nabla\xi(s) = 0, \qquad |p(s)| = 1, \qquad \text{for all } s \in Q_\delta.$$

Thus $p(s)$ is a unit vector normal to the front Φ_o. By the third of (5.2) such a vector is constant along characteristics, and the characteristic system has the explicit integral

$$x(t,s) = tp(s) + \xi(s), \quad z(t,s) = t, \quad p(t,s) = p(s). \tag{5.3}$$

Therefore after a time t, the front Φ_o evolves into the front Φ_t, obtained by transporting each point $\xi(s) \in \Phi_o$, along the normal $p(s)$ with unitary speed, for a time t.

5.1 Wave Fronts, Light Rays, Local Solutions, and Caustics

For a fixed $s \in Q_\delta$ the first of (5.2) are the parametric equations of a straight line in $\mathbb{R}^N$, which we denote by $\ell(t;s)$. Since $p(s)$ is normal to the front Φ_o, such a line can be identified with the light ray through $\xi(s) \in \Phi_o$. By construction such a ray is always normal to the wave front Φ_t that it crosses.

This geometrical interpretation is suggestive on the one hand of the underlying physics, and on the other, it highlights the local nature of the Cauchy problem. Indeed the solution, as constructed, becomes meaningless if two of these rays, say for example $\ell(t;s_1)$ and $\ell(t;s_2)$, intersect at some point, for such a point would have to belong to two distinct wave fronts. To avoid such an occurrence, the number δ that limits the range of the parameters s and t has to be taken sufficiently small.

The possible intersection of the light rays $\ell(s;t)$ might depend also on the initial front. If Φ_o is an $(N-1)$-dimensional hyperplane, then all rays are parallel and normal to Φ_o. In such a case the solution exists for all $s \in \mathbb{R}^{N-1}$ and all $t \in \mathbb{R}$. If Φ_o is an $(N-1)$-dimensional sphere of radius R centered at the origin of $\mathbb{R}^N$, all rays $\ell(t;s)$ intersect at the origin after a time $t = R$. The solution exists for all times, and the integral surfaces are right circular cones with vertex at the origin.

The envelope of the family $\ell(t;s)$ as s ranges over Q_δ, if it exists, is called a *caustic* or *focal curve*. By definition of envelope, the caustic is tangent in any of its points to at least one light ray. Therefore, such a tangency point is instantaneously illuminated, and the caustic can be regarded as a light tracer following the parameter t.

If Φ_o is a hyperplane the caustic does not exists, and if Φ_o is a sphere, the caustic degenerates into its origin.

6 The Initial Value Problem for Hamilton–Jacobi Equations

Denote by $(x; x_{N+1})$ points in $\mathbb{R}^{N+1}$, and for a smooth function u defined in a domain of $\mathbb{R}^{N+1}$, set $\nabla u = (D_x u, u_{x_{N+1}})$. Given a smooth non-linear function

$$(x, x_{N+1}, p) \to \mathcal{H}(x, p; x_{N+1})$$

defined in a domain of $\mathbb{R}^{N+1} \times \mathbb{R}^N$, consider the first-order equation

$$F(x; x_{N+1}, u, D_x u, u_{x_{N+1}}) = u_{x_{N+1}} + \mathcal{H}(x, D_x u; x_{N+1}) = 0. \tag{6.1}$$

The Cauchy problem for (6.1) consists in giving an N-dimensional surface Φ and a smooth function u_o defined on Φ, and seeking a smooth function u that solves (6.1) in a neighborhood of Φ and equals u_o on Φ. If the surface Φ is the hyperplane $x_{N+1} = 0$, it has parametric equations $x = s$ and the characteristic system (3.4) takes the form

$$\begin{aligned}
x_t(t,s) &= D_p\mathcal{H}(x(t,s), p(t,s); x_{N+1}(t,s)) \\
x_{N+1,t}(t,s) &= 1 \\
z_t(x,t) &= p(t,s) \cdot D_p\mathcal{H}(x(t,s), p(t,s); x_{N+1}(t,s)) + p_{N+1}(t,s) \\
p_t(t,s) &= -D_x\mathcal{H}(x(t,s), p(t,s), x_{N+1}(t,s)) \\
p_{N+1,t}(t,s) &= -D_{x_{N+1}}\mathcal{H}(x(t,s), p(t,s); x_{N+1}(t,s))
\end{aligned}$$

with the initial conditions

$$\begin{gathered}
x(0,s) = s, \quad x_{N+1}(0,s) = 0, \quad z(0,s) = u_o(s) \\
p(0,s) = p(s), \quad p_{N+1}(0,s) = p_{N+1}(s).
\end{gathered}$$

The second of these and the corresponding initial datum imply $x_{N+1} = t$. Therefore the $(N+1)$st coordinate may be identified with time, and the Cauchy problem for the surface $[t = 0]$ is the *initial value problem* for the Hamilton–Jacobi equation (6.1). The characteristic system can be written concisely as

$$\begin{aligned}
x_t(t,s) &= D_p\mathcal{H}(x(t,s), p(t,s); t) \qquad & x(0,s) &= s \\
p_t(t,s) &= -D_x\mathcal{H}(x(t,s), p(t,s); t) \qquad & p(0,s) &= p(s)
\end{aligned} \tag{6.2}$$

where the initial data $(p(s), p_{N+1}(s))$ are determined from (3.3) as

$$p(s) = D_s u_o(s), \qquad p_{N+1}(0,s) = -\mathcal{H}(s, p(s); 0). \tag{6.3}$$

Moreover, the functions $(t,s) \to p_{N+1}(t,s), z(t,s)$, satisfy

$$\begin{aligned}
p_{N+1,t}(t,s) &= -\mathcal{H}(x(t,s), p(t,s); t) \\
p_{N+1}(0,s) &= -\mathcal{H}(s, p(s); 0) \\
z_t(x,t) &= p(t,s) \cdot D_p\mathcal{H}(x(t,s), p(t,s); t) + p_{N+1}(t,s) \\
z(0,s) &= u_o(s).
\end{aligned}$$

It is apparent that (6.2) is independent of (6.3), and the latter can be integrated as soon as one determines the functions $(t,s) \to x(t,s), p(t,s)$, solutions of (6.2). Therefore (6.2) is the characteristic system associated with the initial value problem for (6.1).

Consider now a mechanical system with N degrees of freedom governed by a Hamiltonian $\mathcal{H}$. The system (6.2) is precisely the canonical Hamiltonian system that describes the motion of the system, through its Lagrangian coordinates $t \to x(t,s)$ and the kinetic momenta $t \to p(t,s)$, starting from its initial configuration. Therefore the characteristics associated with the initial value problem for the Hamilton–Jacobi equation (6.1) are the dynamic trajectories, in phase space, of the underlying mechanical system.

From now on we will restrict the theory to the case $\mathcal{H}(x,t,p) = \mathcal{H}(p)$, that is, the Hamiltonian depends only on the kinetic momenta p. In such a case the initial value problem takes the form

$$u_t + \mathcal{H}(D_x u) = 0, \qquad u(\cdot, 0) = u_o \tag{6.4}$$

where u_o is a bounded continuous function in $\mathbb{R}^N$. The characteristic curves and initial data are

$$\begin{aligned} x_t(t,s) &= D_p\mathcal{H}(p(t,s) & x(0,s) &= s \\ p_t(t,s) &= 0 & p(0,s) &= D_s u_o(s) \\ p_{N+1,t}(t,s) &= 0, & p_{N+1}(0,s) &= -\mathcal{H}(D_s u_o(s)). \end{aligned} \tag{6.5}$$

Moreover

$$\begin{aligned} z_t(t,s) &= p(t,s) \cdot D_p\mathcal{H}(p(t,s)) + p_{N+1}(t,s) \\ z(0,s) &= u_o(s). \end{aligned} \tag{6.6}$$

7 The Cauchy Problem in Terms of the Lagrangian

Assume that $p \to \mathcal{H}(p)$ is convex and coercive, that is[3]

$$\lim_{|p|\to\infty} \frac{\mathcal{H}(p)}{\|p\|} = \infty.$$

The Lagrangian $q \to \mathcal{L}(q)$, corresponding to the Hamiltonian $\mathcal{H}$, is given by the Legendre transform of $\mathcal{H}$, that is[4]

$$\mathcal{L}(q) = \sup_{p\in\mathbb{R}^N} \left[q \cdot p - \mathcal{H}(p) \right].$$

By the coercivity of $\mathcal{H}$ the supremum is achieved at a vector p satisfying

$$q = D_p\mathcal{H}(p) \quad \text{and} \quad \mathcal{L}(q) = q \cdot p - \mathcal{H}(p). \tag{7.1}$$

---

[3]This occurs, for example, for $\mathcal{H}(p) = |p|^{1+\alpha}$, for all $\alpha > 0$. It does not hold for the Hamiltonian $\mathcal{H}(p) = |p|$ corresponding to the eikonal equation. The Cauchy problem for such non-coercive Hamiltonians is investigated in [86, 87, 88].

[4][32] Chapter 6 Section 5, and [31], Section 13 of the Problems and Complements of Chapter IV.

Moreover, $q \to \mathcal{L}(q)$ is itself convex and coercive, and the Hamiltonian $\mathcal{H}$ is the Legendre transform of the Lagrangian $\mathcal{L}$, that is

$$\mathcal{H}(p) = \sup_{q \in \mathbb{R}^N} \left[p \cdot q - \mathcal{L}(q) \right].$$

Since $\mathcal{L}$ is coercive, the supremum is achieved at a vector q, satisfying

$$p = D_q \mathcal{L}(q) \quad \text{and} \quad \mathcal{H}(p) = q \cdot p - \mathcal{L}(q). \tag{7.2}$$

The equations for the characteristic curves (6.5)–(6.6) can be written in terms of the Lagrangian as follows. The equations in (7.1), written for $q = x_t(t,s)$, and the first of (6.5) imply that the vector $p(s,t)$ for which the supremum in the Legendre transform of $\mathcal{H}$ is achieved is the solution of the second of (6.5). Therefore

$$\mathcal{L}(x_t(t,s)) = x_t(t,s) \cdot p(t,s) - \mathcal{H}(p(t,s)). \tag{7.3}$$

Taking the gradient of $\mathcal{L}$ with respect to x_t and then the derivative with respect to time t, gives

$$D_{\dot{x}} \mathcal{L}(\dot{x}) = p(t,s) \quad \text{and} \quad \frac{d}{dt} \frac{\partial \mathcal{L}(\dot{x})}{\partial \dot{x}_h} = 0, \qquad h = 1, \dots, N.$$

These are the Lagrange equations of motion for a mechanical system of Hamiltonian $\mathcal{H}$.

8 The Hopf Variational Solution

Let u be a smooth solution in $\mathbb{R}^N \times \mathbb{R}^+$ of the Cauchy problem (6.4) for a smooth initial datum u_o. Then for every $x \in \mathbb{R}^N$ and every time $t > 0$, there exists some $s \in \mathbb{R}^N$ such that $x = x(t,s)$, that is the position x is reached in time t by the characteristic $\ell = \{x(t,s)\}$ originating at s. Therefore

$$u(x,t) = u(x(t,s),t) \quad \text{and} \quad D_x u(x(t,s),t) = p(t,s).$$

Equivalently, taking into account that u is a solution of (6.4)

$$\begin{aligned}
u(x,t) - u_o(s) &= \int_\ell \frac{\partial u}{\partial \ell} d\ell \\
&= \int_0^t \big[D_x u(x(\tau,s),\tau) \cdot x_t(\tau,s) + u_t(x(\tau,s),\tau) \big] d\tau \\
&= \int_0^t \big\{ p(\tau,s) \cdot x_t(\tau,s) - \mathcal{H}(p(\tau,s)) \big\} d\tau.
\end{aligned}$$

Using now (7.1), this implies

$$u(x,t) = \int_0^t \mathcal{L}(x_\tau(\tau,s)) d\tau + u_o(s). \tag{8.1}$$

8.1 The First Hopf Variational Formula

The integral on the right-hand side is the Hamiltonian action of a mechanical system with N degrees of freedom, governed by a Lagrangian $\mathcal{L}$, in its motion from a Lagrangian configuration s at time $t = 0$ to a Lagrangian configuration x at time t. Introduce the class of all smooth *synchronous variations*

$$K^s_{\text{sync}} = \left\{ \begin{array}{l} \text{the collection of all smooth paths } q(\cdot) \\ \text{in } \mathbb{R}^N \text{ such that } q(0) = s \text{ and } q(t) = x \end{array} \right\}.$$

By the least action principle ([32], Chapter IX, Section 2)

$$\int_0^t \mathcal{L}(x_t(\tau, s))d\tau = \min_{q\in K^s_{\text{sync}}} \int_0^t \mathcal{L}(\dot{q}(\tau))d\tau.$$

Therefore

$$\begin{aligned} u(x,t) &= \min_{q\in K^s_{\text{sync}}} \int_0^t \mathcal{L}(\dot{q}(\tau))d\tau + u_o(s) \\ &\geq \inf_{y\in\mathbb{R}^N} \inf_{q\in K^y_{\text{sync}}} \left[\int_0^t \mathcal{L}(\dot{q}(\tau))d\tau + u_o(y) \right]. \end{aligned}$$

Such a formula actually holds with the equality sign, since if $u(x,t)$ is known, by (8.1), for each fixed $x \in \mathbb{R}^N$ and $t > 0$ there exist some $s \in \mathbb{R}^N$ and a smooth curve $\tau \to x(\tau, s)$ of extremities s and x such that the infimum is actually achieved. This establishes the first Hopf variational formula, that is, if $(x,t) \to u(x,t)$ is a solution of the Cauchy problem (6.4), then

$$u(x,t) = \min_{y\in\mathbb{R}^N} \min_{q\in K^y_{\text{sync}}} \left[\int_0^t \mathcal{L}(\dot{q}(\tau))d\tau + u_o(y) \right]. \tag{8.2}$$

8.2 The Second Hopf Variational Formula

A drawback of the first Hopf variational formula is that, given x and t, it requires the knowledge of the classes K^y_{sync} for all $y \in \mathbb{R}^N$. The next variational formula dispenses with such classes ([72, 73]).

Proposition 8.1 *Let $(x,t) \to u(x,t)$ be a solution of the minimum problem (8.2). Then for all $x \in \mathbb{R}^N$ and all $t > 0$*

$$u(x,t) = \min_{y\in\mathbb{R}^N} \left[t\mathcal{L}\Big(\frac{x-y}{t}\Big) + u_o(y) \right]. \tag{8.3}$$

Proof For $s \in \mathbb{R}^N$ consider the curve

$$\tau \to q(\tau) = s + \frac{\tau}{t}(x - s) \qquad \tau \in [0,t].$$

If $u(x,t)$ is a solution of (8.2)

$$u(x,t) \le \int_0^t \mathcal{L}(\dot{q}(\tau))d\tau + u_o(s) = t\mathcal{L}\Big(\frac{x-s}{t}\Big) + u_o(s)$$

and since s is arbitrary

$$u(x,t) \le \inf_{y\in\mathbb{R}^N} \Big[t\mathcal{L}\Big(\frac{x-y}{t}\Big) + u_o(y)\Big].$$

Now let $q \in K^s_{\text{sync}}$ for some $s \in \mathbb{R}^N$. Since $\mathcal{L}(\cdot)$ is convex, by Jensen's inequality

$$\mathcal{L}\Big(\frac{x-s}{t}\Big) = \mathcal{L}\left(\frac{1}{t}\int_0^t \dot{q}(\tau)d\tau\right) \le \frac{1}{t}\int_0^t \mathcal{L}(\dot{q}(\tau))d\tau.$$

From this

$$t\mathcal{L}\Big(\frac{x-s}{t}\Big) + u_o(s) \le \int_0^t \mathcal{L}(\dot{q}(\tau))d\tau + u_o(s).$$

Since $s \in \mathbb{R}^N$ is arbitrary, by (8.2)

$$\begin{aligned}\inf_{y\in\mathbb{R}^N} \Big[t\mathcal{L}\Big(\frac{x-y}{t}\Big) + u_o(y)\Big] &\le \min_{y\in\mathbb{R}^N} \min_{q\in K^y_{\text{sync}}} \Big[\int_0^t \mathcal{L}(\dot{q}(\tau))d\tau + u_o(y)\Big] \\ &= u(x,t).\end{aligned}$$

■

9 Semigroup Property of Hopf Variational Solutions

Proposition 9.1 ([5, 13]) *Let $(x,t) \to u(x,t)$ be a solution of the variational problem (8.3). Then for all $x \in \mathbb{R}^N$ and every pair $0 \le \tau < t$*

$$u(x,t) = \min_{y\in\mathbb{R}^N} \Big[(t-\tau)\mathcal{L}\Big(\frac{x-y}{t-\tau}\Big) + u(y,\tau)\Big]. \tag{9.1}$$

Proof Write (8.3) for $x = \eta$ at time τ and let $\xi \in \mathbb{R}^N$ be a point where the minimum is achieved. Thus

$$u(\eta,\tau) = \tau\mathcal{L}\Big(\frac{\eta-\xi}{\tau}\Big) + u_o(\xi).$$

Since $\mathcal{L}(\cdot)$ is convex

$$\mathcal{L}\Big(\frac{x-\xi}{t}\Big) = \Big(1-\frac{\tau}{t}\Big)\mathcal{L}\Big(\frac{x-\eta}{t-\tau}\Big) + \frac{\tau}{t}\mathcal{L}\Big(\frac{\eta-\xi}{\tau}\Big).$$

Therefore

$$\begin{aligned}u(x,t) = \min_{y\in\mathbb{R}^N} \Big[t\mathcal{L}\Big(\frac{x-y}{t}\Big) + u_o(y)\Big] &\le t\mathcal{L}\Big(\frac{x-\xi}{t}\Big) + u_o(\xi) \\ &\le (t-\tau)\mathcal{L}\Big(\frac{x-\eta}{t-\tau}\Big) + \tau\mathcal{L}\left(\frac{\eta-\xi}{\tau}\right) + u_o(\xi)\end{aligned}$$

$$
\begin{aligned}
&= (t-\tau)\mathcal{L}\Big(\frac{x-\eta}{t-\tau}\Big) + u(\eta,\tau) \\
&\le \min_{\eta\in\mathbb{R}^N} \Big[(t-\tau)\mathcal{L}\Big(\frac{x-\eta}{t-\tau}\Big) + u(\eta,\tau)\Big].
\end{aligned}
$$

Now let $\xi \in \mathbb{R}^N$ be a point for which the minimum in (8.3) is achieved, i.e.,

$$u(x,t) = t\mathcal{L}\Big(\frac{x-\xi}{t}\Big) + u_o(\xi).$$

For $\tau \in (0,t)$ write

$$\eta = \frac{\tau}{t}x + \Big(1-\frac{\tau}{t}\Big)\xi \quad\Longrightarrow\quad \frac{x-\eta}{t-\tau} = \frac{x-\xi}{t} = \frac{\eta-\xi}{\tau}.$$

Moreover, by (8.3)

$$u(\eta,\tau) \le \tau\mathcal{L}\Big(\frac{\eta-\xi}{\tau}\Big) + u_o(\xi).$$

Combining these remarks

$$
\begin{aligned}
(t-\tau)\mathcal{L}\Big(\frac{x-\eta}{t-\tau}\Big) + u(\eta,\tau) &\le (t-\tau)\mathcal{L}\Big(\frac{x-\eta}{t-\tau}\Big) + \tau\mathcal{L}\Big(\frac{\eta-\xi}{\tau}\Big) + u_o(\xi) \\
&= t\mathcal{L}\Big(\frac{x-\xi}{t}\Big) + u_o(\xi) = u(x,t).
\end{aligned}
$$

From this

$$u(x,t) \ge \min_{\eta\in\mathbb{R}^N} \Big[(t-\tau)\mathcal{L}\Big(\frac{x-\eta}{t-\tau}\Big) + u(\eta,\tau)\Big]. \qquad \blacksquare$$

10 Regularity of Hopf Variational Solutions

For $(x,t) \to u(x,t)$ to be a solution of the Cauchy problem (6.4), it would have to be differentiable. While this is in general not the case, the next proposition asserts that if the initial datum u_o is Lipschitz continuous, the corresponding Hopf variational solution is Lipschitz continuous. Assume then that there is a positive constant C_o such that

$$|u_o(x) - u_o(y)| \le C_o|x-y| \quad \text{for all } x,y \in \mathbb{R}^N. \tag{10.1}$$

Proposition 10.1 *Let $(x,t) \to u(x,t)$ be a solution of (8.3) for an initial datum u_o satisfying (10.1). Then there exists a positive constant C depending only on C_o and $\mathcal{H}$ such that for all $x,y \in \mathbb{R}^N$ and all $t,\tau \in \mathbb{R}^+$*

$$|u(x,t) - u(y,\tau)| \le C(|x-y| + |t-\tau|). \tag{10.2}$$

Proof For a fixed $t > 0$, let $\xi \in \mathbb{R}^N$ be a vector for which the minimum in (8.3) is achieved. Then for all $y \in \mathbb{R}^N$,

$$
\begin{aligned}
u(y,t) - u(x,t) &= \inf_{\eta \in \mathbb{R}^N} \Big[t\mathcal{L}\Big(\frac{y-\eta}{t}\Big) + u_o(\eta)\Big] - t\mathcal{L}\Big(\frac{x-\xi}{t}\Big) - u_o(\xi) \\
&\le t\mathcal{L}\Big(\frac{y-(y-(x-\xi))}{t}\Big) + u_o(y-(x-\xi)) \\
&\quad - t\mathcal{L}\Big(\frac{x-\xi}{t}\Big) - u_o(\xi) \\
&= u_o(y-(x-\xi)) - u_o(\xi) \le C_o|y-x|.
\end{aligned}
$$

Interchanging the role of x and y gives

$$
|u(x,t) - u(y,t)| \le C_o|x-y| \quad \text{for all } x,y \in \mathbb{R}^N.
$$

This establishes the Lipschitz continuity of u in the space variables uniformly in time. The variational formula (8.3) implies

$$
u(x,t) \le t\mathcal{L}(0) + u_o(x) \quad \text{for all } x \in \mathbb{R}^N
$$

and

$$
\begin{aligned}
u(x,t) &= \min_{y \in \mathbb{R}^N} \Big[t\mathcal{L}\Big(\frac{x-y}{t}\Big) + u_o(y) - u_o(x) + u_o(x)\Big] \\
&\ge u_o(x) + \min_{y \in \mathbb{R}^N} \Big[t\mathcal{L}\Big(\frac{x-y}{t}\Big) - C_o|x-y|\Big] \\
&\ge u_o(x) - \max_{q \in \mathbb{R}^N} [C_o t|q| - t\mathcal{L}(q)] \\
&\ge u_o(x) - t \max_{|p|<C_o} \max_{q \in \mathbb{R}^N} [p \cdot q - \mathcal{L}(q)] \\
&= u_o(x) - t \max_{|p|<C_o} \mathcal{H}(p).
\end{aligned}
$$

Therefore

$$
|u(x,t) - u_o(x)| \le \bar{C}t, \quad \text{where } \bar{C} = \mathcal{L}(0) \wedge \max_{|p|<C_o} \mathcal{H}(p). \tag{10.3}
$$

From this

$$
|u(x,t) - u(x,\tau)| \le \bar{C}|t-\tau| \quad \text{for all } t,\tau \in \mathbb{R}^+. \qquad \blacksquare
$$

Remark 10.1 By the Rademacher theorem $(x,t) \to u(x,t)$ is a.e. differentiable in $\mathbb{R}^N \times \mathbb{R}^+$ ([31], Chapter VII, Section 23).

11 Hopf Variational Solutions (8.3) are Weak Solutions of the Cauchy Problem (6.4)

Assume that u_o is Lipschitz continuous as in (10.1). Then by (10.3), a solution $(x,t) \to u(x,t)$ of the corresponding variational problem (8.3) takes the initial datum u_o in the classical sense. The next proposition asserts that such a variational solution satisfies the Hamilton–Jacobi equation (6.4) at each point (x,t) where it is differentiable.

Proposition 11.1 *Let $(x,t) \to u(x,t)$ be a solution of the minimum problem (8.3). If u is differentiable at $(x,t) \in \mathbb{R}^N \times \mathbb{R}^+$, then*

$$u_t + \mathcal{H}(D_x u) = 0 \qquad \text{at } (x,t).$$

Proof Fix $\eta \in \mathbb{R}^N$ and $h > 0$. By the semigroup property

$$\begin{aligned} u(x+h\eta, t+h) &= \min_{y\in\mathbb{R}^N} \Big[(t+h)\mathcal{L}\Big(\frac{x+h\eta-y}{t+h}\Big) + u_o(y)\Big] \\ &= \min_{y\in\mathbb{R}^N} \Big[h\mathcal{L}\Big(\frac{x+h\eta-y}{h}\Big) + u(y,t)\Big] \le h\mathcal{L}(\eta) + u(x,t). \end{aligned}$$

From this

$$\frac{u(x+h\eta,t+h) - u(x,t)}{h} \le \mathcal{L}(\eta)$$

and letting $h \to 0$

$$\eta \cdot D_x u(x,t) + u_t(x,t) \le \mathcal{L}(\eta).$$

Recalling that $\mathcal{H}$ is the Legendre transform of $\mathcal{L}$

$$u_t(x,t) + \mathcal{H}(D_x u(x,t)) = u_t(x,t) + \max_{q\in\mathbb{R}^N}[q \cdot D_x u(x,t) - \mathcal{L}(q)] \le 0.$$

Let $\xi \in \mathbb{R}^N$ be a point for which the minimum in (8.3) is achieved. Fix $0 < h < t$, set $\tau = t - h$, and let

$$\eta = \frac{\tau}{t}x + \Big(1 - \frac{\tau}{t}\Big)\xi \Longrightarrow \frac{x-\xi}{t} = \frac{\eta-\xi}{\tau}; \quad \eta = x - \frac{h}{t}(x-\xi).$$

Using these definitions, compute

$$\begin{aligned} u(x,t) - u(\eta,\tau) &= t\mathcal{L}\Big(\frac{x-\xi}{t}\Big) + u_o(\xi) - \min_{y\in\mathbb{R}^N}\Big[\tau\mathcal{L}\Big(\frac{\eta-y}{\tau}\Big) + u_o(y)\Big] \\ &\ge t\mathcal{L}\Big(\frac{x-\xi}{t}\Big) + u_o(\xi) - \tau\mathcal{L}\Big(\frac{\eta-\xi}{\tau}\Big) - u_o(\xi) \\ &= (t-\tau)\mathcal{L}\Big(\frac{x-\xi}{t}\Big) = h\mathcal{L}\Big(\frac{x-\xi}{t}\Big). \end{aligned}$$

From this

$$\frac{1}{h}\Big[u(x,t) - u\Big(x - \frac{h}{t}(x-\xi), t-h\Big)\Big] \ge \mathcal{L}\Big(\frac{x-\xi}{t}\Big)$$

and letting $h \to 0$

$$u_t(x,t) + \frac{x-\xi}{t} \cdot D_x u(x,t) \ge \mathcal{L}\Big(\frac{x-\xi}{t}\Big).$$

Since $\mathcal{H}$ is the Legendre transform of the Lagrangian $\mathcal{L}$

$$\begin{aligned} u_t(x,t) + \mathcal{H}(D_x u(x,t)) &= u_t(x,t) + \max_{q\in\mathbb{R}^N}[q \cdot D_x u(x,t) - \mathcal{L}(q)] \\ &\ge u_t(x,t) + \frac{x-\xi}{t} \cdot D_x u(x,t) - \mathcal{L}\Big(\frac{x-\xi}{t}\Big) \ge 0. \end{aligned}$$ ∎

12 Some Examples

Proposition 11.1 ensures that the variational solutions (8.3), satisfy the Hamilton–Jacobi equation in (6.4) only at points of differentiability, as shown by the examples below.

12.1 Example I

$$u_t + \tfrac{1}{2}|D_x u|^2 = 0 \quad \text{in } \mathbb{R}^N \times \mathbb{R}^+, \qquad u(x,0) = |x|. \tag{12.1}$$

The Lagrangian $\mathcal{L}$ corresponding to the Hamiltonian $\mathcal{H}(p) = \frac{1}{2}|p|^2$ is

$$\mathcal{L}(q) = \max_{p\in\mathbb{R}^N} \left[p \cdot q - \tfrac{1}{2}|p|^2\right] = \tfrac{1}{2}|q|^2$$

and the Hopf variational solution is

$$u(x,t) = \min_{y\in\mathbb{R}^N} \left(\frac{|x-y|^2}{2t} + |y|\right).$$

The minimum is computed by setting

$$D_y\left(\frac{|x-y|^2}{2t} + |y|\right) = \frac{y-x}{t} + \frac{y}{|y|} = 0 \quad \Longrightarrow \frac{x}{t} = \left(\frac{1}{t} + \frac{1}{|y|}\right)y.$$

From this compute, for $|x| > t$

$$y = x - \frac{xt}{|x|} \qquad \Longrightarrow \qquad u(x,t) = |x| - \tfrac{1}{2}t \quad \text{for } |x| > t.$$

If $|x| \le t$

$$\begin{aligned}\frac{|x-y|^2}{2t} + |y| &= \frac{|x|^2}{2t} - \frac{x\cdot y}{t} + \frac{|y|^2}{2t} + |y| \\ &\ge \frac{|x|^2}{2t} - \frac{|x|}{t}|y| + \frac{|y|^2}{2t} + |y| \ge \frac{|x|^2 + |y|^2}{2t}.\end{aligned}$$

This holds for all $y \in \mathbb{R}^N$, and equality holds for $y = 0$. Therefore if $|x| \le t$ the minimum is achieved for $y = 0$, and

$$u(x,t) = \begin{cases} \dfrac{|x|^2}{2t} & \text{for } |x| \le t \\ |x| - \frac{1}{2}t & \text{for } |x| > t. \end{cases}$$

One verifies that u satisfies the Hamilton–Jacobi equation in (12.1) in $\mathbb{R}^N \times \mathbb{R}^+$ except at the cone $|x| = t$.

Remark 12.1 For fixed $t > 0$ the graph of $x \to u(x,t)$ is convex for $|x| < t$ and concave for $|x| > t$. In the region of convexity, the Hessian matrix of u is $\mathbb{I}/t$. Therefore for all $\xi \in \mathbb{R}^N$

$$u_{x_i x_j}\xi_i\xi_j \le \frac{|\xi|^2}{t}. \tag{12.2}$$

In the region of concavity $|x| > t$

$$u_{x_i x_j}\xi_i\xi_j = (|x|^2\delta_{ij} - x_i x_j)\frac{\xi_i\xi_j}{|x|^3} \le \frac{|\xi|^2}{t}.$$

Therefore (12.2) holds in the whole of $\mathbb{R}^N \times \mathbb{R}^+$ except for $|x| = t$.

12.2 Example II

$$u_t + \tfrac{1}{2}|D_x u|^2 = 0 \quad \text{in } \mathbb{R}^N \times \mathbb{R}^+, \qquad u(x,0) = -|x|. \tag{12.3}$$

As before, $\mathcal{L}(q) = \frac{1}{2}|q|^2$, and the Hopf variational solution is

$$u(x,t) = \min_{y\in\mathbb{R}^N}\left(\frac{|x-y|^2}{2t} - |y|\right).$$

The minimum is computed by setting

$$D_y\left(\frac{|x-y|^2}{2t} - |y|\right) = \frac{y-x}{t} - \frac{y}{|y|} = 0 \quad\Longrightarrow\quad y = (|x|+t)\frac{x}{|x|}.$$

Therefore

$$u(x,t) = -|x| - \tfrac{1}{2}t \quad \text{in } \mathbb{R}^N \times \mathbb{R}^+.$$

One verifies that this function satisfies the Hamilton–Jacobi equation in (12.3) for all $|x| > 0$.

Remark 12.2 For fixed $t > 0$, the graph of $x \to u(x,t)$ is concave, and

$$u_{x_i x_j}\xi_i\xi_j \le 0 \quad \text{for all } \xi \in \mathbb{R}^N \quad \text{in } \mathbb{R}^N \times \mathbb{R}^+ - \{|x| = 0\}. \tag{12.4}$$

12.3 Example III

The Cauchy problem

$$u_t + (u_x)^2 = 0 \quad \text{in } \mathbb{R} \times \mathbb{R}^+, \qquad u(x,0) = 0 \tag{12.5}$$

has the identically zero solution. However the function

$$u(x,t) = \begin{cases} 0 & \text{for} \quad |x| \ge t \\ x - t & \text{for} \quad 0 \le x \le t \\ -x - t & \text{for} \quad -t \le x \le 0. \end{cases}$$

is Lipschitz continuous in $\mathbb{R}\times\mathbb{R}^+$, and it satisfies the equation (12.5) in $\mathbb{R}\times\mathbb{R}^+$ except on the half-lines $x = \pm t$.

Remark 12.3 For fixed $t > 0$, the graph of $x \to u(x,t)$ is convex for $|x| < t$ and concave for $|x| > t$. In the region of concavity, $u_{xx} = 0$, whereas in the region of convexity, $u(\cdot,t)$ is not of class C^2, and whenever it does exist, the second derivative does not satisfy an upper bound of the type of (12.2). This lack of control on the *convex part* of the graph of $u(\cdot,t)$ is responsible for the lack of uniqueness of the solution of (12.5).

This example raises the issue of identifying a class of solutions of the Cauchy problem (6.4) within which uniqueness holds.

13 Uniqueness

Denote by $\mathcal{C}_o$ the class of solutions $(x,t) \to u(x,t)$ to the Cauchy problem (6.4), of class $C^2(\mathbb{R}^N \times \mathbb{R}^+)$, uniformly Lipschitz continuous in $\mathbb{R}^N \times \mathbb{R}^+$ and such that the graph of $x \to u(x,t)$ is concave for all $t > 0$, that is

$$\mathcal{C}_o = \begin{cases} u \in C^2(\mathbb{R}^N \times \mathbb{R}^+) & \\ u_{x_i x_j} \xi_i \xi_j \le 0 & \text{for all } \xi \in \mathbb{R}^N \quad \text{in } \mathbb{R}^N \times \mathbb{R}^+ \\ |\nabla u| \le C & \text{for some } C > 0 \quad \text{in } \mathbb{R}^N \times \mathbb{R}^+. \end{cases} \tag{13.1}$$

Proposition 13.1 *Let u_1 and u_2 be two solutions of the Cauchy problem (6.4) in the class $\mathcal{C}_o$. Then $u_1 = u_2$.*

Proof Setting $w = u_1 - u_2$, compute

$$\begin{aligned} w_t = \mathcal{H}(D_x u_2) - \mathcal{H}(D_x u_1) &= \int_0^1 \frac{d}{ds} \mathcal{H}(sD_x u_2 + (1-s)D_x u_1) ds \\ &= -\left(\int_0^1 \mathcal{H}_{p_j}(sD_x u_2 + (1-s)D_x u_1) ds \right) w_{x_j} = -\mathbf{V} \cdot D_x w \end{aligned} \tag{13.2}$$

where

$$\mathbf{V} = \int_0^1 D_p \mathcal{H}(sD_x u_2 + (1-s)D_x u_1) ds. \tag{13.3}$$

Multiplying (13.2) by $2w$ gives

$$w_t^2 = -\mathbf{V} \cdot D_x w^2 = -\operatorname{div}(\mathbf{V} w^2) + w^2 \operatorname{div} \mathbf{V}. \tag{13.4}$$

∎

Lemma 13.1 $\operatorname{div} \mathbf{V} \le 0$.

Proof Fix $(x,t) \in \mathbb{R}^N \times \mathbb{R}^+$ and $s \in (0,1)$, set

$$\begin{aligned} p &= sD_x u_2(x,t) + (1-s)D_x u_1(x,t) \\ z_{ij} &= s u_{2,x_i x_j}(x,t) + (1-s) u_{1,x_i x_j}(x,t) \end{aligned}$$

and compute

$$\operatorname{div} \mathbf{V} = \int_0^1 \mathcal{H}_{p_i p_j}(p) z_{ji} ds.$$

The integrand is the trace of the product matrix $(\mathcal{H}_{p_ip_j})(z_{ij})$. Since $\mathcal{H}$ is convex, $(\mathcal{H}_{p_ip_j})$ is symmetric and positive semi-definite, and its eigenvalues $\lambda_h = \lambda_h(x,t,s)$ for $h=1,\dots,N$ are non-negative. Since both matrices (u_{ℓ,x_ix_j}) for $\ell = 1,2$ are negative semi-definite, the same is true for the convex combination

$$(z_{ij}) = s(u_{2,x_ix_j}) + (1-s)(u_{1,x_ix_j}).$$

In particular, its eigenvalues $\mu_h = \mu_h(x,t,s)$ for $h=1,\dots,N$ are non-positive. Therefore

$$\mathcal{H}_{p_ip_j} z_{ji} = \text{trace}(\mathcal{H}_{p_ip_j})(z_{ij}) = \lambda_h\mu_h \le 0. \qquad \blacksquare$$

This in (13.4) gives

$$w_t^2 + \text{div}(\mathbf{V}w^2) \le 0 \quad \text{in } \mathbb{R}^N \times \mathbb{R}^+. \tag{13.5}$$

Fix $x_o \in \mathbb{R}^N$ and $T>0$, and introduce the backward characteristic cone with vertex at (x_o, T)

$$C_M = \big[|x-x_o| \le M(T-\tau);\ 0 \le \tau \le T\big] \tag{13.6}$$

where $M>0$ is to be chosen. The exterior unit normal to the lateral surface of C_M is

$$\nu = \frac{(x/|x|, M)}{\sqrt{1+M^2}} = (\nu_x, \nu_t).$$

For $t \in (0,T)$ introduce also the backward truncated characteristic cone

$$C_M^t = \big[|x-x_o| \le M(T-\tau);\ 0 \le \tau \le t\big]. \tag{13.7}$$

Integrating (13.5) over such a truncated cone gives

$$\begin{aligned}
&\int_{|x-x_o|<M(T-t)} w^2(x,t)dx + \frac{M}{\sqrt{1+M^2}} \int_0^t \int_{|x-x_o|=M(T-\tau)} w^2 d\sigma(\tau)d\tau \\
&\quad \le \int_{|x-x_o|<MT} w^2(x,0)dx - \int_0^t \int_{|x-x_o|=M(T-\tau)} w^2 \mathbf{V}\cdot\nu_x d\sigma(\tau)d\tau
\end{aligned} \tag{13.8}$$

where $d\sigma(\tau)$ is the surface measure on the sphere $[|x-x_o| = M(T-\tau)]$. Using the constant C in (13.1) and the definition (13.3) of $\mathbf{V}$i, choose M from

$$|\mathbf{V}\cdot\nu_x| \le \sup_{|p|<C} |D_p\mathcal{H}(p)| = \delta M \quad \text{for some } \delta \in (0,1). \tag{13.9}$$

This choice of M in (13.8) gives

$$\begin{aligned}
&\int_{|x-x_o|<M(T-t)} w^2(x,t)dx + \frac{(1-\delta)M}{\sqrt{1+M^2}} \int_0^t \int_{|x-x_o|=M(T-\tau)} w^2 d\sigma(\tau)d\tau \\
&\qquad \le \int_{|x-x_o|<MT} w^2(x,0)dx.
\end{aligned}$$

Thus

$$\int_{|x-x_o|<M(T-t)} w^2(x,t)dx \le \int_{|x-x_o|<MT} w^2(x,0)dx. \tag{13.10}$$

14 More on Uniqueness and Stability

Multiply (13.2) by $f'(w)$, for some non-negative $f \in C^1(\mathbb{R})$ and use Lemma 13.1 to get

$$f(w)_t + \operatorname{div}(\mathbf{V} f(w)) \le 0 \qquad \text{in } \mathbb{R}^N \times \mathbb{R}^+.$$

By similar arguments

$$\int_{|x-x_o|<M(T-t)} f(w(x,t))dx \le \int_{|x-x_o|<MT} f(w(x,0))dx. \tag{14.1}$$

By the change of variables $x - x_o = y$

$$\int_{|y|<M(T-t)} f(w(x_o+y,t))d\xi \le \int_{|y|<MT} f(w(x_o+y,0))dy.$$

Integrating this in dx_o over $\mathbb{R}^N$ gives

$$\int_{\mathbb{R}^N} f(w(x,t))dx \le \left(\frac{T}{T-t}\right)^N \int_{\mathbb{R}^N} f(w(x,0))dx$$

provided the integrals are convergent. Fix $0 < t < T$ and let $T \to \infty$ to obtain the stability estimate

$$\int_{\mathbb{R}^N} f(w(x,t))dx \le \int_{\mathbb{R}^N} f(w(x,0))dx. \tag{14.2}$$

14.1 Stability in $L^p(\mathbb{R}^N)$ for All $p \ge 1$

Proposition 14.1 *Let u_1 and u_2 be solutions of (6.4) in the class $\mathcal{C}_o$ introduced in (13.1). If both are in $L^p(\mathbb{R}^N)$ for some $1 \le p \le \infty$, then*

$$\|u_1(\cdot,t) - u_2(\cdot,t)\|_{p,\mathbb{R}^N} \le \|u_1(\cdot,0) - u_2(\cdot,0)\|_{p,\mathbb{R}^N}. \tag{14.3}$$

Proof If $1 \le p < \infty$, the conclusion follows from (14.2) for $f(w) = |w|^p$. If $p = 1$, take $f(w) = \operatorname{sign}(w)$, modulo an approximation process. If $p = \infty$ write (14.1) with $f(w) = |w|^q$ for $1 < q < \infty$, in the form

$$\begin{aligned}\left(\frac{N}{\omega_N[M(T-t)]^N} \int_{|x-x_o|<M(T-t)} |w(x,t)|^q dx\right)^{1/q} \\ \le \left(\frac{T}{T-t}\right)^{N/q} \left(\frac{N}{\omega_N(MT)^N} \int_{|x-x_o|<MT} |w(x,0)|^q dx\right)^{1/q}\end{aligned}$$

where ω_N is the measure of the unit sphere in $\mathbb{R}^N$. Letting $q \to \infty$ gives

$$\|w(\cdot,t)\|_{\infty,[|x-x_o|<M(T-t)]} \le \|w(\cdot,0)\|_{\infty,[|x-x_o|<MT]}.$$

This implies (14.3) for $p = \infty$ since x_o is arbitrary. ∎

14.2 Comparison Principle

Proposition 14.2 *Let u_1 and u_2 be solutions of (6.4) in the class $\mathcal{C}_o$ introduced in (13.1). If $u_{o,1} \le u_{o,2}$, then*

$$u_1(\cdot, t) \le u_2(\cdot, t) \quad \text{in } \mathbb{R}^N \quad \text{for all } t > 0. \tag{14.4}$$

Proof In (14.2) choose $f(w) = w_+$ modulo an approximation process. ∎

15 Semi-Concave Solutions of the Cauchy Problem

Let u be a solution of the Cauchy problem (6.4) of class $C^2(\mathbb{R}^N \times \mathbb{R}^+)$ with no requirement that the graph of $u(\cdot, t)$ be concave. We require, however, that in those regions where such a graph is convex, the "convexity", roughly speaking, be controlled by some uniform bound of the second derivatives of $u(\cdot, t)$. In a precise way it is assumed that there exists a positive constant γ such that

$$(u_{x_i x_j}) - \gamma \mathbb{I} \le 0 \qquad \text{in } \mathbb{R}^N \times \mathbb{R}^+.$$

Since this matrix inequality is invariant by rotations of the coordinate axes, it is equivalent to

$$u_{\nu\nu} \le \gamma \quad \text{in } \mathbb{R}^N \times \mathbb{R}^+ \quad \text{for all } |\nu| = 1. \tag{15.1}$$

A solution of the Cauchy problem (6.4) satisfying such an inequality for all unit vectors $\nu \in \mathbb{R}^N$ is called *semi-concave.*

15.1 Uniqueness of Semi-Concave Solutions

The example in Section 12.3 shows that initial data, however smooth, might give rise to quasi-concave solutions and solutions for which (15.1) is violated. Introduce the class

$$\mathcal{C}_1 = \left\{ \begin{array}{l} u \in C^2(\mathbb{R}^N \times \mathbb{R}^+) \ \text{ satisfies (15.1) and} \\ |\nabla u| \le C \ \text{ for some } \ C > 0 \ \text{ in } \ \mathbb{R}^N \times \mathbb{R}^+. \end{array} \right. \tag{15.2}$$

Proposition 15.1 *Let u_1 and u_2 be two solutions of the Cauchy problem (6.4) in the class $\mathcal{C}_1$. Then $u_1 = u_2$.*

Proof Set $w = u_1 - u_2$ and proceed as in the proof of Proposition 13.1 to arrive at (13.2). Multiplying the latter by $\operatorname{sign} w$ modulo an approximation process gives

$$|w|_t = -\operatorname{div}(\mathbf{V}|w|) + |w| \operatorname{div} \mathbf{V}$$

where $\mathbf{V}$ is defined in (13.3). ∎

Lemma 15.1 *There exists a constant $\bar{\gamma}$ depending only on the constant γ in (15.1) and C in (15.2) such that* $\operatorname{div} \mathbf{V} \le \bar{\gamma}$ *in* $\mathbb{R}^N \times \mathbb{R}^+$.

Proof With the same notation as in Lemma 13.1

$$\begin{aligned}\mathcal{H}_{p_ip_j}(p)z_{ij} &= \text{trace}(\mathcal{H}_{p_ip_j})(z_{ij}) = \text{trace}(\mathcal{H}_{p_ip_j})((z_{ij}) - \gamma\mathbb{I} + \gamma\mathbb{I}) \\ &\le \text{trace}(\mathcal{H}_{p_ip_j})((z_{ij}) - \gamma\mathbb{I}) + \gamma\text{trace}(\mathcal{H}_{p_ip_j}(p)) \le \gamma\mathcal{H}_{p_ip_i}(p).\end{aligned}$$

Since the Hamiltonian is convex, $\mathcal{H}_{p_ip_i}(p) \ge 0$. Moreover, since the solutions u_1 and u_2 are both uniformly Lipschitz in $\mathbb{R}^N \times \mathbb{R}^+$

$$0 \le \gamma\mathcal{H}_{p_ip_i}(p) \le \gamma \sup_{|p|<C} \mathcal{H}_{p_ip_i}(p) = \bar{\gamma}.$$

Combining these estimates

$$|w|_t + \text{div}(\mathbf{V}|w|) \le \bar{\gamma}|w| \quad \text{in } \mathbb{R}^N \times \mathbb{R}^+.$$

Introduce the backward characteristic cone C_M and the truncated backward characteristic cone C_M^t as in (13.6) and (13.7), where the constant M is chosen as in (13.9). Similar calculations yield

$$\int_{|x-x_o|<M(T-t)} |w(x,t)|dx \le \bar{\gamma} \int_0^t \int_{|x-x_o|<M(T-\tau)} |w(x,\tau)|dxd\tau.$$

This implies $w = 0$ by Gronwall's inequality. ∎

16 A Weak Notion of Semi-Concavity

The most limiting requirement of the class $\mathcal{C}_1$ is that solutions have to be of class $C^2(\mathbb{R}^N \times \mathbb{R}^+)$. Such a requirement is not natural, since the equation in (6.4) imposes no conditions on the second derivatives of it solutions. In addition, Proposition 10.1 establishes only that variational solutions of (8.3) are Lipschitz continuous, however smooth the initial datum might be. On the other hand, the example of Section 12.3 shows that uniqueness fails if some assumptions are not formulated on the graph of $u(\cdot,t)$ through the second derivatives of solutions. A condition of semi-concavity can be imposed using a discrete form of second derivatives. A solution of the Cauchy problem (6.4) is weakly semi-concave if there exists a positive constant γ such that for every unit vector $\nu \in \mathbb{R}^N$ and all $h \in \mathbb{R}$

$$u(x+h\nu,t) - 2u(x,t) + u(x-h\nu,t) \le \gamma\Big(1 + \frac{1}{t}\Big)h^2. \tag{16.1}$$

Remark 16.1 The t-dependence on the right-hand side allows for non-semi-concave initial data.

For $\varepsilon > 0$, let k_ε be a mollifying kernel in $\mathbb{R}^N$, and let $u_\varepsilon(\cdot,t)$ be the mollification of $u(\cdot,t)$ with respect to the space variables, i.e.,

$$x \to u_\varepsilon(x,t) = \int_{\mathbb{R}^N} k_\varepsilon(x-y)u(y,t)dy.$$

Lemma 16.1 *Let $u(\cdot,t)$ be weakly semi-concave in the sense of (16.1). Then for every unit vector $\nu\in\mathbb{R}^N$ and all $\varepsilon>0$,*

$$u_{\varepsilon,\nu\nu}\le\gamma\Big(1+\frac{1}{t}\Big)\qquad \textit{in }\ \mathbb{R}^N\times\mathbb{R}^+.$$

Proof Fix $\nu\in\mathbb{R}^N$ and $\varepsilon>0$ and compute

$$\begin{aligned}
u_{\varepsilon,\nu\nu}(x,t)&=\int_{\mathbb{R}^N}k_{\varepsilon,\nu\nu}(x-y)u(y,t)dy\\
&=\lim_{h\to 0}\frac{1}{h^2}\int_{\mathbb{R}^N}\left[k_\varepsilon(x+h\nu-y)-2k_\varepsilon(x-y,t)+k_\varepsilon(x-h\nu-y)\right]u(y,t)dy\\
&=\lim_{h\to 0}\frac{1}{h^2}\int_{\mathbb{R}^N}k_\varepsilon(x-\eta)\left[u(\eta+h\nu,t)-2u(\eta,t)+u(\eta-h\nu,t)\right]d\eta\\
&\le\gamma\Big(1+\frac{1}{t}\Big)\int_{\mathbb{R}^N}k_\varepsilon(x-\eta)d\eta=\gamma\Big(1+\frac{1}{t}\Big).
\end{aligned}$$

∎

Corollary 16.1 *Let $u(\cdot,t)$ be weakly semi-concave in the sense of (16.1). Then for all $\varepsilon>0$*

$$(u_{\varepsilon,x_ix_j})-\gamma\Big(1+\frac{1}{t}\Big)\mathbb{I}\le 0\qquad \textit{in }\ \mathbb{R}^N\times\mathbb{R}^+.$$

17 Semi-Concavity of Hopf Variational Solutions

The semi-concavity condition (16.1) naturally arises from the variational formula (8.3). Indeed, if u_o is weakly semi-concave, the corresponding variational solution is weakly semi-concave. Moreover, if the Hamiltonian $p\to\mathcal{H}(p)$ is strictly convex, then the corresponding variational solution is weakly semi-concave irrespective of whether the initial datum is weakly semi-concave. The next two sections contain these results. Here we stress that they hold for the variational solutions (8.3) and not necessarily for any solution of the Cauchy problem (6.4).

17.1 Weak Semi-Concavity of Hopf Variational Solutions Induced by the Initial Datum u_o

Proposition 17.1 *Let u_o be weakly semi-concave, that is there exists a positive constant γ_o such that for every unit vector $\nu\in\mathbb{R}^N$ and all $h\in\mathbb{R}$*

$$u_o(x+h\nu)-2u_o(x)+u(x-h\nu)\le\gamma_o h^2\qquad \textit{in }\ \mathbb{R}.$$

Then $x\to u(x,t)$ is weakly semi-concave, uniformly in t, for the same constant γ_o.

Proof Let $\xi \in \mathbb{R}^N$ be a vector where the minimum in (8.3) is achieved. Then

$$u(x \pm h\nu, t) = \min_{y\in\mathbb{R}^N} \Big[t\mathcal{L}\Big(\frac{x \pm h\nu - y}{t}\Big) + u_o(y)\Big] \le t\mathcal{L}\Big(\frac{x-\xi}{t}\Big) + u_o(\xi \pm h\nu).$$

From this

$$\begin{aligned} u(x + h\nu, t)-2u(x,t) &+ u(x - h\nu, t) \\ &\le t\mathcal{L}\Big(\frac{x-\xi}{t}\Big) + u_o(\xi + h\nu) - 2t\mathcal{L}\Big(\frac{x-\xi}{t}\Big) \\ &\quad - 2u_o(\xi) + t\mathcal{L}\Big(\frac{x-\xi}{t}\Big) + u_o(\xi - h\nu) \\ &= u_o(\xi + h\nu) - 2u_o(\xi) + u_o(\xi - h\nu) \le \gamma_o h^2. \end{aligned}$$ ■

17.2 Strictly Convex Hamiltonian

The Hamiltonian $p \to \mathcal{H}(p)$ is *strictly* convex, if there exists a positive constant c_o such that $(\mathcal{H}_{p_ip_j}) \ge c_o\mathbb{I}$ in $\mathbb{R}^N$.

Lemma 17.1 *Let $p \to \mathcal{H}(p)$ be strictly convex. then for all $p_1, p_2 \in \mathbb{R}^N$*

$$\mathcal{H}\Big(\frac{p_1 + p_2}{2}\Big) \le \frac{1}{2}\mathcal{H}(p_1) + \frac{1}{2}\mathcal{H}(p_2) - \frac{c_o}{8}|p_1 - p_2|^2.$$

Moreover, if $\mathcal{L}$ is the Lagrangian corresponding to $\mathcal{H}$, then for all $q_1, q_2 \in \mathbb{R}^N$

$$\frac{1}{2}\mathcal{L}(q_1) + \frac{1}{2}\mathcal{L}(q_2) \le \mathcal{L}\Big(\frac{q_1 + q_2}{2}\Big) + \frac{1}{8c_o}|q_1 - q_2|^2.$$

Proof Fix $p_1 \ne p_2$ in $\mathbb{R}^N$ and set

$$\bar{p} = \tfrac{1}{2}(p_1 + p_2), \qquad \ell = |p_2 - p_1|, \qquad \nu = (p_2 - p_1)/2\ell.$$

Consider the two segments $(p_1, \bar{p})$ and $(\bar{p}, p_2)$ with parametric equations

$$\begin{aligned} (p_1, \bar{p}) &= \{y(\sigma) = \bar{p} + \sigma\nu;\ \sigma \in (0, -\tfrac{1}{2}\ell)\} \\ (\bar{p}, p_2) &= \{y(\sigma) = \bar{p} + \sigma\nu;\ \sigma \in (0, \tfrac{1}{2}\ell)\} \end{aligned}$$

and compute

$$\begin{aligned} \mathcal{H}(\bar{p}) &= \mathcal{H}(p_1) + \int_0^{\ell/2} D_\sigma\mathcal{H}(\bar{p} - \sigma\nu)\cdot\nu d\sigma \\ \mathcal{H}(\bar{p}) &= \mathcal{H}(p_2) - \int_0^{\ell/2} D_\sigma\mathcal{H}(\bar{p} + \sigma\nu)\cdot\nu d\sigma. \end{aligned}$$

Adding them up

$$2\mathcal{H}(\bar{p}) = \mathcal{H}(p_1) + \mathcal{H}(p_2) - \int_0^{\ell/2} [D_\sigma \mathcal{H}(\bar{p} + \sigma\nu) - D_\sigma \mathcal{H}(\bar{p} - \sigma\nu)] \cdot \nu d\sigma.$$

By the mean value theorem, there exists some $\sigma' \in (0, \frac{1}{2}\ell)$ such that

$$[D_\sigma \mathcal{H}(\bar{p} + \sigma\nu) - D_\sigma \mathcal{H}(\bar{p} - \sigma\nu)] \cdot \nu = \mathcal{H}_{p_i p_j}(\bar{p} - \sigma'\nu)\nu_i\nu_j 2\sigma \ge c_o 2\sigma.$$

Combining these calculations

$$2\mathcal{H}(\bar{p}) \le \mathcal{H}(p_1) + \mathcal{H}(p_2) - c_o \int_0^{\ell/2} 2\sigma d\sigma.$$

This proves the first statement. To prove the second, recall that $\mathcal{L}$ is the Legendre transform of $\mathcal{H}$. Therefore

$$\mathcal{L}(q_1) \le q_1 \cdot p_1 - \mathcal{H}(p_1), \qquad \mathcal{L}(q_2) \le q_2 \cdot p_2 - \mathcal{H}(p_2)$$

for all $p_1, p_2 \in \mathbb{R}^N$. From this

$$\begin{aligned} \frac{1}{2}\mathcal{L}(q_1) + \frac{1}{2}\mathcal{L}(q_2) &\le \frac{1}{2}(q_1 \cdot p_1 + q_2 \cdot p_2) - \Big(\frac{1}{2}\mathcal{H}(p_1) + \frac{1}{2}\mathcal{H}(p_2)\Big) \\ &\le \frac{1}{2}(q_1 \cdot p_1 + q_2 \cdot p_2) - \mathcal{H}\Big(\frac{p_1 + p_2}{2}\Big) - \frac{c_o}{8}|p_1 - p_2|^2. \end{aligned}$$

Transform

$$\frac{1}{2}(q_1 \cdot p_1 + q_2 \cdot p_2) = \Big(\frac{q_1 + q_2}{2}\Big)\Big(\frac{p_1 + p_2}{2}\Big) + \frac{1}{4}(q_1 - q_2) \cdot (p_1 - p_2)$$

and combine with the previous inequality to obtain

$$\begin{aligned} &\frac{1}{2}\mathcal{L}(q_1) + \frac{1}{2}\mathcal{L}(q_2) \le \Big(\frac{q_1 + q_2}{2}\Big)\Big(\frac{p_1 + p_2}{2}\Big) - \mathcal{H}\Big(\frac{p_1 + p_2}{2}\Big) \\ &- \Big(\frac{c_o}{8}|p_1 - p_2|^2 - \frac{1}{4}(q_1 - q_2) \cdot (p_1 - p_2) + \frac{1}{8c_o}|q_1 - q_2|^2\Big) + \frac{1}{8c_o}|q_1 - q_2|^2 \\ &\le \max_{p \in \mathbb{R}^N} \Big[\Big(\frac{q_1 + q_2}{2}\Big) \cdot p - \mathcal{H}(p)\Big] - \Big[\sqrt{\frac{c_o}{8}}(p_1 - p_2) - \frac{1}{\sqrt{8c_o}}(q_1 - q_2)\Big]^2 \\ &\quad + \frac{1}{8c_o}|q_1 - q_2|^2 \\ &\le \mathcal{L}\Big(\frac{q_1 + q_2}{2}\Big) + \frac{1}{8c_o}|q_1 - q_2|^2. \end{aligned}$$

■

Proposition 17.2 *Let $\mathcal{H}$ be strictly convex. Then every variational solution of (8.3) is weakly semi-concave for all $t > 0$, in the sense of (16.1), for a constant γ independent of u_o.*

Proof Let $\xi \in \mathbb{R}^N$ be a vector where the minimum in (8.3) is achieved. Then

$$\begin{aligned}
u(x+h\nu,t) - 2u(x,t) + u(x-h\nu,t) &= \min_{y\in\mathbb{R}^N}\Big[t\mathcal{L}\Big(\frac{x+h\nu-y}{t}\Big) + u_o(y)\Big] \\
&\quad - 2\Big[t\mathcal{L}\Big(\frac{x-\xi}{t}\Big) + u_o(\xi)\Big] + \min_{y\in\mathbb{R}^N}\Big[t\mathcal{L}\Big(\frac{x-h\nu-y}{t}\Big) + u_o(y)\Big] \\
&\le \Big[t\mathcal{L}\Big(\frac{x+h\nu-\xi}{t}\Big) + u_o(\xi)\Big] - 2\Big[t\mathcal{L}\Big(\frac{x-\xi}{t}\Big) + u_o(\xi)\Big] \\
&\quad + \Big[t\mathcal{L}\Big(\frac{x-h\nu-\xi}{t}\Big) + u_o(\xi)\Big] \\
&= 2t\Big[\frac{1}{2}\mathcal{L}\Big(\frac{x+h\nu-\xi}{t}\Big) + \frac{1}{2}\mathcal{L}\Big(\frac{x-h\nu-\xi}{t}\Big) - \mathcal{L}\Big(\frac{x-\xi}{t}\Big)\Big] \\
&\le 2t\frac{|2h\nu/t|^2}{8c_o} = \frac{h^2}{tc_o}.
\end{aligned}$$

■

18 Uniqueness of Weakly Semi-Concave Variational Hopf Solutions

Introduce the class $\mathcal{C}_2$ of solutions

$$\mathcal{C}_2 = \begin{cases} u \ \text{ is a variational solution of (8.3)} \\ u(\cdot,t) \ \text{ is weakly semi-concave in the sense of (16.1)} \\ |\nabla u| \le C \ \text{ for some } \ C>0 \ \text{ in } \ \mathbb{R}^N\times\mathbb{R}^+. \end{cases} \tag{18.1}$$

Theorem 18.1 *The Cauchy problem has at most one solution within the class* $\mathcal{C}_2$.

Proof Let u_1 and u_2 be two solutions in $\mathcal{C}_2$ and set $w = u_1 - u_2$. Proceeding as in the proof of Proposition 13.1, we arrive at an analogue of (13.2), which in this context holds a.e. in $\mathbb{R}^N\times\mathbb{R}^+$. From the latter, we derive

$$f(w)_t = -\mathbf{V}\cdot D_x f(w) \qquad \text{a.e. in } \ \mathbb{R}^N\times\mathbb{R}^+$$

for any $f\in C^1(\mathbb{R})$, where $\mathbf{V}$ is defined in (13.3). For $\varepsilon>0$, let $u_{\varepsilon,1}$ and $u_{\varepsilon,2}$ be the mollifications of u_1 and u_2 as in Section 16, and set

$$\mathbf{V}_\varepsilon = \int_0^1 D_p\mathcal{H}(sD_x u_{\varepsilon,2} + (1-s)D_x u_{\varepsilon,1})ds.$$

With this notation

$$f(w)_t = -\operatorname{div}(\mathbf{V}_\varepsilon f(w)) + f(w)\operatorname{div}\mathbf{V}_\varepsilon + (\mathbf{V}_\varepsilon - \mathbf{V})\cdot D_x f(w).$$

Lemma 18.1 *There exists a positive constant* $\bar\gamma$, *independent of* ε, *such that*

$$\operatorname{div}\mathbf{V}_\varepsilon \le \bar\gamma\Big(1+\frac{1}{t}\Big) \qquad \text{in } \ \mathbb{R}^N\times\mathbb{R}^+.$$

Proof Same as in Lemma 15.1 with the proper minor modifications. ■

Putting this in the previous expression of $f(w)_t$, and assuming that $f(\cdot)$ is non-negative, gives

$$f(w)_t \le -\operatorname{div}(\mathbf{V}_\varepsilon f(w)) + \bar{\gamma}\Big(1+\frac{1}{t}\Big)f(w) + (\mathbf{V}_\varepsilon - \mathbf{V})\cdot D_x f(w). \tag{18.2}$$

Introduce the backward characteristic cone C_M and the truncated backward characteristic cone C_M^t as in (13.6) and (13.7), where the constant M is chosen as in (13.9). For a fixed $0<\sigma<t<T$, introduce also the truncated cone

$$C_M^{t,\sigma} = \big[|x-x_o| < M(T-\tau);\ 0<\sigma<\tau<t<T\big].$$

Now integrate (18.2) over such a cone. Proceeding as in the proof of Proposition 13.1, and taking into account the choice (13.9) of M, yields

$$\begin{aligned}\int_{|x-x_o|<M(T-t)} f(w(x,t))dx \le& \int_{|x-x_o|<M(T-\sigma)} f(w(x,\sigma))dx \\ &+ \bar{\gamma}\int_\sigma^t \Big(1+\frac{1}{t}\Big)\int_{|x-x_o|<M(T-\tau)} f(w(x,\tau))dxd\tau \\ &+ \int_\sigma^t\int_{|x-x_o|<M(T-\tau)} (\mathbf{V}_\varepsilon - \mathbf{V})\cdot D_x f(w)dxd\tau.\end{aligned} \tag{18.3}$$

By the properties of the class $\mathcal{C}_2$, $|D_x w|$, $|D_x u_1|$, and $|D_x u_2|$ are a.e. bounded in $\mathbb{R}^N\times\mathbb{R}^+$, independent of ε. Therefore

$$\lim_{\varepsilon\to 0}\int_\sigma^t\int_{|x-x_o|<M(T-\tau)} (\mathbf{V}_\varepsilon - \mathbf{V})\cdot D_x w dxd\tau = 0.$$

Observe first that (18.3) continues to hold for non-negative functions $f(\cdot)$, uniformly Lipschitz continuous in $\mathbb{R}$. Choose $f_\delta(w) = (|w|-\delta)_+$, for some fixed $\delta\in(0,1)$. There exists $\sigma>0$ such that $f_\delta(w(\cdot,\tau)) = 0$ for all $\tau\in(0,\sigma]$. Indeed, by virtue of (10.3), for all $\tau\in(0,\sigma]$

$$\begin{aligned}(|w(x,\tau)|-\delta)_+ &\le (|u_1(x,\tau)-u_o(x)| + |u_2(x,\tau)-u_o(x)| - \delta)_+ \\ &\le (2\bar{C}\sigma - \delta)_+ = 0\end{aligned}$$

provided $\sigma < \delta/2\bar{C}$. These remarks in (18.3) yield

$$\begin{aligned}&\int_{|x-x_o|<M(T-t)} f_\delta(w(x,t))dx \\ &\qquad \le \bar{\gamma}\int_\sigma^t \Big(1+\frac{1}{t}\Big)\int_{|x-x_o|<M(T-\tau)} f_\delta(w(x,\tau))dxd\tau \\ &\qquad \le \bar{\gamma}\Big(1+\frac{1}{\sigma}\Big)\int_\sigma^t\int_{|x-x_o|<M(T-\tau)} f_\delta(w(x,\tau))dxd\tau.\end{aligned}$$

Setting

$$\varphi_\delta(t) = \int_\sigma^t \int_{|x-x_o|<M(T-\tau)} f_\delta(w(x,\tau))dxd\tau$$

the previous inequality reads as

$$\varphi'_\delta(t) \le \bar{\gamma}\Big(1+\frac{1}{\sigma}\Big)\varphi_\delta(t) \qquad \text{and } \varphi_\delta(\sigma) = 0.$$

This implies that $\varphi_\delta(\tau) = 0$ for all $\tau \in (0,t)$, and since (x_o, T) is arbitrary, $|w| \le \delta$ in $\mathbb{R}^N \times \mathbb{R}^+$, for all $\delta > 0$. ∎

9

Linear Elliptic Equations with Measurable Coefficients

1 Weak Formulations and Weak Derivatives

Let E be a bounded domain in $\mathbb{R}^N$ with boundary ∂E of class C^1. Denote by (a_{ij}) an $N \times N$ symmetric matrix with entries $a_{ij} \in L^\infty(E)$, and satisfying the *ellipticity condition*

$$\lambda|\xi|^2 \le a_{ij}(x)\xi_i\xi_j \le \Lambda|\xi|^2 \tag{1.1}$$

for all $\xi \in \mathbb{R}^N$ and all $x \in E$, for some $0 < \lambda \le \Lambda$. The number Λ is the least upper bound of the eigenvalues of (a_{ij}) in E, and λ is their greatest lower bound. A vector-valued function $\mathbf{f} = (f_1, \dots, f_N) : E \to \mathbb{R}^N$ is said to be in $L^p_{\text{loc}}(E)$, for some $p \ge 1$, if all the components $f_j \in L^p_{\text{loc}}(E)$. Given a scalar function $f \in L^1_{\text{loc}}(E)$ and a vector-valued function $\mathbf{f} \in L^1_{\text{loc}}(E)$, consider the formal partial differential equation in divergence form (Section 3.1 of the Preliminaries)

$$-\big(a_{ij}u_{x_i}\big)_{x_j} = \operatorname{div}\mathbf{f} - f \quad \text{ in } E. \tag{1.2}$$

Expanding formally the indicated derivatives gives a PDE of the type of (3.1) of Chapter 1, which, in view of the ellipticity condition (1.1), does not admit real characteristic surfaces (Section 3 of Chapter 1). In this formal sense, (1.2) is a *second-order elliptic* equation.

Multiply (1.2) formally by a function $v \in C_o^\infty(E)$ and formally integrate by parts in E to obtain

$$\int_E \big(a_{ij}u_{x_i}v_{x_j} + f_j v_{x_j} + fv\big)dx = 0. \tag{1.3}$$

This is well defined for all $v \in C_o^\infty(E)$, provided $\nabla u \in L^p_{\text{loc}}(E)$, for some $p \ge 1$. In such a case (1.3) is the *weak formulation* of (1.2), and u is a *weak solution*. Such a weak notion of solution coincides with the classical one whenever the various terms in (1.3) are sufficiently regular. Indeed, assume that $f \in C(E)$ and $a_{ij}, \mathbf{f} \in C^1(E)$; if a function $u \in C^2(E)$ satisfies (1.3) for all $v \in C_o^\infty(E)$, integrating by parts gives

E. DiBenedetto, *Partial Differential Equations: Second Edition*,
Cornerstones, DOI 10.1007/978-0-8176-4552-6_10,

$$\int_E \big[\big(a_{ij}u_{x_i}\big)_{x_j} + \operatorname{div}\mathbf{f} + f)\big]v dx = 0 \quad \text{for all } v \in C_o^\infty(E).$$

Thus u satisfies (1.2) in the classical sense. It remains to clarify the meaning of $\nabla u \in L^p_{\rm loc}(E)$ for some $p \ge 1$.

1.1 Weak Derivatives

A function $u \in L^p_{\rm loc}(E)$, for some $p \ge 1$, has a weak partial derivative in $L^p_{\rm loc}(E)$ with respect to the variable x_j if there exists a function $w_j \in L^p_{\rm loc}(E)$ such that

$$\int_E u v_{x_j}\, dx = -\int_E w_j v\, dx \quad \text{for all } v \in C_o^\infty(E). \tag{1.4}$$

If $u \in C^1(E)$, then $w_j = u_{x_j}$ in the classical sense. There are functions admitting weak and not classical derivatives. As an example, $u(x) = |x|$ for $x \in (-1,1)$ does not have a derivative at $x = 0$; however it admits the weak derivative

$$w = \begin{cases} -1 & \text{in } (-1,0) \\ 1 & \text{in } (0,1) \end{cases} \qquad \text{as an element of } L^1(-1,1).$$

With a perhaps improper but suggestive symbolism we set $w_j = u_{x_j}$, warning that in general, u_{x_j} need not be the limit of difference quotients along x_j, and it is meant only in the sense of (1.4). The derivatives u_{x_j} in (1.3) are meant in this weak sense, and solutions of (1.1) are sought as functions in the *Sobolev space* ([142])

$$W^{1,p}(E) = \{\text{the set of } u \in L^p(E) \text{ such that } \nabla u \in L^p(E)\}. \tag{1.5}$$

A norm in $W^{1,p}(E)$ is

$$\|u\|_{1,p} = \|u\|_p + \|\nabla u\|_p. \tag{1.6}$$

Proposition 1.1 *$W^{1,p}(E)$ is a Banach space for the norm (1.6). Moreover, $C^\infty(E)$ is dense in $W^{1,p}(E)$ ([105]).*

Introduce also the two spaces

$$W_o^{1,p}(E) = \{\text{the closure of } C_o^\infty(E) \text{ in the norm (1.6)}\}. \tag{1.7}$$

$$\tilde W^{1,p}(E) = \Big\{\text{all } u \in W^{1,p}(E) \text{ such that } \int_E u\, dx = 0\Big\}. \tag{1.8}$$

Proposition 1.2 *$W_o^{1,p}(E)$ and $\tilde W^{1,p}(E)$ equipped with the norm (1.6) are Banach spaces.*

Functions in $W^{1,p}(E)$ are more "regular" than merely elements in $L^p(E)$, on several accounts. First, they are embedded in $L^q(E)$ for some $q > p$. Second, they form a compact subset of $L^p(E)$. Third, they have boundary values (traces) on ∂E, as elements of $L^p(\partial E)$.

2 Embeddings of $W^{1,p}(E)$

Since ∂E is of class C^1, there is a circular spherical cone $\mathcal{C}$ of height h and solid angle ω such that by putting its vertex at any point of ∂E, it can be properly swung, by a rigid rotation, to remain in E. This is the *cone condition* of ∂E. Denote by $\gamma = \gamma(N,p)$ a constant depending on N and p and independent of E and ∂E.

Theorem 2.1 (Sobolev–Nikol'skii [143]) *If $1 < p < N$ then $W^{1,p}(E) \hookrightarrow L^{p^*}(E)$, where $p^* = \frac{Np}{N-p}$, and there exists $\gamma = \gamma(N,p)$, such that*

$$\|u\|_{p^*} \le \frac{\gamma}{\omega}\Big\{\frac{1}{h}\|u\|_p + \|\nabla u\|_p\Big\}, \quad p^* = \frac{Np}{N-p}, \quad \textit{for all } u \in W^{1,p}(E). \tag{2.1}$$

If $p = 1$ and $|E| < \infty$, then $W^{1,p}(E) \hookrightarrow L^q(E)$ for all $1 \le q < \frac{N}{N-1}$ and there exists $\gamma = \gamma(N,q)$ such that

$$\|u\|_q \le \frac{\gamma}{\omega}|E|^{\frac{1}{q}-\frac{N-1}{N}}\Big\{\frac{1}{h}\|u\|_1 + \|\nabla u\|_1\Big\} \quad \textit{for all } u \in W^{1,1}(E). \tag{2.2}$$

If $p > N$ then $W^{1,p}(E) \hookrightarrow L^\infty(E)$, and there exists $\gamma = \gamma(N,p)$ such that

$$\|u\|_\infty \le \frac{\gamma}{\omega h^{N/p}}(\|u\|_p + h\|\nabla u\|_p) \quad \textit{for all } u \in W^{1,p}(E). \tag{2.3}$$

If $p > N$ and in addition E is convex, then $W^{1,p}(E) \hookrightarrow C^{1-\frac{N}{p}}(\bar{E})$, and there exists $\gamma = \gamma(N,p)$ such that for every pair of points $x, y \in \bar{E}$ with $|x-y| \le h$ ([107])

$$|u(x) - u(y)| \le \frac{\gamma}{\omega}|x-y|^{1-\frac{N}{p}}\|\nabla u\|_p \quad \textit{for all } u \in W^{1,p}(E). \tag{2.4}$$

Remark 2.1 The constants $\gamma(N,p)$ can be computed explicitly, and they tend to infinity as $p \to N$. This is expected as $W^{1,N}(E)$ is not embedded in $L^\infty(E)$. Indeed the function

$$[|x| < e^{-1}] - \{0\} \ni x \to \ln|\ln|x|| \in W^{1,N}(|x| < e^{-1})$$

is not essentially bounded about the origin. In this sense these embeddings are sharp ([148]). In (2.2) the value $q = 1^* = \frac{N}{N-1}$ is not permitted, and the corresponding constant $\gamma(N,q) \to \infty$ as $q \to 1^*$. The limiting embedding for $p = N$ takes a special form ([31], Chapter IX, Section 13).

Remark 2.2 The structure of ∂E enters only through the solid angle ω and the height h of the cone condition of ∂E. Therefore (2.1)–(2.4) continue to hold for domains whose boundaries merely satisfy the cone condition.

Remark 2.3 If E is not convex, the estimate (2.4) can be applied locally. Thus if $p > N$, a function $u \in W^{1,p}(E)$ is locally Hölder continuous in E.

A proof of these embeddings is in Section 2c of the Problems and Complements.

2.1 Compact Embeddings of $W^{1,p}(E)$

Theorem 2.2 (Reillich–Kondrachov [127, 83]) *Let $1 \le p < N$. Then for all $1 \le q < p^*$, the embedding $W^{1,p}(E) \hookrightarrow L^q(E)$ is compact.*

A proof is in 2.2c of the Problems and Complements.

Corollary 2.1 *Let $\{u_n\}$ be a bounded sequence in $W^{1,p}(E)$. If $1 \le p < N$, for each fixed $1 \le q < \frac{Np}{N-p}$ there exist a subsequence $\{u_{n'}\} \subset \{u_n\}$, and $u \in W^{1,p}(E)$ such that*

$$\{u_{n'}\} \to u \;\; \text{weakly in } W^{1,p}(E) \quad \text{and} \quad \{u_{n'}\} \to u \;\; \text{strongly in } L^q(E).$$

3 Multiplicative Embeddings of $W_o^{1,p}(E)$ and $\tilde{W}^{1,p}(E)$

Theorem 3.1 (Gagliardo–Nirenberg [50, 112]) *If $1 \le p < N$, then $W_o^{1,p}(E) \hookrightarrow L^{p^*}(E)$ for $p^* = \frac{Np}{N-p}$, and there exists $\gamma = \gamma(N,p)$ such that*

$$\|u\|_{p^*} \le \gamma \|\nabla u\|_p \qquad p^* = \frac{Np}{N-p} \qquad \text{for all } u \in W_o^{1,p}(E). \tag{3.1}$$

If $p = N$, then $W_o^{1,p}(E) \hookrightarrow L^q(E)$ for all $q > p$, and there exists $\gamma = \gamma(N,q)$ such that

$$\|u\|_q \le \gamma \|\nabla u\|_p^{1-\frac{p}{q}} \|u\|_p^{\frac{p}{q}} \qquad \text{for all } u \in W_o^{1,p}(E). \tag{3.2}$$

If $p > N$, then $W_o^{1,p}(E) \hookrightarrow L^\infty(E)$, and there exists $\gamma = \gamma(N,p)$ such that

$$\|u\|_\infty \le \gamma \|\nabla u\|_p^{\frac{N}{p}} \|u\|_p^{1-\frac{N}{p}} \qquad \text{for all } u \in W_o^{1,p}(E). \tag{3.3}$$

Remark 3.1 The constants $\gamma(N,p)$ can be computed explicitly independent of ∂E, and they tend to infinity as $p \to N$. Unlike (2.2), the value 1^* is permitted in (3.1).

Functions in $W_o^{1,p}(E)$ are limits of functions in $C_o^\infty(E)$ in the norm of $W_o^{1,p}(E)$, and in this sense they vanish on ∂E. This permits embedding inequalities such as (3.1)–(2.3) with constants γ independent of E and ∂E. Inequalities of this kind would not be possible for functions $u \in W^{1,p}(E)$. For example, a constant non-zero function would not satisfy any of them. This suggest that for them to hold some information is required on some values of u. Let

$$u_E = \frac{1}{|E|} \int_E u \, dx$$

denote the integral average of u over E. The multiplicative embeddings (3.1)–(3.3) continue to hold for functions of zero average. Denote by $\gamma = \gamma(N, E, \partial E)$ a constant depending on N, $|E|$, and the C^1-smoothness of ∂E, but invariant under homothetic transformations of E.

Theorem 3.2 (Golovkin–Poincarè [59]) *If $1 < p < N$, then $W^{1,p}(E) \hookrightarrow L^{p^*}(E)$, where $p^* = \frac{Np}{N-p}$, and there exists $\gamma = \gamma(N, p, E, \partial E)$ such that*

$$\|u - u_E\|_{p^*} \le \gamma \|\nabla u\|_p, \qquad p^* = \frac{Np}{N-p} \quad \text{for all } u \in W^{1,p}(E). \tag{3.4}$$

If $p > N$, then $W_o^{1,p}(E) \hookrightarrow L^\infty(E)$, and there exists $\gamma = \gamma(N, p)$ such that

$$\|u - u_E\|_\infty \le \gamma \|\nabla u\|_p^{\frac{N}{p}} \|u - u_E\|_p^{1-\frac{N}{p}} \quad \text{for all } u \in W^{1,p}(E). \tag{3.5}$$

If $p = N$, then $W^{1,p}(E) \hookrightarrow L^q(E)$ for all $q > p$, and there exists $\gamma = \gamma(N, q)$ such that

$$\|u - u_E\|_q \le \gamma \|\nabla u\|_p^{1-\frac{p}{q}} \|u - u_E\|_p^{\frac{p}{q}} \quad \text{for all } u \in W^{1,p}(E). \tag{3.6}$$

Remark 3.2 When E is convex, a simple proof of Theorem 3.2 is due to Poincarè and it is reported in Section 3.2c of the Complements. In such a case the constant $\gamma(N, E, \partial E)$ in (3.4) has the form

$$\gamma(N, E, \partial E) = C \frac{(\operatorname{diam} E)^N}{|E|}$$

for some absolute constants $C > 1$ depending only on N.

3.1 Some Consequences of the Multiplicative Embedding Inequalities

Corollary 3.1 *The norm $\|\cdot\|_{1,p}$ in $W_o^{1,p}(E)$, introduced in (1.6), is equivalent to $\|\nabla u\|_p$; that is, there exists a positive constant $\gamma_o = \gamma_o(N, p, E)$ such that*

$$\gamma_o \|u\|_{1,p} \le \|\nabla u\|_p \le \|u\|_{1,p} \quad \text{for all } u \in W_o^{1,p}(E). \tag{3.7}$$

Corollary 3.2 *The norm $\|\cdot\|_{1,p}$ in $\tilde{W}^{1,p}(E)$, is equivalent to $\|\nabla u\|_p$; that is, there exists a positive constant $\tilde{\gamma}_o = \tilde{\gamma}_o(N, p, E)$ such that*

$$\tilde{\gamma}_o \|u\|_{1,p} \le \|\nabla u\|_p \le \|u\|_{1,p} \quad \text{for all } u \in \tilde{W}^{1,p}(E). \tag{3.8}$$

Corollary 3.3 *$W_o^{1,2}(E)$ and $\tilde{W}^{1,2}(E)$ are Hilbert spaces with equivalent inner products*

$$\langle u, v \rangle + \langle \nabla u, \nabla v \rangle \quad \text{and} \quad \langle \nabla u, \nabla v \rangle \tag{3.9}$$

where $\langle \cdot, \cdot \rangle$ denotes the standard inner product in $L^2(E)$.

4 The Homogeneous Dirichlet Problem

Given $\mathbf{f}, f \in L^\infty(E)$, consider the homogeneous Dirichlet problem

$$\begin{aligned} -\big(a_{ij}u_{x_i}\big)_{x_j} &= \operatorname{div}\mathbf{f} - f && \text{in } E \\ u\big|_{\partial E} &= 0 && \text{on } \partial E. \end{aligned} \tag{4.1}$$

The PDE is meant in the weak sense (1.3) by requiring that $u \in W^{1,p}(E)$ for some $p \ge 1$. The homogeneous boundary datum is enforced, in a weak form, by requiring that u be in the space $W_o^{1,p}(E)$ defined in (1.7). Seeking solutions $u \in W_o^{1,p}(E)$ implies that in (1.3), by density, one may take $v = u$. Thus, by taking into account the ellipticity condition (1.1)

$$\lambda \int_E |\nabla u|^2 dx \le \int_E (|\mathbf{f}||\nabla u| + |f||u|)dx.$$

This forces $p = 2$ and identifies $W_o^{1,2}(E)$ as the natural space where solutions of (4.1) should be sought.

Theorem 4.1 *The homogeneous Dirichlet problem (4.1) admits at most one weak solution $u \in W_o^{1,2}(E)$.*

Proof If $u_1, u_2 \in W_o^{1,2}(E)$ are weak solutions of (4.1)

$$\int_E a_{ij}(u_1 - u_2)_{x_i} v_{x_j} dx = 0 \qquad \text{for all } \ v \in W_o^{1,2}(E).$$

This and the ellipticity condition (1.1) imply $\|\nabla(u_1 - u_2)\|_2 = 0$. Thus $u_1 = u_2$ a.e. in E, by the embedding of Theorem 3.1. ∎

5 Solving the Homogeneous Dirichlet Problem (4.1) by the Riesz Representation Theorem

Regard (1.3) as made out of two pieces

$$a(u,v) = \int_E a_{ij}u_{x_i}v_{x_j}dx \quad \text{and} \quad \ell(v) = -\int_E (f_j v_{x_j} + fv)dx \tag{5.1}$$

for all $v \in W_o^{1,2}(E)$. Finding a solution to (4.1) amounts to finding $u \in W_o^{1,2}(E)$ such that

$$a(u,v) = \ell(v) \quad \text{for all } \ v \in W_o^{1,2}(E). \tag{5.2}$$

The first term in (5.1) is a bilinear form in $W_o^{1,2}(E)$. By the ellipticity condition (1.1) and Corollary 3.1

$$\lambda\gamma_o^2\|u\|_{1,2}^2 \le \lambda\|\nabla u\|_2^2 \le a(u,u) \le \Lambda\|\nabla u\|_2^2 \le \Lambda\|u\|_{1,2}^2.$$

Therefore $a(\cdot,\cdot)$ is an inner product in $W_o^{1,2}(E)$ equivalent to any one of the inner products in (3.9). The second term $\ell(\cdot)$ in (5.1) is a linear functional in $W_o^{1,2}(E)$, bounded in $\|\cdot\|_{1,2}$, and thus bounded in the norm generated by the inner product $a(\cdot,\cdot)$. Therefore by the Riesz representation theorem it is represented as in (5.2) for a unique $u \in W_o^{1,2}(E)$ ([98]).[1]

6 Solving the Homogeneous Dirichlet Problem (4.1) by Variational Methods

Consider the non-linear functional

$$W_o^{1,2}(E) \ni u \to J(u) \stackrel{\text{def}}{=} \int_E \big(\tfrac{1}{2} a_{ij} u_{x_i} u_{x_j} + f_j u_{x_j} + fu\big) dx. \tag{6.1}$$

One verifies that $J(\cdot)$ is strictly convex in $W_o^{1,2}(E)$, that is

$$J\big(tu + (1-t)v\big) < tJ(u) + (1-t)J(v)$$

for every pair (u,v) of non-trivial elements of $W_o^{1,2}(E)$, and all $t \in (0,1)$. Assume momentarily $N > 2$, and let 2^{**} be the Hölder conjugate of 2^*, so that

$$2^* = \frac{2N}{N-2}, \qquad 2^{**} = \frac{2N}{N+2}, \qquad \text{and} \qquad \frac{1}{2^*} + \frac{1}{2^{**}} = 1.$$

The functional $J(u)$ is estimated above using the ellipticity condition (1.1), Hölder's inequality, and the embedding (3.1):

$$\begin{aligned} J(u) &\le \frac{1}{2}\Lambda\|\nabla u\|_2^2 + \|\mathbf{f}\|_2\|\nabla u\|_2 + \|f\|_{2^{**}}\|u\|_{2^*} \\ &\le \frac{1}{2}\Lambda\|\nabla u\|_2^2 + \left(\|\mathbf{f}\|_2 + \gamma\|f\|_{2^{**}}\right)\|\nabla u\|_2 \\ &\le \Lambda\|\nabla u\|_2^2 + \frac{1}{2\Lambda}\left(\|\mathbf{f}\|_2 + \gamma\|f\|_{2^{**}}\right)^2. \end{aligned}$$

Similarly, $J(u)$ is estimated below by

$$J(u) \ge \frac{1}{4}\lambda\|\nabla u\|_2^2 - \frac{1}{\lambda}\left(\|\mathbf{f}\|_2 + \gamma\|f\|_{2^{**}}\right)^2.$$

Therefore

$$-F_o + \tfrac{1}{4}\lambda\|\nabla u\|_2^2 \le J(u) \le \Lambda\|\nabla u\|_2^2 + F_o \tag{6.2}$$

for all $u \in W_o^{1,2}(E)$, where

$$F_o = \frac{1}{\lambda}\left(\|\mathbf{f}\|_2 + \gamma\|f\|_{2^{**}}\right)^2.$$

---

[1] The Riesz representation theorem is in [31], Chapter VI Section 18. This solution method is in [98] and it is referred to as the Lax–Milgram theorem.

Here γ is the constant appearing in the embedding inequality (3.1). These estimates imply that the convex functional $J(\cdot)$ is bounded below, and we denote by J_o its infimum. A minimum can be sought by a minimizing sequence $\{u_n\} \subset W_o^{1,2}(E)$ such that

$$J(u_n) < J(u_{n-1}) < \cdots < J(u_1) \quad \text{and} \quad \lim J(u_n) = J_o. \tag{6.3}$$

By (6.2) and (6.3), the sequence $\{u_n\}$ is bounded in $W_o^{1,2}(E)$. Therefore a subsequence $\{u_{n'}\} \subset \{u_n\}$ can be selected such that $\{u_{n'}\} \to u$ weakly in $W_o^{1,2}(E)$, and $\{J(u_{n'})\} \to J_o$. Since the norm $a(\cdot,\cdot)$ introduced in (5.1) is weakly lower semi-continuous

$$\liminf J(u_{n'}) = J_o \ge J(u).$$

Thus $J(u) = J_o$ and $J(u) \le J(w)$ for all $w \in W_o^{1,2}(E)$. Enforcing this last condition for functions $w = u + \varepsilon v$, for $\varepsilon > 0$ and $v \in C_o^\infty(E)$, gives

$$J(u) \le J(u) + \frac{\varepsilon^2}{2}\int_E a_{ij} v_{x_i} v_{x_j} dx + \varepsilon \int_E \big(a_{ij} u_{x_i} v_{x_j} + f_j v_{x_j} + f v\big)\, dx.$$

Divide by ε and let $\varepsilon \to 0$ to get

$$\int_E \big(a_{ij} u_{x_i} v_{x_j} + f_j v_{x_j} + f v\big)\, dx \ge 0 \quad \text{for all } v \in C_o^\infty(E).$$

Changing v into $-v$ shows that this inequality actually holds with equality for all $v \in C_o^\infty(E)$ and establishes that u is a weak solution of the Dirichlet problem (4.1). In view of its uniqueness, as stated in Theorem 4.1 the whole minimizing sequence $\{u_n\}$ converges weakly to the unique minimizer of $J(\cdot)$. Summarizing, the PDE in divergence form (4.1) is associated with a natural functional $J(\cdot)$ in $W_o^{1,2}(E)$, whose minimum is a solutions of the PDE in $W_o^{1,2}(E)$. Conversely, the functional $J(\cdot)$ in $W_o^{1,2}(E)$ generates naturally a PDE in divergence form whose solutions are the solutions of the homogeneous Dirichlet problem (4.1).

6.1 The Case $N = 2$

The arguments are the same except for a different use of the embedding inequality (3.1) in estimating the term containing fu in (6.1). Pick $1 < p < 2$, define p^* as in (3.1), and let p^{**} be its Hölder conjugate, so that

$$p^* = \frac{Np}{N-p}, \qquad p^{**} = \frac{Np}{N(p-1)+p}, \qquad \frac{1}{p^*} + \frac{1}{p^{**}} = 1.$$

One verifies that $p^{**} > 1$ and estimates

$$\int_E fu\, dx \le \|f\|_{p^{**}} \|u\|_{p^*} \le \gamma \|f\|_{p^{**}} \|\nabla u\|_p \le \gamma |E|^{\frac{2-p}{2}} \|f\|_{p^{**}} \|\nabla u\|_2.$$

The proof now proceeds as before except that the term $\gamma\|f\|_{2^{**}}$ is now replaced by $\gamma|E|^{2-p}2\|f\|_{p^{**}}$.

6.2 Gâteaux Derivative and The Euler Equation of $J(\cdot)$

If u is a minimum for $J(\cdot)$ the function of one variable $t \to J(u+tv)$, for an arbitrary but fixed $v \in C_o^\infty(E)$, has a minimum for $t=0$. Therefore

$$0 = \frac{d}{dt} J(u+tv) \Big|_{t=0} = \int_E \big(a_{ij} u_{x_i} v_{x_j} + f_j v_{x_j} + fv \big) dx. \tag{6.4}$$

This reinforces that solutions of the Dirichlet problem (4.1) are minima of $J(\cdot)$ and vice versa. The procedure leading to (6.4), which is called the Euler equation of $J(\cdot)$, has a broader scope. It can be used to connect stationary points, not necessarily minima, of a functional $J(\cdot)$, for which no convexity information is available, to solutions of its Euler equation.

The derivative of $\varepsilon \to J(u+\varepsilon\varphi)$ at $\varepsilon = 0$ is called the Gâteaux derivative of $J(\cdot)$ at u. Relative variations from $J(u)$ to $J(v)$ are computed along the "line" in $\mathbb{R} \times W_o^{1,2}(E)$ originating at u and "slope" φ. In this sense Gâteaux derivatives are directional derivatives. A more general notion of derivative of $J(\cdot)$ is that of Fréchet derivative, where relative variations from $J(u)$ to $J(v)$ are computed for any v however varying in a $W_o^{1,2}(E)$-neighborhood of u ([6], Chapter I, Section 1.2).

7 Solving the Homogeneous Dirichlet Problem (4.1) by Galerkin Approximations

Let $\{w_n\}$ be a countable, complete, orthonormal system for $W_o^{1,2}(E)$. Such a system exists, since $W_o^{1,2}(E)$ is separable, and it can be constructed, for example, by the Gram–Schmidt procedure, for the natural inner product $a(u,v)$ introduced in (5.1). Thus in particular

$$A_{hk} = \int_E a_{ij} w_{h,x_i} w_{k,x_j} dx = \delta_{hk} \tag{7.1}$$

where δ_{hk} is the Kronecker delta. Every $v \in W_o^{1,2}(E)$ admits the representation as $v = \sum v_k w_k$ for constants v_k. The Galerkin method consists in constructing the solution u of (4.1) as the weak $W_o^{1,2}(E)$ limit, as $n \to \infty$, of the finite-dimensional approximations

$$u_n = \sum_{h=1}^{n} c_{n,h} w_h$$

where the coefficients $c_{n,h}$ are computed by enforcing an n-dimensional version of the PDE in its weak form (1.3). Precisely, u_n is sought as the solution of

$$\int_E \Big(a_{ij} u_{n,x_i} \sum_{k=1}^{n} v_k w_{k,x_j} + f_j \sum_{k=1}^{n} v_k w_{k,x_j} + f \sum_{k=1}^{n} v_k w_k \Big) dx = 0 \tag{7.2}$$

for all $v \in \overset{o}{W}{}^{1,2}(E)$. From this

$$\sum_{k=1}^{n} v_k \Big\{ \sum_{h=1}^{n} c_{n,h} \int_E a_{ij} w_{h,x_i} w_{k,x_j} dx + \int_E (f_j w_{k,x_j} + f w_k)\, dx \Big\} = 0$$

Setting

$$A_{hk} = \int_E a_{ij} w_{h,x_i} w_{k,x_j} dx, \qquad \phi_k = - \int_E (f_j w_{k,x_j} + f w_k)\, dx$$

and taking into account that $v \in \overset{o}{W}{}^{1,2}(E)$ is arbitrary, the coefficients $c_{n,h}$ are computed as the unique solution of the linear algebraic system

$$\sum_{h=1}^{n} c_{n,h} A_{hk} = \phi_k \qquad \text{for } k = 1, \dots, n. \tag{7.3}$$

By (7.1), A_{hk} is the identity matrix, and therefore $c_{n,h} = \phi_h$ for all $n, h \in \mathbb{N}$. With u_n so determined, put $v = u_n$ in (7.2), and assuming momentarily that $N > 2$, estimate

$$\lambda \|\nabla u_n\|_2^2 \le \|\mathbf{f}\|_2 \|\nabla u_n\|_2 + \|f\|_{2_{**}} \|u_n\|_{2^*} \le \big(\|\mathbf{f}\|_2 + \gamma \|f\|_{2_{**}} \big) \|\nabla u_n\|_2$$

where γ is the constant of the embedding inequality (3.1). Therefore $\{u_n\}$ is bounded in $\overset{o}{W}{}^{1,2}(E)$, and a subsequence $\{u_{n'}\} \subset \{u_n\}$ can be selected, converging weakly to some $u \in \overset{o}{W}{}^{1,2}(E)$. Letting $n \to \infty$ in (7.2) shows that such a u is a solution of the homogeneous Dirichlet problem (4.1). In view of its uniqueness, the whole sequence $\{u_n\}$ of finite-dimensional approximations converges weakly to u. The same arguments hold true for $N = 2$, by the minor modifications indicated in Section 6.1.

7.1 On the Selection of an Orthonormal System in $\overset{o}{W}{}^{1,2}(E)$

The selection of the complete system $\{w_n\}$ is arbitrary. For example the Gram–Schmidt procedure could be carried on starting from a countable collection of linearly independent elements of $\overset{o}{W}{}^{1,2}(E)$, and using anyone of the equivalent inner products in (3.9), or the construction could be completely independent of Gram–Schmidt procedure. The Galerkin method continues to hold, except that the coefficients A_{hk} defined in (7.1) are no longer identified by the Kronecker symbol. This leads to the determination of the coefficients $c_{n,h}$ as solutions of the linear algebraic system (7.3), whose leading $n \times n$ matrix (A_{hk}) is still invertible, for all n, because of the ellipticity condition (1.1). The corresponding unique solutions might depend on the nth approximating truncation, and therefore are labeled by $c_{n,h}$.

While the method is simple and elegant, it hinges on a suitable choice of complete system in $\overset{o}{W}{}^{1,2}(E)$. Such a choice is suggested by the specific geometry of E and the structure of the matrix (a_{ij}) ([21]).

7.2 Conditions on f and f for the Solvability of the Dirichlet Problem (4.1)

Revisiting the proofs of these solvability methods shows that the only required conditions on $\mathbf{f}$ and f are

$$\mathbf{f} \in L^2(E) \qquad f \in L^q(E) \text{ where } \begin{cases} q = 2^{**} & \text{if } N > 2 \\ \text{any } q > 1 & \text{if } N = 2. \end{cases} \tag{7.4}$$

Theorem 7.1 *Let (7.4) hold. Then the homogeneous Dirichlet problem (4.1) admits a unique solution.*

8 Traces on ∂E of Functions in $W^{1,p}(E)$

8.1 The Segment Property

The boundary ∂E has the *segment property* if there exist a locally finite open covering of ∂E with balls $\{B_t(x_j)\}$ centered at $x_j \in \partial E$ with radius t, a corresponding sequence of unit vectors $\mathbf{n}_j$, and a number $t^* \in (0,1)$ such that

$$x \in \bar{E} \cap B_t(x_j) \quad \Longrightarrow \quad x + t\mathbf{n}_j \in E \quad \text{for all } t \in (0, t^*). \tag{8.1}$$

Such a requirement forces, in some sense, the domain E to lie locally on one side of its boundary. However no smoothness is required on ∂E. As an example for $x \in \mathbb{R}$ let

$$h(x) = \begin{cases} \sqrt{|x|}\sin^2 \dfrac{1}{x} & \text{for } |x| > 0 \\ 0 & \text{for } x = 0. \end{cases} \quad \text{and} \quad E = [y > h]. \tag{8.2}$$

The set E satisfies the segment property. The cone property does not imply the segment property. For example, the unit disc from which a radius is removed satisfies the cone property and does not satisfy the segment property. The segment property does not imply the cone property. For example the set in (8.2), does not satisfy the cone property. The segment property does not imply that ∂E is of class C^1. Conversely, ∂E of class C^1 does not imply the segment property.

A remarkable fact about domains with the segment property is that functions in $W^{1,p}(E)$ can be extended "outside" E to be in $W^{1,p}(E')$, for a larger open set E' containing E. A consequence of such an extension is that functions in $W^{1,p}(E)$ can be approximated in the norm (1.6) by functions smooth up to ∂E. Precisely

Proposition 8.1 *Let E be a bounded open set in $\mathbb{R}^N$ with boundary ∂E of class C^1 and with the segment property. Then $C_o^\infty(\mathbb{R}^N)$ is dense in $W^{1,p}(E)$.*

Proof 8.1c of the Problems and Complements. ∎

8.2 Defining Traces

Denote by $\gamma = \gamma(N, p, \partial E)$ a constant that can be quantitatively determined a priori in terms of N, p, and the structure of ∂E only.

Proposition 8.2 *Let ∂E be of class C^1 and satisfy the segment property. If $1 \le p < N$, there exists $\gamma = \gamma(N, p, \partial E)$ such that for all $\varepsilon > 0$*

$$\|u\|_{p^* \frac{N-1}{N}; \partial E} \le \varepsilon \|\nabla u\|_p + \gamma\Big(1 + \frac{1}{\varepsilon}\Big)\|u\|_p \quad \textit{for all } u \in C_o^\infty(\mathbb{R}^N). \tag{8.3}$$

If $p = N$, then for all $q \ge 1$, there exists $\gamma = \gamma(N, q, \partial E)$ such that for all $\varepsilon > 0$

$$\|u\|_{q; \partial E} \le \varepsilon \|\nabla u\|_p + \gamma\Big(1 + \frac{1}{\varepsilon}\Big)\|u\|_p \quad \textit{for all } u \in C_o^\infty(\mathbb{R}^N). \tag{8.4}$$

If $p > N$, there exists $\gamma = \gamma(N, p, \partial E)$ such that for all $\varepsilon > 0$

$$\begin{aligned} \|u\|_{\infty, \partial E} &\le \varepsilon \|\nabla u\|_p + \gamma\Big(1 + \frac{1}{\varepsilon}\Big)\|u\|_p \\ |u(x) - u(y)| &\le \gamma |x - y|^{1 - \frac{N}{p}} \|u\|_{1,p} \quad \textit{for all } x, y \in \bar{E}. \end{aligned} \tag{8.5}$$

Proof Section 8.2c of the Problems and Complements. ∎

Remark 8.1 The constants γ in (8.3) and (8.5) tend to infinity as $p \to N$, and the constant γ in (8.4) tends to infinity as $q \to \infty$.

Since $u \in C_o^\infty(\mathbb{R}^N)$, the values of u on ∂E are meant in the classical sense. By Proposition 8.1, given $u \in W^{1,p}(E)$ there exists a sequence $\{u_n\} \subset C_o^\infty(\mathbb{R}^N)$ such that $\{u_n\} \to u$ in $W^{1,p}(E)$. In particular, $\{u_n\}$ is Cauchy in $W^{1,p}(E)$ and from (8.3) and (8.4)

$$\begin{aligned} \|u_n - u_m\|_{p^* \frac{N-1}{N}; \partial E} &\le \gamma \|u_n - u_m\|_{1,p} \quad \text{if } 1 \le p < N \\ \|u_n - u_m\|_{q; \partial E} &\le \gamma \|u_n - u_m\|_{1,p} \quad \text{for fixed } q \ge 1 \text{ if } p \ge N > 1. \end{aligned}$$

By the completeness of the spaces $L^p(\partial E)$, for $p \ge 1$

$$\{u_n\big|_{\partial E}\} \to \operatorname{tr}(u) \ \text{ in } \begin{cases} L^{p^* \frac{N-1}{N}}(\partial E) & \text{if } 1 \le p < N \\ L^q(\partial E) & \text{for fixed } q \ge 1 \text{ if } p \ge N > 1. \end{cases} \tag{8.6}$$

One verifies that $\operatorname{tr}(u)$ is independent of the particular sequence $\{u_n\}$. Therefore given $u \in W^{1,p}(E)$, this limiting process identifies its "boundary values" $\operatorname{tr}(u)$, called the *trace* of u on ∂E, as an element of $L^r(E)$, with r specified by (8.6). With perhaps an improper but suggestive symbolism we write $\operatorname{tr}(u) = u|_{\partial E}$.

8.3 Characterizing the Traces on ∂E of Functions in $W^{1,p}(E)$

The trace of a function in $W^{1,p}(E)$ is somewhat more regular that merely an element in $L^p(\partial E)$ for some $p \ge 1$. For $v \in C^\infty(\partial E)$ and $s \in (0,1)$, set

$$|||v|||_{s,p;\partial E}^p = \int_{\partial E}\int_{\partial E} \frac{|v(x)-v(y)|^p}{|x-y|^{(N-1)+sp}} d\sigma(x)d\sigma(y) < \infty \tag{8.7}$$

where $d\sigma(\cdot)$ is the surface measure on ∂E. Denote by $W^{s,p}(\partial E)$ the collections of functions v in $L^p(\partial E)$ with finite norm

$$\|v\|_{s,p;\partial E} = \|v\|_{p,\partial E} + |||v|||_{s,p;\partial E}. \tag{8.8}$$

The next theorem characterizes the traces of functions in $W^{1,p}(E)$ in terms of the spaces $W^{s,p}(\partial E)$.

Theorem 8.1 *Let ∂E be of class C^1 and satisfy the segment property. If $u \in W^{1,p}(E)$, then $tr(u) \in W^{s,p}(\partial E)$, with $s = 1 - \frac{1}{p}$. Conversely, given $v \in W^{s,p}(\partial E)$, with $s = 1 - \frac{1}{p}$, there exists $u \in W^{1,p}(E)$ such that $tr(u) = v$.*

Proof Section 8.3c of the Problems and Complements. ∎

9 The Inhomogeneous Dirichlet Problem

Assume that ∂E is of class C^1 and satisfies the segment property. Given $\mathbf{f}$ and f satisfying (7.4) and $\varphi \in W^{\frac{1}{2},2}(\partial E)$, consider the Dirichlet problem

$$\begin{aligned} -\left(a_{ij}u_{x_i}\right)_{x_j} &= \operatorname{div}\mathbf{f} - f \qquad \text{in } E \\ u\big|_{\partial E} &= \varphi \qquad \text{on } \partial E. \end{aligned} \tag{9.1}$$

By Theorem 8.1 there exists $v \in W^{1,2}(E)$ such that $\mathrm{tr}(v) = \varphi$. A solution of (9.1) is sought of the form $u = w + v$, where $w \in W_o^{1,2}(E)$ is the unique weak solution of the auxiliary, homogeneous Dirichlet problem

$$\begin{aligned} -\left(a_{ij}w_{x_i}\right)_{x_j} &= \operatorname{div}\tilde{\mathbf{f}} - f \quad \text{in } E \\ w\big|_{\partial E} &= 0 \quad \text{on } \partial E \end{aligned} \qquad \text{where } \tilde{f}_j = f_j + a_{ij}v_{x_i}. \tag{9.2}$$

Theorem 9.1 *Assume that ∂E is of class C^1 and satisfies the segment property. For every $\mathbf{f}$ and f satisfying (7.4) and $\varphi \in W^{\frac{1}{2},2}(\partial E)$, the Dirichlet problem (9.1) has a unique weak solution $u \in W^{1,2}(E)$.*

Remark 9.1 The class $W^{1,2}(E)$ where a weak solution is sought characterizes the boundary data φ on ∂E that ensure solvability.

10 The Neumann Problem

Assume that ∂E is of class C^1 and satisfies the segment property. Given $\mathbf{f}$ and f satisfying (7.4), consider the formal Neumann problem

$$\begin{aligned} -\big(a_{ij}u_{x_i}\big)_{x_j} &= \operatorname{div}\mathbf{f} - f && \text{in } E \\ \big(a_{ij}u_{x_i} + f_j\big)\, n_j &= \psi && \text{on } \partial E \end{aligned} \tag{10.1}$$

where $\mathbf{n} = (n_1, \dots, n_N)$ is the outward unit normal to ∂E and $\psi \in L^p(\partial E)$ for some $p \ge 1$. If $a_{ij} = \delta_{ij}$, $\mathbf{f} = f = 0$, and ψ are sufficiently regular, this is precisely the Neumann problem (1.3) of Chapter 2. Since $a_{ij} \in L^\infty(E)$ and $\mathbf{f} \in L^2(E)$, neither the PDE nor the boundary condition in (10.1) are well defined, and they have to be interpreted in some weak form. Multiply formally the first of (10.1) by $v \in C_o^\infty(\mathbb{R}^N)$ and integrate by parts over E, as if both the PDE and the boundary condition were satisfied in the classical sense. This gives formally

$$\int_E \big(a_{ij}u_{x_i}v_{x_j} + f_j v_{x_j} + fv\big)dx = \int_{\partial E} \psi v \, d\sigma. \tag{10.2}$$

If $v \in C_o^\infty(\mathbb{R}^N)$ is constant in a neighborhood of E, this implies the necessary condition of solvability[2]

$$\int_E f\, dx = \int_{\partial E} \psi \, d\sigma. \tag{10.3}$$

It turns out that this condition linking the data f and ψ is also sufficient for the solvability of (10.1), provided a precise class for ψ is identified. By Proposition 8.1, if (10.2) holds for all $v \in C_o^\infty(\mathbb{R}^N)$, it must hold for all $v \in W^{1,2}(E)$, provided a solution u is sought in $W^{1,2}(E)$. In such a case, the right-hand side is well defined if ψ is in the conjugate space of integrability of the traces of functions in $W^{1,2}(E)$. Therefore the natural class for the Neumann datum is

$$\psi \in L^q(\partial E), \qquad \text{where} \quad \begin{cases} q = \dfrac{2(N-1)}{N} & \text{if } N > 2 \\[2ex] \text{any } q > 1 & \text{if } N = 2. \end{cases} \tag{10.4}$$

Theorem 10.1 *Let ∂E be of class C^1 and satisfy the segment property. Let $\mathbf{f}$ and f satisfy (7.4) and ψ satisfy (10.4) and be linked by the compatibility condition (10.3). Then the Neumann problem (10.1) admits a solution in the weak form (10.2) for all $v \in W^{1,2}(E)$. The solution is unique up to a constant.*

Proof Consider the non-linear functional in $W^{1,2}(E)$

[2]This is a version of the compatibility condition (1.4) of Chapter 2; see also Theorem 6.1 of Chapter 3, and Section 6 of Chapter 4.

$$J(u) \stackrel{\text{def}}{=} \int_E \left(\tfrac{1}{2} a_{ij} u_{x_i} u_{x_j} + f_j u_{x_j} + f u\right) dx - \int_{\partial E} \psi \operatorname{tr}(u) d\sigma. \tag{10.5}$$

By the compatibility condition (10.3), $J(u) = J(u - u_E)$, where u_E is the integral average of u over E. Therefore $J(\cdot)$ can be regarded as defined in the space $\tilde{W}^{1,2}(E)$ introduced in (1.8). One verifies that $J(\cdot)$ is strictly convex in $\tilde{W}^{1,2}(E)$. Assume momentarily that $N > 2$, let 2^{**} be the Hölder conjugate of 2^*, and estimate

$$\left| \int_E f_j u_{x_j} dx \right| \le \|\mathbf{f}\|_2 \|\nabla u\|_2$$

$$\left| \int_E f u dx \right| \le \|f\|_{2^{**}} \|u\|_{2^*} \le \gamma \|f\|_{2^{**}} \|\nabla u\|_2$$

where γ is the constant in the embedding inequality (3.4). Similarly, using the trace inequality (8.3)

$$\begin{aligned} \left| \int_{\partial E} \psi \operatorname{tr}(u) d\sigma \right| &\le \|\psi\|_{q;\partial E} \|\operatorname{tr}(u)\|_{2^* \frac{N-1}{N};\partial E} \\ &\le \gamma \|\psi\|_{q;\partial E} \left(\|u\|_2 + \|\nabla u\|_2\right) \le 2\gamma^2 \|\psi\|_{q;\partial E} \|\nabla u\|_2 \end{aligned}$$

where γ is the largest of the constants in (3.4) and (8.3). Therefore

$$-F_1 + \tfrac{1}{4}\lambda \|\nabla u\|_2^2 \le J(u) \le \Lambda \|\nabla u\|_2^2 + F_1 \tag{10.6}$$

for all $u \in \tilde{W}^{1,2}(E)$, where

$$F_1 = \frac{1}{\lambda} \left(\|\mathbf{f}\|_2 + \gamma \|f\|_{2^{**}} + 2\gamma^2 \|\psi\|_{q;\partial E} \right)^2.$$

By Corollary 3.2, the $\|u\|_{1,2}$ norm of $\tilde{W}^{1,2}(E)$ is equivalent to $\|\nabla u\|_2$. With these estimates in hand the proof can now be concluded by a minimization process in $\tilde{W}^{1,2}(E)$, similar to that of Section 6. The minimum $u \in \tilde{W}^{1,2}(E)$ satisfies (10.2), for all $v \in W^{1,2}(E)$ and the latter can be characterized as the Euler equation of $J(\cdot)$. ∎

Essentially the same arguments continue to hold for $N = 2$, modulo minor variants that can be modeled after those in Section 6.1.

10.1 A Variant of (10.1)

The compatibility condition (10.3) has the role of estimating $J(\cdot)$ above and below as in (10.6), via the multiplicative embeddings of Theorem 3.2. Consider next the Neumann problem

$$\begin{aligned} -\left(a_{ij} u_{x_i}\right)_{x_j} + \mu u &= \operatorname{div} \mathbf{f} - f \quad && \text{in } E \\ \left(a_{ij} u_{x_i} + f_j\right) n_j &= \psi && \text{on } \partial E \end{aligned} \tag{10.7}$$

where $\mu > 0$, $\mathbf{f}, f \in L^2(E)$, and ψ satisfies (10.4). The problem is meant in its weak form

$$\int_E (a_{ij} u_{x_i} v_{x_j} + f_j v_{x_j} + \mu uv + fv) dx = \int_{\partial E} \psi v \, d\sigma \tag{10.8}$$

for all $v \in W^{1,2}(E)$. No compatibility conditions are needed on the data f and ψ for a solution to exist, and in addition, the solution is unique.

Theorem 10.2 *Let ∂E be of class C^1 and satisfy the segment property. Let $\mathbf{f}, f \in L^2(E)$, and let ψ satisfy (10.4). Then the Neumann problem (10.7) with $\mu > 0$ admits a unique solution in the weak form (10.8).*

Proof If u_1 and u_2 are two solutions in $W^{1,2}(E)$, their difference w satisfies

$$\int_E (a_{ij} w_{x_i} w_{x_j} + \mu w^2) dx = 0.$$

The non-linear functional in $W^{1,2}(E)$

$$J(u) \stackrel{\text{def}}{=} \int_E \left(\tfrac{1}{2} a_{ij} u_{x_i} u_{x_j} + \tfrac{1}{2}\mu u^2 + f_j u_{x_j} + fu\right) dx - \int_{\partial E} \psi \operatorname{tr}(u) d\sigma \tag{10.9}$$

is strictly convex. Then a solution can be constructed by the variational method of Section 6, modulus establishing an estimate analogous to (6.2) or (10.6), with $\|\nabla u\|_2$ replaced by the norm $\|u\|_{1,2}$ of $W^{1,2}(E)$. Estimate

$$\Big| \int_E f_j u_{x_j} dx \Big| \le \|\mathbf{f}\|_2 \|\nabla u\|_2, \qquad \Big| \int_E fu dx \Big| \le \|f\|_2 \|u\|_2$$
$$\Big| \int_{\partial E} \psi \operatorname{tr}(u) d\sigma \Big| \le \gamma \|\psi\|_{q;\partial E} \left(\|u\|_2 + \|\nabla u\|_2\right)$$

where γ is the constant of the trace inequality (8.3). Then the functional $J(\cdot)$ is estimated above and below by

$$-F_\mu + \tfrac{1}{4} \min\{\lambda; \mu\} \|u\|_{1,2}^2 \le J(u) \le \tfrac{3}{2} \max\{\Lambda; \mu\} \|u\|_{1,2}^2 + F_\mu \tag{10.10}$$

for all $u \in W^{1,2}(E)$, where

$$F_\mu = \frac{1}{\min\{\lambda; \mu\}} \left(\|\mathbf{f}\|_2 + \|f\|_2 + \gamma \|\psi\|_{q;\partial E}\right)^2. \qquad \blacksquare$$

11 The Eigenvalue Problem

Consider the problem of finding a non-trivial pair (μ, u) with $\mu \in \mathbb{R}$ and $u \in W_o^{1,2}(E)$, a solution of

$$\begin{aligned} -\left(a_{ij}u_{x_i}\right)_{x_j} &= \mu u \qquad \text{in } E \\ u &= 0 \qquad \text{on } \partial E. \end{aligned} \tag{11.1}$$

This is meant in the weak sense

$$\int_E \left(a_{ij}u_{x_i}v_{x_j} - \mu uv\right)dx = 0 \qquad \text{for all } v \in W_o^{1,2}(E). \tag{11.2}$$

If (μ, u) is a solution pair, μ is an eigenvalue and u is an eigenfunction of (11.1). In principle, the pair (μ, u) is sought for $\mu \in \mathbb{C}$, and for u in the complex-valued Hilbert space $W_o^{1,2}(E)$, with complex inner product as in Section 1 of Chapter 4. However by considerations analogous to those of Proposition 7.1 of that chapter, eigenvalues of (11.1) are real, and eigenfunctions can be taken to be real-valued. Moreover, any two distinct eigenfunctions corresponding to two distinct eigenvalues are orthogonal in $L^2(E)$.

Proposition 11.1 *Eigenvalues of (11.1) are positive. Moreover, to each eigenvalue μ there correspond at most finitely many eigenfunctions, linearly independent, and orthonormal in $L^2(E)$.*

Proof If $\mu \le 0$, the functional

$$W_o^{1,2}(E) \ni u \to \int_E (a_{ij}u_{x_i}u_{x_j} - \mu u^2)\,dx \tag{11.3}$$

is strictly convex and bounded below by $\lambda\|\nabla u\|_2^2$. Therefore it has a unique minimum, which is the unique solution of its Euler equation (11.1). Since $u = 0$ is a solution, it is the only one. Let $\{u_n\}$ be a sequence of eigenfunctions linearly independent in $L^2(E)$, corresponding to μ. Without loss of generality we may assume they are orthonormal. Then from (11.2), $\|\nabla u_n\|_2 \le \sqrt{\mu/\lambda}$ for all n, and $\{u_n\}$ is equi-bounded in $W_o^{1,2}(E)$. If $\{u_n\}$ is infinite, a subsequence can be selected, and relabeled with n, such that $\{u_n\} \to u$ weakly in $W_o^{1,2}(E)$ and strongly in $L^2(E)$. However, $\{u_n\}$ cannot be a Cauchy sequence in $L^2(E)$, since $\|u_n - u_m\|_2 = \sqrt{2}$ for all n, m. ■

Let $\{u_{\mu,1}, \dots, u_{\mu,n_\mu}\}$ be the linearly independent eigenfunctions corresponding to the eigenvalue μ. The number n_μ is the multiplicity of μ. If $n_\mu = 1$, then μ is said to be simple.

12 Constructing the Eigenvalues of (11.1)

Minimize the strictly convex functional $a(\cdot,\cdot)$ on the unit sphere S_1 of $L^2(E)$, that is

$$\min_{\substack{u \in W_o^{1,2}(E) \\ \|u\|_2 = 1}} a(u,u) = \min_{\substack{u \in W_o^{1,2}(E) \\ \|u\|_2 = 1}} \int_E a_{ij}u_{x_i}u_{x_j}dx$$

and let $\mu_1 \ge 0$ be its minimum value. A minimizing sequence $\{u_n\} \subset S_1$ is bounded in $W_o^{1,2}(E)$, and a subsequence $\{u_{n'}\} \subset \{u_n\}$ can be selected such that $\{u_{n'}\} \to w_1$ weakly in $W_o^{1,2}(E)$ and strongly in $L^2(E)$. Therefore $w_1 \in S_1$, it is non-trivial, $\mu_1 > 0$, and

$$\mu_1 = \lim a(u_{n'}, u_{n'}) \ge \lim a(w_1, w_1) \ge \mu_1. \tag{12.1}$$

Thus

$$\mu_1 \le \frac{a(w_1 + v, w_1 + v)}{\|w_1 + v\|_2^2} \qquad \text{for all } v \in W_o^{1,2}(E). \tag{12.2}$$

It follows that the functional

$$\begin{aligned} W_o^{1,2}(E) \ni v \to I_{\mu_1}(v) = &\int_E \big(a_{ij} w_{1,x_i} v_{x_j} - \mu_1 w_1 v\big) dx \\ &+ \frac{1}{2} \int_E \big(a_{ij} v_{x_i} v_{x_j} - \mu_1 v^2\big) dx \end{aligned}$$

is non-negative, its minimum is zero, and the minimum is achieved for $v = 0$. Thus

$$\frac{d}{dt} I_{\mu_1}(tv)\Big|_{t=0} = 0 \qquad \text{for all } v \in W_o^{1,2}(E).$$

The latter is precisely (11.2) for the pair (μ_1, w_1). While this process identifies μ_1 uniquely, the minimizer $w_1 \in S_1$ depends on the choice of subsequence $\{u_{n'}\} \subset \{u_n\}$. Thus a priori, to μ_1 there might correspond several eigenfunctions in $W_o^{1,2}(E) \cap S_1$.

If μ_n and the set of its linearly independent eigenfunctions have been found, set

$$\begin{aligned} \mathcal{E}_n &= \{\text{span of the eigenfunctions of } \mu_n\} \\ W_n^{1,2}(E) &= W_o^{1,2}(E) \cap [\mathcal{E}_1 \cup \cdots \cup \mathcal{E}_n]^{\perp} \end{aligned}$$

consider the minimization problem

$$\min_{\substack{u \in W_n^{1,2}(E) \\ \|u\|_2 = 1}} a(u, u) = \min_{\substack{u \in W_n^{1,2}(E) \\ \|u\|_2 = 1}} \int_E a_{ij} u_{x_i} u_{x_j} dx$$

and let $\mu_{n+1} > \mu_n$ be its minimum value. A minimizing sequence $\{u_n\} \subset S_1$ is bounded in $W_n^{1,2}(E)$, and a subsequence $\{u_{n'}\} \subset \{u_n\}$ can be selected such that $\{u_{n'}\} \to w_{n+1}$ weakly in $W_n^{1,2}(E)$ and strongly in $L^2(E)$. Therefore $w_{n+1} \in S_1$ is non-trivial, and

$$\mu_{n+1} = \lim a(u_{n'}, u_{n'}) \ge \lim a(w_{n+1}, w_{n+1}) \ge \mu_{n+1}.$$

Thus

$$\mu_{n+1} \le \frac{a(w_{n+1} + v, w_{n+1} + v)}{\|w_{n+1} + v\|_2^2} \qquad \text{for all } v \in W_n^{1,2}(E).$$

It follows that the functional

$$W_n^{1,2}(E) \ni v \to I_{\mu_{n+1}}(v) = \int_E \big(a_{ij} w_{n+1,x_i} v_{x_j} - \mu_{n+1} w_{n+1} v\big)dx + \frac{1}{2}\int_E \big(a_{ij} v_{x_i} v_{x_j} - \mu_{n+1} v^2\big)dx$$

is non-negative, its minimum is zero, and the minimum is achieved for $v = 0$. Thus

$$\frac{d}{dt} I_{\mu_{n+1}}(tv) \Big|_{t=0} = 0 \quad \text{for all } v \in W_n^{1,2}(E).$$

This implies

$$\int_E \big(a_{ij} w_{n+1,x_i} v_{x_j} - \mu_{n+1} w_{n+1} v\big)dx = 0 \tag{12.3}$$

for all $v \in W_n^{1,2}(E)$. The latter coincides with (11.2), except that the test functions v are taken out of $W_n^{1,2}(E)$ instead of the entire $W_o^{1,2}(E)$. Any $v \in W_o^{1,2}(E)$ can be written as $v = v^\perp + v_o$, where $v \in W_n^{1,2}(E)$ and v_o has the form

$$v_o = \sum_{j=1}^{k_n} v_j w_j$$

where v_j are constants, and w_j are eigenfunctions of (11.1) corresponding to eigenvalues μ_j, for $j \le n$. By construction

$$\int_E \big(a_{ij} w_{n+1,x_i} v_{o,x_j} - \mu_{n+1} w_{n+1} v_o\big)dx = 0.$$

Hence (11.2) holds for all $v \in W_o^{1,2}(E)$, and (μ_{n+1}, w_{n+1}) is a non-trivial solution pair of (11.1). While this process identifies μ_{n+1} uniquely, the minimizer $w_{n+1} \in S_1$ depends on the choice of subsequence $\{u_{n'}\} \subset \{u_n\}$. Thus a priori, to μ_{n+1} there might correspond several eigenfunctions.

13 The Sequence of Eigenvalues and Eigenfunctions

This process generates a sequence of eigenvalues $\mu_n < \mu_{n+1}$ each with its own multiplicity. The linearly independent eigenfunctions $\{w_{\mu_n,1}, \dots, w_{\mu_n,n_{\mu_n}}\}$ corresponding to μ_n can be chosen to be orthonormal. The eigenfunctions are relabeled with n to form an orthonormal sequence $\{w_n\}$, and each is associated with its own eigenvalue, which in this reordering remains the same as the index of the corresponding eigenfunctions ranges over its own multiplicity. We then write

$$\begin{array}{ccccccc} \mu_1 \le & \mu_2 \le & \cdots \le & \mu_n \le & \cdots \\ w_1 & w_2 & \cdots & w_n & \cdots . \end{array} \tag{13.1}$$

Proposition 13.1 *Let $\{\mu_n\}$ and $\{w_n\}$ be as in (13.1). Then $\{\mu_n\} \to \infty$ as $n \to \infty$. The orthonormal system $\{w_n\}$ is complete in $L^2(E)$. The system $\{\sqrt{\mu_n} w_n\}$ is orthonormal and complete in $W_o^{1,2}(E)$ with respect to the inner product $a(\cdot,\cdot)$.*

Proof If $\{\mu_n\} \to \mu_\infty < \infty$, the sequence $\{w_n\}$ will be bounded in $W_o^{1,2}(E)$, and by compactness, a subsequence $\{w_{n'}\} \subset \{w_n\}$ can be selected such that $\{w_{n'}\} \to w$ strongly in $L^2(E)$. Since $\{w_n\}$ is orthonormal in $L^2(E)$

$$2 = \lim_{n',m' \to \infty} \|w_{n'} - w_{m'}\|^2 \to 0.$$

Let $f \in L^2(E)$ be non-zero and orthogonal to the $L^2(E)$-closure of $\{w_n\}$. Let $u_f \in W_o^{1,2}(E)$ be the unique solution of the homogeneous Dirichlet problem (4.1) with $\mathbf{f} = 0$, for such a given f. Since $f \neq 0$, the solution $u_f \neq 0$ can be renormalized so that $\|u_f\|_2 = 1$. Then, for all $n \in \mathbb{N}$

$$\mu_n = \inf_{\substack{u \in W_n^{1,2}(E) \\ \|u\|_2 = 1}} a(u,u) \le a(u_f,u_f) \le \|f\|_2.$$

It is apparent that $\{\sqrt{\mu_n} w_n\}$ is an orthogonal system in $W_o^{1,2}(E)$ with respect to the inner product $a(\cdot,\cdot)$. To establish its completeness in $W_o^{1,2}(E)$ it suffices to verify that $a(w_n, u) = 0$ for all w_n implies $u = 0$. This in turn follows from the completeness of $\{w_n\}$ in $L^2(E)$. ■

Proposition 13.2 *μ_1 is simple and $w_1 > 0$ in E.*

Proof Let (μ_1, w) be a solution pair for (11.1) for the first eigenvalue μ_1. Since $w \in W_o^{1,2}(E)$, also $w^\pm \in W_o^{1,2}(E)$, and either of these can be taken as a test function in the corresponding weak form (11.2) for the pair (μ_1, w). This gives

$$a(w^\pm, w^\pm) = \mu_1 \|w^\pm\|_2^2.$$

Therefore in view of the minimum problem (12.2), the two functions $w^\pm$ are both non-negative solutions of

$$\begin{aligned} -\big(a_{ij} w_{x_i}^\pm\big)_{x_j} &= \mu_1 w^\pm && \text{weakly in } E \\ w^\pm &= 0 && \text{on } \partial E. \end{aligned} \tag{13.2}$$

Lemma 13.1 *The functions $w^\pm$ are Hölder continuous in E, and if $w^+(x_o) > 0$ ($w^-(x_o) > 0$), for some $x_o \in E$, then $w^+ > 0$ ($w^- > 0$) in E.*

Assuming the lemma for the moment, either $w^+ \equiv 0$ or $w^- \equiv 0$ in E. Therefore, since $w = w^+ - w^-$, the eigenfunction w can be chosen to be strictly positive in E. If v and w are two linearly independent eigenfunctions corresponding to μ_1, they can be selected to be both positive in E and thus cannot be orthogonal. Thus $v = \gamma w$ for some $\gamma \in \mathbb{R}$, and μ_1 is simple. ■

The proof of Lemma 13.1 will follow from the Harnack Inequality of Section 9 of Chapter 10.

14 A Priori $L^\infty(E)$ Estimates for Solutions of the Dirichlet Problem (9.1)

A weak sub(super)-solution of the Dirichlet problem (9.1) is a function $u \in W^{1,2}(E)$, whose trace on ∂E satisfies $\mathrm{tr}(u) \le (\ge)\varphi$ and such that

$$\int_E \big(a_{ij}u_{x_i}v_{x_j} + f_j v_{x_j} + fv\big)dx \le (\ge)0 \tag{14.1}$$

for all non-negative $v \in W_o^{1,2}(E)$. A function $u \in W^{1,2}(E)$ is a weak solution of the Dirichlet problem (9.1), if and only if is both a weak sub- and super-solution of that problem.

Proposition 14.1 *Let $u \in W^{1,2}(E)$ be a weak sub-solution of (9.1) for $N \ge 2$. Assume*

$$\varphi_+ \in L^\infty(\partial E), \quad \mathbf{f} \in L^{N+\varepsilon}(E), \quad f_+ \in L^{\frac{N+\varepsilon}{2}}(E) \tag*{$(14.2)_+$}$$

for some $\varepsilon > 0$. Then $u_+ \in L^\infty(E)$ and there exists a constant C_ε that can be determined a priori only in terms of λ, Λ, N, ε, and the constant γ in the Sobolev embedding (3.1)–(3.2), such that

$$\operatorname*{ess\,sup}_E u_+ \le \max\Big\{ \operatorname*{ess\,sup}_{\partial E} \varphi_+ ; C_\varepsilon\big[\|\mathbf{f}\|_{N+\varepsilon}; |E|^\delta \|f_+\|_{\frac{N+\varepsilon}{2}}\big]|E|^\delta\Big\} \tag*{$(14.3)_+$}$$

where

$$\delta = \frac{\varepsilon}{N(N+\varepsilon)}. \tag{14.4}$$

A similar statement holds for super-solutions. Precisely

Proposition 14.2 *Let $u \in W^{1,2}(E)$ be a weak super-solution of (9.1) for $N \ge 2$. Assume*

$$\varphi_- \in L^\infty(\partial E), \quad \mathbf{f} \in L^{N+\varepsilon}(E), \quad f_- \in L^{\frac{N+\varepsilon}{2}}(E) \tag*{$(14.2)_-$}$$

for some $\varepsilon > 0$. Then $u_- \in L^\infty(E)$ and

$$\operatorname*{ess\,sup}_E u_- \le \max\Big\{ \operatorname*{ess\,sup}_{\partial E} \varphi_- ; C_\varepsilon\big[\|\mathbf{f}\|_{N+\varepsilon}; |E|^\delta \|f_-\|_{\frac{N+\varepsilon}{2}}\big]|E|^\delta\Big\} \tag*{$(14.3)_-$}$$

for the same constants C_ε and δ.

Remark 14.1 The constant C_ε in $(14.3)_\pm$ is "stable" as $\varepsilon \to \infty$, in the sense that if $\mathbf{f}$ and $f_\pm$ are in $L^\infty(E)$, then $u_\pm \in L^\infty(E)$ and there exists a constant C_∞ depending on the indicated quantities except ε, such that

$$\operatorname*{ess\,sup}_E u_\pm \le \max\Big\{ \operatorname*{ess\,sup}_{\partial E} \varphi_\pm ; C_\infty\big[\|\mathbf{f}\|_\infty; |E|^{\frac{1}{N}} \|f_\pm\|_\infty\big]|E|^{\frac{1}{N}}\Big\}. \tag{14.5}$$

Remark 14.2 The constant C_ε tends to infinity as $\varepsilon \to 0$. Indeed, the propositions are false for $\varepsilon = 0$, as shown by the following example. For $N > 2$, the two equations

$$\begin{array}{ll} \Delta u = f & \text{where } f = \dfrac{N-2}{|x|^2 \ln|x|} - \dfrac{1}{|x|^2 \ln^2|x|} \\ \Delta u = f_{j,x_j} & \text{where } f_j = \dfrac{x_j}{|x|^2 \ln|x|} \end{array}$$

are both solved, in a neighborhood E of the origin by $u(x) = \ln|\ln|x||$. One verifies that

$$\begin{array}{lll} f \in L^{\frac{N}{2}}(E) & \text{and } f \notin L^{\frac{N+\varepsilon}{2}}(E) & \text{for any } \varepsilon > 0 \\ \mathbf{f} \in L^N(E) & \text{and } \mathbf{f} \notin L^{N+\varepsilon}(E) & \text{for any } \varepsilon > 0. \end{array}$$

Remark 14.3 The propositions can be regarded as a weak form of the maximum principle (Section 4.1 of Chapter 2). Indeed, if u is a weak sub(super)-solution of the Dirichlet problem (9.1), with $\mathbf{f} = f = 0$, then $u_+ \le \text{tr}(u)_+$ ($u_- \le \text{tr}(u)_-$).

15 Proof of Propositions 14.1–14.2

It suffices to establish Proposition 14.1. Let $u \in W^{1,2}(E)$ be a weak sub-solution of the Dirichlet problem (9.1), in the sense of (14.1) for all non-negative $v \in W_o^{1,2}(E)$. Let $k \ge \|\varphi_+\|_{\infty,\partial E}$ to be chosen, and set

$$k_n = k\Big(2 - \frac{1}{2^{n-1}}\Big), \qquad A_n = [u > k_n], \qquad n = 1, 2, \dots . \tag{15.1}$$

Then $(u - k_n)_+ \in W_o^{1,2}(E)$ for all $n \in \mathbb{N}$, and it can be taken as a test function in the weak formulation (14.1) to yield

$$\int_E \big[[a_{ij} u_{x_i} + f_j]\, (u - k_n)_{+x_j} + f_+ (u - k_n)_+ \big] dx \le 0.$$

From this, estimate

$$\begin{aligned} \lambda \|\nabla (u - k_n)_+\|_2^2 &\le \|\mathbf{f}\chi_{A_n}\|_2 \|\nabla(u - k_n)_+\|_2 + \int_E f_+ (u - k_n)_+ dx \\ &\le \frac{\lambda}{4} \|\nabla(u - k_n)_+\|_2^2 + \frac{1}{\lambda} \|\mathbf{f}\chi_{A_n}\|_2^2 + \int_E f_+ (u - k_n)_+ dx \\ &\le \frac{\lambda}{4} \|\nabla(u - k_n)_+\|_2^2 + \frac{1}{\lambda} \|\mathbf{f}\|_{N+\varepsilon}^2 |A_n|^{1 - \frac{2}{N+\varepsilon}} \\ &\quad + \int_E f_+ (u - k_n)_+ dx. \end{aligned}$$

The last term is estimated by Hölder's inequality as

$$\int_E f_+(u-k_n)_+ dx \le \|f_+\|_{\frac{p^*}{p^*-1}} \|(u-k_n)_+\|_{p^*}$$

where

$$\frac{p^*}{p^*-1} = \frac{N+\varepsilon}{2} \quad \text{and} \quad p^* = \frac{Np}{N-p}. \tag{15.2}$$

For these choices one verifies that $1 < p < N$ for all $N \ge 2$. Therefore by (3.1) of the embedding of Theorem 3.1

$$\begin{aligned}\int_E f_+(u-k_n)_+ dx &\le \|f_+\|_{\frac{N+\varepsilon}{2}} \|\nabla(u-k_n)_+\|_p \\ &\le \|\nabla(u-k_n)_+\|_2 \|f_+\|_{\frac{N+\varepsilon}{2}} |A_n|^{\frac{1}{p}(1-\frac{p}{2})} \\ &\le \frac{\lambda}{4}\|\nabla(u-k_n)_+\|_2^2 + \frac{1}{\lambda}\|f_+\|^2_{\frac{N+\varepsilon}{2}} |A_n|^{\frac{2}{p}-1}.\end{aligned}$$

Combining these estimates gives

$$\|\nabla(u-k_n)_+\|_2^2 \le C_o^2 |A_n|^{1-\frac{2}{N}+2\delta} \tag{15.3}$$

where

$$C_o^2 = \lambda^{-2} \max\left\{ \|\mathbf{f}\|^2_{N+\varepsilon}; |E|^{2\delta} \|f_+\|^2_{\frac{N+\varepsilon}{2}} \right\}$$

and δ is defined in (14.4).

15.1 An Auxiliary Lemma on Fast Geometric Convergence

Lemma 15.1 *Let $\{Y_n\}$ be a sequence of positive numbers linked by the recursive inequalities*

$$Y_{n+1} \le b^n K Y_n^{1+\sigma} \tag{15.4}$$

for some $b > 1$, $K > 0$, and $\sigma > 0$. If

$$Y_1 \le b^{-1/\sigma^2} K^{-1/\sigma} \tag{15.5}$$

Then $\{Y_n\} \to 0$ as $n \to \infty$.

Proof By direct verification by applying (15.5) recursively. ∎

15.2 Proof of Proposition 14.1 for $N > 2$

By the embedding inequality (3.1) for $W_o^{1,2}(E)$ and (15.3)

$$\begin{aligned}\frac{k^2}{4^n}|A_{n+1}| &\le \int_E (u-k_n)_+^2 \chi_{[u>k_{n+1}]} dx \le \|(u-k_n)_+\|_2^2 \\ &\le \|(u-k_n)_+\|^2_{2^*} |A_n|^{\frac{2}{N}} \le \gamma^2 \|\nabla(u-k_n)_+\|_2^2 |A_n|^{\frac{2}{N}} \\ &\le \gamma^2 C_o^2 |A_n|^{1+2\delta}.\end{aligned} \tag{15.6}$$

From this

$$|A_{n+1}| \le \frac{4^n \gamma^2 C_o^2}{k^2} |A_n|^{1+2\delta} \qquad \text{for all } n \in \mathbb{N} \tag{15.7}$$

where γ is the constant of the embedding inequality (3.1). If $\{|A_n|\} \to 0$ as $n \to \infty$, then $u \le 2k$ a.e. in E. By Lemma 15.1, this occurs if

$$|A_1| \le |E| \le 2^{-1/2\delta^2} C_o^{-1/\delta} k^{1/\delta}.$$

This in turn is satisfied if k is chosen from $k = 2^{1/2\delta} C_o |E|^\delta$. ∎

15.3 Proof of Proposition 14.1 for $N = 2$

The main difference is in the application of the embedding inequality in (15.6), leading to the recursive inequalities (15.7). Let $q > 2$ to be chosen, and modify (15.6) by applying the embedding inequality (3.2) of Theorem 3.1, as follows. First

$$\begin{aligned} \|(u-k_n)_+\|_2 &\le \|(u-k_n)_+\|_q |A_n|^{\frac{1}{2}(1-\frac{2}{q})} \\ &\le \gamma(q) \|\nabla(u-k_n)_+\|_2^{1-\frac{2}{q}} \|(u-k_n)_+\|_2^{\frac{2}{q}} |A_n|^{\frac{1}{2}(1-\frac{2}{q})} \\ &\le \frac{1}{2}\|(u-k_n)_+\|_2 + \gamma(q)\|\nabla(u-k_n)_+\|_2 |A_n|^{\frac{1}{2}}. \end{aligned}$$

Therefore

$$\begin{aligned} \frac{k^2}{4^n}|A_{n+1}| &\le \int_E (u-k_n)_+^2 \chi_{[u>k_{n+1}]} dx \\ &\le \|(u-k_n)_+\|_2^2 \le 2\gamma(q) C_o^2 |A_n|^{1+2\delta}. \end{aligned}$$

∎

16 A Priori $L^\infty(E)$ Estimates for Solutions of the Neumann Problem (10.1)

A weak sub(super)-solution of the Neumann problem (10.1) is a function $u \in W^{1,2}(E)$ satisfying

$$\int_E (a_{ij} u_{x_i} v_{x_j} + f_j v_{x_j} + f v) dx \le (\ge) \int_{\partial E} \psi v \, d\sigma \tag{16.1}$$

for all non-negative test functions $v \in W^{1,2}(E)$. A function $u \in W^{1,2}(E)$ is a weak solution of the Neumann problem (10.1), if and only if is both a weak sub- and super-solution of that problem.

Proposition 16.1 *Let ∂E be of class C^1 and satisfying the segment property. Let $u \in W^{1,2}(E)$ be a weak sub-solution of (10.1) for $N \ge 2$, and assume that*

$$\psi_+ \in L^{N-1+\sigma}(\partial E), \quad \mathbf{f} \in L^{N+\varepsilon}(E), \quad f_+ \in L^{\frac{N+\varepsilon}{2}}(E) \tag{16.2$_+$}$$

for some $\sigma > 0$ and $\varepsilon > 0$. Then $u_+ \in L^\infty(E)$, and there exists a positive constant C_ε that can be determined quantitatively a priori only in terms of the set of parameters $\{N, \lambda, \Lambda, \varepsilon, \sigma\}$, the constant γ in the embeddings of Theorem 2.1, the constant γ of the trace inequality of Proposition 8.2, and the structure of ∂E through the parameters h and ω of its cone condition such that

$$\operatorname*{ess\,sup}_E u_+ \le C_\varepsilon \max\left\{\|u_+\|_2; \|\psi_+\|_{q;\partial E}; \|\mathbf{f}\|_{N+\varepsilon}; |E|^\delta \|f_+\|_{\frac{N+\varepsilon}{2}}\right\} \qquad (16.3)_+$$

where

$$q = N - 1 + \sigma, \qquad \sigma = \varepsilon \frac{N-1}{N}, \qquad \textit{and} \qquad \delta = \frac{\varepsilon}{N(N+\varepsilon)}. \qquad (16.4)$$

Proposition 16.2 *Let ∂E be of class C^1 and satisfying the segment property. Let $u \in W^{1,2}(E)$ be a weak super-solution of (10.1) for $N \ge 2$, and assume that*

$$\psi_- \in L^{N-1+\sigma}(\partial E), \quad \mathbf{f} \in L^{N+\varepsilon}(E), \quad f_- \in L^{\frac{N+\varepsilon}{2}}(E) \qquad (16.2)_-$$

for some $\sigma > 0$ and $\varepsilon > 0$. Then $u_- \in L^\infty(E)$

$$\operatorname*{ess\,sup}_E u_- \le C_\varepsilon \max\left\{\|u_-\|_2; \|\psi_-\|_{q;\partial E}; \|\mathbf{f}\|_{N+\varepsilon}; |E|^\delta \|f_-\|_{\frac{N+\varepsilon}{2}}\right\} \qquad (16.3)_-$$

where the parameters q, σ, δ and C_ε are the same as in $(16.3)_+$ and (16.4).

Remark 16.1 The dependence on some norm of u, for example $\|u_\pm\|_2$, is expected, since the solutions of (10.1) are unique up to constants.

Remark 16.2 The constant C_ε in $(16.3)_\pm$ is "stable" as $\varepsilon \to \infty$, in the sense that if

$$\psi_\pm \in L^\infty(\partial E), \quad \mathbf{f} \in L^\infty(E), \quad f_\pm \in L^\infty(E), \qquad (16.5)$$

then $u_\pm \in L^\infty(E)$ and there exists a constant C_∞ depending on the indicated quantities except ε and σ such that

$$\operatorname*{ess\,sup}_E u_\pm \le C_\infty \max\left\{\|u_\pm\|_2; \operatorname*{ess\,sup}_{\partial E} \psi_\pm; \|\mathbf{f}\|_\infty; |E|^{\frac{1}{N}} \|f_\pm\|_\infty\right\}. \qquad (16.6)$$

Remark 16.3 The constant C_ε in $(16.3)_\pm$ tends to infinity as $\varepsilon \to 0$. The order of integrability of $\mathbf{f}$ and f required in $(16.2)_\pm$ is optimal for $u_\pm$ to be in $L^\infty(E)$. This can be established by the same local solutions in Remark 14.2. Also, the order of integrability of $\psi_\pm$ is optimal for $u_\pm$ to be in $L^\infty(E)$, even if $\mathbf{f} = f = 0$. Indeed, the propositions are false for $\sigma = 0$ and $\mathbf{f} = f = 0$, as shown by the following counterexample. Consider the family of functions parametrized by $\eta > 0$ (Section 8 of Chapter 2)

$$\mathbb{R} \times \mathbb{R}^+ = E \ni (x, y) \to F_\eta(x, y) = \frac{1}{2\pi} \ln \sqrt{x^2 + (y+\eta)^2}.$$

One verifies that F_η are harmonic in E, and on the boundary $y = 0$ of E

$$F_{\eta,y}\Big|_{y=0} = \frac{\eta}{2\pi(x^2+\eta^2)}.$$

One also verifies that

$$\int_{\mathbb{R}} F_{\eta,y}(x,0)dx = \frac{1}{2} \qquad \text{for all } \ \eta > 0.$$

Therefore if an estimate of the type of $(16.3)_\pm$ were to exist for $\sigma = 0$, and with C independent of σ, we would have, for all (x, y) in a neighborhood E_o of the origin

$$|F_\eta(x,y)| \le C(1 + \|F_{\eta,y}(\cdot,0)\|_{1,\partial E}) = \tfrac{3}{2}C \qquad \text{for all } \ \eta > 0.$$

Letting $\eta \to 0$ gives a contradiction. While the counterexample is set in $\mathbb{R}\times\mathbb{R}^+$, it generates a contradiction in a subset of E_o about the origin of $\mathbb{R}^2$.

Remark 16.4 The constants C_ε in $(16.3)_\pm$ and C_∞ in (16.6) depend on the embedding constants of Theorem 2.1. As such, they depend on the structure of ∂E through the parameters h and ω of its cone condition. Because of this dependence, C_ε and C_∞ tend to ∞ as either $h \to 0$ or $\omega \to 0$.

Remark 16.5 The propositions are a priori estimates assuming that a sub(super)-solution exists. Sufficient conditions for the existence of a solution require that f and ψ must be linked by the compatibility condition (10.3). Such a requirement, however, plays no role in the a priori $L^\infty(E)$ estimates.

17 Proof of Propositions 16.1–16.2

It suffices to establish Proposition 16.1. Let $u \in W^{1,2}(E)$ be a sub-solution of the Neumann problem (10.1), in the sense of (16.1), and for $k > 0$ to be chosen, define k_n and A_n as in (15.1). In the weak formulation (16.1) take $v = (u - k_{n+1})_+ \in W^{1,2}(E)$, to obtain

$$\int_E [(a_{ij}u_{x_i} + f_j)(u-k_{n+1})_{+x_j} + f_+(u-k_{n+1})_+]dx \le \int_{\partial E} \psi_+(u-k_{n+1})_+ d\sigma.$$

Estimate the various terms by making use of the embedding (2.1), as follows

$$\begin{aligned}\Big|\int_E f_j(u-k_{n+1})_{+x_j}dx\Big| &\le \|\nabla(u-k_{n+1})_+\|_2\|\mathbf{f}\chi_{A_{n+1}}\|_2 \\ &\le \frac{\lambda}{4}\|\nabla(u-k_{n+1})_+\|_2^2 + \frac{1}{\lambda}\|\mathbf{f}\|_{N+\varepsilon}^2|A_{n+1}|^{1-\frac{2}{N+\varepsilon}}.\end{aligned}$$

Next, for the same choices of p_* as in (15.2), by the embedding (2.1) of Theorem 2.1

$$\begin{aligned}\Big|\int_E f_+(u-k_{n+1})_+dx\Big| &\le \|f_+\|_{\frac{N+\varepsilon}{2}}\|(u-k_{n+1})_+\|_{p^*}\\ &\le \frac{\gamma}{\omega}\|f_+\|_{\frac{N+\varepsilon}{2}}\Big(\frac{1}{h}\|(u-k_{n+1})_+\|_p+\|\nabla(u-k_{n+1})_+\|_p\Big)\\ &\le \frac{\lambda}{4}\|\nabla(u-k_{n+1})_+\|_2^2+\|(u-k_{n+1})_+\|_2^2\\ &\quad+\frac{\gamma^2}{\omega^2}\big(\frac{1}{\lambda}+\frac{1}{4h^2}\big)|E|^{\delta}\|f_+\|^2_{\frac{N+\varepsilon}{2}}|A_{n+1}|^{1-\frac{2}{N}+2\delta}\end{aligned}$$

where γ is the constant of the embedding inequality (2.1), and δ is defined in (14.4). Setting

$$F^2=\|\mathbf{f}\|^2_{N+\varepsilon}+|E|^{\delta}\|f_+\|^2_{\frac{N+\varepsilon}{2}},\qquad C_1=\frac{\gamma^2}{\omega^2}\big(\frac{1}{\lambda}+\frac{1}{4h^2}\big)$$

the previous remarks imply

$$\begin{aligned}\frac{\lambda}{2}\|\nabla(u-k_{n+1})_+\|_2^2\le \|(u-k_{n+1})_+\|_2^2+C_1F^2|A_{n+1}|^{1-\frac{2}{N}+2\delta}\\ +\Big|\int_{\partial E}\psi_+(u-k_{n+1})_+d\sigma\Big|.\end{aligned}$$

The last integral is estimated by means of the trace inequality (8.3). Let

$$q=N-1+\sigma=(N-1)\Big(\frac{N+\varepsilon}{N}\Big)$$

be the order of integrability of ψ on ∂E and determine p^* and p from

$$1-\frac{1}{q}=\frac{1}{p^*}\frac{N}{N-1},\qquad p^*=\frac{Np}{N-p},\qquad \frac{1}{q}=\frac{p-1}{p}\frac{N}{N-1}.$$

One verifies that for these choices, $1<p<2<N$, and the trace inequality (8.3) can be applied. Therefore

$$\begin{aligned}&\Big|\int_{\partial E}\psi_+(u-k_{n+1})_+d\sigma\Big|\le \|\psi\|_{q;\partial E}\|(u-k_{n+1})_+\|_{p^*\frac{N-1}{N};\partial E}\\ &\quad\le \|\psi_+\|_{q;\partial E}\big[\|\nabla(u-k_{n+1})_+\|_p+2\gamma\|(u-k_{n+1})_+\|_p\big]\\ &\quad\le \|\psi_+\|_{q;\partial E}\big[\|\nabla(u-k_{n+1})_+\|_2+2\gamma\|(u-k_{n+1})_+\|_2\big]|A_{n+1}|^{\frac{1}{p}-\frac{1}{2}}\\ &\quad\le \frac{\lambda}{4}\|\nabla(u-k_{n+1})_+\|_2^2+\|(u-k_{n+1})_+\|_2^2\\ &\qquad+\Big(\gamma^2+\frac{1}{\lambda}\Big)\|\psi_+\|^2_{q;\partial E}|A_{n+1}|^{\frac{2}{p}-1}.\end{aligned}$$

Denote by C_ℓ, $\ell=1,2,\dots$ generic positive constants that can be determined quantitatively a priori, only in terms of the set of parameters $\{N,\lambda,\Lambda\}$, the constant γ in the embeddings of Theorem 2.1, the constant γ of the trace inequality of Proposition 8.2, and the structure of ∂E through the parameters h and ω of the cone condition. Then combining the previous estimates yields the existence of constants C_2 and C_3 such that

$$\|\nabla(u-k_{n+1})_+\|_2^2 \le C_2\|(u-k_{n+1})_+\|_2^2 + C_3F_*^2|A_{n+1}|^{1-\frac{2}{N}+2\delta} \tag{17.1}$$

where we have set

$$F_*^2 = \max\big\{\|\psi\|_{q;\partial E}^2\,;\ \|\mathbf{f}\|_{N+\varepsilon}^2\,;\ |E|^{2\delta}\|f_+\|_{\frac{N+\varepsilon}{2}}^2\big\}.$$

17.1 Proof of Proposition 16.1 for $N>2$

By the embedding inequality (2.1) of Theorem 2.1 for $W^{1,2}(E)$ and (17.1)

$$\begin{aligned}\|(u-k_{n+1})_+\|_2^2 &\le \|(u-k_{n+1})_+\|_{2^*}^2|A_{n+1}|^{\frac{2}{N}}\\ &\le \frac{2\gamma^2}{\omega^2}\big(\|\nabla(u-k_{n+1})_+\|_2^2 + \frac{1}{h^2}\|(u-k_{n+1})_+\|_2^2\big)|A_{n+1}|^{\frac{2}{N}}\\ &\le C_4\|(u-k_n)_+\|_2^2|A_{n+1}|^{\frac{2}{N}} + C_5F_*^2|A_{n+1}|^{1+2\delta}.\end{aligned} \tag{17.2}$$

For all $n\in\mathbb{N}$

$$Y_n \overset{\text{def}}{=} \int_E (u-k_n)_+^2dx \ge \int_{A_{n+1}}(u-k_{n+1})_+^2dx \ge \frac{k^2}{4^n}|A_{n+1}|.$$

Therefore the previous inequality yields

$$Y_{n+1} \le \frac{4^nC_4}{k^{4\delta}}Y_n^{1+2\delta}\left(\frac{1}{k^2}\int_E u^2dx\right)^{\frac{2}{N+\varepsilon}} + \frac{F_*^2}{k^2}\frac{4^{2n}C_5}{k^{4\delta}}Y_n^{1+2\delta}.$$

Take $k\ge\max\{\|u\|_2;F_*\}$, so that

$$\frac{1}{k^2}\int_E u^2dx \le 1 \qquad \text{and} \qquad \frac{F_*^2}{k^2}\le 1.$$

This choice leads to the recursive inequalities

$$Y_{n+1} \le \frac{4^{2n}C_6}{k^{4\delta}}Y_n^{1+2\delta} \tag{17.3}$$

for a constant C_6 that can be determined a priori only in terms of $\{N,\lambda,\Lambda\}$, the constants γ in the embedding inequalities of Theorem 2.1, the trace inequalities of Proposition 8.2, the smoothness of ∂E through the parameters ω and h of its cone condition, and is otherwise independent of $\mathbf{f}$, f, and ψ. By the fast geometric convergence Lemma 15.1, $\{Y_n\}\to 0$ as $n\to\infty$, provided

$$Y_1 \le 2^{-1/\delta^2}C_6^{-1/2\delta}k^2.$$

We conclude that by choosing

$$k = 2^{1/2\delta^2}C_6^{1/4\delta}\max\{\|u\|_2\,;\ F_*\}$$

then $Y_\infty = \|(u-2k)_+\|_2 = 0$, and therefore $u\le 2k$ in E. ∎

17.2 Proof of Proposition 16.1 for $N = 2$

The only differences occur in the application of the embedding inequalities of Theorem 2.1, in the inequalities (17.2), leading to the recursive inequalities (17.3). Inequality (17.2) is modified by fixing $1 < p < 2$ and applying the embedding inequality (2.1) of Theorem 2.1 for $1 < p < N$. This gives

$$\begin{aligned}\|(u-k_{n+1})_+\|_2^2 &\le \|(u-k_{n+1})_+\|_{p^*}^2 |A_{n+1}|^{2\frac{p-1}{p}} \\ &\le \gamma(p,h,\omega)\big(\|\nabla(u-k_{n+1})_+\|_p^2 + \|(u-k_{n+1})_+\|_p^2\big)|A_{n+1}|^{2\frac{p-1}{p}} \\ &\le \gamma(p,h,\omega)\big(\|\nabla(u-k_{n+1})_+\|_2^2 + \|(u-k_{n+1})_+\|_2^2\big)|A_{n+1}| \\ &\le \gamma\|(u-k_n)_+\|_2^2|A_{n+1}| + C_5F_*^2|A_{n+1}|^{1+2\delta}. \qquad \blacksquare\end{aligned}$$

18 Miscellaneous Remarks on Further Regularity

A function $u \in W^{1,2}_{\text{loc}}(E)$ is a *local* weak solution of (1.2), irrespective of possible boundary data, if it satisfies (1.3) for all $v \in W^{1,2}_o(E_o)$ for all open sets E_o such that $\bar{E}_o \subset E$. On the data $\mathbf{f}$ and f assume

$$\mathbf{f} \in L^{N+\varepsilon}(E), \qquad f_\pm \in L^{\frac{N+\varepsilon}{2}}(E), \qquad \text{for some } \varepsilon > 0. \tag{18.1}$$

The set of parameters $\{N, \lambda, \Lambda, \varepsilon, \|\mathbf{f}\|_{N+\varepsilon}, \|f\|_{\frac{N+\varepsilon}{2}}\}$ are the data, and we say that a constant $C, \gamma, \dots$ depends on the data if it can be quantitatively determined a priori in terms of only these quantities. Continue to assume that the boundary ∂E is of class C^1 and with the segment property. For a compact set $\mathcal{K} \subset \mathbb{R}^N$ and $\eta \in (0,1)$ continue to denote by $|\!|\!| \cdot |\!|\!|_{\eta;\mathcal{K}}$ the Hölder norms introduced in (9.3) of Chapter 2.

Theorem 18.1 *Let $u \in W^{1,2}_{\text{loc}}(E)$ be a local weak solution of (1.2) and let (18.1) hold. Then u is locally bounded and locally Hölder continuous in E, and for every compact set $K \subset E$, there exist positive constants γ_K, and C_K depending upon the data and* $\text{dist}\{K;\partial E\}$, *and $\alpha \in (0,1)$ depending only on the data and independent of* $\text{dist}\{K;\partial E\}$, *such that*

$$|\!|\!|u|\!|\!|_{\alpha;K} \le \gamma_K(\text{data}, \text{dist}\{K;\partial E\}). \tag{18.2}$$

Theorem 18.2 *Let $u \in W^{1,2}(E)$ be a solution of the Dirichlet problem (9.1), with $\mathbf{f}$ and f satisfying (18.1) and $\varphi \in C^\epsilon(\partial E)$ for some $\epsilon \in (0,1)$. Then u is Hölder continuous in $\bar{E}$ and there exist constants $\gamma > 1$ and $\alpha \in (0,1)$, depending upon the data, the C^1 structure of ∂E, and the Hölder norm $|\!|\!|\varphi|\!|\!|_{\epsilon;\partial E}$, such that*

$$|\!|\!|u|\!|\!|_{\alpha,\bar{E}} \le \gamma(\text{data}, \varphi, \partial E). \tag{18.3}$$

Theorem 18.3 *Let $u \in W^{1,2}(E)$ be a solution of the Neumann problem (10.1), with $\mathbf{f}$ and f satisfying (16.1) and $\psi \in L^{N-1+\sigma}(\partial E)$ for some*

$\sigma \in (0,1)$. *Then u is Hölder continuous in $\bar{E}$, and there exist constants γ and $\alpha \in (0,1)$, depending on the data, the C^1 structure of ∂E, and $\|\psi\|_{N-1+\sigma;\partial E}$, such that*

$$|||u|||_{\alpha,\bar{E}} \le \gamma(\text{data}, \psi, \partial E). \tag{18.4}$$

The precise structure of these estimates in terms of the Dirichlet data φ or the Neumann ψ, as well as the dependence on the structure of ∂E is specified in more general theorems for functions in the DeGiorgi classes (Theorem 7.1 and Theorem 8.1 of the next chapter). These are the key, seminal facts in the theory of regularity of solutions of elliptic equations. They can be used, by boot-strap arguments, to establish further regularity on the solutions, whenever further regularity is assumed on the data.

Problems and Complements

1c Weak Formulations and Weak Derivatives

1.1c The Chain Rule in $W^{1,p}(E)$

Proposition 1.1c *Let $u \in W^{1,p}(E)$ for some $p \ge 1$, and let $f \in C^1(\mathbb{R})$ satisfy $\sup|f'| \le M$, for some positive constant M. Then $f(u) \in W^{1,p}(E)$ and $\nabla f(u) = f'(u)\nabla u$.*

Proposition 1.2c *Let $u \in W^{1,p}(E)$ for some $p \ge 1$. Then $u^{\pm} \in W^{1,p}(E)$ and*

$$\nabla u^{\pm} = \begin{cases} \text{sign}(u)\nabla u & \text{a.e. in } [u^{\pm} > 0] \\ 0 & \text{a.e. in } [u = 0]. \end{cases}$$

Proof (Hint) To prove the statement for u^+, for $\varepsilon > 0$, apply the previous proposition with

$$f_\varepsilon(u) = \begin{cases} \sqrt{u^2+\varepsilon^2} - \varepsilon & \text{for } u > 0 \\ 0 & \text{for } u \le 0. \end{cases}$$

Then let $\varepsilon \to 0$. ■

Corollary 1.1c *Let $u \in W^{1,p}(E)$ for some $p \ge 1$. Then $|u-k| \in W^{1,p}(E)$, for all $k \in \mathbb{R}$, and $\nabla u = 0$ a.e. on any level set of u.*

Corollary 1.2c *Let $f, g \in W^{1,p}(E)$ for some $p \ge 1$. Then $f \wedge g$ and $f \vee g$ are in $W^{1,p}(E)$ and*

$$\nabla f \wedge g = \begin{cases} \nabla f & a.e.\ in\ [f > g] \\ \nabla g & a.e.\ in\ [f < g] \\ 0 & a.e.\ in\ [f = g]. \end{cases}$$

A similar formula holds for $f \vee g$.

1.2. Prove Proposition 1.2 and the first part of Proposition 1.1.

2c Embeddings of $W^{1,p}(E)$

It suffices to prove the various assertions for $u \in C^\infty(E)$. Fix $x \in E$ and let $\mathcal{C}_x \subset \bar{E}$ be a cone congruent to the cone $\mathcal{C}$ of the cone property. Let $\mathbf{n}$ be the unit vector exterior to $\mathcal{C}_x$, ranging over its same solid angle, and compute

$$|u(x)| = \Big| \int_0^h \frac{\partial}{\partial \rho} \Big(1 - \frac{\rho}{h}\Big) u(\rho \mathbf{n}) d\rho \Big| \le \int_0^h |\nabla u(\rho \mathbf{n})| d\rho + \frac{1}{h} \int_0^h |u(\rho \mathbf{n})| d\rho.$$

Integrating over the solid angle of $\mathcal{C}_x$ gives

$$\begin{aligned} \omega |u(x)| &\le \int_{\mathcal{C}_x} \frac{|\nabla u(y)|}{|x-y|^{N-1}} dy + \frac{1}{h} \int_{\mathcal{C}_x} \frac{|u(y)|}{|x-y|^{N-1}} dy \\ &\le \int_E \frac{|\nabla u(y)|}{|x-y|^{N-1}} dy + \frac{1}{h} \int_E \frac{|u(y)|}{|x-y|^{N-1}} dy. \end{aligned} \tag{2.1c}$$

The right-hand side is the sum of two Riesz potentials of the form (10.1) of Chapter 2. The embeddings (2.1)–(2.3) are now established from this and the estimates of Riesz potentials (10.2) of Proposition 10.1 of Chapter 2. Complete the estimates and compute the constants γ explicitly.

2.1c Proof of (2.4)

Let $\mathcal{C}_{x,\rho}$ be the cone of vertex at x, radius $0 < \rho \le h$, coaxial with $\mathcal{C}_x$ and with the same solid angle ω. Denote by $(u)_{x,\rho}$ the integral average of u over $\mathcal{C}_{x,\rho}$.

Lemma 2.1c *For every pair $x, y \in E$ such that $|x - y| = \rho \le h$*

$$|u(y) - (u)_{x,\rho}| \le \frac{\gamma(N,p)}{\omega} \rho^{1-\frac{N}{p}} \|\nabla u\|_p.$$

Proof For all $\xi \in \mathcal{C}_{x,\rho}$

$$|u(y) - u(\xi)| = \Big| \int_0^1 \frac{\partial}{\partial t} u\big(y + t(\xi - y)\big) dt \Big|.$$

Integrate in $d\xi$ over $\mathcal{C}_{x,\rho}$, and then in the resulting integral perform the change of variables $y + t(\xi - y) = \eta$. The Jacobian is t^{-N}, and the new domain of integration is transformed into those η for which $|y - \eta| = t|\xi - y|$ as ξ ranges over $\mathcal{C}_{x,\rho}$. Such a transformed domain is contained in the ball $B_{2\rho t}(y)$. These operations give

$$\begin{aligned}\frac{\omega}{N}\rho^N|u(y) - (u)_{x,\rho}| &\le \int_0^1 \Big(\int_{\mathcal{C}_{x,\rho}} |\xi - y|\,|\nabla u(y + t(\xi - y))|d\xi\Big)dt \\ &\le \int_0^1 t^{-(N+1)} \int_{E\cap B_{2\rho t}(y)} |\eta - y||\nabla u(\eta)|d\eta\, dt \\ &\le \gamma(N,p)\int_0^1 t^{-(N+1)}(2\rho t)^{N(1-\frac{1}{p})+1}\|\nabla u\|_p t. \qquad \blacksquare\end{aligned}$$

To conclude the proof of (2.4), fix $x, y \in E$, let $z = \frac{1}{2}(x + y)$, $\rho = \frac{1}{2}|x - y|$, and estimate

$$|u(x) - u(y)| \le |u(x) - (u)_{z,\rho}| + |u(y) - (u)_{z,\rho}| \le \frac{\gamma(N,p)}{\omega}|x - y|^{1-\frac{N}{p}}\|\nabla u\|_p. \quad \blacksquare$$

2.2c Compact Embeddings of $W^{1,p}(E)$

The proof consists in verifying that a bounded subset of $W^{1,p}(E)$ satisfies the conditions for a subset of $L^q(E)$ to be compact ([31], Chapter V). For $\delta > 0$ let

$$E_\delta = \big\{x \in E \big| \operatorname{dist}\{x; \partial E\} > \delta\big\}.$$

For $q \in [1, p^*)$ and $u \in W^{1,p}(E)$

$$\|u\|_{q,E-E_\delta} \le \|u\|_{p^*}\mu(E - E_\delta)^{\frac{1}{q}-\frac{1}{p^*}}.$$

Next, for $h \in \mathbb{R}^N$ of length $|h| < \delta$ compute

$$\begin{aligned}\int_{E_\delta} |u(x + h) - u(x)|dx &\le \int_{E_\delta}\int_0^1 \Big|\frac{d}{dt}u(x + th)\Big|dtdx \\ &\le |h|\int_0^1\int_{E_\delta} |\nabla u(x + th)|dx\, dt \le |h||E|^{\frac{p-1}{p}}\|\nabla u\|_p.\end{aligned}$$

Therefore for all $\sigma \in (0, \frac{1}{q})$

$$\begin{aligned}\int_{E_\delta} |T_h u - u|^q dx &= \int_{E_\delta} |T_h u - u|^{q\sigma + q(1-\sigma)}dx \\ &\le \Big(\int_{E_\delta} |T_h u - u|dx\Big)^{q\sigma}\Big(\int_{E_\delta} |T_h u - u|^{\frac{q(1-\sigma)}{1-q\sigma}}dx\Big)^{1-q\sigma}.\end{aligned}$$

Choose σ so that

$$\frac{q(1-\sigma)}{1-q\sigma} = p^*, \qquad \text{that is,} \qquad \sigma q = \frac{p^*-q}{p^*-1}.$$

Such a choice is possible if $1 < q < p^*$. Applying the embedding Theorem 2.1 gives

$$\int_{E_\delta} |T_h u - u|^q dx \le \gamma^{(1-\sigma)q} \Big(\int_{E_\delta} |T_h u - u| dx \Big)^{\frac{p^*-q}{p^*-1}} \|u\|_{1,p}^{(1-\sigma)q}$$

for a constant γ depending only on N, p and the geometry of the cone property of E. Combining these estimates

$$\|T_h u - u\|_{q,E_\delta} \le \gamma_1 |h|^\sigma \|u\|_{1,p}.$$

■

3c Multiplicative Embeddings of $W_o^{1,p}(E)$ and $\tilde{W}^{1,p}(E)$

3.1c Proof of Theorem 3.1 for $1 \le p < N$

Lemma 3.1c *Let $u \in C_o^\infty(E)$ and $N > 1$. Then*

$$\|u\|_{\frac{N}{N-1}} \le \prod_{j=1}^N \|u_{x_j}\|_1^{1/N}.$$

Proof If $N = 2$

$$\begin{aligned}
\iint_E u^2(x_1,x_2)dx_1dx_2 &= \iint_E u(x_1,x_2)u(x_1,x_2)dx_1dx_2 \\
&\le \iint_E \max_{x_2} u(x_1,x_2) \max_{x_1} u(x_1,x_2)dx_1dx_2 \\
&= \int_{\mathbb{R}} \max_{x_2} u(x_1,x_2)dx_1 \int_{\mathbb{R}} \max_{x_1} u(x_1,x_2)dx_2 \\
&\le \iint_E |u_{x_1}|dx \iint_E |u_{x_2}|dx.
\end{aligned}$$

Thus the lemma holds for $N = 2$. Assuming that it does hold for N, set

$$\overline{x} = (x_1, \dots, x_N) \quad \text{and} \quad x = (\bar{x}, x_{N+1}).$$

By repeated application of Hölder's inequality and the induction

$$
\begin{aligned}
\|u\|_{\frac{N+1}{N}}^{\frac{N+1}{N}} &= \int_{\mathbb{R}}\int_{\mathbb{R}^N} |u(\overline{x}, x_{N+1})|^{\frac{N+1}{N}} d\overline{x} dx_{N+1} \\
&= \int_{\mathbb{R}} dx_{N+1} \int_{\mathbb{R}^N} |u(\overline{x}, x_{N+1})||u(\overline{x}, x_{N+1})|^{\frac{1}{N}} d\overline{x} \\
&\le \int_{\mathbb{R}} dx_{N+1} \Big(\int_{\mathbb{R}^N} |u(\overline{x}, x_{N+1})| d\overline{x} \Big)^{\frac{1}{N}} \Big(\int_{\mathbb{R}^N} |u(\overline{x}, x_{N+1})|^{\frac{N}{N-1}} d\overline{x} \Big)^{\frac{N-1}{N}} \\
&\le \Big(\int_E |u_{x_{N+1}}| dx \Big)^{\frac{1}{N}} \int_{\mathbb{R}} \prod_{j=1}^{N} \Big(\int_{\mathbb{R}^N} |u_{x_j}(\bar{x}, x_{N+1})| d\bar{x} \Big)^{\frac{1}{N}} dx_{N+1} \\
&\le \Big(\int_E |u_{x_{N+1}}| dx \Big)^{\frac{1}{N}} \Big(\prod_{j=1}^{N} \int_E |u_{x_j}| dx \Big)^{\frac{1}{N}} = \Big(\prod_{j=1}^{N+1} \int_E |u_{x_j}| dx \Big)^{\frac{1}{N}} \qquad \blacksquare
\end{aligned}
$$

Next, for $1 \le p < N$ write

$$
\|u\|_{\frac{Np}{N-p}} = \Big(\int_E w^{\frac{N}{N-1}} dx \Big)^{\frac{N-1}{N} \frac{N-p}{p(N-1)}} \qquad \text{where } w = |u|^{\frac{p(N-1)}{N-p}}
$$

and apply Lemma 3.1c to the function w. This gives

$$
\begin{aligned}
\|u\|_{\frac{Np}{N-p}} &\le \Big[\prod_{j=1}^{N} \Big(\int_E |w_{x_j}| dx \Big)^{\frac{1}{N}} \Big]^{\frac{N-p}{p(N-1)}} \\
&= \gamma(N,p) \prod_{j=1}^{N} \Big(\int_E |u|^{\frac{p(N-1)}{N-p} - 1} |u_{x_j}| dx \Big)^{\frac{N-p}{Np(N-1)}}
\end{aligned}
$$

where

$$
\gamma = \Big(\frac{p(N-1)}{N-p} \Big)^{\frac{N-p}{p(N-1)}}.
$$

Now for all $j = 1, \dots, N$, by Hölder's inequality

$$
\int_E |u|^{\frac{p(N-1)}{N-p} - 1} |u_{x_j}| dx \le \Big(\int_E |u_{x_i}|^p dx \Big)^{\frac{1}{p}} \Big(\int_E |u|^{\frac{Np}{N-p}} dx \Big)^{\frac{p-1}{p}}.
$$

Therefore

$$
\begin{aligned}
\prod_{j=1}^{N} \Big(\int_E |u|^{\frac{p(N-1)}{N-p} - 1} |u_{x_j}| dx \Big)^{\frac{N-p}{Np(N-1)}} &= \prod_{j=1}^{N} \|u_{x_i}\|_p^{\frac{N-p}{Np(N-1)}} \|u\|_{\frac{Np}{N-p}}^{\frac{p-1}{p(N-1)}} \\
&\le \|\nabla u\|_p^{\frac{N-p}{p(N-1)}} \|u\|_{\frac{Np}{N-p}}^{\frac{p-1}{p} \frac{N}{N-1}}. \qquad \blacksquare
\end{aligned}
$$

3.2c Proof of Theorem 3.1 for $p \ge N > 1$

Let $F(x;y)$ be the fundamental solution of the Laplacian. Then for $u \in C_o^\infty(E)$, by the Stokes formula (2.3)–(2.4) of Chapter 2

$$\begin{aligned} u(x) &= -\int_{\mathbb{R}^N} F(x;y)\Delta u(y)dy = \int_{\mathbb{R}^N} \nabla u(y)\cdot\nabla_y F(x;y)dy \\ &= \int_{|x-y|<\rho} \nabla u(y)\cdot\nabla_y F(x;y)dy + \int_{|x-y|>\rho} \nabla u(y)\cdot\nabla_y F(x;y)dy. \end{aligned}$$

The last integral can be computed by an integration by parts, and equals

$$\int_{|x-y|>\rho} \nabla u(y)\cdot\nabla_y F(x;y)dy = \frac{1}{\omega_N\rho^{N-1}}\int_{|x-y|=\rho} u(y)d\sigma$$

since $F(x;\cdot)$ is harmonic in $\mathbb{R}^N - \{x\}$. Here $d\sigma$ denotes the surface measure on the sphere $|x-y|=\rho$. Put this in the previous expression of $u(x)$, multiply by $N\omega_N\rho^{N-1}$, and integrate in $d\rho$ over $(0,R)$, where R is a positive number to be chosen later. This gives

$$\begin{aligned} \omega_N R^N|u(x)| \le N&\int_0^R\left(\int_{|x-y|<\rho}\frac{|\nabla u(y)|}{|x-y|^{N-1}}dy\right)\rho^{N-1}d\rho \\ &+ N\int_0^R\left(\int_{|x-y|=\rho}|u(y)|d\sigma\right)d\rho. \end{aligned}$$

From this, for all $x \in E$

$$\begin{aligned} \omega_N|u(x)| &\le \int_{B_R(x)}\frac{|\nabla u(y)|}{|x-y|^{N-1}}dy + \frac{N}{R^N}\int_{B_R(x)}|u(y)|dy \\ &= I_1(x,R) + NI_2(x,R). \end{aligned}$$

3.2.1c Estimate of $I_1(x,R)$

Choose two positive numbers $a, b < N$ such that

$$\frac{a}{q} + b\left(1-\frac{1}{p}\right) = N-1.$$

Since $p \ge N$, this choice is possible for the indicated range of q. Now write

$$\frac{|\nabla u|}{|x-y|^{N-1}} = |\nabla u|^{p(\frac{1}{p}-\frac{1}{q})}\frac{|\nabla u|^{\frac{p}{q}}}{|x-y|^{\frac{a}{q}}}\frac{1}{|x-y|^{b(1-\frac{1}{p})}}$$

and apply Hölder's inequality with the conjugate exponents

$$\left(\frac{1}{p}-\frac{1}{q}\right)+\frac{1}{q}+\left(1-\frac{1}{p}\right)=1.$$

This gives

$$I_1(x,R) \le \|\nabla u\|_p^{1-\frac{p}{q}} \Big(\int_{B_R(x)} \frac{|\nabla u(y)|^p}{|x-y|^a} dy \Big)^{\frac{1}{q}} \Big(\int_{B_R(x)} \frac{1}{|x-y|^b} dy \Big)^{1-\frac{1}{p}}.$$

Taking the qth power and integrating over E gives

$$\|I_1(R)\|_q \le \frac{\omega_N^{1-\frac{1}{p}+\frac{1}{q}} R^{N(\frac{1}{N}-\frac{1}{p}+\frac{1}{q})}}{(N-a)^{\frac{1}{q}}(N-b)^{1-\frac{1}{p}}} \|\nabla u\|_p.$$

3.2.2c Estimate of $I_2(x,R)$

$$\begin{aligned} I_2(x,R) &\le R^{-N} \Big(\int_{|x-y|<R} |u(y)|^p dy \Big)^{\frac{1}{p}} \Big(\int_{|x-y|<R} 1 dy \Big)^{1-\frac{1}{p}} \\ &\le \Big(\frac{\omega_N}{N} \Big)^{1-\frac{1}{p}} R^{-\frac{N}{p}} \|u\|_p^{1-\frac{p}{q}} \Big(\int_{|\xi|<R} |u(x+\xi)|^p d\xi \Big)^{\frac{1}{q}}. \end{aligned}$$

Take the qth power and integrate in dx over $\mathbb{R}^N$ to obtain

$$\|I_2\|_q \le \Big(\frac{\omega_N}{N} \Big)^{1+\frac{1}{q}-\frac{1}{p}} R^{-N(\frac{1}{p}-\frac{1}{q})} \|u\|_p.$$

3.2.3c Proof of Theorem 3.1 for $p \ge N > 1$ (Concluded)

Combining these estimates yields

$$\|u\|_q \le \gamma \big(R^{1-\delta} \|\nabla u\|_p + R^{-\delta} \|u\|_p \big), \qquad \delta = N \Big(\frac{1}{p} - \frac{1}{q} \Big)$$

for a constant $\gamma(N,p,q,a,b)$. Minimizing the right-hand side with respect to the parameter R proves the estimate.

3.3c Proof of Theorem 3.2 for $1 \le p < N$ and E Convex

Having fixed $x, y \in E$, let $R(x,y)$ be the distance from x to ∂E along $y - x$ and write

$$|u(x) - u(y)| \le \int_0^{R(x,y)} \Big| \frac{\partial}{\partial \rho} u(x + \rho \mathbf{n}) \Big| d\rho, \qquad \mathbf{n} = \frac{y-x}{|y-x|}.$$

Integrate in dy over E to obtain

$$|E| |u(x) - u_E| \le \int_E \Big(\int_0^{R(x,y)} |\nabla u(x + \rho \mathbf{n})| d\rho \Big) dy.$$

The integral in dy is calculated by introducing polar coordinates with pole at x. Therefore if $\mathbf{n}$ is the angular variable spanning the sphere $|\mathbf{n}|=1$, the right-hand side is majorized by

$$(\operatorname{diam} E)^{N-1}\int_0^{\operatorname{diam} E}\int_{|\mathbf{n}|=1}\int_0^{R(x,y)}\rho^{N-1}\frac{|\nabla u(x+\rho\mathbf{n})|}{|x-y|^{N-1}}d\rho d\mathbf{n}dr$$
$$\leq(\operatorname{diam} E)^N\int_E\frac{|\nabla u|}{|x-y|^{N-1}}dy.$$

Therefore

$$|u(x)-u_E|\leq\frac{(\operatorname{diam} E)^N}{|E|}\int_E\frac{|\nabla u(y)|}{|x-y|^{N-1}}dy.$$

The proof is now concluded using the estimates of the Riesz potentials in Section 10 of Chapter 2. The remaining cases for $p\geq N$ are left as an exercise following similar arguments in the analogous multiplicative embeddings of $W_o^{1,p}(E)$.

5c Solving the Homogeneous Dirichlet Problem (4.1) by the Riesz Representation Theorem

Consider formally the linear operator with variable coefficients

$$\mathcal{L}(u)=-\big(a_{ij}u_{x_i}+a_ju\big)_{x_j}+b_iu_{x_i}+cu \tag{5.1c}$$

and the associated, formal bilinear form

$$a(u,v)=\int_E\big(a_{ij}u_{x_i}v_{x_j}+a_juv_{x_j}+b_iu_{x_i}v+cuv\big)\,dx. \tag{5.2c}$$

Assuming that (a_{ij}) satisfies the ellipticity condition (1.1), all the various terms are well defined for $u,v\in W_o^{1,2}(E)$, provided $c\in L^\infty(E)$ and

$$\left.\begin{array}{l}\mathbf{a}=(a_1,\dots,a_N)\\ \mathbf{b}=(b_1,\dots,b_N)\end{array}\right\}\in\left\{\begin{array}{ll}L^N(E) & \text{if } N>2\\ L^q(E) & \text{for some } q>2 \text{ if } N=2.\end{array}\right. \tag{5.3c}$$

The homogeneous Dirichlet problem (4.1) takes the form

$$\mathcal{L}(u)=\operatorname{div}\mathbf{f}-f \text{ in } E,\qquad\text{and } u\big|_{\partial E}=0 \text{ on } \partial E. \tag{5.4c}$$

The latter is meant in the weak form of seeking $u\in W_o^{1,2}(E)$ such that

$$a(u,v)=-\int_E\big(\mathbf{f}\cdot\nabla v+fv\big)dx \tag{5.5c}$$

for all $v\in W_o^{1,2}(E)$. The unique solvability of this problem can be established almost verbatim by any one of the methods of Sections 5–7, provided the bilinear form $a(\cdot,\cdot)$ introduced in (5.2c) generates an inner product in $W_o^{1,2}(E)$, equivalent to any one of the inner products in (3.9), precisely, if there are constants $0 < \lambda_o < \Lambda_o$ such that

$$\lambda_o\|\nabla u\|_2^2 \le a(u,u) \le \Lambda_o\|\nabla u\|_2^2 \qquad \text{for all } u \in W_o^{1,2}(E).$$

This can be ensured by a number of conditions on $\mathbf{a}$, $\mathbf{b}$, and c, and on the size of E. Let $c = c^+ - c^-$ be partitioned into its positive and negative parts. Prove that if $N > 2$, the following condition is sufficient for the unique solvability of (5.4c):

$$\gamma\left(\|\mathbf{a}\|_N + \|\mathbf{b}\|_N + \gamma\|c^-\|_\infty\right) \le (1-\varepsilon)\lambda \tag{5.6c}$$

for some $\varepsilon \in (0,1)$, where γ is the constant appearing in the embedding inequality (3.1). The latter occurs, for example, if $c \ge 0$, $\mathbf{b} \in L^q(E)$ for some $q > N$, and $|E|$ is sufficiently small. Prove that another sufficient condition is

$$c \ge c_o > 0 \quad \text{and} \quad \frac{1}{4(1-\varepsilon)\lambda}\left(\|\mathbf{a}\|_\infty + \|\mathbf{b}\|_\infty\right) \le c_o. \tag{5.7c}$$

6c Solving the Homogeneous Dirichlet Problem (4.1) by Variational Methods

The homogeneous Dirichlet problem (5.4c), can also be solved by variational methods. The corresponding functional is

$$2J(u) = \int_E \{[a_{ij}u_{x_i} + b_i u_{x_i} + (a_j + b_j)u + 2f_j]u_{x_j} + (\mathbf{b}\cdot\nabla u + cu + 2f)u\}dx.$$

The same minimization procedure can be carried out, provided $\mathbf{b}$ and c satisfy either (5.6c) or (5.7c).

6.1c More General Variational Problems

More generally one might consider minimizing functionals of the type

$$W_o^{1,p}(E) \ni u \to J(u) = \int_E F(x,u,\nabla u)dx, \qquad p > 1 \tag{6.1c}$$

where the function

$$E \times \mathbb{R} \times \mathbb{R}^N \ni (x,z,\mathbf{q}) \to F(x,z,\mathbf{q})$$

is measurable in x for a.e. $(z,\mathbf{q}) \in \mathbb{R}^{N+1}$, differentiable in z and $\mathbf{q}$ for a.e. $x \in E$, and satisfies the structure condition

$$\lambda|\mathbf{q}|^p - f(x) \le F(x,z,\mathbf{q}) \le \Lambda|\mathbf{q}|^p + f(x) \tag{6.2c}$$

for a given non-negative $f \in L^1(E)$. On F impose also the convexity (ellipticity) condition, that is, $F(x,z,\cdot) \in C^2(\mathbb{R}^N)$ for a.e. $(x,z) \in E \times \mathbb{R}$, and

$$F_{q_iq_j}\xi_i\xi_j \ge \lambda|\xi|^p \quad \text{for all } \xi \in \mathbb{R}^N \text{ for a.e. } (x,z) \in E \times \mathbb{R}. \tag{6.3c}$$

A Prototype Example

Let (a_{ij}) denote a symmetric $N \times N$ matrix with entries $a_{ij} \in L^\infty(E)$ and satisfying the ellipticity condition (1.1), and consider the functional

$$W_o^{1,p}(E) \ni u \to pJ(u) = \int_E (|\nabla u|^{p-2} a_{ij} u_{x_i} u_{x_j} + pfu)dx \tag{6.4c}$$

for a given $f \in L^q(E)$, where $q \geq 1$ satisfies

$$\frac{1}{p} + \frac{1}{q} = \frac{1}{N} + 1 \quad \text{if } 1 < p < N, \qquad \text{and } q \geq 1 \text{ if } p \geq N. \tag{6.5c}$$

Let $\mathbf{v} = (v_1, \ldots, v_N)$ be a vector-valued function defined in E. Verify that the map

$$[L^p(E)]^N \ni \mathbf{v} = \int_E |\mathbf{v}|^{p-2} a_{ij} v_i v_j dx$$

defines a norm in $[L^p(E)]^N$ equivalent to $\|\mathbf{v}\|_p$. Since the norm is weakly lower semi-continuous, for every sequence $\{\mathbf{v}_n\} \subset [L^p(E)]^N$ weakly convergent to some $\mathbf{v} \in [L^p(E)]^N$

$$\liminf \int_E |\mathbf{v}_n|^{p-2} a_{ij} v_{i,n} v_{j,n} dx \geq \int_E |\mathbf{v}|^{p-2} a_{ij} v_i v_j dx.$$

The convexity condition (6.3c), called also the Legendre condition, ensures that a similar notion of semi-continuity holds for the functional $J(\cdot)$ in (6.1c) (see [108]).

Lower Semi-Continuity

A functional J from a topological space X into $\mathbb{R}$ is lower semi-continuous if $[J > a]$ is open in X for all $a \in \mathbb{R}$. Prove the following.

Proposition 6.1c *Let X be a topological space satisfying the first axiom of countability. A functional $J : X \to \mathbb{R}$ is lower semi-continuous if and only if for every sequence $\{u_n\} \subset X$ convergent to some $u \in X$*

$$\liminf J(u_n) \geq J(u).$$

6.3. The epigraph of J is the set

$$\mathcal{E}_J = \{(x, a) \in X \times \mathbb{R} \mid J(x) \leq a\}.$$

Assume that X satisfies the first axiom of countability and prove that J is lower semi-continuous if and only if its epigraph is closed.

6.4. Prove that $J : X \to \mathbb{R}$ is convex if and only if its epigraph is convex.

6.5. Prove the following:

Proposition 6.2c *Let* $J : W_o^{1,p}(E) \to \mathbb{R}$ *be the functional in (6.1c) where* F *satisfies (6.2c)–(6.3c). Then* J *is weakly lower semi-continuous.*

Hint: Assume first that F is independent of x and z and depends only on $\mathbf{q}$. Then J may be regarded as a convex functional from $\tilde{J} : [L^p(E)]^N \to \mathbb{R}$. Prove that its epigraph is (strongly and hence weakly) closed in $[L^p(E)]^N$.

6.6. Prove the following:

Proposition 6.3c *Let* $J : W_o^{1,p}(E) \to \mathbb{R}$ *be the functional in (6.1c) where* F *satisfies (6.2c)–(6.3c). Then* J *has a minimum in* $W_o^{1,p}(E)$.

Hint: Parallel the procedure of Section 6.

6.7. The minimum claimed by Proposition 6.3c need not be unique. Provide a counterexample. Formulate sufficient assumptions on F to ensure uniqueness of the minimum.

6.8c Gâteaux Derivatives, Euler Equations, and Quasi-Linear Elliptic Equations

Let X be a Hausdorff space. A functional $J : X \to \mathbb{R}$ is Gâteaux differentiable at $w \in X$ in the direction of some $v \in X$ if there exists an element $J'(w; v) \in \mathbb{R}$ such that

$$\lim_{t\to 0} \frac{J(w + tv) - J(w)}{t} = J'(w; v).$$

The equation

$$J'(w; v) = 0 \quad \text{for all } \; v \in X$$

is called the Euler equation of J. In particular, (6.4) is the Euler equation of the functional in (6.1). The Euler equation of the functional in (6.4c) is

$$-\left(|\nabla u|^{p-2} a_{ij} u_{x_i}\right)_{x_j} = f. \tag{6.6c}$$

In the special case $(a_{ij}) = \mathbb{I}$ this is the p-Laplacian equation

$$-\operatorname{div} |\nabla u|^{p-2} \nabla u = f. \tag{6.7c}$$

The Euler equation of the functional in (6.1c) is

$$-\operatorname{div} \mathbf{A}(x, u, \nabla u) + B(x, u, \nabla u) = 0, \qquad u \in W_o^{1,p}(E) \tag{6.8c}$$

where $\mathbf{A} = \nabla_{\mathbf{q}} F$ and $B = F_z$. The equation is elliptic in the sense that $(a_{ij}) = (F_{q_i q_j})$ satisfies (6.3c). Thus the functional in (6.1c) generates the PDE in (6.8c) as its Euler equation, and minima of J are solutions of (6.8c).

6.8.1c Quasi-Linear Elliptic Equations

Consider now (6.8c) independently of its variational origin, where

$$E \times \mathbb{R} \times \mathbb{R}^N \ni (x, z, \mathbf{q}) \to \begin{cases} \mathbf{A}(x, z, \mathbf{q}) \in \mathbb{R}^N \\ B(x, z, \mathbf{q}) \in \mathbb{R} \end{cases} \tag{6.9c}$$

are continuous functions of their arguments and subject to the structure conditions

$$\begin{cases} \mathbf{A}(x, z, \mathbf{q}) \cdot \mathbf{q} \geq \lambda |\mathbf{q}|^p - C^p \\ |\mathbf{A}(x, z, \mathbf{q})| \leq \Lambda |\mathbf{q}|^{p-1} + C^{p-1} \\ |B(x, z, \mathbf{q})| \leq C |\mathbf{q}|^{p-1} + C^p \end{cases} \tag{6.10c}$$

for all $(x, z, \mathbf{q}) \in E \times \mathbb{R} \times \mathbb{R}^N$, for given positive constants $\lambda \leq \Lambda$ and non-negative constant C. A local solution of (6.8c)–(6.10c), irrespective of possible prescribed boundary data, is a function $u \in W^{1,p}_{\mathrm{loc}}(E)$ satisfying

$$\int_E \big[\mathbf{A}(x, u, \nabla u) \nabla v + B(x, u, \nabla u) v\big] dx = 0 \quad \text{for all } v \in W^{1,p}_o(E_o) \tag{6.11c}$$

for every open set E_o such that $\bar{E}_o \subset E$. In general, there is not a function F satisfying (6.2c)–(6.3c) and a corresponding functional as in (6.1c) for which (6.8c) is its Euler equation. It turns out, however, that local solutions of (6.8c)–(6.10c), whenever they exist, possess the same local behavior, regardless of their possible variational origin (Chapter 10).

6.8.2c Quasi-Minima

Let $J : W^{1,p}_{\mathrm{loc}}(E) \to \mathbb{R}$ be given by (6.1c), where F satisfies (6.2c) but not necessarily (6.3c). A function $u \in W^{1,p}_{\mathrm{loc}}(E)$ is a Q-minimum for J if there is a number $Q \geq 1$ such that

$$J(u) \leq J(u + v) \qquad \text{for all } v \in W^{1,p}_o(E_o)$$

for every open set E_o such that $\bar{E}_o \subset E$. The notion is of local nature. Minima are Q-minima, but the converse is false. Every functional of the type (6.1c)–(6.3c) generates a quasi-linear elliptic PDE of the type of (6.8c)–(6.10c). The converse is in general false. However every *local* solution $u \in W^{1,p}_{\mathrm{loc}}(E)$ of (6.8c)–(6.10c) is a Q-minimum, in the sense that there exists some F satisfying (6.2c), but not necessarily (6.3c), such that u is a Q-minimum for the function J in (6.1c) for such a F ([53]).

8c Traces on ∂E of Functions in $W^{1,p}(E)$

8.1c Extending Functions in $W^{1,p}(E)$

Establish Proposition 8.1 by the following steps. Let $\mathbb{R}^N_+$ be the upper-half space $x_N > 0$ and denote its coordinates by $x = (\bar{x}, x_N)$, where

$\bar{x} = (x_1, \dots, x_{N-1})$. Assume first that $E = \mathbb{R}^N_+$ so that ∂E is the hyperplane $x_N = 0$. Given $u \in W^{1,p}(\mathbb{R}^N_+)$, set ([102])

$$\tilde{u}(\bar{x}, x_N) = \begin{cases} u(\bar{x}, x_N) & \text{if } x_N > 0 \\ -3u(\bar{x}, -x_N) + 4u(\bar{x}, -\frac{1}{2}x_N) & \text{if } x_N < 0. \end{cases}$$

Prove that $\tilde{u} \in W^{1,p}(\mathbb{R}^N)$, and that $C_o^\infty(\mathbb{R}^N)$ is dense in $W^{1,p}(\mathbb{R}^N_+)$.

If ∂E is of class C^1 and has the segment property, it admits a finite covering with balls $B_t(x_j)$ for some $t > 0$, and $x_j \in \partial E$ for $j = 1, \dots, m$. Let then

$$U = \{B_o, B_{2t}(x_1), \dots, B_{2t}(x_m)\}, \qquad B_o = E - \bigcup_{j=1}^m \bar{B}_t(x_j)$$

be an open covering of E, and let Φ be a partition of unity subordinate to U. Set

$$\psi_j = \{\text{the sum of the } \varphi \in \Phi \text{ supported in } B_{2t}(x_j)\}$$

so that

$$u = \sum_{j=1}^m u_j \quad \text{where} \quad u_j = \begin{cases} u\psi_j & \text{in } E \\ 0 & \text{otherwise.} \end{cases}$$

By construction, $u_j \in W^{1,p}(B_{2t}(x_j))$ with bounds depending on t. By choosing t sufficiently small, the portion $\partial E \cap B_{2t}(x_j)$ can be mapped, in a local system of coordinates, into a portion of the hyperplane $x_N = 0$. Denote by U_j an open ball containing the image of $B_{2t}(x_j)$ and set $U_j^+ = U_j \cap [x_N > 0]$. The transformed functions $\bar{u}_j$ belong to $W^{1,p}(U_j^+)$. Perform the extension as indicated earlier, return to the original coordinates and piece together the various integrals each relative to the balls $B_{2t}(x_j)$ of the covering U. This technique is refereed to as local "flattening of the boundary."

8.2c The Trace Inequality

Proposition 8.1c *Let $u \in C_o^\infty(\mathbb{R}^N)$. If $1 \le p < N$, there exists a constant $\gamma = \gamma(N, p)$ such that*

$$\|u(\cdot, 0)\|_{p^* \frac{N-1}{N}, \mathbb{R}^{N-1}} \le \gamma \|\nabla u\|_{p, \mathbb{R}^N_+}. \tag{8.1c}$$

If $p > N$, there exist constants $\gamma = \gamma(N, p)$ such that

$$\|u(\cdot, 0)\|_{\infty, \mathbb{R}^{N-1}} \le \gamma \|u\|_{p, \mathbb{R}^N_+}^{1 - \frac{N}{p}} \|\nabla u\|_{p, \mathbb{R}^N_+}^{\frac{N}{p}} \tag{8.2c}$$

$$|u(\bar{x}, 0) - u(\bar{y}, 0)| \le \gamma |\bar{x} - \bar{y}|^{1 - \frac{N}{p}} \|\nabla u\|_{p, \mathbb{R}^N_+} \tag{8.3c}$$

for all $\bar{x}, \bar{y} \in \mathbb{R}^{N-1}$.

Proof For all $\bar{x} \in \mathbb{R}^{N-1}$ and all $r \ge 1$

$$|u(\bar{x},0)|^r \le r \int_0^\infty |u(\bar{x},x_N)|^{r-1} |u_{x_N}(\bar{x},x_N)| dx_N.$$

Integrate both sides in $d\bar{x}$ over $\mathbb{R}^{N-1}$ and apply *Hölder's inequality* to the resulting integral on the right-hand side to obtain

$$\|u(\cdot,0)\|_{r,\mathbb{R}^{N-1}}^r \le r \|\nabla u\|_{p,\mathbb{R}_+^N} \|u\|_{q,\mathbb{R}_+^N}^{r-1}, \qquad \text{where } q = \frac{p}{p-1}(r-1). \tag{8.4c}$$

Apply this with $r = p^* \frac{N-1}{N}$, and use the embedding (3.1) of Theorem 3.1 to get

$$\|u(\cdot,0)\|_{p^*\frac{N-1}{N},\mathbb{R}^{N-1}} \le \gamma \|u\|_{p^*,\mathbb{R}_+^N}^{1-\frac{1}{r}} \|\nabla u\|_{p,\mathbb{R}_+^N}^{\frac{1}{r}} \le \gamma \|\nabla u\|_{p,\mathbb{R}_+^N}.$$

The domain $\mathbb{R}_+^N$ satisfies the cone condition with cone $\mathcal{C}$ of solid angle $\frac{1}{2}\omega_N$ and height $h \in (0,\infty)$. Then (8.3c) follows from (2.4) of Theorem 2.1, whereas (8.2c) follows from (2.3) of the same theorem, by minimizing over $h \in (0,\infty)$. ∎

Prove Proposition 8.2 by a local flattening of ∂E.

8.3c Characterizing the Traces on ∂E of Functions in $W^{1,p}(E)$

Set $\mathbb{R}_+^{N+1} = \mathbb{R}^N \times \mathbb{R}_+$ and denote the coordinates in $\mathbb{R}_+^{N+1}$ by (x,t) where $x \in \mathbb{R}^N$ and $t \ge 0$. Also set

$$\nabla_N = \Big(\frac{\partial}{\partial x_1}, \dots, \frac{\partial}{\partial x_N}\Big), \qquad \nabla = \Big(\nabla_N, \frac{\partial}{\partial t}\Big).$$

Proposition 8.2c *Let $u \in C_o^\infty(\mathbb{R}_+^{N+1})$. Then*

$$|\!|\!|u(\cdot,0)|\!|\!|_{1-\frac{1}{p},p;\mathbb{R}^N} \le \gamma \|u_t\|_{p,\mathbb{R}_+^{N+1}}^{\frac{1}{p}} \|\nabla_N u\|_{p,\mathbb{R}_+^{N+1}}^{1-\frac{1}{p}} \tag{8.5c}$$

where $\gamma = \gamma(p)$ depends only on p and $\gamma(p) \to \infty$ as $p \to 1$.

Proof For every pair $x, y \in \mathbb{R}^N$, set $2\xi = x - y$ and consider the point $z \in \mathbb{R}_+^{N+1}$ of coordinates $z = (\frac{1}{2}(x+y), \lambda|\xi|)$, where λ is a positive parameter to be chosen. Then

$$\begin{aligned}
|u(x,0) - u(y,0)| &\le |u(z) - u(x,0)| + |u(z) - u(y,0)| \\
&\le |\xi| \int_0^1 |\nabla_N u(x - \rho\xi, \lambda\rho|\xi|)| d\rho + |\xi| \int_0^1 |\nabla_N u(y + \rho\xi, \lambda\rho|\xi|)| d\rho \\
&\quad + \lambda|\xi| \int_0^1 |u_t(x - \rho\xi, \lambda\rho|\xi|)| d\rho + \lambda|\xi| \int_0^1 |u_t(y + \rho\xi, \lambda\rho|\xi|)| d\rho.
\end{aligned}$$

From this

$$\begin{aligned}\frac{|u(x,0)-u(y,0)|^p}{|x-y|^{N+(p-1)}} &\le \frac{1}{2^p}\Big(\int_0^1 \frac{|\nabla_N u(x-\rho\xi,\lambda\rho|\xi|)|}{|x-y|^{\frac{N-1}{p}}}d\rho\Big)^p \\ &+\frac{1}{2^p}\Big(\int_0^1 \frac{|\nabla_N u(y+\rho\xi,\lambda\rho|\xi|)|}{|x-y|^{\frac{N-1}{p}}}d\rho\Big)^p \\ &+\frac{1}{2^p}\lambda^p\Big(\int_0^1 \frac{|u_t(x-\rho\xi,\lambda\rho|\xi|)|}{|x-y|^{\frac{N-1}{p}}}d\rho\Big)^p \\ &+\frac{1}{2^p}\lambda^p\Big(\int_0^1 \frac{|u_t(y+\rho\xi,\lambda\rho|\xi|)|}{|x-y|^{\frac{N-1}{p}}}d\rho\Big)^p.\end{aligned}$$

Next integrate both sides over $\mathbb{R}^N\times\mathbb{R}^N$. In the resulting inequality take the $\frac{1}{p}$ power and estimate the various integrals on the right-hand side by the continuous version of Minkowski's inequality. This gives

$$\begin{aligned}|||u(\cdot,0)|||_{1-\frac{1}{p},\mathbb{R}^N} &\le \int_0^1\Big(\int_{\mathbb{R}^N}\int_{\mathbb{R}^N}\frac{|\nabla_N u(x-\rho\xi,\lambda\rho|\xi|)|^p}{|x-y|^{N-1}}dxdy\Big)^{\frac{1}{p}}d\rho \\ &+\lambda\int_0^1\Big(\int_{\mathbb{R}^N}\int_{\mathbb{R}^N}\frac{|u_t(x-\rho\xi,\lambda\rho|\xi|)|^p}{|x-y|^{N-1}}dxdy\Big)^{\frac{1}{p}}d\rho.\end{aligned}$$

Compute the first integral by integrating first in dy and perform such integration in polar coordinates with pole at x. Denoting by $\mathbf{n}$ the unit vector spanning the unit sphere in $\mathbb{R}^N$ and recalling that $2|\xi|=|x-y|$, we obtain

$$\begin{aligned}&\int_{\mathbb{R}^N}\int_{\mathbb{R}^N}\frac{|\nabla_N u(x-\rho\xi,\lambda\rho|\xi|)|^p}{|x-y|^{N-1}}dxdy \\ &\qquad = 2\int_{|\mathbf{n}|=1}d\mathbf{n}\int_0^\infty d|\xi|\int_{\mathbb{R}^N}|\nabla_N u(x+\rho\mathbf{n}|\xi|,\lambda\rho|\xi|)|^p dx \\ &\qquad = 2\frac{\omega_N}{\lambda\rho}\int_{\mathbb{R}^{N+1}_+}|\nabla_N u|^p dx.\end{aligned}$$

Compute the second integral in a similar fashion and combine them into

$$\begin{aligned}|||u(\cdot,0)|||_{1-\frac{1}{p},p;\mathbb{R}^N} &\le 2^{1/p}\lambda^{-\frac{1}{p}}\|\nabla_N u\|_{p,\mathbb{R}^{N+1}_+}\int_0^1\rho^{-\frac{1}{p}}d\rho \\ &+2^{1/p}\lambda^{1-\frac{1}{p}}\|u_t\|_{p,\mathbb{R}^{N+1}_+}\int_0^1\rho^{-\frac{1}{p}}d\rho \\ &=2^{1/p}\frac{p}{p-1}\Big(\lambda^{-\frac{1}{p}}\|\nabla_N u\|_{p,\mathbb{R}^{N+1}_+}+\lambda^{1-\frac{1}{p}}\|u_t\|_{p,\mathbb{R}^{N+1}_+}\Big).\end{aligned}$$

The proof is completed by minimizing with respect to λ. ■

Prove Theorem 8.1 by the following steps:

8.4. Proposition 8.2c shows that a function in $W^{1,p}(\mathbb{R}^{N+1}_+)$ has a trace on $\mathbb{R}^N = [x_{N+1} = 0]$ in $W^{1-\frac{1}{p},p}(\mathbb{R}^N)$. Prove the direct part of the theorem for general ∂E of class C^1 and with the segment property by a local flattening technique.

8.5. Every $v \in W^{1-\frac{1}{p},p}(\mathbb{R}^N)$ admits an extension $u \in W^{1,p}(\mathbb{R}^{N+1}_+)$ such that $v = \mathrm{tr}(u)$. To construct such an extension, assume first that v is continuous and bounded in $\mathbb{R}^N$. Let $H_v(x,t)$ be its harmonic extension in $\mathbb{R}^{N+1}_+$ constructed in Section 8 of Chapter 2, and in particular in (8.3), and set

$$u(x, x_{N+1}) = H_v(x, x_{N+1})e^{-x_{N+1}}.$$

Verify that $u \in W^{1,p}(\mathbb{R}^{N+1}_+)$ and that $\mathrm{tr}(u) = v$. Modify the construction to remove the assumption that v is bounded and continuous in $\mathbb{R}^N$.

8.6. Prove that the Poisson kernel $K(\cdot;\cdot)$ in $\mathbb{R}^N \times \mathbb{R}^+$, constructed in (8.2) of Section 8 of Chapter 2, is not in $W^{1,p}(\mathbb{R}^N \times \mathbb{R}^+)$ for any $p \ge 1$. Argue indirectly by examining its trace on $x_{N+1} = 0$.

8.7. Prove a similar fact for the kernel in the Poisson representation of harmonic functions in a ball B_R (formula (3.9) of Section 3 of Chapter 2).

9c The Inhomogeneous Dirichlet Problem

9.1c The Lebesgue Spike

The segment property on ∂E is required to ensure an extension of φ into E by a function $v \in W^{1,2}(E)$. Whence such an extension is achieved, the structure of ∂E does not play any role. Indeed, the problem is recast into one with homogeneous Dirichlet data on ∂E whose solvability by either methods of Sections 5–7 use only the embeddings of $W^{1,2}_o(E)$ of Theorem 3.1, whose constants are independent of ∂E. Verify that the domain of Section 7.2 of Chapter 2 does not satisfy the segment property. Nevertheless the Dirichlet problem (7.3), while not admitting a classical solution, has a unique weak solution given by (7.2). Specify in what sense such a function is a weak solution.

9.2c Variational Integrals and Quasi-Linear Equations

Consider the quasi-linear Dirichlet problem

$$\begin{aligned} -\operatorname{div}\mathbf{A}(x,u,\nabla u) + B(x,u,\nabla u) &= 0 && \text{in } E \\ u\big|_{\partial E} &= \varphi \in W^{1-\frac{1}{p},p}(\partial E) && \text{on } \partial E \end{aligned} \tag{9.1c}$$

where the functions $\mathbf{A}$ and B satisfy the structure condition (6.10c). Assume moreover, that (9.1c) has a variational structure, that is, there exists a function F, as in Section 6.1c, and satisfying (6.2c)–(6.3c), such that $\mathbf{A} = \nabla_{\mathbf{q}} F$

and $B = F_z$. A weak solution is a function $u \in W^{1,p}(E)$ such that $\text{tr}(u) = \varphi$ and satisfying (6.11c). Introduce the set

$$K_\varphi = \{u \in W^{1,p}(E) \;\text{ such that }\; \text{tr}(u) = \varphi\} \tag{9.2c}$$

and the functional

$$K_\varphi \ni u \to J(u) = \int_E F(x, u, \nabla u)dx \qquad p > 1. \tag{9.3c}$$

Prove the following:

9.3. K_φ is convex and weakly (and hence strongly) closed. *Hint:* Use the trace inequalities (8.3)–(8.5).

9.4. Subsets of K_φ, bounded in $W^{1,p}(E)$ are weakly sequentially compact.

9.5. The functional J in (9.3c) has a minimum in K_φ. Such a minimum is a solution of (9.1c) and the latter is the Euler equation of J. *Hint:* Use Proposition 6.2c.

9.6. Solve the inhomogeneous Dirichlet problem for the more general linear operator in (5.1c).

9.7. Explain why the method of extending the boundary datum φ and recasting the problem as a homogeneous Dirichlet problem might not be applicable for quasi-linear equations of the form (9.1c). *Hint:* Examine the functionals in (6.4c) and their Euler equations (6.6c)–(6.7c).

10c The Neumann Problem

Consider the quasi-linear Neumann problem

$$\begin{aligned} -\operatorname{div} \mathbf{A}(x, u, \nabla u) + B(x, u, \nabla u) &= 0 \qquad \text{in } E \\ \mathbf{A}(x, u, \nabla u) \cdot \mathbf{n} &= \psi \qquad \text{on } \partial E \end{aligned} \tag{10.1c}$$

where $\mathbf{n}$ is the outward unit normal to ∂E and ψ satisfies (10.4). The functions $\mathbf{A}$ and B satisfy the structure condition (6.10c) and have a variational structure in the sense of Section 9.2c. Introduce the functional

$$W^{1,p}(E) \ni u \to J(u) = \int_E F(x, u, \nabla u)dx - \int_{\partial E} \psi \text{tr}(u) d\sigma. \tag{10.2c}$$

In dependence of various assumptions on F, identify the correct weakly closed subspace of $W^{1,p}(E)$, where the minimization of J should be set, and find such a minimum, to coincide with a solution of (10.1c).

As a starting point, formulate sufficient conditions on the various parts of the operators in (5.1c), (6.6c), and (6.7c) that would ensure solvability of the corresponding Neumann problem. Discuss uniqueness.

11c The Eigenvalue Problem

11.1. Formulate the eigenvalue problem for homogeneous Dirichlet data as in (11.1) for the more general operator (5.1c). Formulate conditions on the coefficients for an analogue of Proposition 11.1 to hold.

11.2. Formulate the eigenvalue problem for homogeneous Neumann data. State and prove a proposition analogous to Proposition 11.1. Extend it to the more general operator (5.1c).

12c Constructing the Eigenvalues

12.1. Set up the proper variational functionals to construct the eigenvalues for homogeneous Dirichlet data for the more general operator (5.1c). Formulate conditions on the coefficients for such a variational problem to be well-posed.

12.2. Set up the proper variational functionals to construct the eigenvalues for homogeneous Neumann data. Extend these variational integrals and formulate sufficient conditions to include the more general operator (5.1c).

13c The Sequence of Eigenvalues and Eigenfunctions

13.1. It might seem that the arguments of Proposition 13.2 would apply to all eigenvalues and eigenfunctions. Explain where the argument fails for the eigenvalues following the first.

13.2. Formulate facts analogous to Proposition 13.1 for the sequence of eigenvalues and eigenfunctions for homogeneous Dirichlet data for the more general operator (5.1c).

13.3. Formulate facts analogous to Proposition 13.1 for the sequence of eigenvalues and eigenfunctions for homogeneous Neumann data.

14c A Priori $L^\infty(E)$ Estimates for Solutions of the Dirichlet Problem (9.1)

The proof of Propositions 14.1–14.2 shows that the $L^\infty(E)$-estimate stems only from the recursive inequalities (15.3), and a sup-bound would hold for any function satisfying them. For these inequalities to hold the linearity of the PDE in (9.1) is immaterial. As an example, consider the quasi-linear Dirichlet problem (9.1c) where $B = 0$ and $\mathbf{A}$ is subject to the structure condition (6.10c). In particular the problem is not required to have a variational structure.

14.1. Prove that weak solutions of such a quasi-linear Dirichlet problem satisfy recursive inequalities analogous to (15.3). Prove that they are essentially bounded with an upper of the form (14.5), with $\mathbf{f} = 0$ and $f = C^p$, where C is the constant in the structure conditions (6.10c).

14.2. Prove that the boundedness of u continues to hold, if $\mathbf{A}$ and B satisfy the more general conditions

$$\begin{aligned} \mathbf{A}(x,z,\mathbf{q})\cdot\mathbf{q} &\geq \lambda|\mathbf{q}|^p - f(x) \\ |B(x,z,\mathbf{q})| &\leq f(x) \end{aligned} \qquad \text{for some } f \in L^{\frac{N+\varepsilon}{p}}(E).$$

Prove that an upper bound for $\|u\|_\infty$ has the same form as $(14.3)_\pm$ with $\mathbf{f} = 0$ and the same value of δ.

15c A Priori $L^\infty(E)$ Estimates for Solutions of the Neumann Problem (10.1)

The estimates $(16.3)_\pm$ and (16.6) are a sole consequence of the recursive inequalities (17.3) and therefore continue to hold for weak solutions of equations from which they can be derived.

15.1. Prove that they can be derived for weak solutions of the quasi-linear Neumann problem (10.1c), where $\mathbf{A}$ and B satisfy the structure conditions (6.10c) and are not required to be variational. Prove that the estimate takes the form

$$\|u\|_\infty \leq C_\sigma \max\left\{\|u\|_2\,;\ \|\psi\|_{N-1+\sigma}\,;\ |E|^{\frac{1}{N}}\right\}.$$

15.2. Prove that $L^\infty(E)$ estimates continue to hold if the constant C in the structure conditions (6.10c) is replaced by a non-negative function $f \in L^{\frac{N+\varepsilon}{p}}$ for some $\varepsilon > 0$. In such a case the estimate takes exactly the form (16.6) with $\mathbf{f} = 0$.

15.3. Establish $L^\infty(E)$ estimates for weak solutions to the Neumann problem for the operator $\mathcal{L}(\cdot)$ in (5.1c).

15.4. The estimates deteriorate if either the opening or the height of the circular spherical cone of the cone condition of ∂E tend to zero (Remark 16.4). Generate examples of such occurrences for the Laplacian in dimension $N = 2$.

15.1c Back to the Quasi-Linear Dirichlet Problem (9.1c)

The main difference between the estimates $(14.3)_\pm$ and $(16.3)_\pm$ is that the right-hand side contains the norm $\|u_\pm\|_2$ of the solution. Having the proof of Proposition 14.1 as a guideline, establish $L^\infty(E)$ bounds for solutions of the quasi-linear Dirichlet (9.1c) where $\mathbf{A}$ and B satisfy the full quasi-linear structure (6.10c), where, in addition, C may be replaced by a non-negative function $f \in L^{\frac{N+\varepsilon}{p}}$ for some $\varepsilon > 0$. Prove that the resulting estimate has the form

$$\|u\|_\infty \le \max\{\|\varphi\|_{\infty,\partial E}; C_\varepsilon[\|u\|_2; |E|^{p\delta}\|f\|_{\frac{N+\varepsilon}{p}}]\}.$$

10

DeGiorgi Classes

1 Quasi-Linear Equations and DeGiorgi Classes

A quasi-linear elliptic equation in an open set $E \subset \mathbb{R}^N$ is an expression of the form

$$- \operatorname{div} \mathbf{A}(x, u, \nabla u) + B(x, u, \nabla u) = 0 \tag{1.1}$$

where for $u \in W^{1,p}_{\mathrm{loc}}(E)$, the functions

$$E \ni x \to \begin{cases} \mathbf{A}\big(x, u(x), \nabla u(x)\big) \in \mathbb{R}^N \\ B\big(x, u(x), \nabla u(x)\big) \in \mathbb{R} \end{cases}$$

are measurable and satisfy the structure conditions

$$\begin{aligned} \mathbf{A}\big(x, u, \nabla u\big) \cdot \nabla u &\ge \lambda |\nabla u|^p - f^p \\ |\mathbf{A}\big(x, u, \nabla u\big)| &\le \Lambda |\nabla u|^{p-1} + f^{p-1} \\ |B\big(x, u, \nabla u\big)| &\le \Lambda_o |\nabla u|^{p-1} + f_o \end{aligned} \tag{1.2}$$

for given constants $0 < \lambda \le \Lambda$ and $\Lambda_o > 0$, and given non-negative functions

$$f \in L^{N+\varepsilon}(E), \qquad f_o \in L^{\frac{N+\varepsilon}{p}}(E), \qquad \text{for some } \varepsilon > 0. \tag{1.3}$$

The Dirichlet and Neumann problems for these equations were introduced in Sections 9.2c and 10c of the Problems and Complements of Chapter 9, their solvability was established for a class of functions $\mathbf{A}$ and B, and $L^\infty(E)$ bounds were derived for suitable data. Here we are interested in the local behavior of these solutions irrespective of possible prescribed boundary data. A function $u \in W^{1,p}_{\mathrm{loc}}(E)$ is a local weak sub(super)-solution of (1.1), if

$$\int_E [\mathbf{A}(x, u, \nabla u)\nabla v + B(x, u, \nabla u)v]\, dx \le (\ge) 0 \tag{1.4}$$

for all non-negative test functions $v \in W^{1,p}_o(E_o)$, for every open set E_o such that $\bar{E}_o \subset E$. A local weak solution to (1.1) is a function $u \in W^{1,p}_{\mathrm{loc}}(E)$

E. DiBenedetto, *Partial Differential Equations: Second Edition*,
Cornerstones, DOI 10.1007/978-0-8176-4552-6_11,

satisfying (1.4) with the equality sign, for all $v \in W_o^{1,p}(E_o)$. No further requirements are placed on $\mathbf{A}$ and B other than the structure conditions (1.2). Specific examples of these PDEs are those introduced in the previous chapter. In particular they include the class of linear equations (1.2), those in (5.1c)–(5.4c), and the non-linear p-Laplacian-type equations in (6.6c)–(6.7c) of the Complements of Chapter 9. In all these examples the coefficients of the principal part are only measurable. Nevertheless local weak solutions of (1.1) are locally Hölder continuous in E. If $p > N$, this follows from the embedding inequality (2.4) of Theorem 2.1 of Chapter 9. If $1 < p \le N$, this follows from their membership in more general classes of functions called DeGiorgi classes, which are introduced next. Let $B_\rho(y) \subset E$ denote a ball of center y and radius ρ; if y is the origin, write $B_\rho(0) = B_\rho$. For $\sigma \in (0,1)$, consider the concentric ball $B_{\sigma\rho}(y)$ and denote by ζ a non-negative, piecewise smooth cutoff function that equals 1 on $B_{\sigma\rho}(y)$, vanishes outside $B_\rho(y)$ and such that $|\nabla\zeta| \le [(1-\sigma)\rho]^{-1}$. Let u be a local sub(super)-solution of (1.1). For $k \in \mathbb{R}$, the localized truncations $\pm\zeta^p(u-k)_\pm$ belong to $W_o^{1,p}(E)$ and can be taken as test functions v in (1.4). Using the structure conditions (1.2) yields

$$\begin{aligned}
\lambda \int_{B_\rho(y)} & |\nabla(u-k)_\pm|^p \zeta^p dx \\
&\le \int_{B_\rho(y)} |\nabla(u-k)_\pm|^{p-1}\zeta^{p-1}(p\Lambda|\nabla\zeta| + \Lambda_o\zeta)(u-k)_\pm dx \\
&\quad + \int_{B_\rho(y)} \big\{ f^p\zeta^p\chi_{[(u-k)_\pm>0]} + pf^{p-1}\zeta^{p-1}(u-k)_\pm|\nabla\zeta|)\big\} dx \\
&\quad + \int_{B_\rho(y)} f_o(u-k)_\pm\zeta^p dx \\
&\le \frac{\lambda}{2}\int_{B_\rho(y)} |\nabla(u-k)_\pm|^p\zeta^p dx + \frac{\gamma(\Lambda,p)}{(1-\sigma)^p\rho^p}\int_{B_\rho(y)} (u-k)_\pm^p dx \\
&\quad + \int_{B_\rho(y)} f^p\chi_{[(u-k)_\pm>0]} dx + \int_{B_\rho(y)} f_o(u-k)_\pm\zeta^p dx
\end{aligned}$$

where ρ has been taken so small that $\rho \le \max\{1;\Lambda_o\}^{-1}$. Next estimate

$$\int_{B_\rho(y)} f^p\chi_{[(u-k)_\pm>0]} dx \le \|f\|_{N+\varepsilon}^p |A_{k,\rho}^\pm|^{1-\frac{p}{N}+p\delta}$$

where we have assumed $1 < p \le N$, and

$$A_{k,\rho}^\pm = [(u-k)_\pm > 0] \cap B_\rho(y) \quad \text{and} \quad \delta = \frac{\varepsilon}{N(N+\varepsilon)}. \tag{1.5}$$

The term involving f_o is estimated by Hölder's inequality with conjugate exponents

$$\frac{N+\varepsilon}{p} = \frac{q^*}{q^*-1}, \qquad q^* = \frac{Nq}{N-q}.$$

Continuing to assume $1 < p \le N$, one checks that $1 < q < p < N$ for all $N \ge 2$ and the Sobolev embedding of Theorem 3.1 of Chapter 9, can be applied since $(u-k)_\pm\zeta \in W_o^{1,q}(B_\rho(y))$. Therefore

$$\begin{aligned}\int_{B_\rho(y)} f_o(u-k)_\pm\zeta^p dx &\le \|f_o\|_{\frac{N+\varepsilon}{p}} \|(u-k)_\pm\zeta\|_{q^*} \\ &\le \gamma(N,p)\|f_o\|_{\frac{N+\varepsilon}{p}} \|\nabla[(u-k)_\pm\zeta]\|_q \\ &\le \gamma(N,p)\|\nabla[(u-k)_\pm\zeta]\|_p \|f_o\|_{\frac{N+\varepsilon}{p}} |A_{k,\rho}^\pm|^{\frac{1}{q}-\frac{1}{p}} \\ &\le \frac{\lambda}{4}\int_E |\nabla(u-k)_\pm|^p\zeta^p dx + \int_E (u-k)_\pm^p |\nabla\zeta|^p dx \\ &\quad + \gamma(N,p,\lambda)\|f_o\|_{\frac{N+\varepsilon}{p}}^{\frac{p}{p-1}} |A_{k,\rho}^\pm|^{1-\frac{p}{N}+\frac{p}{p-1}p\delta}.\end{aligned}$$

Continue to assume that $\rho \le \max\{1; \Lambda_o\}^{-1}$ and combine these estimates to conclude that there exists a constant $\gamma = \gamma(N,p,\lambda,\Lambda)$ dependent only on the indicated quantities and independent of ρ, y, k, and σ such that for $1 < p \le N$

$$\begin{aligned}\|\nabla(u-k)_\pm\|_{p,B_{\sigma\rho}(y)}^p \le \frac{\gamma}{(1-\sigma)^p\rho^p} \|(u-k)_\pm\|_{p,B_\rho(y)}^p \\ + \gamma_*^p |A_{k,\rho}^\pm|^{1-\frac{p}{N}+p\delta}\end{aligned} \tag{1.6}$$

where δ is given by (1.5) and

$$\gamma_*^p = \gamma(N,p)\big(\|f\|_{N+\varepsilon}^p + \|f_o\|_{\frac{N+\varepsilon}{p}}^{\frac{p}{p-1}}\big). \tag{1.7}$$

1.1 DeGiorgi Classes

Let E be an open subset of $\mathbb{R}^N$, let $p \in (1,N]$, and let γ, γ_*, and δ be given positive constants. The DeGiorgi class $\mathrm{DG}^+(E,p,\gamma,\gamma_*,\delta)$ is the collection of all functions $u \in W_{\mathrm{loc}}^{1,p}(E)$ such that $(u-k)_+$ satisfy (1.6) for all $k \in \mathbb{R}$, and for all pair of balls $B_{\sigma\rho}(y) \subset B_\rho(y) \subset E$. Local weak sub-solutions of (1.1) belong to DG^+, for the constants γ, γ_* and δ identified in (1.5)–(1.7). The DeGiorgi class $\mathrm{DG}^-(E,p,\gamma,\gamma_*,\delta)$ are defined similarly, with $(u-k)_+$ replaced by $(u-k)_-$. Local weak super-solutions of (1.1) belong to DG^-. The DeGiorgi classes $\mathrm{DG}(E,p,\gamma,\gamma_*,\delta)$ are the intersection of $\mathrm{DG}^+ \cap \mathrm{DG}^-$, or equivalently the collection of all functions $u \in W_o^{1,p}(E)$ satisfying (1.6) for all pair of balls $B_{\sigma\rho}(y) \subset B_\rho(y) \subset E$ and all $k \in \mathbb{R}$. We refer to these classes as homogeneous if $\gamma_* = 0$. In such a case the choice of the parameter δ is immaterial. The set of parameters $\{N,p,\gamma\}$ are the homogeneous data of the DG classes, whereas γ_* and δ are the inhomogeneous parameters. This terminology stems from the structure of (1.6) versus the structure of the quasi-linear elliptic equations in (1.1), and is evidenced by (1.7).

Functions in DG have remarkable properties, irrespective of their connection with the quasi-linear equations (1.1). In particular, they are locally bounded, and locally Hölder continuous in E. Even more striking is that non-negative functions in DG satisfy the Harnack inequality of Section 5.1 of Chapter 2, which is typical of non-negative harmonic functions.

2 Local Boundedness of Functions in the DeGiorgi Classes

We say that constants $C, \gamma, \dots$ depend only on the data, and are independent of γ_* and δ, if they can be quantitatively determined a priori only in terms of the inhomogeneous parameters $\{N, p, \gamma\}$. The dependence on the homogeneous parameters $\{\gamma_*, \delta\}$ will be traced, as a way to identify those additional properties afforded by homogeneous structures.

Theorem 2.1 (DeGiorgi [26]) *Let $u \in DG^{\pm}$ and $\tau \in (0,1)$. There exists a constant C depending only on the data such that for every pair of concentric balls $B_{\tau\rho}(y) \subset B_\rho(y) \subset E$*

$$\operatorname*{ess\,sup}_{B_{\tau\rho}(y)} u_\pm \le \max\left\{\gamma_* \rho^{N\delta}\,;\ \frac{C}{(1-\tau)^{\frac{1}{\delta}}}\left(\fint_{B_\rho} u_\pm^p\,dx\right)^{\frac{1}{p}}\right\}. \tag{2.1}$$

For homogeneous $DG^{\pm}$ classes, $\gamma_* = 0$ and δ can be taken $\delta = \frac{1}{N}$.

Proof Having fixed the pair of balls $B_{\tau\rho}(y) \subset B_\rho(y) \subset E$ assume $y = 0$ and consider the sequences of nested concentric balls $\{B_n\}$ and $\{\tilde{B}_n\}$, and the sequences of increasing levels $\{k_n\}$

$$\begin{aligned} B_n &= B_{\rho_n}(0) \quad \text{where } \rho_n = \tau\rho + \frac{1-\tau}{2^{n-1}}\rho \\ \tilde{B}_n &= B_{\tilde{\rho}_n}(0) \quad \text{where } \tilde{\rho}_n = \frac{\rho_n + \rho_{n+1}}{2} = \tau\rho + \frac{3}{2}\frac{1-\tau}{2^n}\rho \\ k_n &= k - \frac{1}{2^{n-1}}k \end{aligned} \tag{2.2}$$

where $k > 0$ is to be chosen. Introduce also non-negative piecewise smooth cutoff functions

$$\zeta_n(x) = \begin{cases} 1 & \text{for } x \in B_{n+1} \\ \dfrac{\tilde{\rho}_n - |x|}{\tilde{\rho}_n - \rho_{n+1}} = \dfrac{2^{n+1}}{(1-\tau)\rho}(\tilde{\rho}_n - |x|) & \text{for } \rho_{n+1} \le |x| \le \tilde{\rho}_n \\ 0 & \text{for } |x| \ge \tilde{\rho}_n \end{cases} \tag{2.3}$$

for which

$$|\nabla\zeta_n| \le \frac{2^{n+1}}{(1-\tau)\rho}.$$

Write down the inequalities (1.6) for $(u - k_{n+1})_+$, for the levels k_{n+1} over the pair of balls $\tilde{B}_n \subset B_n$ for which $(1-\sigma) = 2^{-(n+1)}(1-\tau)$, to get

$$\|\nabla(u-k_{n+1})_+\|^p_{p,\tilde{B}_n} \le \frac{2^{(n+1)p}\gamma}{(1-\tau)^p\rho^p}\|(u-k_{n+1})_+\|^p_{p,B_n} + \gamma_*^p|A^+_{k_{n+1},\rho_n}|^{1-\frac{p}{N}+p\delta}.$$

In the arguments below, γ is a positive constant depending only on the data and that might be different in different contexts.

2.1 Proof of Theorem 2.1 for $1 < p < N$

Apply the embedding inequality (3.1) of Theorem 3.1 of Chapter 9 to the functions $(u-k_{n+1})_+\zeta_n$ over the balls $\tilde{B}_n$ to get

$$\begin{aligned} \|(u-k_{n+1})_+\|^p_{p,B_{n+1}} &\le \|(u-k_{n+1})_+\zeta_n\|^p_{p,\tilde{B}_n} \\ &\le \|(u-k_{n+1})_+\zeta_n\|^{\frac{p}{p^*}}_{p^*,\tilde{B}_n}|A^+_{k_{n+1},\tilde{\rho}_n}|^{\frac{p}{N}} \\ &\le \gamma\|\nabla[(u-k_{n+1})_+\zeta_n]\|^p_{p,\tilde{B}_n}|A^+_{k_{n+1},\tilde{\rho}_n}|^{\frac{p}{N}} \\ &\le \gamma\bigg(\frac{2^{pn}}{(1-\tau)^p\rho^p}\|(u-k_{n+1})_+\|^p_{p,B_n} \\ &\quad + \gamma_*^p|A^+_{k_{n+1},\rho_n}|^{1-\frac{p}{N}+p\delta}\bigg)|A^+_{k_{n+1},\rho_n}|^{\frac{p}{N}}. \end{aligned} \tag{2.4}$$

Next

$$\begin{aligned} \|(u-k_n)_+\|^p_{p,B_n} &= \int_{B_n}(u-k_n)^p_+dx \ge \int_{B_n\cap[u>k_{n+1}]}(u-k_n)^p_+dx \\ &\ge \int_{B_n\cap[u>k_{n+1}]}(k_{n+1}-k_n)^pdx \ge \frac{k^p}{2^{np}}|A^+_{k_{n+1},\rho_n}|. \end{aligned} \tag{2.5}$$

Therefore

$$|A^+_{k_{n+1},\rho_n}| \le \frac{2^{np}}{k^p}\|(u-k_n)_+\|^p_{p,B_n}. \tag{2.6}$$

Combining these estimates yields

$$\begin{aligned} \|(u-k_{n+1})_+\|^p_{p,B_{n+1}} &\le \gamma\frac{2^{np\frac{N+p}{N}}}{(1-\tau)^p\rho^p}\frac{1}{k^{p\frac{p}{N}}}\|(u-k_n)_+\|^{p(1+\frac{p}{N})}_{p,B_n} \\ &\quad + \gamma\gamma_*^p\frac{2^{np(1+p\delta)}}{k^{p(1+p\delta)}}\|(u-k_n)_+\|^{p(1+p\delta)}_{p,B_n}. \end{aligned} \tag{2.7}$$

Set

$$Y_n = \frac{1}{k^p}\fint_{B_n}(u-k_n)^p_+dx = \frac{\|(u-k_n)_+\|^p_{p,B_n}}{k^p\,|B_n|}, \qquad b = 2^{\frac{N+p}{N}}$$

and rewrite the previous recursive inequalities as

$$Y_{n+1} \le \frac{\gamma b^{pn}}{(1-\tau)^p} \left(Y_n^{1+\frac{p}{N}} + \gamma_*^p \frac{\rho^{Np\delta}}{k^p} Y_n^{1+p\delta} \right). \tag{2.8}$$

Stipulate to take k so large that

$$k \ge \gamma_* \rho^{N\delta}, \qquad k > \left(\fint_{B_\rho} u_+^p dx \right)^{\frac{1}{p}}. \tag{2.9}$$

Then $Y_n \le 1$ for all n and $Y_n^{\frac{p}{N}} \le Y_n^{p\delta}$. With these remarks and stipulations, the previous recursive inequalities take the form

$$Y_{n+1} \le \frac{\gamma b^{pn}}{(1-\tau)^p} Y_n^{1+p\delta} \quad \text{for all } n = 1, 2, \dots \tag{2.10}$$

From the fast geometric convergence Lemma 15.1 of Chapter 9, it follows that $\{Y_n\} \to 0$ as $n \to \infty$, provided

$$Y_1 = \frac{1}{k^p} \fint_{B_\rho} u_+^p dx \le b^{-\frac{1}{p\delta^2}} \gamma^{-\frac{1}{p\delta}} (1-\tau)^{\frac{1}{\delta}}.$$

Therefore, taking also into account (2.9), choosing

$$k = \max \left\{ \gamma_* \rho^{N\delta} ; \ \frac{b^{\frac{1}{(p\delta)^2}} \gamma^{\frac{1}{p^2\delta}}}{(1-\tau)^{\frac{1}{p\delta}}} \left(\fint_{B_\rho} u_+^p dx \right)^{\frac{1}{p}} \right\}$$

one derives

$$Y_\infty = \frac{1}{k^p} \fint_{B_{\tau\rho}} (u-k)_+^p dx = 0 \quad \Longrightarrow \quad \operatorname*{ess\,sup}_{B_{\tau\rho}} u_+ \le k.$$

If $\gamma_* = 0$, then (2.8) are already in the form (2.10) with $\delta = \frac{1}{N}$. ∎

2.2 Proof of Theorem 2.1 for $p = N$

The main difference occurs in the application of the embedding inequality (3.2) of Theorem 3.1 of Chapter 9 to the functions $(u - k_{n+1})_+ \zeta_n$ over the balls $\tilde{B}_n$ to derive inequalities analogous to (2.4). Let $q > N$ to be chosen and estimate

$$\begin{aligned}
\|(u-k_{n+1})_+\|_{p,B_{n+1}}^p &\le \|(u-k_{n+1})_+\zeta_n\|_{p,\tilde{B}_n}^p \\
&\le \|(u-k_{n+1})_+\zeta_n\|_{q,\tilde{B}_n}^{\frac{p}{q}} |A_{k_{n+1},\tilde{\rho}_n}^+|^{1-\frac{p}{q}} \\
&\le \gamma(N,q) \left(\|\nabla[(u-k_{n+1})_+\zeta_n]\|_{p,\tilde{B}_n}^{1-\frac{p}{q}} \|(u-k_{n+1})_+\zeta\|_{p,\tilde{B}_n}^{\frac{p}{q}} \right)^p |A_{k_{n+1},\tilde{\rho}_n}^+|^{1-\frac{p}{q}} \\
&\le \gamma(N,q) \left(\frac{2^{pn}}{(1-\tau)^p \rho^p} \|(u-k_{n+1})_+\|_{p,B_n}^p + \gamma_* |A_{k_{n+1},\rho_n}^+|^{p\delta} \right) |A_{k_{n+1},\rho_n}^+|^{1-\frac{p}{q}}.
\end{aligned}$$

Choose $q = 2/\delta$, estimate $|A^+_{k_{n+1},\rho_n}|$ as in (2.5)–(2.6), and arrive at the analogues of (2.7), which now take the form

$$\begin{aligned}\|(u-k_{n+1})_+\|^p_{p,B_{n+1}} &\le \gamma\frac{2^{np(2-\frac{p}{2}\delta)}}{(1-\tau)^p\rho^p}\frac{1}{k^{p(2-\frac{p}{2}\delta)}}\|(u-k_n)_+\|^{p(2-\frac{p}{2}\delta)}_{p,B_n}\\ &\quad+\gamma\gamma_*\frac{2^{np(1+\frac{p}{2}\delta)}}{k^{p(1+\frac{p}{2}\delta)}}\|(u-k_n)_+\|^{p(1+\frac{p}{2}\delta)}_{p,B_n}.\end{aligned}$$

Set

$$Y_n = \frac{1}{k^p}\fint_{B_n}(u-k_n)^p_+\,dx \qquad \text{and} \qquad b = 2^{2-\frac{p}{2}\delta}$$

and rewrite the previous recursive inequalities as

$$Y_{n+1} \le \frac{\gamma b^{pn}}{(1-\tau)^p}\left(Y_n^{1+\frac{p}{2}\delta+(1-\frac{p}{2}\delta)} + \gamma_*^p\frac{\rho^{N\frac{p}{2}\delta}}{k^p}Y_n^{1+\frac{p}{2}\delta}\right).$$

Stipulate to take k as in (2.9) with δ replaced by $\frac{1}{2}\delta$, and recast these recursive inequalities in the form (2.10) with δ replaced by $\frac{1}{2}\delta$. ∎

3 Hölder Continuity of Functions in the DG Classes

For a function $u \in \mathrm{DG}(E,p,\gamma,\gamma_*,\delta)$ and $B_{2\rho}(y) \subset E$ set

$$\mu^+ = \operatorname*{ess\,sup}_{B_{2\rho}(y)} u, \quad \mu^- = \operatorname*{ess\,inf}_{B_{2\rho}(y)} u, \quad \omega(2\rho) = \mu^+ - \mu^- = \operatorname*{ess\,osc}_{B_{2\rho}(y)} u. \tag{3.1}$$

These quantities are well defined since $u \in L^\infty_{\mathrm{loc}}(E)$.

Theorem 3.1 (DeGiorgi [26]) *Let $u \in DG(E,p,\gamma,\gamma_*,\delta)$. There exist constants $C > 1$ and $\alpha \in (0,1)$ depending only upon the data and independent of u, such that for every pair of balls $B_\rho(y) \subset B_R(y) \subset E$*

$$\omega(\rho) \le C\max\left\{\omega(R)\left(\frac{\rho}{R}\right)^\alpha;\ \gamma_*\rho^{N\delta}\right\}. \tag{3.2}$$

The Hölder continuity is local to E, with Hölder exponent $\alpha_o = \min\{\alpha; N\delta\}$. An upper bound for the Hölder constant is

$$\{\text{Hölder constant}\} \le C\max\{2MR^{-\alpha};\gamma_*\}, \qquad \text{where} \qquad M = \|u\|_\infty.$$

This implies that the local Hölder estimates deteriorate near ∂E. Indeed, fix $x, y \in E$ and let

$$R = \min\{\mathrm{dist}\{x;\partial E\}\,;\,\mathrm{dist}\{y;\partial E\}\}.$$

If $|x-y| < R$, then (3.2) implies

$$|u(x)-u(y)| \le C\max\{\omega(R)R^{-\alpha_o};\gamma_*\}|x-y|^{\alpha_o}.$$

If $|x-y| \ge R$, then

$$|u(x)-u(y)| \le 2MR^{-\alpha_o}|x-y|^{\alpha_o}.$$

Corollary 3.1 *Let u be a local weak solution of (1.1)–(1.4). Then for every compact subset $K \subset E$, and for every pair $x, y \in K$*

$$|u(x) - u(y)| \le C \max\left\{\frac{2M_K}{\operatorname{dist}\{K;\partial E\}^{\alpha}};\gamma_*\right\}|x-y|^{\alpha_o}$$

where $M_K = \operatorname{ess\,sup}_K |u|$.

3.1 On the Proof of Theorem 3.1

Although the parameters δ and p are fixed, in view of the value of δ in (1.5), which naturally arises from quasi-linear equations, we will assume $\delta \le \frac{1}{N}$. The value $\delta = \frac{1}{N}$ would occur if $\varepsilon \to \infty$ in the integrability requirements (1.3). For homogeneous DG classes $\gamma_* = 0$, while immaterial, we take $\delta = 1/N$. The proof will be carried on for $1 < p < N$. The case $p = N$ only differs in the application of the embedding Theorem 3.1 of Chapter 9, and the minor modifications needed to cover this case can be modeled after almost identical arguments in Section 2.2 above. In what follows we assume that $u \in \mathrm{DG}$ is given, the ball $B_{2\rho}(y) \subset E$ is fixed, $\mu^{\pm}$ and $\omega(2\rho)$ are defined as in (3.1), and denote by ω any number larger than $\omega(2\rho)$.

4 Estimating the Values of u by the Measure of the Set where u is Either Near μ^+ or Near μ^-

Proposition 4.1 *For every $a \in (0,1)$, there exists $\nu \in (0,1)$ depending only on the data and a, but independent of ω, such that if for some $\varepsilon \in (0,1)$*

$$\big|[u > \mu^+ - \varepsilon\omega] \cap B_\rho(y)\big| \le \nu |B_\rho| \tag{4.1$_+$}$$

then either $\varepsilon\omega \le \gamma_* \rho^{N\delta}$ or

$$u \le \mu^+ - a\varepsilon\omega \qquad \text{a.e. in } B_{\frac{1}{2}\rho}(y). \tag{4.2$_+$}$$

Similarly, if

$$\big|[u < \mu^- + \varepsilon\omega] \cap B_\rho(y)\big| \le \nu |B_\rho| \tag{4.1$_-$}$$

then either $\varepsilon\omega \le \gamma_* \rho^{N\delta}$ or

$$u \ge \mu^- + a\varepsilon\omega \qquad \text{a.e. in } B_{\frac{1}{2}\rho}(y). \tag{4.2$_-$}$$

Proof We prove only (4.1)$_+$–(4.2)$_+$, the arguments for (4.1)$_-$–(4.2)$_-$ being analogous. Set $y = 0$ and consider the sequence of balls $\{B_n\}$ and $\{\tilde{B}_n\}$ introduced in (2.2) for $\tau = \frac{1}{2}$ and the cutoff functions ζ_n introduced in (2.3). For $n \in \mathbb{N}$, introduce also the increasing levels $\{k_n\}$, the nested sets $\{A_n\}$, and their relative measure $\{Y_n\}$ by

$$k_n = \mu^+ - a\varepsilon\omega - \frac{1-a}{2^n}\varepsilon\omega, \quad A_n = [u > k_n] \cap B_n, \quad Y_n = \frac{|A_n|}{|B_n|}.$$

Apply (1.6) to $(u - k_n)_+$ over the pair of concentric balls $\tilde{B}_n \subset B_n$, for which $(1-\sigma) = 2^{-n}$, to get

$$\|\nabla(u - k_n)_+\|^p_{p,\tilde{B}_n} \le \frac{\gamma 2^{np}}{\rho^p}\|(u - k_n)_+\|^p_{p,B_n} + \gamma_*^p |A_n|^{1-\frac{p}{N}+p\delta}.$$

If $1 < p < N$, by the embedding (3.1) of Theorem 3.1

$$\begin{aligned}
\left[\frac{(1-a)\varepsilon\omega}{2^{n+1}}\right]^p |A_{n+1}| &= (k_{n+1} - k_n)^p |A_{n+1}| \le \|(u - k_n)_+\zeta_n\|^p_{p,\tilde{B}_n} \\
&\le \|(u - k_n)_+\zeta_n\|^p_{p^*,\tilde{B}_n} |A_n|^{\frac{p}{N}} \le \|\nabla[(u - k_n)_+\zeta_n]\|^p_{p,\tilde{B}_n} |A_n|^{\frac{p}{N}} \\
&\le \left(\frac{\gamma 2^{np}}{\rho^p}\|(u - k_n)_+\|^p_{p,B_n} + \gamma\gamma_*^p |A_n|^{1-\frac{p}{N}+p\delta}\right)|A_n|^{\frac{p}{N}} \\
&\le \frac{\gamma 2^{np}}{\rho^p}\left(\frac{\varepsilon\omega}{2^n}\right)^p |A_n|^{1+\frac{p}{N}} + \gamma\gamma_*^p |A_n|^{1+p\delta}.
\end{aligned}$$

From this, in dimensionless form, in terms of Y_n one derives

$$Y_{n+1} \le \frac{\gamma 2^{np}}{(1-a)^p}\left[Y_n^{1+\frac{p}{N}} + \left(\frac{\gamma_*\rho^{N\delta}}{\varepsilon\omega}\right)^p Y_n^{1+p\delta}\right] \le \frac{\gamma 2^{np}}{(1-a)^p} Y_n^{1+p\delta}$$

provided $\varepsilon\omega > \gamma_*\rho^{N\delta}$. It follows from these recursive inequalities that $\{Y_n\} \to 0$ as $n \to \infty$, provided (Lemma 15.1 of Chapter 9)

$$Y_1 = \frac{\big|[u > \mu^+ - \varepsilon\omega] \cap B_\rho\big|}{|B_\rho|} \le \frac{(1-a)^{1/\delta}}{\gamma^{1/p\delta} 2^{1/p\delta^2}} \stackrel{\text{def}}{=} \nu. \tag{4.3}$$

∎

Remark 4.1 This formula provides a precise dependence of ν on a and the data. In particular, ν is independent of ε.

5 Reducing the Measure of the Set where u is Either Near μ^+ or Near μ^-

Proposition 5.1 *Assume that*

$$\big|[u \le \mu^+ - \tfrac{1}{2}\omega] \cap B_\rho\big| \ge \theta |B_\rho| \tag{5.1$_+$}$$

for some $\theta \in (0,1)$. Then for every $\nu \in (0,1)$ there exists $\varepsilon \in (0,1)$ that can be determined a priori only in terms of the data and θ, and independent of ω, such that either $\varepsilon\omega \le \gamma_*\rho^{N\delta}$ or

$$\big|[u > \mu^+ - \varepsilon\omega] \cap B_\rho\big| \le \nu \big|B_\rho\big|. \tag{5.2$_+$}$$

Similarly, if

$$\big|[u \geq \mu^- + \tfrac{1}{2}\omega] \cap B_\rho\big| \geq \theta |B_\rho| \tag{5.1$_-$}$$

for some $\theta \in (0,1)$*, then for every* $\nu \in (0,1)$ *there exists* $\varepsilon \in (0,1)$ *depending only on the data and* θ*, and independent of* ω*, such that either* $\varepsilon\omega \leq \gamma_* \rho^{N\delta}$ *or*

$$\big|[u < \mu^- + \varepsilon\omega] \cap B_\rho\big| \leq \nu \big|B_\rho\big|. \tag{5.2$_-$}$$

5.1 The Discrete Isoperimetric Inequality

Proposition 5.2 *Let* E *be a bounded convex open set in* $\mathbb{R}^N$*, let* $u \in W^{1,1}(E)$*, and assume that* $|[u = 0]| > 0$*. Then*

$$\|u\|_1 \leq \gamma(N) \frac{(\operatorname{diam} E)^{N+1}}{|[u=0]|} \|\nabla u\|_1. \tag{5.3}$$

Proof For almost all $x \in E$ and almost all $y \in [u = 0]$

$$|u(x)| = \Big| \int_0^{|y-x|} \frac{\partial}{\partial \rho} u(x + \mathbf{n}\rho) d\rho \Big| \leq \int_0^{|y-x|} |\nabla u(x + \mathbf{n}\rho)| d\rho, \qquad \mathbf{n} = \frac{x-y}{|x-y|}.$$

Integrating in dx over E and in dy over $[u = 0]$ gives

$$|[u=0]|\, \|u\|_1 \leq \int_E \Big\{ \int_{[u=0]} \int_0^{|y-x|} |\nabla u(x + \mathbf{n}\rho)| d\rho dy \Big\} dx.$$

The integral over $[u = 0]$ is computed by introducing polar coordinates with center at x. Denoting by $R(x, y)$ the distance from x to ∂E along $\mathbf{n}$

$$\begin{aligned} &\int_{[u=0]} \int_0^{|y-x|} |\nabla u(x + \mathbf{n}\rho)| d\rho dy \\ &\qquad \leq \int_0^{R(x,y)} s^{N-1} ds \int_{|\mathbf{n}|=1} \int_0^{R(x,y)} |\nabla u(x + \mathbf{n}\rho)| d\rho d\mathbf{n}. \end{aligned}$$

Combining these remarks, we arrive at

$$|[u=0]|\, \|u\|_1 \leq \frac{1}{N} (\operatorname{diam} E)^N \int_E \int_E \frac{|\nabla u(y)|}{|x-y|^{N-1}} dy dx.$$

Inequality (5.3) follows from this, since

$$\sup_{y \in E} \int_E \frac{dx}{|x-y|^{N-1}} \leq \omega_N \operatorname{diam} E. \qquad \blacksquare$$

For a real number ℓ and $u \in W^{1,1}(E)$, set

$$u_\ell = \begin{cases} \ell & \text{if } u > \ell \\ u & \text{if } u \leq \ell. \end{cases}$$

Apply (5.3) to the function $(u_\ell - k)_+$ for $k < \ell$ to obtain

$$(\ell - k)|[u > \ell]| \le \gamma(N)\frac{(\operatorname{diam} E)^{N+1}}{|[u < k]|}\int_{[k<u<\ell]} |\nabla u| dx. \tag{5.4}$$

This is referred to as a discrete version of the isoperimetric inequality ([26]). A continuous version is in [41].

5.2 Proof of Proposition 5.1

We will establish $(5.2)_+$ starting from $(5.1)_+$. Set

$$k_s = \mu^+ - \frac{1}{2^s}\omega, \quad A_s = [u > k_s] \cap B_\rho, \quad \text{for } s = 1, 2, \dots, s_* \tag{5.5}$$

where s_* is a positive integer to be chosen. Apply (5.4) for the levels $k_s < k_{s+1}$ over the ball B_ρ. By virtue of $(5.1)_+$

$$\big|[u < k_s] \cap B_\rho\big| \ge \theta |B_\rho| \quad \text{for all } s \in \mathbb{N}.$$

Therefore

$$\begin{aligned}
\frac{\omega}{2^{s+1}}|A_{s+1}| &\le \frac{\gamma(N)}{\theta}\rho \int_{A_s - A_{s+1}} |\nabla u| dx \\
&\le \frac{\gamma(N)}{\theta}\rho \Big(\int_{B_\rho} |\nabla (u - k_s)_+|^p dx\Big)^{\frac{1}{p}} \big|A_s - A_{s+1}\big|^{\frac{p-1}{p}}
\end{aligned}$$

Take the p-power of both sides and estimate the term involving $\nabla(u - k_s)_+$ by making use of the DG classes (1.6) over the pair of balls $B_\rho \subset B_{2\rho}$ for which $(1-\sigma) = \frac{1}{2}$. This gives

$$\begin{aligned}
\frac{\omega^p}{2^{sp}}|A_{s+1}|^p &\le \frac{\gamma^p \rho^p}{\theta^p}\bigg(\frac{\|(u - k_s)_+\|^p_{p,B_{2\rho}}}{\rho^p} + \gamma_*^p \rho^{N-p+Np\delta}\bigg)\big|A_s - A_{s+1}\big|^{p-1} \\
&\le \frac{\gamma^p \rho^N}{\theta^p}\frac{\omega^p}{2^{sp}}\bigg[1 + \Big(\frac{2^s}{\omega}\gamma_* \rho^{N\delta}\Big)^p\bigg]\big|A_s - A_{s+1}\big|^{p-1}.
\end{aligned}$$

Let $\varepsilon = 2^{-(s_*+1)}$ and stipulate that the term in $[\cdots]$ is majorized by 2. Then, after we divide through by $(\omega/2^s)^p$ and take the $\frac{1}{p-1}$ power of both sides, this inequality yields

$$|A_{s+1}|^{\frac{p}{p-1}} \le \Big(\frac{\gamma}{\theta}\Big)^{\frac{p}{p-1}} \rho^{\frac{N}{p-1}} \big|A_s - A_{s+1}\big|.$$

Add both sides over $s = 1, \dots, s_*$ and observe that the sum on the right-hand side can be majorized by a telescopic series, which in turn is majorized by $|B_\rho|$. On the left-hand side the sum is carried over the constant minorizing term $|A_{s_*+1}|$. Thus

$$\begin{aligned} s_*|A_{s_*+1}|^{\frac{p}{p-1}} &\le \sum_{s=1}^{s_*}|A_s|^{\frac{p}{p-1}} \\ &\le \left(\frac{\gamma}{\theta}\right)^{\frac{p}{p-1}} \rho^{\frac{N}{p-1}} \sum_{s=1}^{\infty} \big|A_s - A_{s+1}\big| \le \left(\frac{\gamma}{\theta}\right)^{\frac{p}{p-1}} |B_\rho|^{\frac{p}{p-1}}. \end{aligned}$$

From this

$$|A_{s_*}| \le \frac{1}{s_*^{\frac{p-1}{p}}}\frac{\gamma}{\theta}|B_\rho| \overset{\text{def}}{=} \nu|B_\rho|.$$

■

6 Proof of Theorem 3.1

Consider the assumption $(5.1)_\pm$ with $\theta = \frac{1}{2}$. Since $\omega \ge \omega(2\rho)$, by the definitions (3.1)

$$\left([u \le \mu^+ - \tfrac{1}{2}\omega] \cap B_\rho\right) \bigcup \left([u \ge \mu^- + \tfrac{1}{2}\omega] \cap B_\rho\right) \supset B_\rho.$$

Therefore not both of $(5.1)_\pm$ can be violated. Assuming the first is in force, fix the number ν as the one claimed by Proposition 4.1 for the choice $a = \frac{1}{2}$, and then, such a number being fixed, determine s_* and hence $\varepsilon = 2^{-(s_*+1)}$ by the procedure of Proposition 5.1. Then by Proposition 4.1, either $\varepsilon\omega \le \gamma\rho^{N\delta}$, or $(4.2)_+$ holds. The latter implies

$$\operatorname*{ess\,sup}_{B_{\frac{1}{2}\rho}} u \le \operatorname*{ess\,sup}_{B_{2\rho}} - \tfrac{1}{2}\varepsilon \operatorname*{ess\,osc}_{B_{2\rho}} u. \tag{6.1}$$

Now

$$-\operatorname*{ess\,inf}_{B_{\frac{1}{2}\rho}} u \le -\operatorname*{ess\,inf}_{B_{2\rho}} u.$$

Adding these inequalities gives

$$\omega(\tfrac{1}{2}\rho) \le \eta\omega(2\rho), \qquad \text{where } \eta = 1 - \frac{1}{2}\varepsilon. \tag{6.2}$$

Let $B_R(y) \subset E$ be fixed and set $\rho_n = 4^{-n}R$. The previous remarks imply that

$$\omega(\rho_{n+1}) \le \max\{\eta\omega(\rho_n)\,;\, \varepsilon^{-1}\gamma_*\rho_n^N\delta\} \tag{6.3}$$

and by iteration

$$\omega(\rho_{n+1}) \le \max\{\eta^n\omega(R)\,;\, \varepsilon^{-1}\gamma_*\rho_n^{N\delta}\}.$$

Compute

$$\rho_n = 4^{-n}R \implies -n = \ln\left(\frac{\rho_n}{R}\right)^{\frac{1}{\ln 4}} \implies \eta^n = \left(\frac{\rho_n}{R}\right)^\alpha \quad \text{for } \alpha = -\frac{\ln\eta}{\ln 4}.$$

■

7 Boundary DeGiorgi Classes: Dirichlet Data

Let ∂E be the finite union of portions of $(N-1)$-dimensional surfaces of class C^1, so that the trace of a function $u \in W^{1,p}(E)$ can be defined except possibly on an $(N-2)$-dimensional subset of ∂E. Given $\varphi \in W^{1-\frac{1}{p},p}(\partial E)$, the Dirichlet problem for the quasi-linear equation (1.1) consists in finding $u \in W^{1,p}(E)$ such that $\text{tr}(u) = \varphi$ and u satisfies the PDE in the weak form (1.4), with the equality sign, for all $v \in W_o^{1,p}(E)$. Weak sub(super)-solutions of the Dirichlet problem are functions $u \in W^{1,p}(E)$ with $\text{tr}(u) \le (\ge)\varphi$ and satisfying (1.4) for all non-negative $v \in W_o^{1,p}(E)$. If $\varphi \in C(\partial E)$, it is natural to ask whether a solution of the Dirichlet problem, whenever it exists, is continuous up the boundary ∂E. The issue builds on the Lebesgue counterexample of Section 7.1 of Chapter 2, and can be rephrased by asking what requirements are needed on ∂E for the interior continuity of functions in the DG classes to extend up to ∂E. Assume that ∂E satisfies the property of *positive geometric density*, that is, there exist $\beta \in (0,1)$ and $R > 0$ such that for all $y \in \partial E$

$$\big|B_\rho(y) \cap (\mathbb{R}^N - E)\big| \ge \beta\big|B_\rho\big| \quad \text{for all } 0 < \rho \le R. \tag{7.1}$$

Fix $y \in \partial E$, assume up to a possible translation that it coincides with the origin, and consider nested concentric balls $B_{\sigma\rho} \subset B_\rho$ for some $\rho > 0$ and $\sigma \in (0,1)$. Let $\varphi \in C(\partial E)$ and set

$$\begin{aligned} \varphi^+(\rho) &= \sup_{\partial E \cap B_\rho} \varphi, \qquad \varphi^-(\rho) = \inf_{\partial E \cap B_\rho} \varphi \\ \omega_\varphi(\rho) &= \varphi^+(\rho) - \varphi^-(\rho) = \operatorname*{osc}_{\partial E \cap B_\rho} \varphi. \end{aligned} \tag{7.2}$$

Let ζ be a non-negative, piecewise smooth cutoff function, that equals 1 on $B_{\sigma\rho}(y)$, vanishes outside $B_\rho(y)$, and such that $|\nabla\zeta| \le [(1-\sigma)\rho]^{-1}$, and let u be a local sub(super)-solution of the Dirichlet problem associated to (1.1) for the given φ. In the weak formulation (1.4), take as test functions v, the localized truncations $\pm\zeta^p(u-k)_\pm$. While ζ vanishes on ∂B_ρ, it does not vanish of $\partial E \cap B_\rho$; however

$$\begin{aligned} &\zeta^p(u-k)_+ \ \text{ is admissible if } \ k \ge \varphi^+(\rho) \\ &\zeta^p(u-k)_- \ \text{ is admissible if } \ k \le \varphi^-(\rho). \end{aligned} \tag{7.3}$$

Putting these choices in (1.4), all the calculations and estimates of Section 1 can be reproduced verbatim, with the understanding that the various integrals are now extended over $B_\rho \cap E$. However, since $\zeta^p(u-k)_\pm \in W_o^{1,p}(B_\rho \cap E)$, we may regard them as elements of $W_o^{1,p}(B_\rho)$ by defining them to be zero outside E. Then the same calculations lead to the inequalities (1.6), with the same stipulations that the various functions vanish outside E and the various integrals are extended over the full ball B_ρ. Given $\varphi \in C(\partial E)$, the boundary DeGiorgi classes $\text{DG}_\varphi^\pm = \text{DG}_\varphi^\pm(\partial E, p, \gamma, \gamma_*, \delta)$ are the collection of all $u \in W^{1,p}(E)$ such that for all $y \in \partial E$ and all pairs of balls $B_{\sigma\rho}(y) \subset B_\rho(y)$ the localized truncations $(u-k)_\pm$ satisfy (1.6) for all levels k subject to the restrictions (7.3). We further define $\mathrm{DG}_\varphi = \mathrm{DG}_\varphi^+ \cap \mathrm{DG}_\varphi^-$ and refer to these classes as homogeneous if $\gamma_* = 0$.

7.1 Continuity up to ∂E of Functions in the Boundary DG Classes (Dirichlet Data)

Let R be the parameter in the condition of positive geometric density (7.1). For $y \in \partial E$ consider concentric balls $B_\rho(y) \subset B_{2\rho}(y) \subset B_R(y)$ and set

$$\begin{gathered} \mu^+ = \operatorname*{ess\,sup}_{B_{2\rho(y)}\cap E} u, \quad \mu^- = \operatorname*{ess\,inf}_{B_{2\rho}(y)\cap E} u \\ \omega(2\rho) = \mu^+ - \mu^- = \operatorname*{ess\,osc}_{B_{2\rho}(y)\cap E} u. \end{gathered} \tag{7.4}$$

Let also $\omega_\varphi(2\rho)$ be defined as in (7.2).

Theorem 7.1 *Let ∂E satisfy the condition of positive geometric density (7.1), and let $\varphi \in C(\partial E)$. Then every $u \in DG_\varphi$ is continuous up to ∂E, and there exist constants $C > 1$ and $\alpha \in (0,1)$, depending only on the data defining the DG_φ classes and the parameter β in (7.1), and independent of φ and u, such that for all $y \in \partial E$ and all balls $B_\rho(y) \subset B_R(y)$*

$$\omega(\rho) \le C \max\left\{\omega(R)\left(\frac{\rho}{R}\right)^\alpha ; \omega_\varphi(2\rho) ; \gamma_* \rho^{N\delta}\right\}. \tag{7.5}$$

The proof of this theorem is almost identical to that of the interior Hölder continuity, except for a few changes, which we outline next. First, Proposition 4.1 and its proof continue to hold, provided the levels $\varepsilon\omega$ satisfy (7.3). Next, Proposition 5.1 and its proof continue to be in force, provided the levels k_s in (5.5) satisfy the restriction (7.3) for all $s \ge 1$. Now either one of the inequalities

$$\mu^+ - \tfrac{1}{2}\omega \ge \varphi^+, \qquad \mu^- + \tfrac{1}{2}\omega \le \varphi^-$$

must be satisfied. Indeed, if both are violated

$$\mu^+ - \tfrac{1}{2}\omega \le \varphi^+ \quad \text{and} \quad -\mu^- - \tfrac{1}{2}\omega \le -\varphi^-.$$

Adding these inequalities gives

$$\omega(\rho) \le 2\omega_\varphi(2\rho)$$

and there is nothing to prove. Assuming the first holds, then all levels k_s as defined in (5.5) satisfy the first of the restrictions (7.3) and thus are admissible. Moreover, $(u-k_1)_+$ vanishes outside E, and therefore

$$\left|[u \le \mu^+ - \tfrac{1}{2}\omega] \cap B_\rho\right| \ge \beta |B_\rho|$$

where β is the parameter in the positive geometric density condition (7.1). From this, the procedure of Proposition 5.1 can be repeated with the understanding that $(u-k_s)_+$ are defined in the full ball B_ρ and are zero outside E. Proposition 5.1 now guarantees the existence of ε as in $(5.2)_+$ and then Proposition 4.1 ensures that (6.1) holds. ∎

Remark 7.1 If φ is Hölder continuous, then u is Hölder continuous up to ∂E.

Remark 7.2 The arguments are local in nature and as such they require only local assumptions. For example, the positive geometric density (7.1) could be satisfied on only a portion of ∂E, open in the relative topology of ∂E, and φ could be continuous only on that portion of ∂E. Then the boundary continuity of Theorem 7.1 continues to hold only locally, on that portion of ∂E.

Corollary 7.1 *Let ∂E satisfy (7.1). A solution u of the Dirichlet problem for (1.1) for a datum $\varphi \in C(\partial E)$ is continuous in $\bar{E}$. If φ is Hölder continuous in ∂E, then u is Hölder continuous in $\bar{E}$. Analogous statements hold if ∂E satisfies (7.1) on an open portion of ∂E and if φ is continuous (Hölder continuous) on that portion of ∂E.*

8 Boundary DeGiorgi Classes: Neumann Data

Consider the quasi-linear Neumann problem

$$\begin{aligned} -\operatorname{div}\mathbf{A}(x,u,\nabla u) + B(x,u,\nabla u) &= 0 \qquad \text{in } E \\ \mathbf{A}(x,u,\nabla u)\cdot\mathbf{n} &= \psi \qquad \text{on } \partial E \end{aligned} \tag{8.1}$$

where $\mathbf{n}$ is the outward unit normal to ∂E. The functions $\mathbf{A}$ and B satisfy the structure (1.2), and the Neumann datum ψ satisfies

$$\psi \in L^q(\partial E), \qquad \text{where} \qquad \begin{cases} q = \dfrac{p}{p-1}\dfrac{N-1}{N} & \text{if } 1<p<N \\ \\ \text{any } q>1 & \text{if } p=N. \end{cases} \tag{8.2}$$

A weak sub(super)-solution to (8.1) is a function $u \in W^{1,p}(E)$ such that

$$\int_E [\mathbf{A}(x,u,\nabla u)\nabla v + B(x,u,\nabla u)v]\,dx \le (\ge) \int_{\partial E} \psi v\,d\sigma \tag{8.3}$$

for all non-negative $v \in W^{1,p}(E)$, where $d\sigma$ is the surface measure on ∂E. All terms on the left-hand side are well defined by virtue of the structure conditions (1.2), whereas the boundary integral on the right-hand side is well defined by virtue of the trace inequalities of Proposition 8.2 of Chapter 9.

In defining boundary DG classes for the Neumann data ψ, fix $y \in \partial E$, assume without loss of generality that $y=0$, and introduce a local change of coordinates by which $\partial E \cap B_R$ for some fixed $R > 0$ coincides with the hyperplane $x_N = 0$, and E lies locally in $\{x_N > 0\}$. Setting

$$B_\rho^+ = B_\rho \cap [x_N > 0] \quad \text{for all } 0 < \rho \le R$$

we require that all "concentric" $\frac{1}{2}$-balls $B_{\sigma\rho}^+ \subset B_\rho^+ \subset B_R^+$ be contained in E. Denote by ζ a non-negative piecewise smooth cutoff function that equals 1 on $B_{\sigma\rho}(y)$, vanishes outside $B_\rho(y)$, and such that $|\nabla\zeta| \le [(1-\sigma)\rho]^{-1}$. Notice that ζ vanishes on ∂B_ρ and not on ∂B_ρ^+. Let u be a local sub(super)-solution of (8.1) in the sense of (8.3), and in the latter take the test functions $v = \pm\zeta^p(u-k)_\pm \in W^{1,p}(E)$. Carrying on the same estimations as in Section 1, we arrive at integral inequalities analogous to (1.6) with the only difference that the various integrals are extended over $B_{\sigma\rho}^+$ and B_ρ^+, and that the right-hand side contains the boundary term arising from the right-hand side of (8.3). Precisely

$$\begin{aligned} \|\nabla(u-k)_\pm\zeta\|_{p,B_\rho^+}^p &\le \frac{\gamma}{(1-\sigma)^p\rho^p}\|(u-k)_\pm\|_{p,B_\rho^+}^p \\ &\quad + \gamma_*^p|A_{k,\rho}^\pm|^{1-\frac{p}{N}+p\delta} + \left|\int_{x_N=0}\psi(u-k)_\pm\zeta^p d\bar{x}\right| \end{aligned} \tag{8.4}$$

where δ is given by (1.5), γ_*^p is defined in (1.7), the sets $A_{k,\rho}^\pm$ are redefined accordingly, and $\bar{x} = (x_1, \dots, x_{N-1})$. The requirement (8.2) merely ensures that (8.3) is well defined. The boundary DG classes for Neumann data ψ require a higher order of integrability of ψ. We assume that

$$\psi \in L^q(\partial E), \qquad \text{where} \qquad \begin{cases} q = \dfrac{N-1}{p-1}\left(\dfrac{N+\varepsilon}{N}\right) & \text{if } 1 < p < N \\ \text{any } q > 1 & \text{if } p = N \end{cases} \tag{8.5}$$

for some $\varepsilon > 0$. Using such a q, define $\bar{p} > 1$ by

$$1 - \frac{1}{q} = \frac{1}{\bar{p}^*}\frac{N}{N-1}, \qquad \bar{p}^* = \frac{N\bar{p}}{N-\bar{p}}, \qquad \frac{1}{q} = \frac{\bar{p}-1}{\bar{p}}\frac{N}{N-1}.$$

One verifies that for these choices, $1 < \bar{p} < p \le N$ and the trace inequality (8.3) of Chapter 9 can be applied. With this stipulation, estimate the last integral as

$$\begin{aligned} &\left|\int_{x_N=0}\psi(u-k)_\pm\zeta^p d\sigma\right| \\ &\quad \le \|\psi\|_{q;\partial E}\|(u-k)_\pm\zeta\|_{\bar{p}^*\frac{N-1}{N};\partial E} \\ &\quad \le \|\psi\|_{q;\partial E}\big[\|\nabla[(u-k)_\pm\zeta]\|_{\bar{p}} + 2\gamma\|(u-k)_\pm\zeta\|_{\bar{p}}\big] \\ &\quad \le \|\psi\|_{q;\partial E}\big[\|\nabla[(u-k)_\pm\zeta]\|_p + 2\gamma\|(u-k)_\pm\zeta\|_p\big]|A_{k,\rho}^\pm|^{\frac{1}{\bar{p}}-\frac{1}{p}} \\ &\quad \le \frac{1}{2}\|\nabla(u-k)_\pm\zeta\|_{p,B_\rho^+}^p + \|(u-k)_\pm(\zeta+|\nabla\zeta|)\|_{p,B_\rho^+}^p \\ &\qquad + \gamma(N,p)\|\psi\|_{q;\partial E}^{\frac{p}{p-1}}|A_{k,\rho}^\pm|^{(\frac{1}{\bar{p}}-\frac{1}{p})\frac{p}{p-1}}. \end{aligned}$$

Combining this with (8.4) and stipulating $\rho \le 1$ gives

$$\|\nabla(u-k)_\pm\|^p_{p,B^+_{\sigma\rho}} \le \frac{\gamma}{(1-\sigma)^p\rho^p}\|(u-k)_\pm\|^p_{p,B^+_\rho} + \gamma^p_{**}|A^\pm_{k,\rho}|^{1-\frac{p}{N}+p\delta} \tag{8.6}$$

where δ is given by (1.5), and

$$\gamma^p_{**} = \gamma(N,p)\big(\|f\|^p_{N+\varepsilon} + \|f_o\|^{\frac{p}{p-1}}_{\frac{N+\varepsilon}{p}} + \|\psi\|^{\frac{p}{p-1}}_{q;\partial E}\big). \tag{8.7}$$

Given $\psi \in L^q(\partial E)$ as in (8.5), the boundary DeGiorgi classes $\mathrm{DG}^\pm_\psi = \mathrm{DG}^\pm_\psi(\partial E, p, \gamma, \gamma_{**}, \delta)$ are the collection of all $u \in W^{1,p}(E)$ such that for all $y \in \partial E$ and all pairs of $\frac{1}{2}$-balls $B^+_{\sigma\rho}(y) \subset B^+_\rho(y)$ for $\rho < R$, the localized truncations $(u-k)_\pm$ satisfy (8.6). We further define $\mathrm{DG}_\psi = \mathrm{DG}^+_\psi \cap \mathrm{DG}^-_\psi$ and refer to these classes as homogeneous if $\gamma_{**} = 0$.

8.1 Continuity up to ∂E of Functions in the Boundary DG Classes (Neumann Data)

Having fixed $y \in \partial E$, assume after a flattening of ∂E about y that ∂E coincides with the hyperplane $x_N = 0$ within a ball $B_R(y)$. Consider the "concentric" $\frac{1}{2}$-balls $B^+_\rho(y) \subset B^+_{2\rho}(y) \subset B^+_R(y)$ and set

$$\mu^+ = \operatorname*{ess\,sup}_{B^+_{2\rho}(y)} u, \quad \mu^- = \operatorname*{ess\,inf}_{B^+_{2\rho}(y)} u, \quad \omega(2\rho) = \mu^+ - \mu^- = \operatorname*{ess\,osc}_{B^+_{2\rho}(y)} u. \tag{8.8}$$

Theorem 8.1 *Let ∂E be of class C^1 satisfying the segment property. Then every $u \in DG_\psi$ is continuous up to ∂E, and there exist constants $C > 1$ and $\alpha \in (0,1)$, depending only on the data defining the DG_ψ classes and the C^1 structure of ∂E, and independent of ψ and u, such that for all $y \in \partial E$ and all $\frac{1}{2}$-balls $B^+_\rho(y) \subset B^+_R(y)$*

$$\omega(\rho) \le C \max\left\{\omega(R)\left(\frac{\rho}{R}\right)^\alpha ; \gamma_{**}\rho^{N\delta}\right\}. \tag{8.9}$$

The proof of this theorem is almost identical to that of the interior Hölder continuity, the only difference being that we are working with "concentric" $\frac{1}{2}$-balls instead of balls. Proposition 4.1 and its proof continue to hold. Since $(u-k)_\pm\zeta$ do not vanish on ∂B^+_ρ, the embedding Theorem 2.1 of Chapter 9 is used instead of the multiplicative embedding. Next, Proposition 5.1 relies on the discrete isoperimetric inequality of Proposition 5.2, which holds for convex domains, and thus for $\frac{1}{2}$-balls. The rest of the proof is identical with the indicated change in the use of the embedding inequalities. ∎

Remark 8.1 The regularity of ψ enters only in the requirement (8.5) through the constant γ_{**}.

Remark 8.2 The arguments are local in nature, and as such they require only local assumptions.

Corollary 8.1 *Let ∂E be of class C^1 satisfying the segment property. A weak solution u of the Neumann problem for (8.1) for a datum ψ satisfying (8.5), is Hölder continuous in $\bar{E}$. Analogous local statements are in force, if the assumptions on ∂E and ψ hold on portions of ∂E.*

9 The Harnack Inequality

Theorem 9.1 ([27, 29]) *Let u be a non-negative element of $DG(E,p,\gamma,\gamma_*,\delta)$. There exists a positive constant c_* that can be quantitatively determined a priori in terms of only the parameters N,p,γ and independent of u, γ_*, and δ such that for every ball $B_{4\rho}(y)\subset E$, either $u(y)\le c_*^{-1}\gamma_*\rho^{N\delta}$ or*

$$c_*\,u(y)\le \inf_{B_\rho(y)} u. \tag{9.1}$$

This inequality was first proved for non-negative harmonic functions (Section 5.1 of Chapter 2). Then it was shown to hold for non-negative solutions of quasi-linear elliptic equations of the type of (1.1) ([109, 134, 154]). It is quite remarkable that they continue to hold for non-negative functions in the DG classes, and it raises the still unsettled question of the structure of these classes, versus Harnack estimates, and weak forms of the maximum principle.

The first proof of Theorem 9.1 is in [27]. A different proof that avoids coverings is in [29]. This is the proof presented here, in view of its relative flexibility.

9.1 Proof of Theorem 9.1 (Preliminaries)

Fix $B_{4\rho}(y)\subset E$, assume $u(y)>0$, and introduce the change of function and variables

$$w=\frac{u}{u(y)},\qquad x\to\frac{x-y}{\rho}.$$

Then $w(0)=1$, and w belongs to the DG classes relative to the ball B_4, with the same parameters as the original DG classes, except that γ_* is now replaced by

$$\Gamma_*=(2\rho)^{N\delta}\frac{\gamma_*}{u(y)}. \tag{9.2}$$

In particular, the truncations $(w-k)_\pm$ satisfy

$$\|\nabla(w-k)_\pm\|^p_{p,B_{\sigma r}(x_*)}\le\frac{\gamma}{(1-\sigma)^p r^p}\|(w-k)_\pm\|^p_{p,B_r(x_*)}+\Gamma_*^p|A^-_{k,r}|^{1-\frac{p}{N}+p\delta} \tag{9.3}$$

for all $B_r(x_*)\subset B_4$ and for all $k>0$. By these transformations, (9.1) reduces to finding a positive constant c_* that can be determined a priori in terms of only the parameters of the original DG classes, such that

$$c_*\le\max\{\inf_{B_1} w\,;\,\Gamma_*\}. \tag{9.4}$$

9.2 Proof of Theorem 9.1. Expansion of Positivity

Proposition 9.1 *Let $M > 0$ and $B_{4r}(x_*) \subset B_4$. If*

$$\big|[w \ge M] \cap B_r(x_*)\big| \ge \tfrac{1}{2}|B_r| \tag{9.5}$$

then for every $\nu \in (0,1)$ there exists $\varepsilon \in (0,1)$ depending only on the data and ν, and independent of Γ_*, such that either $\varepsilon M \le \Gamma_* r^{N\delta}$ or

$$\big|[w < 2\varepsilon M] \cap B_{4r}(x_*)\big| \le \nu \big|B_{4r}\big|. \tag{9.6}$$

As a consequence, either $\varepsilon M \le \Gamma_* r^{N\delta}$ or

$$w \ge \varepsilon M \qquad \text{in } \ B_{2r}(x_*). \tag{9.7}$$

Proof The assumption (9.5) implies that

$$\big|[w \ge M] \cap B_{4r}(x_*)\big| \ge \theta |B_{4r}|, \qquad \text{where } \theta = \frac{1}{2}\frac{1}{4^N}.$$

Then Proposition 5.1 applied for such a θ and for ρ replaced by $4r$ implies that (9.6) holds, for any prefixed $\nu \in (0,1)$. This in turn implies (9.7), by virtue of Proposition 4.1, applied with ρ replaced by $4r$. ∎

Remark 9.1 Proposition 4.1 is a "shrinking" proposition, in that information on a ball B_ρ, yields information on a smaller ball $B_{\frac{1}{2}\rho}$. Proposition 9.1 is an "expanding" proposition in the sense that information on a ball $B_r(x_*)$ yields information on a larger ball $B_{2r}(x_*)$. This "expansion of positivity" is at the heart of the Harnack inequality (9.1).

9.3 Proof of Theorem 9.1

For $s \in [0,1)$ consider the balls B_s and the increasing families of numbers

$$M_s = \sup_{B_s} u, \qquad N_s = (1-s)^{-\beta}$$

where $\beta > 0$ is to be chosen. Since $w \in L^\infty(B_2)$, the net $\{M_s\}$ is bounded. One verifies that

$$M_o = N_o = 1, \qquad \lim_{s\to 1} M_s < \infty, \qquad \text{and} \qquad \lim_{s \to 1} N_s = \infty.$$

Therefore the equation $M_s = N_s$ has roots, and we denote by s_* the largest of these roots. Since w is continuous in B_2, there exists $x_* \in B_{s_*}$ such that

$$\sup_{B_{s_*}(x_*)} w = w(x_*) = (1 - s_*)^{-\beta}.$$

Also, since s_* is the largest root of $M_s = N_s$

$$\sup_{B_R(x_*)} w \le \left(\frac{1-s_*}{2}\right)^{-\beta}, \qquad \text{where} \qquad R = \frac{1-s_*}{2}.$$

By virtue of the Hölder continuity of w, in the form (3.2), for all $0 < r < R$ and for all $x \in B_r(x_*)$

$$\begin{aligned} w(x) - w(x_*) &\ge -C\Big\{\Big[\sup_{B_R(x_*)} w - \inf_{B_R(x_*)} w\Big]\left(\frac{r}{R}\right)^{\alpha} + \Gamma_* r^{N\delta}\Big\} \\ &\ge -C\Big[2^{\beta}(1-s_*)^{-\beta}\left(\frac{r}{R}\right)^{\alpha} + \Gamma_* r^{N\delta}\Big]. \end{aligned} \tag{9.8}$$

Next take $r = \epsilon_* R$, and then ϵ_* so small that

$$C\left\{2^{\beta}(1-s_*)^{-\beta}\epsilon_*^{\alpha} + \Gamma_*\epsilon_*^{N\delta}\right\} \le \frac{1}{2}(1-s_*)^{-\beta}.$$

The choice of ϵ_* depends on $C, \alpha, \Gamma_*, N, \delta$, which are quantitatively determined parameters; it depends also on β, which is still to be chosen; however the choice of ϵ_* can be made independent of s_*. For these choices

$$w(x) \ge w(x_*) - \frac{1}{2}(1-s_*)^{-\beta} = \frac{1}{2}(1-s_*)^{-\beta} \stackrel{\text{def}}{=} M$$

for all $x \in B_r(x_*)$. Therefore

$$\big|[w \ge M] \cap B_r(x_*)\big| \ge \tfrac{1}{2}\big|B_r\big|. \tag{9.9}$$

From this and Proposition 9.1, there exists $\varepsilon \in (0,1)$ that can be quantitatively determined in terms of only the non-homogeneous parameters in the DG classes and is independent of β, r, Γ_*, and w such that either

$$\varepsilon M \le \Gamma_* r^{N\delta}, \qquad \text{where} \qquad r = \tfrac{1}{2}\epsilon_*(1-s_*)$$

or

$$w \ge \varepsilon M \qquad \text{on} \qquad B_{2r}(x_*).$$

Iterating this process from the ball $B_{2^j r}(x_*)$ to the ball $B_{2^{j+1}r}(x_*)$ gives the recursive alternatives, either

$$\varepsilon^j M \le \Gamma_*(2^j r)^{N\delta} \qquad \text{or} \qquad w > \varepsilon^j M \quad \text{on } B_{2^{j+1}r}(x_*). \tag{9.10}$$

After n iterations, the ball $B_{2^{n+1}r}(x_*)$ will cover B_1 if n is so large that

$$2 \le 2^{n+1} r = 2^{n+1}\tfrac{1}{2}\epsilon_*(1-s_*) \le 4 \tag{9.11}$$

from which

$$2\varepsilon^n M = \varepsilon^n(1-s_*)^{-\beta} \le (2^{\beta}\varepsilon)^n \epsilon_*^{\beta} \le 2^{\beta}\varepsilon^n(1-s_*)^{-\beta} = 2^{\beta+1}\varepsilon^n M.$$

In these inequalities, all constants except s_* and β are quantitatively determined a priori in terms of only the non-homogeneous parameters of the DG classes. The parameter ϵ_* depends on β but is independent of s_*. The latter is determined only qualitatively. The remainder of the proof consists in selecting β so that the qualitative parameter s_* is eliminated. Select β so large that $\varepsilon 2^\beta = 1$. Such a choice determines ϵ_*, and

$$\varepsilon^n M = \varepsilon^n \tfrac{1}{2}(1-s_*)^{-\beta} \geq 2^{-(\beta+1)}\epsilon_*^\beta \stackrel{\text{def}}{=} c_*.$$

Returning to (9.10), if the first alternative is violated for all $j = 1, 2, \dots, n$, then the second alternative holds recursively and gives

$$w \geq \varepsilon^n M \geq c_* \quad \text{in } B_1.$$

If the first alternative holds for some $j \in \{1, \dots, n\}$, then a fortiori it holds for $j = n$, which, taking into account the definition (9.2) of Γ_* and (9.11), implies

$$c_* u(y) \leq \gamma_*(2\rho)^{N\delta}.$$ ■

10 Harnack Inequality and Hölder Continuity

The Hölder continuity of a function u in the DG classes in the form (3.2) has been used in an essential way in the proof of Theorem 9.1. For non-negative solutions of elliptic equations, the Harnack estimate can be established independent of the Hölder continuity, and indeed, the former implies the latter ([109]).

Let $\mu^\pm$ and $\omega(2\rho)$ be defined as in (3.1). Applying Theorem 9.1 to the two non-negative functions $w^+ = \mu^+ - u$ and $w^- = u - \mu^-$, gives either

$$\begin{aligned} \operatorname*{ess\,sup}_{B_\rho(y)} w^+ &= \mu^+ - \operatorname*{ess\,inf}_{B_\rho(y)} u \leq c_*^{-1}\gamma_*\rho^{N\delta} \\ \operatorname*{ess\,sup}_{B_\rho(y)} w^- &= \operatorname*{ess\,sup}_{B_\rho(y)} u - \mu^- \leq c_*^{-1}\gamma_*\rho^{N\delta} \end{aligned} \tag{10.1}$$

or

$$\begin{aligned} c_*(\mu^+ - \operatorname*{ess\,inf}_{B_\rho(y)} u) &\leq \mu^+ - \operatorname*{ess\,sup}_{B_\rho(y)} u \\ c_*(\operatorname*{ess\,sup}_{B_\rho(y)} u - \mu^-) &\leq \operatorname*{ess\,inf}_{B_\rho(y)} u - \mu^-. \end{aligned} \tag{10.2}$$

If either one of (10.1) holds, then

$$\omega(\rho) \leq \omega(2\rho) \leq c_*^{-1}\gamma_*\rho^{N\delta}. \tag{10.3}$$

Otherwise, both inequalities in (10.2) are in force. Adding them gives

$$c_*\omega(2\rho) + c_*\omega(\rho) \leq \omega(2\rho) - \omega(\rho).$$

From this

$$\omega(\rho) \le \eta\omega(2\rho), \qquad \text{where} \qquad \eta = \frac{1-c_*}{1+c_*}. \tag{10.4}$$

The alternatives (10.3)–(10.4) yield recursive inequalities of the same form as (6.3), from which the Hölder continuity follows. These remarks raise the question whether the Harnack estimate for non-negative functions in the DG classes can be established independently of the Hölder continuity. The link between these two facts rendering them essentially equivalent, is the next lemma of real analysis.

11 Local Clustering of the Positivity Set of Functions in $W^{1,1}(E)$

For $R > 0$, denote by $K_R(y) \subset \mathbb{R}^N$ a cube of edge R centered at y and with faces parallel to the coordinate planes. If y is the origin on $\mathbb{R}^N$, write $K_R(0) = K_R$.

Lemma 11.1 ([33]) *Let $v \in W^{1,1}(K_R)$ satisfy*

$$\|v\|_{W^{1,1}(K_R)} \le \gamma R^{N-1} \qquad \text{and} \qquad |[v > 1]| \ge \nu |K_R| \tag{11.1}$$

for some $\gamma > 0$ and $\nu \in (0,1)$. Then for every $\nu_* \in (0,1)$ and $0 < \lambda < 1$, there exist $x_* \in K_R$ and $\epsilon_* = \epsilon_*(\nu, \nu_*, \varepsilon, \gamma, N) \in (0,1)$ such that

$$|[v > \lambda] \cap K_{\epsilon_* R}(x_*)| > (1-\nu_*)|K_{\epsilon_* R}|. \tag{11.2}$$

Remark 11.1 Roughly speaking, the lemma asserts that if the set where u is bounded away from zero occupies a sizable portion of K_R, then there exists at least one point x_* and a neighborhood $K_{\epsilon_* R}(x_*)$ where u remains large in a large portion of $K_{\epsilon_* R}(x_*)$. Thus the set where u is positive clusters about at least one point of K_R.

Proof (Lemma 11.1) It suffices to establish the lemma for u continuous and $R = 1$. For $n \in \mathbb{N}$ partition K_1 into n^N cubes, with pairwise disjoint interior and each of edge $1/n$. Divide these cubes into two finite sub-collections $\mathbf{Q}^+$ and $\mathbf{Q}^-$ by

$$\begin{aligned} Q_j \in \mathbf{Q}^+ \quad &\iff \quad |[v > 1] \cap Q_j| > \tfrac{1}{2}\nu |Q_j| \\ Q_i \in \mathbf{Q}^- \quad &\iff \quad |[v > 1] \cap Q_i| \le \tfrac{1}{2}\nu |Q_i| \end{aligned}$$

and denote by $\#(\mathbf{Q}^+)$ the number of cubes in $\mathbf{Q}^+$. By the assumption

$$\sum_{Q_j \in \mathbf{Q}^+} |[v > 1] \cap Q_j| + \sum_{Q_i \in \mathbf{Q}^-} |[v > 1] \cap Q_i| > \nu |K_1| = \nu n^N |Q|$$

where $|Q|$ is the common measure of the Q_ℓ. From the definitions of $\mathbf{Q}^\pm$

$$\nu n^N \sum_{Q_j \in \mathbf{Q}^+} \frac{|[v>1]\cap Q_j|}{|Q_j|} + \sum_{Q_i\in\mathbf{Q}^-} \frac{|[v>1]\cap Q_i|}{|Q_i|} < \#(\mathbf{Q}^+) + \tfrac{1}{2}\nu(n^N - \#(\mathbf{Q}^+)).$$

Therefore

$$\#(\mathbf{Q}^+) > \frac{\nu}{2-\nu} n^N. \tag{11.3}$$

Fix $\nu_*, \lambda \in (0,1)$. The integer n can be chosen depending on $\nu, \nu_*, \lambda, \gamma$, and N, such that

$$|[v>\lambda]\cap Q_j| \ge (1-\nu_*)|Q_j| \quad \text{for some } \; Q_j \in \mathbf{Q}^+. \tag{11.4}$$

This would establish the lemma for $\epsilon_* = 1/n$. We first show that if Q is a cube in $\mathbf{Q}^+$ for which

$$|[v>\lambda]\cap Q| < (1-\nu_*)|Q|, \tag{11.5}$$

then there exists a constant $c = c(\nu, \nu_*, \lambda, N)$ such that

$$\|v\|_{W^{1,1}(Q)} \ge c(\nu,\nu_*,\lambda,N)\frac{1}{n^{N-1}}. \tag{11.6}$$

From (11.5)

$$|[v\le\lambda]\cap Q| \ge \nu_*|Q| \qquad \text{and} \qquad \left|\left[v > \frac{1+\lambda}{2}\right]\cap Q\right| > \frac{1}{2}\nu|Q|.$$

For fixed $x \in [v\le\lambda]\cap Q$ and $y\in[v>(1+\lambda)/2]\cap Q$

$$\frac{1-\lambda}{2} \le v(y) - v(x) = \int_0^{|y-x|} \nabla u(x+t\mathbf{n})\cdot \mathbf{n}dt, \quad \mathbf{n} = \frac{y-x}{|x-y|}.$$

Let $R(x,\mathbf{n})$ be the polar representation of ∂Q with pole at x for the solid angle $\mathbf{n}$. Integrate the previous relation in dy over $[v>(1+\lambda)/2]\cap Q$. Minorize the resulting left-hand side, by using the lower bound on the measure of such a set, and majorize the resulting integral on the right-hand side by extending the integration over Q. Expressing such integration in polar coordinates with pole at $x\in[v\le\lambda]\cap Q$ gives

$$\begin{aligned}\frac{\nu(1-\lambda)}{4}|Q| &\le \int_{|\mathbf{n}|=1}\int_0^{R(x,\mathbf{n})} r^{N-1}\int_0^{|y-x|} |\nabla v(x+t\mathbf{n})|\,dt\,dr\,d\mathbf{n}\\ &\le N^{N/2}|Q|\int_{|\mathbf{n}|=1}\int_0^{R(x,\mathbf{n})} |\nabla v(x+t\mathbf{n})|\,dt\,d\mathbf{n}\\ &= N^{N/2}|Q|\int_Q \frac{|\nabla v(z)|}{|z-x|^{N-1}}dz.\end{aligned}$$

Now integrate in dx over $[u \le \lambda] \cap Q$. Minorize the resulting left-hand side using the lower bound on the measure of such a set, and majorize the resulting right-hand side, by extending the integration to Q. This gives

$$\begin{aligned}\frac{\nu\nu_*(1-\lambda)}{4N^{N/2}}|Q| &\le \|v\|_{W^{1,1}(Q)} \sup_{z\in Q}\int_Q \frac{1}{|z-x|^{N-1}}dx \\ &\le C(N)|Q|^{1/N}\|v\|_{W^{1,1}(Q)}\end{aligned}$$

for a constant $C(N)$ depending only on N, thereby proving (11.6).

If (11.4) does not hold for any cube $Q_j \in \mathbf{Q}^+$, then (11.6) is verified for all such Q_j. Adding (11.6) over such cubes and taking into account (11.3)

$$\frac{\nu}{2-\nu}c(\nu,\nu_*,\lambda,N)n \le \|u\|_{W^{1,1}(K_1)} \le \gamma. \qquad \blacksquare$$

Remark 11.2 While the lemma has been proved for cubes, by reducing the number ϵ_* if needed, we may assume without loss of generality that it continues to hold for balls.

12 A Proof of the Harnack Inequality Independent of Hölder Continuity

Introduce the same transformations of Section 9.1 and reduce the proof to establishing (9.4). Following the same arguments and notation of Section 9.3, for $\beta > 0$ to be chosen, let s_* be the largest root of $M_s = N_s$ and set

$$M_* = (1-s_*)^{-\beta} \qquad M^* = 2^\beta(1-s_*)^{-\beta}$$
$$R = \frac{1-s_*}{4} \qquad R_o = \frac{1+s_*}{2}$$

so that $M^* = 2^\beta M_*$ and $B_R(x) \subset B_{R_o}$ for all $x \in B_{s_*}$, and

$$\operatorname*{ess\,sup}_{B_R(x)} w \le M^* \quad \text{for all } x \in B_{s_*}.$$

Proposition 12.1 *There exists a ball $B_R(x) \subset B_{R_o}$ such that either*

$$M^* \le \Gamma_* R^{N\delta} \quad \Longrightarrow \quad u(y) \le \gamma_* \rho^{N\delta} \tag{12.1}$$

or

$$|[w > M^* - \varepsilon M^*] \cap B_R(x)| > \nu_a |B_R| \tag{12.2}$$

where

$$\nu_a = \frac{(1-a)^{1/\delta}}{\gamma^{1/\delta}2^{1/p\delta^2}}, \qquad a = \varepsilon = \sqrt{1 - \frac{1}{2^\beta}} \tag{12.3}$$

and where γ is the quantitative constant appearing in (4.3) and dependent only on the inhomogeneous data of the DG classes.

Proof If the first alternative (12.1) holds for some $B_R(x) \subset B_{R_o}$, there is nothing to prove. Thus assuming (12.1) fails for all such balls, if (12.2) holds for some of these balls, there is nothing to prove. Thus we may assume that (12.1) and (12.2) are both violated for all $B_R(x) \subset B_{R_o}$. Apply Proposition 4.1 with $\varepsilon = a$ and conclude, by the choice of a and ν_a, that

$$w < (1 - a^2)M^* = M_* \quad \text{in all balls } B_{\frac{1}{2}R}(x) \subset B_{R_o}.$$

Thus $w < M_*$ in B_{s_*}, contradicting the definition of M_*. ■

A consequence is that there exists $B_R(x) \subset B_{R_o}$ such that

$$|[v > 1] \cap B_R(x)| > \nu |B_R|, \qquad \text{where} \quad v = \frac{w}{(1-a)M^*}. \tag{12.4}$$

Write the inequalities (9.3) for w_+ $(k = 0)$ over the pair of balls $B_R(x) \subset B_{2R}(x) \subset B_{R_o}$, and then divide the resulting inequalities by $[(1-a)M^*]^p$. Taking into account the definition (9.2) of Γ_*, and (12.2), this gives

$$\|\nabla v\|^p_{p, B_R(x)} \leq \gamma 2^{p\beta/2} R^{N-p}.$$

From this

$$\|\nabla v\|_{1, B_R(x)} \leq \gamma(\beta) R^{N-1}. \tag{12.5}$$

Thus the function v satisfies the assumptions of Lemma 11.1, with given and fixed constants $\gamma = \gamma(\text{data}, \beta)$ and $\nu = \nu_a(\text{data}, \beta)$. The parameter $\beta > 1$ has to be chosen. Applying the lemma for $\lambda = \nu = \frac{1}{2}$ yields the existence of $x_* \in B_R(x)$ and $\epsilon_* = \epsilon_*(\text{data}, \beta)$ such that

$$|[w > M] \cap B_r(x_*)| \geq \tfrac{1}{2} |B_r|, \qquad \text{where} \quad r = \epsilon_* R$$

and where $M = \frac{1}{4} M_*$. This is precisely (9.9), and the proof can now be concluded as before. ■

Remark 12.1 The Hölder continuity was used to ensure, starting from (9.8) that w is bounded below by M in a sizable portion of a small ball $B_{\epsilon_* R}(x_*)$ about x_*. In that process, the parameter ϵ_* had to be chosen in terms of the data and the still to be determined parameter β. Thus $\epsilon_* = \epsilon_*(\text{data}, \beta)$. The alternative proof based on Lemma 11.1, is intended to achieve the same lower bound on a sizable portion of $B_{\epsilon_* R}(x_*)$. The discussion has been conducted in order to trace the dependence of the various parameters on the unknown β. Indeed, also in this alternative argument, $\epsilon_* = \epsilon_*(\text{data}, \beta)$, but this is the only parameter dependent on β, whose choice can then be made by the very same argument, which from (9.9) leads to the conclusion of the proof.

References

1. *Œuvres complètes de N. H. Abel mathématicien*, M.M. L. Sylow and S. Lie, Eds. 2 Vols. (Oslo 1881).
2. N. Abel, Solution de quelques problèmes à l'aide d'intégrales définies, *Œuvres*, #1, 11–27.
3. N. Abel, Résolution d'un problème de mécanique, *Œuvres*, #1, 97–101.
4. S. Agmon, *Lectures on Elliptic Boundary Value Problems*, Van Nostrand, Princeton NJ, 1965.
5. S. Aizawa, A semigroup Treatment of the Hamilton–Jacobi Equations in Several Space Variables, *Hiroshima Math. J.*, No. 6, (1976), 15–30.
6. A. Ambrosetti and G. Prodi, *Analisi Non-Lineare*, Quaderni della Scuola Normale Superiore di Pisa, 1973.
7. P. Appell, Sur l'équation $(\partial^2 z/\partial x^2) - (\partial z/\partial y) = 0$ et la théorie de chaleur, *J. Math. Pures Appl.*, **8**, (1892), 187–216.
8. G. Barenblatt, On some unsteady motions of a liquid or a gas in a porous medium, *Prikl. Mat. Meh.* **16**, (1952), 67–78.
9. D. Bernoulli, Réflexions et éclaircissements sur les nouvelles vibrations des cordes, *Mémoires de l'Academie Royale des Sciences et belles lettres*, Berlin, (1755).
10. D. Bernoulli, Commentatio physico-mechanica generalior principii de coexistentia vibrationum simplicium haud perturbaturum in systemate compositio, *Novi Commentarii Academiae Scientiarum Imperialis Petropolitanae*, **19**, (1775), 239.
11. D. Bernoulli, Mémoire sur les vibrations des cordes d'une épaisseur inégale, *Mémoires de l'Academie royale des Sciences et belles lettres*, Berlin, (1767).
12. F. Bowman, *Introduction to Bessel Functions*, Dover, New York, 1958.
13. C. Burch, A Semigroup Treatment of the Hamilton–Jacobi Equations in Several Space Variables, *J. of Diff. Eq.*, **23**, (1977), 107–124.
14. J.M. Burgers, Application of a Model System to Illustrate Some Points of the Statistical Theory of Free Turbulence, *Proc. Acad. Sci. Amsterdam*, **43**, (1940), 2–12.
15. J. M. Burgers, A Mathematical Model Illustrating the Theory of Turbulence, in *Advances in Applied Mechanics, Ed. R. von Mises and T. von Kármán*, Vol. 1, Academic Press, New York, 1948, 171–199.

16. A.P. Calderón and A. Zygmund, On the existence of certain singular integrals, *Acta Math.* **88**, (1952), 85–139.
17. H.S. Carlslaw and J.C. Jaeger, *Conduction of Heat in Solids*, Oxford Univ. Press, Oxford, 1959.
18. H. Cartan, *Functions Analytiques d'une Variable Complexe*, Dunod, Paris, 1961.
19. A. Cauchy, Mémoire sur les systèmes d'équations aux derivées partielles d'ordre quelconque, et sur leur réduction à des systèmes d'équations linéaires du premier ordre, *C.R. Acad. Sci. Paris*, **40**, (1842), 131–138.
20. A. Cauchy, Mémoire sur les intégrales des systèmes d'équations différentielles at aux derivées partielles, et sur le développement de ces intégrales en séries ordonnés suivant les puissances ascendentes d'un paramètre que renferment les équations proposées, *C.R. Acad. Sci. Paris*, **40**, (1842), 141–146.
21. P.G. Ciarlet *The Finite Element Method for Elliptic Problems*, SIAM, Classics in Analysis, No. 40, Philadelphia, 2002.
22. R. Courant and D. Hilbert, *Methods of Mathematical Physics*, Vols. I and II, Interscience, New York, 1953, 1962.
23. C.M. Dafermos, Hyperbolic Systems of Conservation Laws, Proceedings of *Systems of non-linear Partial Differential Equations*, (Oxford, 1982), 25–70, NATO ASI Sci. Inst. Ser. C: Math. Phys. Sci. **111**, Reidel, Dordrecht–Boston, Mass, 1983.
24. G. Darboux, Sur l'existence de l'intégrale dans les équations aux derivées partielles d'ordre quelconque, *C. R. Acad. Sci. Paris*, **80**, (1875), 317–318.
25. G. Darboux, *Leçons sur la Théorie Générale des Surfaces et les Applications Géometriques du Calcul Infinitesimal*, Gauthiers–Villars, Paris, 1896.
26. E. DeGiorgi, Sulla differenziabilità e l'analiticità delle estremali degli integrali multipli regolari, *Mem. Acc. Sc. Torino, Cl. Sc. Mat. Fis. Nat.* **3**(3), (1957), 25–43.
27. E. DiBenedetto and N.S. Trudinger, Harnack Inequalities for Quasi–Minima of Variational Integrals, *Ann. Inst. H. Poincaré, Analyse Non-Linéaire*, **14**, (1984), 295–308.
28. E. DiBenedetto and A. Friedman, Bubble growth in porous media, *Indiana Univ. Math. J.*, **35**(3) (1986), 573–606.
29. E. DiBenedetto, Harnack Estimates in Certain Function Classes, *Atti Sem. Mat. Fis. Univ. Modena*, **XXXVII**, (1989), 173–182.
30. E. DiBenedetto, *Degenerate Parabolic Equations*, Springer-Verlag, New York, Series, *Universitext*, 1993.
31. E. DiBenedetto, *Real Analysis*, Birkhäuser, Boston, 2002.
32. E. DiBenedetto, *Classical Mechanics*, Birkhäuser, Boston, 2010.
33. E. DiBenedetto, U. Gianazza, V. Vespri, Local Clustering of the Non-Zero Set of Functions in $W^{1,1}(E)$, *Rend. Acc. Naz. Lincei, Mat. Appl.* **17**, (2006), 223–225.
34. E. DiBenedetto, U. Gianazza and V.Vespri, Intrinsic Harnack Estimates for Non-Negative Solutions of Quasi-Linear Degenerate Parabolic Equations, *Acta Mathematica* **200**, 2008, 181–209.
35. E. DiBenedetto, U. Gianazza and V. Vespri, Alternative Forms of the Harnack Inequality for Non-Negative Solutions to Certain Degenerate and Singular Parabolic Equations, *Rend. Acc. Naz. Lincei, (in press).*
36. J. Dieudonné, *Treatise on Analysis*, Academic Press, New York.

37. P. Dive, Attraction d'ellipsoides homogènes et réciproque d'un théorème de Newton, *Bull. Soc. Math. France*, No. 59 (1931), 128–140.
38. J.M. Duhamel, Sur les vibrations d'une corde flexible chargée d'un ou plusieurs curseurs, *J. de l'École Polytechnique*, No. 17, (1843), cahier 29, 1–36.
39. L.C. Evans, *Partial Differential Equations*, American Mathematical Society, Graduate Studies in Mathematics **19**, Providence R.I., 1998.
40. G. Fichera, *Linear Elliptic Differential Systems and Eigenvalue Problems*, Lecture Notes in Mathematics, No. 8, Springer-Verlag, Berlin, 1965.
41. W.H. Fleming and R. Rischel, An integral formula for total gradient variation, *Arch. Math.* **XI**, (1960), 218–222.
42. J.B. Fourier, *Théorie Analytique de la Chaleur*, Chez Firmin Didot, Paris, 1822.
43. I. Fredholm, Sur une nouvelle méthode pour la résolution du problème de Dirichlet, *Kong. Vetenskaps-Akademiens Fröh. Stockholm*, (1900), 39–46.
44. I. Fredholm, Sur une classe d'équations fonctionnelles, *Acta Math.* **27**, (1903), 365–390.
45. A. Friedman, *Partial Differential Equations of Parabolic Type*, Prentice Hall, 1964.
46. A. Friedman, *Partial Differential Equations*, Holt, Rinehart and Winston, New York, 1969.
47. A. Friedman, *Variational Principles and Free Boundary Problems*, John Wiley & Sons, New York, 1982.
48. A. Friedman, Interior Estimates for Parabolic Systems of Partial Differential equations, *J. Math. Mech.*, **7**, (1958), 393–417.
49. A. Friedman, A new proof and generalizations of the Cauchy–Kowalewski Theorem, *Trans. Amer. Math. Soc.*, No. 98, (1961), 1–20.
50. E. Gagliardo, Proprietà di Alcune Funzioni in n Variabili, *Ricerche Mat.* **7**, (1958), 102–137.
51. P.R. Garabedian, *Partial Differential Equations*, John Wiley & Sons, New York, 1964.
52. M. Gevrey, Sur les équations aux dérivées partielles du type parabolique, *J. Math. Pures et Appl.* **9**(VI), (1913), 305–471; ibidem **10**(IV), (1914), 105–148.
53. M. Giaquinta and E. Giusti, Quasi–Minima, *Ann. Inst. Poincaré, Analyse non Linéaire*, **1**, (1984), 79–107.
54. D. Gilbarg and N.S. Trudinger, *Elliptic Partial Differential Equations of Second-Order*, (2nd ed.), Die Grundlehren der Mathematischen Wissenschaften, No. 224, Springer-Verlag, Berlin, 1983.
55. D. Gilbarg and J. Serrin, On isolated singularities of solutions of second-order elliptic differential equations, *J. d'Analyse Math.* **4**, (1955), 309–340.
56. E. Giusti *Functions of Bounded Variation*, Birkhäuser, Basel, 1983.
57. J. Glimm and P.D. Lax, *Decay of Solutions of Systems of Non-Linear Hyperbolic Conservation Laws, Mem. Amer. Math. Soc.* **10**(101), (1970).
58. G. Green, *An Essay on the Applications of Mathematical Analysis to the Theories of Electricity and Magnetism*, Nottingham, 1828.
59. K.K. Golovkin, Embedding Theorems, *Doklady Akad. Nauk, USSR*, **134**(1), (1960), 19–22.
60. E. Goursat, *Cours d'Analyse Mathématique*, Gauthiers–Villars, Paris, 1913.
61. J. Hadamard, *Le problème de Cauchy et les Équations aux Deriveées Partielles Linéaires Hyperboliques*, Hermann et Cie, Paris, 1932.

62. J. Hadamard, *Lectures on Cauchy's Problem in Linear Partial Differential Equations*, Dover, New York, 1952.
63. J. Hadamard, Extension à l'équation de la chaleur d'un théorème de A. Harnack, *Rend. Circ. Mat. di Palermo*, **Ser. 2 3**, (1954), 37–346.
64. A. Hammerstein, Nichtlineare Integralgleichungen nebst Angewendungen, *Acta Math.* **54** (1930), 117–176.
65. H.H. Hardy, Note on a theorem of Hilbert, *Math. Z.* **6**, (1920), 314–317.
66. G.H. Hardy, J.E. Littlewood, G. Pólya, *Inequalities*, Cambridge Univ. Press, 1963.
67. A. Harnack, *Grundlagen der Theorie des Logarithmischen Potentials*, Leipzig, 1887.
68. A. Harnack, Existenzbeweise zur Theorie des Potentiales in der Ebene und in Raume, *Math. Ann.* **35**(1–2), (1889), 19–40.
69. D. Hilbert, *Grundzüge einer allgemeinen Theorie der linearen Integralgleichungen*, Leipzig, 1912.
70. E.W. Hobson, *The Theory of Spherical and Ellipsoidal Harmonics*, Chelsea Pub. Co., New York, 1965.
71. E. Hopf, The Partial Differential Equation $u_t + uu_x = \mu_{xx}$, *Comm. Pure Appl. Math.* **3**, (1950), 201–230.
72. E. Hopf, Generalized Solutions of Non-Linear Equations of First Order, *J. Math. Mech.* **14**, (1965), 951–974.
73. E. Hopf, On the right Weak Solution of the Cauchy Problem for a Quasi-Linear Equation of First Order, *J. Mech. Math.*, **19**, (1969/1970), 483–487.
74. L. Hörmander, *Linear Partial Differential Operators*, Springer-Verlag, Berlin–Heidelberg–New York, 1963.
75. L. Hörmander, *An Introduction to Complex Analysis in Several Variables*, D. Van Nostrand Co. Inc., Princeton, New Jersey, 1966.
76. H. Hugoniot, Sur la Propagation du Mouvement dans les Corps et Spécialement dans les Gaz Parfaits, *Journ. de l'École Polytechnique*, **58**, (1889), 1–125.
77. F. John, *Partial Differential Equations*, Springer-Verlag, New York, 1986.
78. F. John, *Plane Waves and Spherical Means Applied to Partial Differential Equations*, Interscience Publishers, 1955.
79. A.S. Kalashnikov, Construction of Generalized Solutions of Quasi-Linear Equations of First Order Without Convexity Conditions as Limits of Solutions of Parabolic Equations With a Small Parameter, *Dokl. Akad. Nauk SSSR*, **127**, (1959), 27–30.
80. O.D. Kellogg, *Foundations of Potential Theory*, Springer-Verlag, Berlin, 1929, reprinted 1967.
81. J.L. Kelley, *General Topology*, Van Nostrand, New York, 1961.
82. D. Kinderlehrer and G. Stampacchia, *An Introduction to Variational Inequalities and Their Applications*, Academic Press, New York, 1980.
83. V.I. Kondrachov, Sur certaines propriétés des fonctions dans les espaces L^p, *C.R. (Dokl.) Acad. Sci. USSR (N.S.)*, **48**, (1945), 535–539.
84. S. Kowalewski, Zur Theorie der Partiellen Differentialgleichungen, *J. Reine Angew. Math.* **80**, (1875), 1–32.
85. S.N. Kruzhkov, First Order Quasi-linear Equations in Several Independent Variables, *Mat. USSR Sbornik* **10**(2), (1970), 217–243.
86. S.N. Kruzhkov, Generalized Solutions of the Hamilton–Jacobi Equations of Eikonal Type-I, *Soviet Mat. Dokl.*, **16**, (1975), 1344–1348.

87. S.N. Kruzhkov, Generalized Solutions of the Hamilton–Jacobi Equations of Eikonal Type-II, *Soviet Mat. Dokl.*, **27**, (1975), 405–446.
88. S.N. Kruzhkov, Non-local Theory for Hamilton–Jacobi Equations, *Proceeding of a meeting in Partial Differential Equations* Edited by P. Alexandrov and O. Oleinik, Moscow Univ., 1978.
89. N.S. Krylov, *Controlled Diffusion Processes*, Springer-Verlag, Series Applications of Mathematics **14**, New York, 1980.
90. O.A. Ladyzenskajia, New equations for the description of motion of viscous incompressible fluids and solvability in the large of boundary value problems for them, *Proc. Steklov Inst. Math.*, **102**, (1967), 95–118 (transl. Trudy Math. Inst. Steklov, **102**, (1967), 85–104).
91. O.A. Ladyzenskajia, N.N. Ural'tzeva, *Linear and Quasi-linear Elliptic Equations*, Academic Press, London–New York, 1968.
92. O.A. Ladyzenskajia, N.A. Solonnikov, N.N. Ural'tzeva, *Linear and Quasi-linear Equations of Parabolic Type*, Transl. Math. Monographs, **23**, AMS, Providence R.I., 1968.
93. E.M. Landis, *Second Order Equations of Elliptic and Parabolic Type*, Transl. Math. Monographs, **171**, AMS Providence R.I., 1997 (Nauka, Moscow, 1971).
94. N.S. Landkof, *Foundations of Modern Potential Theory*, Springer-Verlag, Berlin, 1972.
95. M.M. Lavrentiev, *Some Improperly Posed Problems of Mathematical Physics*, Springer-Verlag, Tracts in Nat. Philosophy, **11**, Berlin–Heidelberg–New York, 1967.
96. M.M. Lavrentiev, V.G. Romanov, and S.P. Sisatskij, *Problemi Non Ben Posti in Fisica Matematica e Analisi*, C.N.R. Istit. Analisi Globale **12**, Firenze 1983. Italian transl. of *Nekorrektnye zadachi Matematicheskoi Fisiki i Analisa* Akad. Nauka Moscou, 1980.
97. P.D. Lax, Non-Linear Hyperbolic Equations, *Comm. Pure Appl. Math.*, **4**, (1953), 231–258.
98. P.D. Lax and A.M. Milgram, *Parabolic Equations*, Contrib. to the Theory of P.D.E.'s, Princeton Univ. Press, 1954, 167–190.
99. P. Lax, On the Cauchy Problem for Partial Differential Equations with Multiple Characteristics, *Comm. Pure Appl. Math.* **9**, (1956), 135–169.
100. P. Lax, *Hyperbolic Systems of Conservation Laws and the Mathematical Theory of Shock Waves*, SIAM, Philadelphia, 1973.
101. H.L. Lebesgue, Sur des cas d'impossibilité du problème de Dirichlet, *Comptes Rendus Soc. Math. de France*, **17**, (1913).
102. L. Lichenstein, Eine elementare Bemerkung zur reellen Analysis, *Math. Z.*, **30**, (1929), 794–795.
103. A. Majda, *Compressible Fluid Flow and Systems of Conservation Laws in Several Space Variables*, Appl. Math. **53**, Springer-Verlag, New York, 1984.
104. V.G. Mazja, *Sobolev Spaces*, Springer-Verlag, New York, 1985.
105. N.Meyers and J. Serrin, $H = W$, *Proc. Nat. Acad. Sci.* **51**, (1964), 1055–1056.
106. S.G. Mikhlin, *Integral Equations*, Pergamon Press **4**, 1964.
107. C.B. Morrey, On the Solutions of Quasi-linear Elliptic Partial Differential Equations, *Trans. AMS* **43**, (1938), 126–166.
108. C.B. Morrey, *Multiple Integrals in the Calculus of Variations*, Springer-Verlag, New York, 1966.
109. J. Moser, On Harnack's Theorem for Elliptic Differential Equations, *Comm. Pure Appl. Math.* **14**, (1961), 577–591.

110. J. Moser, A Harnack inequality for parabolic differential equations, *Comm. Pure and Appl. Math.*, **17**, (1964), 101–134.
111. C.G. Neumann, *Untersuchungen über das logarithmische und Newtonsche Potential*, Leipzig, 1877.
112. L. Nirenberg, On Elliptic Partial Differential Equations, *Ann. Scuola Norm. Sup. Pisa*, **3**(13), (1959), 115–162.
113. O.A. Oleinik, Discontinuous Solutions of Non-Linear Differential Equations, *Uspekhi Mat. Nauk (N.S.)*, **12**(3), (1957), 3–73; (*Amer. Math. Soc. Transl.*, **2**(26), 95–172).
114. O.A. Oleinik, Uniqueness and Stability of the Generalized Solution of the Cauchy Problem for a Quasi-Linear Equation, *Uspekhi Mat. Nauk (N.S.)*, **14**(6), (1959), 165–170.
115. R.E. Pattle, Diffusion from an instantaneous point source with a concentration-dependent coefficient, *Quarterly J. of Appl. Math.* **12**, (1959), 407–409.
116. L.E. Payne, *Improperly Posed Problems in Partial Differential Equations*, SIAM, Vol. 22, Philadelphia, 1975.
117. O. Perron, Eine neue Behandlung der Randewertaufgabe für $\Delta u = 0$. *Math. Z.*, No. 18 (1923), 42–54.
118. B. Pini, Sulla soluzione generalizzata di Wiener per il primo problema di valori al contorno nel caso parabolico, *Rend. Sem. Mat. Univ. Padova*, **23**, (1954) 422–434.
119. S.D. Poisson, Mémoire sur la théorie du son, *J. de l'École Polytechnique, Tome VII, Cahier XIV*^{ème}, (1808), 319–392.
120. S.D. Poisson, Mouvement d'une corde vibrante composée de deux parties de matières différentes, *J. de l'École Polytechnique, Tome XI, Cahier XVIII*^{ème}, (1820), 442–476.
121. M.H. Protter and H.F. Weinberger, *Maximum Principles in Differential Equations*, Prentice Hall, 1967.
122. T. Radó, *Sub-harmonic Functions*, Chelsea Pub. Co., New York, 1940.
123. J. Radon, Über lineare Funktionaltransformationen und Funktionalgleichungen, *Sitzsber. Akad. Wiss. Wien*, **128**, (1919), 1083–1121.
124. W.J.M. Rankine, On the Thermodynamic Theory of Waves of Finite Longitudinal Disturbance, *Trans. Royal Soc. of London*, **160**, (1870), 277–288.
125. F. Riesz, Über lineare Functionalgleichungen, *Acta Math.* **41**, (1918), 71–98.
126. F. Riesz, Sur les fonctions subharmoniques at leur rapport à la théorie du potentiel, part I and II, *Acta Math.* **48**, (1926), 329–343, and ibidem **54**, (1930), 321–360.
127. R. Reillich, Ein Satz über mittlere Konvergenz, *Nach. Akad. Wiss. Göttingen Math. Phys. Kl.*, (1930), 30–35.
128. F. Reillich, Spektraltheorie in nicht-separablen Räumen, *Math. Annalen*, Band 110, (1934), 342–356.
129. B. Riemann, Ueber die Fortpflanzung ebener Luftwellen von endlicher Schwingungsweite, *Abhandlung der Königlichen Gesellschaft der Wissenschaften zu Göttingen*, Band 8, 1860.
130. F. Riesz and B. Nagy, *Functional Analysis*, Dover, New York, 1990.
131. P.C. Rosenbloom and D. V. Widder, A temperature function which vanishes initially, *Amer. Math. Monthly*, **65**, (1958), 607–609.
132. H.L. Royden, *Real Analysis*, 3rd ed., Macmillan, New York, 1988.
133. J.B.Serrin, On the Phragmen-Lindelöf principle for elliptic differential equations, *J. Rat. Mech. Anal.* **3**, (1954), 395–413.

134. J.B. Serrin, Local behavior of Solutions of Quasi-linear Elliptic Equations, *Acta Math.* **111**, (1964), 101–134.
135. J.P. Schauder, Über lineare elliptische Differentialgleichungen zweiter Ordnung, *Math. Z.* **38**, (1934), 257–282.
136. J.P. Schauder, Numerische Abschaetzungen in elliptischen linearen Differentialgleichungen, *Studia Math.* **5**, (1935), 34–42.
137. A.E. Scheidegger, *The Physics of Flow Through Porous Media*, Univ. of Toronto Press, Toronto, 1974.
138. E. Schmidt, Anflösung der allgemeinen linearen Integralgleichungen, *Math. Annalen*, Band 64, (1907), 161–174.
139. L. Schwartz, *Théorie des Distributions*, Herman et Cie, Paris, 1966.
140. H. Shahgolian, On the Newtonian potential of a heterogeneous ellipsoid, *SIAM J. Math. Anal.*, **22**(5), (1991), 1246–1255.
141. M. Shimbrot and R.E. Welland, The Cauchy–Kowalewski Theorem, *J. Math. Anal. and Appl.*, **25**(3), (1976), 757–772.
142. S.L. Sobolev, On a Theorem of Functional Analysis, *Math. Sbornik* **46**, (1938), 471–496.
143. S.L. Sobolev and S.M. Nikol'skii, Embedding Theorems, *Izdat. Akad. Nauk SSSR, Leningrad* (1963), 227–242.
144. E.M. Stein, *Singular Integrals and Differentiability Properties of Functions*, Princeton Univ. Press, 1970.
145. W. Strauss, *Non-Linear Wave Equations*, Conference Board of the Mathematical Sciences Regional Conf. Series in Math. **73**, AMS, Providence R.I. 1989.
146. S. Tacklind, Sur les classes quasianalytiques des solutions des équations aux deriveées partielles du type parabolique, *Acta Reg. Sc. Uppsaliensis*, (Ser. 4), **10**, (1936), 3–55.
147. G. Talenti, Sulle equazioni integrali di Wiener–Hopf, *Boll. Un. Mat. Ital.*, **7**(4), Suppl. fasc. 1, (1973), 18–118.
148. G. Talenti, Best Constants in Sobolev Inequalities, *Ann. Mat. Pura Appl.* **110**, (1976), 353–372.
149. G. Talenti, Sui problemi mal posti, *Boll. U.M.I. (5)*, **15**, (1978), 1–29.
150. L. Tartar, *Une introduction à la Théorie Mathématique des Systèmes Hyperboliques des Lois de Conservation*, Pubblic. **682**, Ist. Analisi Numerica, C.N.R. Pavia, 1989.
151. W. Thomson Kelvin, Extraits de deux lettres adressées à M. Liouville, *J. de Mathématiques Pures et Appliquées*, **12**, (1847), 256.
152. F.G. Tricomi, Sulle equazioni lineari alle derivate parziali di tipo misto, *Atti Accad. Naz. Lincei*, **14**, (1923), 218–270.
153. F. Tricomi, *Integral Equations*, Dover, New York, 1957.
154. N.S. Trudinger, On Harnack Type Inequalities and their Application to Quasi-Linear Elliptic Partial Differential Equations, *Comm. Pure Appl. Math.* **20**, (1967), 721–747.
155. A.N.Tychonov, Théorèmes d'unicité pour l'équation de la chaleur, *Math. Sbornik*, **42**, (1935), 199–216.
156. A.N. Tychonov and V.Y. Arsenin, *Solutions of Ill-Posed Problems*, Winston/ Wiley, 1977.
157. A.I. Vol'pert, The Spaces BV and Quasi-Linear Equations, *Math. USSR Sbornik*, **2**, (1967), 225–267.

158. V.Volterra, Sulla inversione degli integrali definiti, *Rend. Accad. Lincei, Ser. 5* (1896), 177–185.
159. V.Volterra, Sopra alcune questioni di inversione di integrali definiti, *Ann. di Mat.* **25**(2), (1897), 139–178.
160. R. Von Mises and K.O. Friedrichs, *Fluid Dynamics*, Appl. Math. Sc., **5**, Springer-Verlag, New York, 1971.
161. N.Wiener, Certain notions in potential theory, *J. Math. Phys. Mass. Inst. Tech.* **III**, 24–51, (1924).
162. N. Wiener, Une condition nécessaire et suffisante de possibilité pour le problème de Dirichlet, *Comptes Rendus Acad. Sci. Paris*, **178**, (1924), 1050–1054.
163. N. Wiener and E. Hopf, Über eine Klasse singulärer Integralgleichungen, *Sitzungsber. Preuss. Akad. der Wiss.*, 1931, 696–706.
164. D. V. Widder, Positive temperatures in an infinite rod, *Trans. Amer. Math. Soc.*, **55**, (1944), 85–95.
165. K. Yoshida, *Functional Analysis*, Springer-Verlag, New York, 1974.

Index

Breinigsville, PA USA
03 February 2010
231848BV00005B/46/P

9 780817 645519